Numerical Methods for Chemical Engineers with MATLAB Applications

PRENTICE HALL INTERNATIONAL SERIES
IN THE PHYSICAL AND CHEMICAL ENGINEERING SCIENCES

BALZHISER, SAMUELS, AND ELIASSEN *Chemical Engineering Thermodynamics*
BEQUETTE *Process Dynamics*
BIEGLER, GROSSMAN, AND WESTERBERG *Systematic Methods of Chemical Process Design*
CONSTANTINIDES AND MOSTOUFI *Numerical Methods for Chemical Engineers with MATLAB Applications*
CROWL AND LOUVAR *Chemical Process Safety*
CUTLIP AND SHACHAM *Problem Solving in Chemical Engineering with Numerical Methods*
DENN *Process Fluid Mechanics*
ELLIOT AND LIRA *Introductory Chemical Engineering Thermodynamics*
FOGLER *Elements of Chemical Reaction Engineering, 3rd edition*
HANNA AND SANDALL *Computational Methods in Chemical Engineering*
HIMMELBLAU *Basic Principles and Calculations in Chemical Engineering, 6th edition*
HINES AND MADDOX *Mass Transfer*
KYLE *Chemical and Process Thermodynamics, 3rd edition*
NEWMAN *Electrochemical Systems, 2nd edition*
PRAUSNITZ, LICHTENTHALER, AND DE AZEVEDO *Molecular Thermodynamics of Fluid-Phase Equilibria, 3rd edition*
PRENTICE *Electrochemical Engineering Principles*
SHULER AND KARGI *Bioprocess Engineering*
STEPHANOPOULOS *Chemical Process Control*
TESTER AND MODELL *Thermodynamics and Its Applications, 3rd edition*
TURTON, BAILIE, WHITING, AND SHAEIWITZ *Analysis, Synthesis and Design of Chemical Processes*
WILKES *Fluid Mechanics for Chemical Engineering*

Numerical Methods for Chemical Engineers with MATLAB Applications

Alkis Constantinides
Department of Chemical and Biochemical Engineering
Rutgers, The State University of New Jersey

Navid Mostoufi
Department of Chemical Engineering
University of Tehran

Prentice Hall PTR
Upper Saddle River, New Jersey 07458
http://www.prenhall.com

Library of Congress Cataloging-in-Publication Data

Constantinides, A.
Numerical methods for chemical engineers with MATLAB applications / Alkis Constantinides, Navid Mostoufi.
p. cm. -- (Prentice Hall international series in the physical and chemical engineering sciences)
ISBN 0-13-013851-7 (alk. paper)
1. Numerical analysis--Data processing. 2. Chemical engineering--Mathematics--Data processing. 3. MATLAB. I. Mostoufi, Navid. II. Title. III. Series.
QA297.C6494 1999
660'.01'5194--dc21 99-22296
CIP

Editorial/Production Supervision: Craig Little
Acquisitions Editor: Bernard Goodwin
Manufacturing Manager: Alan Fischer
Marketing Manager: Lisa Konzelmann
Cover Design Director: Jerry Votta
Cover Design: Talar Agasyan

Reprinted with corrections March, 2000.

Printed in the United States of America
10 9 8 7 6 5

ISBN 0-13-013851-7

Prentice-Hall International (UK) Limited, *London*
Prentice-Hall of Australia Pty. Limited, *Sydney*
Prentice-Hall Canada Inc., *Toronto*
Prentice-Hall Hispanoamericana, S.A., *Mexico*
Prentice-Hall of India Private Limited, *New Delhi*
Prentice-Hall of Japan, Inc., *Tokyo*
Prentice-Hall (Singapore) Pte. Ltd., *Singapore*
Editora Prentice-Hall do Brasil, Ltda., *Rio de Janeiro*

Dedicated to our wives
Melody Richards Constantinides
and
Fereshteh Rashchi (Mostoufi)
and our children
Paul Constantinides
Kourosh and Soroush Mostoufi

Contents

[1] Sections marked with an asterisk (*) may be omitted in an undergraduate course.

Preface

This book emphasizes the derivation of a variety of numerical methods and their application to the solution of engineering problems, with special attention to problems in the chemical engineering field. These algorithms encompass linear and nonlinear algebraic equations, eigenvalue problems, finite difference methods, interpolation, differentiation and integration, ordinary differential equations, boundary value problems, partial differential equations, and linear and nonlinear regression analysis. MATLAB[2] is adopted as the calculation environment throughout the book. MATLAB is a high-performance language for technical computing. It integrates computation, visualization, and programming in an easy-to-use environment. MATLAB is distinguished by its ability to perform all the calculations in matrix form, its large library of built-in functions, its strong structural language, and its rich graphical visualization tools. In addition, MATLAB is available on all three operating platforms: WINDOWS, Macintosh, and UNIX. The reader is expected to have a basic knowledge of using MATLAB. However, for those who are not familiar with MATLAB, it is recommended that they cover the subjects discussed in Appendix A: Introduction to MATLAB prior to studying the numerical methods.

Several worked examples are given in each chapter to demonstrate the numerical techniques. Most of these examples require computer programs for their solution. These programs were written in the MATLAB language and are compatible with MATLAB 5.0 or higher. In all the examples, we tried to present a general MATLAB function that implements

[2] MATLAB is a registered trademark of the MathWorks, Inc.

the method and that may be applied to the solution of other problems that fall in the same category of application as the worked example. The general algorithm for these programs is illustrated in the section entitled, "General Algorithm for the Software Developed in this Book." All the programs that appear in the text are included on the CD-ROM that accompanies this book. There are three versions of these programs on the CD-ROM, one for each of the major operating systems in which MATLAB exists: WINDOWS, Macintosh, and UNIX. Installation procedures, a complete list, and brief descriptions of all the programs are given in the section entitled "Programs on the CD-ROM" that immediately follows this Preface. In addition, the programs are described in detail in the text in order to provide the reader with a thorough background and understanding of how MATLAB is used to implement the numerical methods.

It is important to mention that the main purpose of this book is to teach the student numerical methods and problem solving, rather than to be a MATLAB manual. In order to assure that the student develops a thorough understanding of the numerical methods and their implementation, new MATLAB functions have been written to demonstrate each of the numerical methods covered in this text. Admittedly, MATLAB already has its own built-in functions for some of the methods introduced in this book. We mention and discuss the built-in functions, whenever they exist.

The material in this book has been used in undergraduate and graduate courses in the Department of Chemical and Biochemical Engineering at Rutgers University. Basic and advanced numerical methods are covered in each chapter. Whenever feasible, the more advanced techniques are covered in the last few sections of each chapter. A one-semester graduate level course in applied numerical methods would cover all the material in this book. An undergraduate course (junior or senior level) would cover the more basic methods in each chapter. To facilitate the professor teaching the course, we have marked with an asterisk (*) in the Table of Contents those sections that may be omitted in an undergraduate course. Of course, this choice is left to the discretion of the professor.

Future updates of the software, revisions of the text, and other news about this book will be listed on our web site at http://sol.rutgers.edu/~constant.

Prentice Hall and the authors would like to thank the reviewers of this book for their constructive comments and suggestions. NM is grateful to Professor Jamal Chaouki of Ecole Polytechnique de Montréal for his support and understanding.

Alkis Constantinides
Navid Mostoufi

Programs on the CD-ROM

Brief Description

The programs contained on the CD-ROM that accompanies this book have been written in the MATLAB 5.0 language and will execute in the MATLAB command environment in all three operating systems (WINDOWS, Macintosh, and UNIX). There are 21 examples, 29 methods, and 13 other function scripts on this CD-ROM. A list of the programs is given later in this section. Complete discussions of all programs are given in the corresponding chapters of the text.

MATLAB is a high-performance language for technical computing. It integrates computation, visualization, and programming in an easy-to-use environment. It is assumed that the user has access to MATLAB. If not, MATLAB may be purchased from:

The MathWorks, Inc.
24 Prime Park Way
Natick, MA 07160-1500
Tel: 508-647-7000 Fax: 508-647-7001
E-mail: info@mathworks.com
http://www.mathworks.com

The Student Edition of MATLAB may be obtained from:

Prentice Hall PTR, Inc.
One Lake Street
Upper Saddle River, NJ 07458
http://www.prenhall.com

An introduction to MATLAB fundamentals is given in Appendix A of this book.

Program Installation for WINDOWS

To start the installation, do the following:

1. Insert the CD-ROM in your CD-ROM drive (usually *d*: or *e*:)
2. Choose Run from the WINDOWS Start menu, type *d:\setup* (or *e:\setup*) and click OK.

3. Follow the instructions on screen.
4. When the installation is complete, run MATLAB and set the MATLAB search path as described below.

This installation procedure copies all the MATLAB files to the user's hard disk (the default destination folder is C:\Program Files\Numerical Methods). It also places a shortcut, called Numerical Methods, on the Start Programs menu of WINDOWS (see Fig. 1). This shortcut accesses all twenty-one examples from the seven chapters of the book (but not the methods). In addition, the shortcut provides access to the *readme* file (in three different formats: *pdf*, *html*, and *doc*). Choosing an example from the shortcut enables the user to view the MATLAB script of that example with the MATLAB Editor. Files have been installed on the hard disk with the "read-only" attribute in order to prevent the user from inadvertently modifying the program files (see **Editing the Programs**, below). To execute any of the examples, see **Executing the Programs,** below.

Program Installation for Macintosh

The CD-ROM is in ISO format, therefore it can be read by Macintosh computers that have File Exchange (for System 8 or higher) or PC Exchange (for System 7). If you have not activated the File Exchange, please do so via the Control Panels before using this CD-ROM. To start the installation, do the following:

1. Insert the CD-ROM in your CD-ROM drive on a Macintosh computer.
2. Open the folder named *MAC* on the CD-ROM. This contains a compressed file (zip file) named *NUMMETH*.ZIP.
3. Copy the file *NUMMETH*.ZIP to your computer and uncompress it using *zipit* or *StuffIt Expander*. This will create a folder named *Numerical Methods* which contains all the programs of this book.
4. When the installation is complete, run MATLAB and set the MATLAB search path as described below.

Program Installation for UNIX Systems

To start the installation, do the following:

1. Insert the CD-ROM in your CD-ROM drive on a UNIX workstation.
2. Open the folder named *UNIX on the CD-ROM*. This contains a compressed file (tar file) named *nummeth.tar*.
3. Copy the file *nummeth.tar* to your computer and uncompress it using the *tar command*:

   tar xf *nummeth.tar*

This will create a folder named *Numerical Methods* which contains all the programs of this book.

4. When the installation is complete, run MATLAB and set the MATLAB search path as described below.

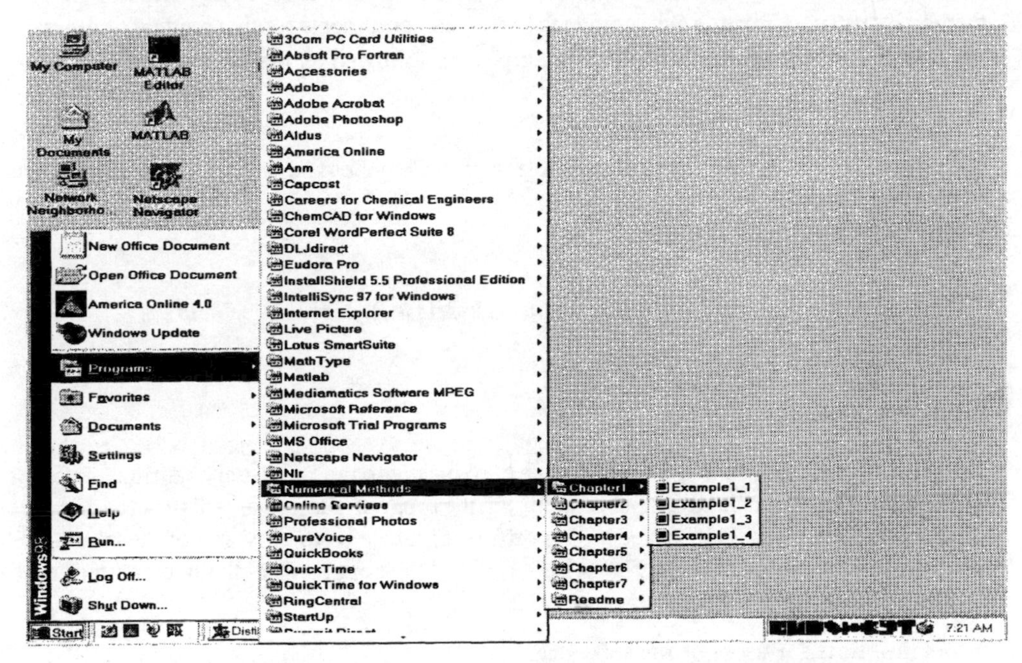

Figure 1 Arrangement of the Numerical Methods programs in the Start menu.

Setting the MATLAB Search Path

It is important that the search path used by MATLAB is set correctly so that the files may be found from any directory that MATLAB may be running. In the MATLAB Command Window choose File, Set Path. This will open the Path Browser. From the menu of the Path Browser choose Path, Add to Path. Add the directories of your hard disk where the Numerical Methods programs have been installed (the default directory for the WINDOWS installation is C:\Program Files\Numerical Methods\Chapter1, etc.). The path should look as in Fig. 2, provided that the default directory was not modified by the user during setup.

Executing the Programs

Once the search path is set as described above, any of the examples, methods, and functions in this book may be used from anywhere within the MATLAB environment. To execute one of the examples, simply enter the name of that example in the MATLAB Command Window:

»Example1_1

To use any of the methods or functions from within another MATLAB script, invoke the method by its specific name and provide the necessary arguments for that method or function. To get a brief description of any program, type *help* followed by the name of the program:

»*help* Example1_1

To get descriptions of the programs available in each Chapter, type *help* followed by the name of the Chapter:

»*help* Chapter1

To find out what topics of help are available in MATLAB, simply type *help*:

»*help*

Editing the Programs

The setup procedure installed the files on the hard disk with the "read-only" attribute in order to prevent the user from inadvertently modifying the program files. If any of the program files are modified, they should be saved with a different name. To modify any of the MATLAB language programs, use the MATLAB Editor. Read the comments at the beginning of each program before making changes.

Important note for users of the software

Last-minute changes have been made to the software; however, these changes do not appear in the text or on the CD-ROM that accompanies this book. To download the latest version of the software, please visit our website:

http://sol.rutgers.edu/~constant

Note for users of MATLAB 5.2

The original MATLAB Version 5.2 had a "bug." The command

linspace(0,0,100)

which is used in *LI.m*, *NR.m*, and *Nrpoly.m* in Chapter 1, would not work properly in the MATLAB installations of Version 5.2 which have not yet been corrected. A patch which corrects this problem is available on the website of Math Works, Inc.:

http://www.mathworks.com

If you have Version 5.2, you are strongly encouraged to download and install this patch, if you have not done so already.

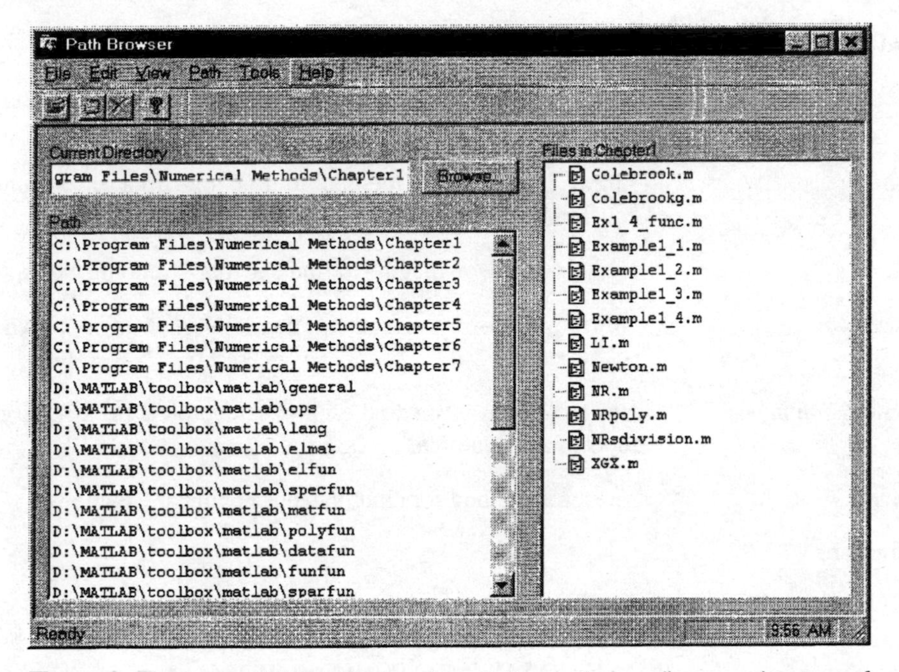

Figure 2 The correct MATLAB search path that includes all seven chapters of the Numerical Methods software.

LISTING AND DESCRIPTION OF PROGRAMS

CHAPTER 1

Program Name	Description
Examples	
Example1_1.m	Calculates the friction factor from the Colebrook equation using the Successive Substitution (*XGX.m*), the Linear Interpolation (*LI.m*), and the Newton-Raphson (*NR.m*) methods.
Example1_2.m	Solves the Soave-Redlich-Kwong equation of state using the Newton-Raphson method for polynomial equations (*NRpoly.m*).
Example1_3.m	Solves *n*th-degree polynomials and transfer functions using the Newton-Raphson method with synthetic division (*NRsdivision.m*).
Example1_4.m	Solves simultaneous reactions in chemical equilibrium using Newton's method for simultaneous nonlinear equations (*Newton.m*).

Methods

XGX.m — Successive Substitution method to find one root of a nonlinear equation.

LI.m — Linear Interpolation method to find one root of a nonlinear equation.

NR.m — Newton-Raphson method to find one root of a nonlinear equation.

NRpoly.m — Newton-Raphson method to find one root of a polynomial equation.

NRsdivision.m — Newton-Raphson method with synthetic division to find all the roots of a polynomial equation.

Newton.m — Newton's method for simultaneous nonlinear equations.

Functions

Colebrookg.m — Contains the Colebrook equation in a form so that it can be solved by Successive Substitution (used in *Example1_1.m*).

Colebrook.m — Contains the Colebrook equation in a form so that it can be solved by Linear Interpolation and/or Newton-Raphson (used in *Example1_1.m*).

Ex1_4_func.m — Contains the set of simultaneous nonlinear equations (used in *Example1_4.m*).

CHAPTER 2

Examples

Example2_1.m — Solves a set of simultaneous linear algebraic equations that model the heat transfer in a steel pipe using the Gauss Elimination method (*Gauss.m*).

Example2_2.m — Solves a set of simultaneous linear algebraic equations that model the steam distribution system of a chemical plant using the Gauss-Jordan Reduction method (*Jordan.m*).

Example2_3.m — Solves a set of simultaneous linear algebraic equations that represent the material balances for a set of continuous stirred tank reactors using the Jacobi Iterative method (*Jacobi.m*).

Methods

Gauss.m	Gauss Elimination method for solution of simultaneous linear algebraic equations.
Jordan.m	Gauss-Jordan Reduction method for solution of simultaneous linear algebraic equations.
Jacobi.m	Jacobi Iterative method for solution of predominantly diagonal sets of simultaneous linear algebraic equations.

CHAPTER 3

Examples

Example3_1.m	Interpolates equally spaced points using the Gregory-Newton forward interpolation formula (*GregoryNewton.m*).
Example3_2.m	Interpolates unequally spaced points using Lagrange polynomials (*Lagrange.m*) and cubic splines (*NaturalSPLINE.m*).

Methods

GregoryNewton.m	Gregory-Newton forward interpolation method.
Lagrange.m	Lagrange polynomial interpolation method.
NaturalSPLINE.m	Cubic splines interpolation method.

CHAPTER 4

Examples

Example4_1.m	Calculates the unsteady flux of water vapor from the open top of a vessel using numerical differentiation of a function (*fder.m*).
Example4_2.m	Calculates the solids volume fraction profile in the riser of a gas-solid fluidized bed using differentiation of tabulated data (*deriv.m*).
Example4_3.m	Integrates a vector of experimental data using the trapezoidal rule (*trapz.m*) and the Simpson's 1/3 rule (*Simpson.m*).
Example4_4.m	Integrates a function using the Gauss-Legendre quadrature (*GaussLegendre.m*).

Methods

fder.m	Differentiation of a function.
deriv.m	Differentiation of tabulated data.
Simpson.m	Integration of tabulated data by the Simpson's 1/3 rule.
GaussLegendre.m	Integration of a function by the Gauss-Legendre quadrature.

Functions

Ex4_1_phi.m	Contains the nonlinear equation for calculation of phi (used in *Example4_1.m*).
Ex4_1_profile.m	Contains the function of concentration profile (used in *Example4_1.m*).
Ex4_4_func.m	Contains the function to be integrated (used in *Example4_4.m*).

Chapter 5

Examples

Example5_2.m	Calculates the concentration profile of a system of first-order chemical reactions by solving the set of linear ordinary differential equations (*LinearODE.m*).
Example5_3.m	Calculates the concentration and temperature profiles of a nonisothermal reactor by solving the mole and energy balances (*Euler.m, MEuler.m, RK.m, Adams.m, AdamsMoulton.m*).
Example5_4.m	Calculates the velocity profile of a non-Newtonian fluid flowing in a circular pipe by solving the momentum balance equation (*shooting.m*).
Example5_5.m	Calculates the optimum concentration and temperature profiles in a batch penicillin fermentor (*collocation.m*).

Methods

LinearODE.m	Solution of a set of linear ordinary differential equations.
Euler.m	Solution of a set of nonlinear ordinary differential equations by the explicit Euler method.

MEuler.m Solution of a set of nonlinear ordinary differential equations by the modified Euler (predictor-corrector) method.

RK.m Solution of a set of nonlinear ordinary differential equations by the Runge-Kutta methods of order 2 to 5.

Adams.m Solution of a set of nonlinear ordinary differential equations by the Adams method.

AdamsMoulton.m Solution of a set of nonlinear ordinary differential equations by the Adams-Moulton predictor-corrector method.

shooting.m Solution of a boundary-value problem in the form of a set of ordinary differential equations by the shooting method using Newton's technique.

collocation.m Solution of a boundary-value problem in the form of a set of ordinary differential equations by the orthogonal collocation method.

Functions

Ex5_3_func.m Contains the mole and energy balances (used in *Example5_3.m*).

Ex5_4_func.m Contains the set of differential equations obtained from the momentum balance (used in *Example5_4.m*).

Ex5_5_func.m Contains the set of system and adjoint equations (used in *Example5_5.m*).

Ex5_5_theta.m Contains the necessary condition for maximum as a function of temperature (used in *Example5_5.m*).

CHAPTER 6

Examples

Example6_1.m Calculates the temperature profile of a rectangular plate solving the two-dimensional heat balance (*elliptic.m*).

Example6_2.m Calculates the unsteady-state one-dimensional concentration profile of gas A diffusing in liquid B (*parabolic1D.m*).

Example6_3.m Calculates the unsteady-state two-dimensional temperature profile in a furnace wall (*parabolic2D.m*).

Methods

elliptic.m Solution of two-dimensional elliptic partial differential equation.

parabolic1D.m Solution of parabolic partial differential equation in one space dimension by the implicit Crank-Nicolson method.

parabolic2D.m Solution of parabolic partial differential equation in two space dimensions by the explicit method.

Functions

Ex6_2_func.m Contains the equation of the rate of chemical reaction (used in *Example6_2.m*).

CHAPTER 7

Examples

Example7_1.m Uses the nonlinear regression program (*NLR.m* and *statistics.m*) to determine the parameters of two differential equations that represent the kinetics of penicillin fermentation. The equations are fitted to experimental data.

Methods

NLR.m Least squares multiple nonlinear regression using the Marquardt and Gauss-Newton methods. The program can fit simultaneous ordinary differential equations and/or algebraic equations to multiresponse data.

statistics.m Performs a series of statistical tests on the data being fitted and on the regression results.

Functions

Ex7_1_func.m Contains the model equations for cell growth and penicillin formation used in *Example7_1.m*.

stud.m Evaluates the Student *t* distribution.

Data

Ex7_1_data.mat The MATLAB workspace containing the data for Example 7.1

General Algorithm for the Software Developed in this Book

The Algorithm

Example.m

This is a program that solves the specific example described in the text. It is interactive with the user. It asks the user to enter, from the keyboard, the parameters that will be used by the method (such as the name of the function that contains the equations, constants, initial guesses, convergence criterion).

This program calls the *method.m* function, passes the parameters to it, and receives back the results. It writes out the results in a formatted form and generates plots of the results, if needed.

Method.m

This is a general function that implements a method (such as the Newton-Raphson, Linear Interpolation, Gauss Elimination). This function is portable so that it can be called by other input-output programs and/or from the MATLAB work space (with parameters).

It may call the *function.m* that contains the specific equations to be solved. It may also call any of the built-in MATLAB functions. The results of the method may be printed out (or plotted) here, if they are generic.

Function.m

This function contains the specific equations to be solved. It may also contain some or all constants that are particular to these equations.

This function must be provided by the user.

MATLAB *functions*

Any of the built-in functions and plotting routines that may be needed.

CHAPTER 1

Numerical Solution of Nonlinear Equations

1.1 Introduction

Many problems in engineering and science require the solution of nonlinear equations. Several examples of such problems drawn from the field of chemical engineering and from other application areas are discussed in this section. The methods of solution are developed in the remaining sections of the chapter, and specific examples of the solutions are demonstrated using the MATLAB software.

In thermodynamics, the pressure-volume-temperature relationship of real gases is described by the equation of state. There are several semitheoretical or empirical equations, such as Redlich-Kwong, Soave-Redlich-Kwong, and the Benedict-Webb-Rubin equations, which have been used extensively in chemical engineering. For example, the Soave-Redlich-Kwong equation of state has the form

$$P = \frac{RT}{V - b} - \frac{a\alpha}{V(V + b)} \tag{1.1}$$

where P, V, and T are the pressure, specific volume, and temperature, respectively. R is the gas constant, α is a function of temperature, and a and b are constants, specific for each gas. Eq. (1.1) is a third-degree polynomial in V and can be easily rearranged into the canonical form for a polynomial, which is

$$Z^3 - Z^2 + (A - B - B^2)Z - AB = 0 \tag{1.2}$$

where $Z = PV/RT$ is the compressibility factor, $A = \alpha a P/R^2T^2$ and $B = bP/RT$. Therefore, the problem of finding the specific volume of a gas at a given temperature and pressure reduces to the problem of finding the appropriate root of a polynomial equation.

In the calculations for multicomponent separations, it is often necessary to estimate the minimum reflux ratio of a multistage distillation column. A method developed for this purpose by Underwood [1], and described in detail by Treybal [2], requires the solution of the equation

$$\sum_{j=1}^{n} \frac{\alpha_j z_{jF} F}{\alpha_j - \phi} - F(1 - q) = 0 \tag{1.3}$$

where F is the molar feed flow rate, n is the number of components in the feed, z_{jF} is the mole fraction of each component in the feed, q is the quality of the feed, α_j is the relative volatility of each component at average column conditions, and ϕ is the root of the equation. The feed flow rate, composition, and quality are usually known, and the average column conditions can be approximated. Therefore, ϕ is the only unknown in Eq. (1.3). Because this equation is a polynomial in ϕ of degree n, there are n possible values of ϕ (roots) that satisfy the equation.

The friction factor f for turbulent flow of an incompressible fluid in a pipe is given by the nonlinear Colebrook equation

$$\sqrt{\frac{1}{f}} = -0.86 \ln\left(\frac{\epsilon/D}{3.7} + \frac{2.51}{N_{Re}\sqrt{f}}\right) \tag{1.4}$$

where ϵ and D are roughness and inside diameter of the pipe, respectively, and N_{Re} is the

Reynolds number. This equation does not readily rearrange itself into a polynomial form; however, it can be arranged so that all the nonzero terms are on the left side of the equation as follows:

$$\sqrt{\frac{1}{f}} + 0.86\ln\left(\frac{\epsilon/D}{3.7} + \frac{2.51}{N_{Re}\sqrt{f}}\right) = 0 \tag{1.5}$$

The method of differential operators is applied in finding analytical solutions of nth-order linear homogeneous differential equations. The general form of an nth-order linear homogeneous differential equation is

$$a_n\frac{d^n y}{dx^n} + a_{n-1}\frac{d^{n-1}y}{dx^{n-1}} + \ldots + a_1\frac{dy}{dx} + a_0 y = 0 \tag{1.6}$$

By defining D as the differentiation with respect to x:

$$D = \frac{d}{dx} \tag{1.7}$$

Eq. (1.6) can be written as

$$[a_n D^n + a_{n-1}D^{n-1} + \ldots + a_1 D + a_0]y = 0 \tag{1.8}$$

where the bracketed term is called the differential operator. In order for Eq. (1.8) to have a nontrivial solution, the differential operator must be equal to zero:

$$a_n D^n + a_{n-1}D^{n-1} + \ldots + a_1 D + a_0 = 0 \tag{1.9}$$

This, of course, is a polynomial equation in D whose roots must be evaluated in order to construct the complementary solution of the differential equation.

The field of process dynamics and control often requires the location of the roots of transfer functions that usually have the form of polynomial equations. In kinetics and reactor design, the simultaneous solution of rate equations and energy balances results in mathematical models of simultaneous nonlinear and transcendental equations. Methods of solution for these and other such problems are developed in this chapter.

1.2 Types of Roots and Their Approximation

All the nonlinear equations presented in Sec. 1.1 can be written in the general form

$$f(x) = 0 \tag{1.10}$$

where x is a single variable that can have multiple values (roots) that satisfy this equation. The function $f(x)$ may assume a variety of nonlinear functionalities ranging from that of a polynomial equation whose canonical form is

$$f(x) = a_n x^n + a_{n-1}x^{n-1} + \ldots + a_1 x + a_0 = 0 \tag{1.11}$$

to the transcendental equations, which involve trigonometric, exponential, and logarithmic terms. The roots of these functions could be

1. Real and distinct
2. Real and repeated
3. Complex conjugates
4. A combination of any or all of the above.

The real parts of the roots may be positive, negative, or zero.

Fig. 1.1 graphically demonstrates all the above cases using fourth-degree polynomials. Fig. 1.1*a* is a plot of the polynomial equation (1.12):

$$x^4 + 6x^3 + 7x^2 - 6x - 8 = 0 \tag{1.12}$$

which has four real and distinct roots at -4, -2, -1, and 1, as indicated by the intersections of the function with the x axis. Fig. 1.1*b* is a graph of the polynomial equation (1.13):

$$x^4 + 7x^3 + 12x^2 - 4x - 16 = 0 \tag{1.13}$$

which has two real and distinct roots at -4 and 1 and two real and repeated roots at -2. The point of tangency with the x axis indicates the presence of the repeated roots. At this point $f(x) = 0$ and $f'(x) = 0$. Fig. 1.1*c* is a plot of the polynomial equation (1.14):

$$x^4 - 6x^3 + 18x^2 - 30x + 25 = 0 \tag{1.14}$$

which has only complex roots at $1 \pm 2i$ and $2 \pm i$. In this case, no intersection with the x axis of the Cartesian coordinate system occurs, as all of the roots are located in the complex plane. Finally, Fig. 1.1*d* demonstrates the presence of two real and two complex roots with

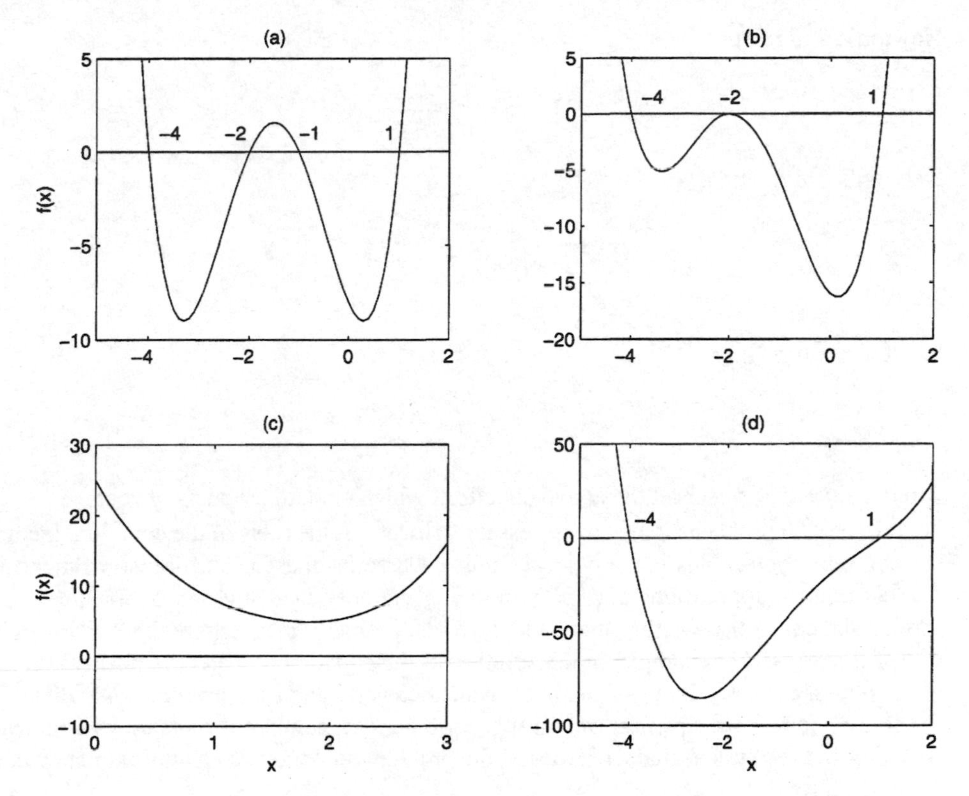

Figure 1.1 Roots of fourth-degree polynomial equations. (*a*) Four real distinct. (*b*) Two real and two repeated. (*c*) Four complex. (*d*) Two real and two complex.

the polynomial equation (1.15):

$$x^4 + x^3 - 5x^2 + 23x - 20 = 0 \tag{1.15}$$

whose roots are -4, 1, and $1 \pm 2i$. As expected, the function crosses the x axis only at two points: -4 and 1.

The roots of an nth-degree polynomial, such as Eq. (1.11), may be verified using Newton's relations, which are:

Newton's 1st relation:

$$\sum_{i=1}^{n} x_i = -\frac{a_{n-1}}{a_n} \tag{1.16}$$

where x_i are the roots of the polynomial.

Newton's 2nd relation:

$$\sum_{i,j=1}^{n} x_i x_j = \frac{a_{n-2}}{a_n} \tag{1.17}$$

Newton's 3rd relation:

$$\sum_{i,j,k=1}^{n} x_i x_j x_k = -\frac{a_{n-3}}{a_n} \tag{1.18}$$

Newton's *n*th relation:

$$x_1 x_2 x_3 \ldots x_n = (-1)^n \frac{a_0}{a_n} \tag{1.19}$$

where $i \neq j \neq k \neq \ldots$ for all the above equations which contain products of roots.

In certain problems it may be necessary to locate all the roots of the equation, including the complex roots. This is the case in finding the zeros and poles of transfer functions in process control applications and in formulating the analytical solution of linear *n*th-order differential equations. On the other hand, different problems may require the location of only one of the roots. For example, in the solution of the equation of state, the positive real root is the one of interest. In any case, the physical constraints of the problem may dictate the feasible region of search where only a subset of the total number of roots may be indicated. In addition, the physical characteristics of the problem may provide an approximate value of the desired root.

The most effective way of finding the roots of nonlinear equations is to devise iterative algorithms that start at an initial estimate of a root and converge to the exact value of the desired root in a finite number of steps. Once a root is located, it may be removed by synthetic division if the equation is of the polynomial form. Otherwise, convergence on the same root may be avoided by initiating the search for subsequent roots in different region of the feasible space.

For equations of the polynomial form, Descartes' rule of sign may be used to determine the number of positive and negative roots. This rule states: The number of positive roots is equal to the number of sign changes in the coefficients of the equation (or less than that by an even integer); the number of negative roots is equal to the number of sign repetitions in the coefficients (or less than that by an even integer). Zero coefficients are counted as positive [3]. The purpose of the qualifier, "less than that by an even integer," is to allow for the existence of conjugate pairs of complex roots. The reader is encouraged to apply Descartes' rule to Eqs. (1.12)-(1.15) to verify the results already shown.

If the problem to be solved is a purely mathematical one, that is, the model whose roots are being sought has no physical origin, then brute-force methods would have to be used to establish approximate starting values of the roots for the iterative technique. Two categories of such methods will be mentioned here. The first one is a truncation method applicable to equation of the polynomial form. For example, the following polynomial

$$a_4 x^4 + a_3 x^3 + a_2 x^2 + a_1 x + a_0 = 0 \tag{1.20}$$

may have its lower powered terms truncated

$$a_4 x^4 + a_3 x^3 \simeq 0 \tag{1.21}$$

to yield an approximation of one of the roots

$$x \simeq -\frac{a_3}{a_4} \tag{1.22}$$

Alternatively, if the higher powered terms are truncated

$$a_1 x + a_0 \simeq 0 \tag{1.23}$$

the approximate root is

$$x \simeq -\frac{a_0}{a_1} \tag{1.24}$$

This technique applied to Soave-Redlich-Kwong equation [Eq. (1.2)] results in

$$Z = \frac{PV}{RT} \simeq 1 \tag{1.25}$$

This, of course, is the well-known *ideal gas law*, which is an excellent approximation of the pressure-volume-temperature relationship of real gases at low pressures. On the other end of the polynomial, truncation of the higher powered terms results in

$$Z \simeq \frac{AB}{A - B - B^2} \tag{1.26}$$

giving a value of Z very close to zero which is the case for liquids. In this case, the physical considerations of the problem determine that Eq. (1.25) or Eq. (1.26) should be used for gas phase or liquid phase, respectively, to initiate the iterative search technique for the real root.

Another method of locating initial estimates of the roots is to scan the entire region of search by small increments and to observe the steps in which a change of sign in the function $f(x)$ occurs. This signals that the function $f(x)$ crosses the x axis within the particular step. This search can be done easily in MATLAB environment using *fplot* function. Once the function $f(x)$ is introduced in a MATLAB function *file_name.m*, the statement *fplot*('*file_name*',[*a*, *b*]) shows the plot of the function from $x=a$ to $x=b$. The values of a and b may be changed until the plot crosses the x axis.

The scan method may be a rather time-consuming procedure for polynomials whose roots lie in a large region of search. A variation of this search is the method of bisection that divides the interval of search by 2 and always retains that half of the search interval in which the change of sign has occurred. When the range of search has been narrowed down sufficiently, a more accurate search technique would then be applied within that step in order to refine the value of the root.

More efficient methods based on rearrangement of the function to $x = g(x)$ (*method of successive substitution*), linear interpolation of the function (*method of false position*), and the tangential descent of the function (*Newton-Raphson method*) will be described in the next three sections of this chapter.

MATLAB has its own built-in function *fzero* for root finding. The statement *fzero*('*file_name*',x_0) finds the root of the function $f(x)$ introduced in the user-defined MATLAB function *file_name.m*. The second argument x_0 is a starting guess. Starting with this initial value, the function *fzero* searches for change in the sign of the function $f(x)$. The calculation then continues with either bisection or linear interpolation method until the convergence is achieved.

1.3 The Method of Successive Substitution

The simplest one-point iterative root-finding technique can be developed by rearranging the function $f(x)$ so that x is on the left-hand side of the equation

$$x = g(x) \tag{1.27}$$

The function $g(x)$ is a formula to predict the root. In fact, the root is the intersection of the line $y = x$ with the curve $y = g(x)$. Starting with an initial value of x_1, as shown in Fig. 1.2*a*, we obtain the value of x_2:

$$x_2 = g(x_1) \tag{1.28}$$

which is closer to the root than x_1 and may be used as an initial value for the next iteration. Therefore, general iterative formula for this method is

$$x_{n+1} = g(x_n) \tag{1.29}$$

which is known as the method of successive substitution or the method of $x = g(x)$.

A sufficient condition for convergence of Eq. (1.29) to the root x^* is that $|g'(x)| < 1$ for all x in the search interval. Fig. 1.2*b* shows the case when this condition is not valid and the method diverges. This analytical test is often difficult in practice. In a computer program it is easier to determine whether $|x_3 - x_2| < |x_2 - x_1|$ and, therefore, the successive x_n values converge. The advantage of this method is that it can be started with only a single point, without the need for calculating the derivative of the function.

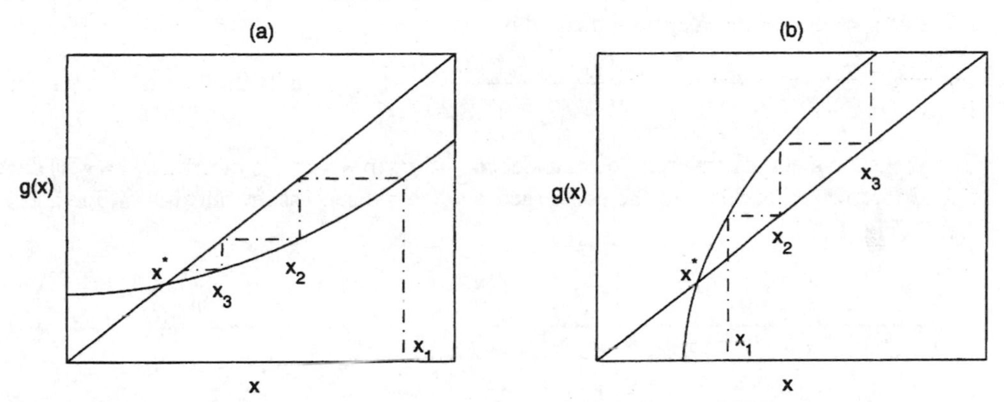

Figure 1.2 Use of $x = g(x)$ method. (*a*) Convergence. (*b*) Divergence.

1.4 The Wegstein Method

The Wegstein method may also be used for the solution of the equations of the form

$$x = g(x) \tag{1.27}$$

Starting with an initial value of x_1, we first obtain another estimation of the root from

$$x_2 = g(x_1) \tag{1.28}$$

As shown in Fig. 1.3, x_2 does not have to be closer to the root than x_1. At this stage, we estimate the function $g(x)$ with a line passing from the points $(x_1, g(x_1))$ and $(x_2, g(x_2))$

$$\frac{y - g(x_1)}{x - x_1} = \frac{g(x_2) - g(x_1)}{x_2 - x_1} \tag{1.30}$$

and find the next estimation of the root, x_3, from the intersection of the line (1.30) and the line $y = x$:

$$x_3 = \frac{x_1 g(x_2) - x_2 g(x_1)}{x_1 - g(x_1) - x_2 + g(x_2)} \tag{1.31}$$

It can be seen from Fig. 1.3*a* that x_3 is closer to the root than either x_1 and x_2. In the next iteration we pass the line from the points $(x_2, g(x_2))$ and $(x_3, g(x_3))$ and again evaluate the next estimation of the root from the intersection of this line with $y = x$. Therefore, the general iterative formula for the Wegstein method is

$$x_{n+1} = \frac{x_{n-1} g(x_n) - x_n g(x_{n-1})}{x_{n-1} - g(x_{n-1}) - x_n + g(x_n)} \qquad n \geq 2 \tag{1.32}$$

The Wegstein method converges, even under conditions in which the method of $x = g(x)$ does not. Moreover, it accelerates the convergence when the successive substitution method is stable (Fig. 1.3*b*).

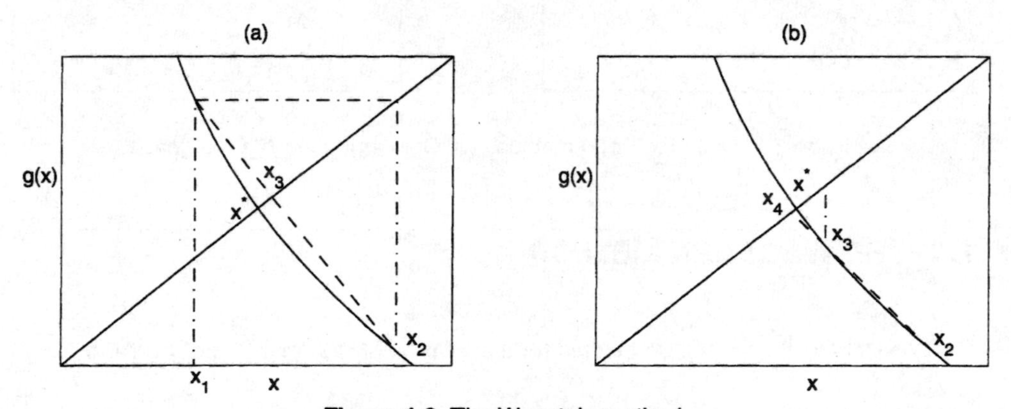

Figure 1.3 The Wegstein method.

1.5 The Method of Linear Interpolation (Method of False Position)

This technique is based on linear interpolation between two points on the function that have been found by a scan to lie on either side of a root. For example, x_1 and x_2 in Fig. 1.4*a* are positions on opposite sides of the root x^* of the nonlinear function $f(x)$. The points $(x_1, f(x_1))$ and $(x_2, f(x_2))$ are connected by a straight line, which we will call a chord, whose equation is

$$y(x) = ax + b \tag{1.33}$$

Because this chord passes through the two points $(x_1, f(x_1))$ and $(x_2, f(x_2))$, its slope is

$$a = \frac{f(x_2) - f(x_1)}{x_2 - x_1} \tag{1.34}$$

and its y intercept is

$$b = f(x_1) - ax_1 \tag{1.35}$$

Eq. (1.33) then becomes

$$y(x) = \left[\frac{f(x_2) - f(x_1)}{x_2 - x_1}\right]x + \left\{f(x_1) - \left[\frac{f(x_2) - f(x_1)}{x_2 - x_1}\right]x_1\right\} \tag{1.36}$$

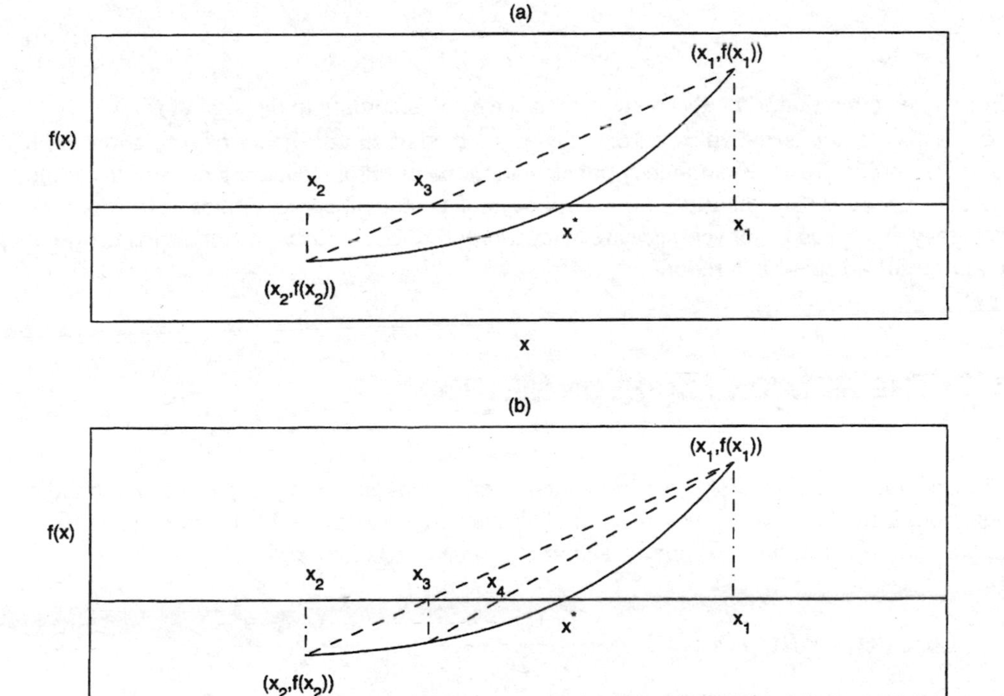

Figure 1.4 Method of linear interpolation.

Locating x_3 using Eq. (1.36), where $y(x_3) = 0$:

$$x_3 = x_1 - \frac{f(x_1)(x_2 - x_1)}{f(x_2) - f(x_1)} \tag{1.37}$$

Note that for the shape of curve chosen on Fig. 1.4, x_3 is nearer to the root x^* than either x_1 or x_2. This, of course, will not always be the case with all functions. Discussion of criteria for convergence will be given in the next section.

According to Fig. 1.4, $f(x_3)$ has the same sign as $f(x_2)$; therefore, x_2 may be replaced by x_3. Now repeating the above operation and connecting the points $(x_1, f(x_1))$ and $(x_3, f(x_3))$ with a new chord, as shown in Fig. 1.4*b*, we obtain the value of x_4:

$$x_4 = x_1 - \frac{f(x_1)(x_3 - x_1)}{f(x_3) - f(x_1)} \tag{1.38}$$

which is nearer to the root than x_3. For general formulation of this method, consider x^+ to be the value at which $f(x^+) > 0$ and x^- to be the value at which $f(x^-) < 0$. Next improved approximation of the root of the function may be calculated by successive application of the general formula

$$x_n = x^+ - \frac{f(x^+)(x^+ - x^-)}{f(x^+) - f(x^-)} \tag{1.39}$$

For the next iteration, x^+ or x^- should be replaced by x_n according to the sign of $f(x_n)$.

This method is known by several names: method of chords, linear interpolation, false position (*regula falsi*). Its simplicity of calculation (no need for evaluating derivatives of the function) gave it its popularity in the early days of numerical computations. However, its accuracy and speed of convergence are hampered by the choice of x_1, which forms the pivot point for all subsequent iterations.

1.6 The Newton-Raphson Method

The best known, and possibly the most widely used, technique for locating roots of nonlinear equations is the *Newton-Raphson* method. This method is based on a Taylor series expansion of the nonlinear function $f(x)$ around an initial estimate (x_1) of the root:

$$f(x) = f(x_1) + f'(x_1)(x - x_1) + \frac{f''(x_1)(x - x_1)^2}{2!} + \frac{f'''(x_1)(x - x_1)^3}{3!} + \dots \tag{1.40}$$

Because what is being sought is the value of x that forces the function $f(x)$ to assume zero value, the left side of Eq. (1.40) is set to zero, and the resulting equation is solved for x.

However, the right-hand side is an infinite series. Therefore, a finite number of terms must be retained and the remaining terms must be truncated. Retaining only the first two terms on the right-hand side of the Taylor series is equivalent to linearizing the function $f(x)$. This operation results in

$$x = x_1 - \frac{f(x_1)}{f'(x_1)} \tag{1.41}$$

that is, the value of x is calculated from x_1 by correcting this initial guess by $f(x_1)/f'(x_1)$. The geometrical significance of this correction is shown in Fig. 1.5*a*. The value of x is obtained by moving from x_1 to x in the direction of the tangent $f'(x_1)$ of the function $f(x)$.

Because the Taylor series was truncated, retaining only two terms, the new value x will not yet satisfy Eq. (1.10). We will designate this value as x_2 and reapply the Taylor series linearization at x_2 (shown in Fig. 1.5*b*) to obtain x_3. Repetitive application of this step converts Eq. (1.41) to an iterative formula:

$$x_{n+1} = x_n - \frac{f(x_n)}{f'(x_n)} \tag{1.42}$$

In contrast to the method of linear interpolation discussed in Sec. 1.5, the Newton-Raphson method uses the newly found position as the starting point for each subsequent iteration.

In the discussion for both linear interpolation and Newton-Raphson methods, a certain shape of the function was used to demonstrate how these techniques converge toward a root in the space of search. However, the shapes of nonlinear functions may vary drastically, and convergence is not always guaranteed. As a matter of fact, divergence is more likely to occur, as shown in Fig. 1.6, unless extreme care is taken in the choice of the initial starting points.

To investigate the convergence behavior of the Newton-Raphson method, one has to examine the term $[-f(x_n)/f'(x_n)]$ in Eq. (1.42). This is the error term or correction term applied to the previous estimate of the root at each iteration. A function with a strong vertical trajectory near the root will cause the denominator of the error term to be large; therefore, the

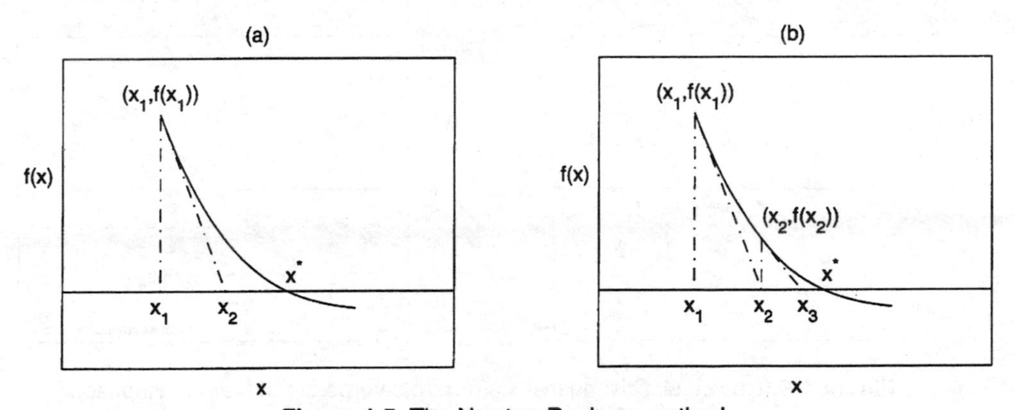

Figure 1.5 The Newton-Raphson method.

convergence will be quite fast. If, however, $f(x)$ is nearly horizontal near the root, the convergence will be slow. If at any point during the search, $f'(x) = 0$, the method would fail due to division by zero. Inflection points on the curve, within the region of search, are also troublesome and may cause the search to diverge.

A sufficient, but not necessary, condition for convergence of the Newton-Raphson method was stated by Lapidus [4] as follows: "If $f'(x)$ and $f''(x)$ do not change sign in the interval (x_1, x^*) and if $f(x_1)$ and $f''(x_1)$ have the same sign, the iteration will always converge to x^*." These convergence criteria may be easily programmed as part of the computer program which performs the Newton-Raphson search, and a warning may be issued or other appropriate action may be taken by the computer if the conditions are violated.

A more accurate extension of the Newton-Raphson method is Newton's 2nd-order method, which truncates the right-hand side of the Taylor series [Eq. (1.40)] after the third term to yield the equation:

$$\frac{f''(x_1)}{2!}(\Delta x_1)^2 + f'(x_1)\,\Delta x_1 + f(x_1) = 0 \tag{1.43}$$

where $\Delta x_1 = x - x_1$. This is a quadratic equation in Δx_1 whose solution is given by

$$\Delta x_1 = \frac{-f'(x_1) \pm \sqrt{[f'(x_1)]^2 - 2f''(x_1)f(x_1)}}{f''(x_1)} \tag{1.44}$$

The general iterative formula for this method would be

$$x_{n+1} \overset{+}{=} x_n - \frac{f'(x_n)}{f''(x_n)} + \frac{\sqrt{[f'(x_n)]^2 - 2f''(x_n)f(x_n)}}{f''(x_n)} \tag{1.45a}$$

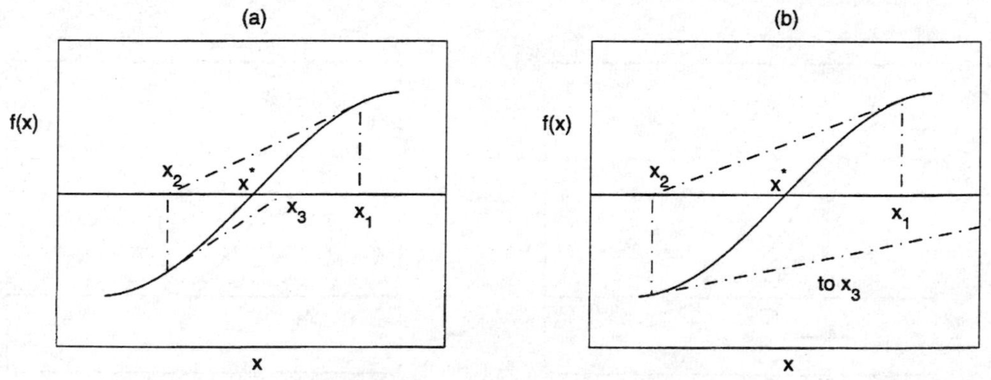

Figure 1.6 Choice of initial guesses affects convergence of Newton-Raphson method. (*a*) Convergence. (*b*) Divergence.

or

$$x_{n+1}^{-} = x_n - \frac{f'(x_n)}{f''(x_n)} - \frac{\sqrt{[f'(x_n)]^2 - 2f''(x_n)f(x_n)}}{f''(x_n)} \tag{1.45b}$$

The choice between (1.45*a*) and (1.45*b*) will be determined by exploring both values of x_{n+1}^{+} and x_{n+1}^{-} and determining which one results in the function $f(x_{n+1}^{+})$ or $f(x_{n+1}^{-})$ being closer to zero.

An alternative to the above exploration will be to treat Eq. (1.43) as another nonlinear equation in Δx and to apply the Newton-Raphson method for its solution:

$$F(\Delta x) = \frac{f''(x_1)}{2!}(\Delta x)^2 + f'(x_1)\,\Delta x + f(x_1) = 0 \tag{1.46}$$

where

$$\Delta x_{n+1} = \Delta x_n - \frac{F(\Delta x_n)}{F'(\Delta x_n)} \tag{1.47}$$

Two nested Newton-Raphson algorithms would have to be programmed together as follows:

1. Assume a value of x_1.
2. Calculate Δx_1 from Eq. (1.44).
3. Calculate Δx_2 from Eq. (1.47).
4. Calculate x_2 from $x_2 = x_1 + \Delta x_2$.
5. Repeat steps 2 to 4 until convergence is achieved.

Example 1.1: Solution of the Colebrook Equation by Successive Substitution, Linear Interpolation, and Newton-Raphson Methods. Develop MATLAB functions to solve nonlinear equations by the successive substitution method, the linear interpolation, and the Newton-Raphson root-finding techniques. Use these functions to calculate the friction factor from the Colebrook equation [Eq. (1.4)] for flow of a fluid in a pipe with $\epsilon/D = 10^{-4}$ and $N_{Re} = 10^5$. Compare these methods with each other.

Method of Solution: Eqs. (1.29), (1.39), and (1.42) are used for the method of $x = g(x)$, linear interpolation, and Newton-Raphson, respectively. The iterative procedure stops when the difference between two succeeding approximations of the root is less than the convergence criterion (default value is 10^{-6}), or when the number of iterations reaches 100, whichever is satisfied first. The program may show the convergence results numerically and/or graphically, if required, to illustrate how each method arrives at the answer.

Program Description: Three MATLAB functions called *XGX.m*, *LI.m*, and *NR.m* are developed to find the root of a general nonlinear equation using successive substitution [the method of $x = g(x)$], linear interpolation, and Newton-Raphson methods, respectively. The name of the nonlinear function subject to root finding is introduced in the input arguments; therefore, these MATLAB functions may be applied to any problem.

Successive substitution method (XGX.m): This function starts with initialization section in which input arguments are evaluated and initial values for the main iteration are introduced. The first argument is the name of the MATLAB function in which the function $g(x)$ is described. The second argument is a starting value and has to be a scalar. By default, convergence is assumed when two succeeding iterations result in root approximations with less than 10^{-6} in difference. If another value for convergence is desired, it may be introduced to the function by the third argument. A value of 1 as the fourth argument makes the function show the results of each iteration step numerically. If this value is set to 2, the function shows the results numerically and graphically. The third and fourth arguments are optional. Every additional argument that is introduced after the fourth argument is passed directly to the function $g(x)$. In this case, if it is desired to use the default values for the third and fourth arguments, an empty matrix should be entered in their place. For solution of the problem, the values of N_{Re} and ϵ/D are passed to the Colebrook function by introducing them in fifth and sixth arguments.

The next section in the function is the main iteration loop, in which the iteration according to Eq. (1.29) takes place and the convergence is checked. In the case of the Colebrook equation, Eq. (1.4) is rearranged to solve for f: The right-hand side of this equation is taken as $g(f)$ and is introduced in the MATLAB function *Colebrookg.m*. Numerical results of the calculations are also shown, if requested, in each iteration of this section.

$$f = \frac{1}{\left[0.86 \ln\left(\frac{\epsilon/D}{3.7} + \frac{2.51}{N_{Re}\sqrt{f}}\right)\right]^2} = g(f)$$

At the end, the MATLAB function plots the function as well as the results of the calculation, if required, to illustrate the convergence procedure.

Linear interpolation method (LI.m): This function consists of the same parts as the *XGX.m* function. The number of input arguments is one more than that of *XGX.m*, because the linear interpolation method needs two starting points. Special care should be taken to introduce two starting values in which the function have opposite signs. Eq. (1.5) is used without change as the function the root of which is to be located. This function is contained in a MATLAB function called *Colebrook.m*.

Newton-Raphson method (NR.m): The structure of this function is the same as that of the two previous functions. The derivative of the function is taken numerically to reduce the inputs. It is also more applicable for complicated functions. The reader may simply introduce the derivative function in another MATLAB function and use it instead of numerical derivation. In the case of the Colebrook equation, the same MATLAB function *Colebrook.m*, which represents Eq. (1.5), may be used with this function to calculate the value of the friction factor.

The MATLAB program *Example1_1.m* finds the friction factor from the Colebrook equation by three different methods of root finding. The program first asks the user to input the values for N_{Re} and ϵ/D. It then asks for the method of solution of the Colebrook equation, name of the *m*-file that contains the Colebrook equation, and the initial value(s) to start the method. The program then calculates the friction factor by the selectcd method and continues asking for the method of solution until the user enters 0.

WARNING: The original MATLAB Version 5.2 had a "bug." The command

linspace(0,0,100)

which is used in *LI.m*, *NR.m*, and *Nrpoly.m* in Chapter 1, would not work properly in the MATLAB installations of Version 5.2 which have not yet been corrected. A patch which corrects this problem is available on the website of Math Works, Inc.:

http://www.mathworks.com

If you have Version 5.2, you are strongly encouraged to download and install this patch, if you have not done so already.

Program

Example1_1.m

```
% Example1_1.m
% This program solves the problem posed in Example 1.1.
% It calculates the friction factor from the Colebrook equation
% using the Successive Substitution, the Linear Interpolation,
% and the Newton-Raphson methods.

clear
clc
clf

disp('Calculating the friction factor from the Colebrook equation')

% Input
Re = input('\n Reynolds No.         = ');
e_over_D = input(' Relative roughness = ');

method = 1;
while method
   fprintf('\n')
   disp('  1 ) Successive substitution')
   disp('  2 ) Linear Interpolation')
   disp('  3 ) Newton Raphson')
   disp('  0 ) Exit')
   method = input('\n Choose the method of solution : ');
   if method
```

```
 fname = input('\n Function containing the Colebrook equation : ');
  end
  switch method
  case 1                                          % Successive substitution
     x0 = input(' Starting value = ');
     f = xgx(fname,x0,[],2,Re,e_over_D);
     fprintf('\n f = %6.4f\n',f)
  case 2                                          % Linear interpolation
     x1 = input(' First starting value = ');
     x2 = input(' Second starting value = ');
     f = LI(fname,x1,x2,[],2,Re,e_over_D);
     fprintf('\n f = %6.4f\n',f)
  case 3                                          % Newton-Raphson
     x0 = input(' Starting value = ');
     f = NR(fname,x0,[],2,Re,e_over_D);
     fprintf('\n f = %6.4f\n',f)
  end
end
```

XGX.m

```
function x = XGX(fnctn,x0,tol,trace,varargin)
%XGX Finds a zero of a function by x=g(x) method.
%
%   XGX('G',X0) finds the intersection of the curve y=g(x)
%   with the line y=x. The function g(x) is  described by the
%   M-file G.M. X0 is a starting guess.
%
%   XGX('G',X0,TOL,TRACE) uses tolerance TOL for convergence
%   test. TRACE=1 shows the calculation steps numerically and
%   TRACE=2 shows the calculation steps both numerically and
%   graphically.
%
%   XGX('G',X0,TOL,TRACE,P1,P2,...) allows for additional
%   arguments which are passed to the function G(X,P1,P2,...).
%   Pass an empty matrix for TOL or TRACE to use the default
%   value.
%
%   See also FZERO, ROOTS, NR, LI

% (c) by N. Mostoufi & A. Constantinides
% January 1, 1999

% Initialization
if nargin < 3 | isempty(tol)
   tol = 1e-6;
end
if nargin < 4 | isempty(trace)
   trace = 0;
end
if tol == 0
   tol = 1e-6;
```

```
end
if (length(x0) > 1) | (~isfinite(x0))
   error('Second argument must be a finite scalar.')
end
if trace
   header = ' Iteration        x              g(x)';
   disp(' ')
   disp(header)
   if trace == 2
      xpath = [x0];
      ypath = [0];
   end
end

x = x0;
x0 = x + 1;
iter = 1;
itermax = 100;

% Main iteration loop
while abs(x - x0) > tol & iter <= itermax
   x0 = x;
   fnk = feval(fnctn,x0,varargin{:});

   % Next approximation of the root
   x = fnk;

   % Show the results of calculation
   if trace
      fprintf('%5.0f   %13.6g %13.6g\n',iter, [x0 fnk])
      if trace == 2
         xpath = [xpath x0 x];
         ypath = [ypath fnk x];
      end
   end
   iter = iter + 1;
end

if trace == 2
   % Plot the function and path to the root
   xmin = min(xpath);
   xmax = max(xpath);
   dx = xmax - xmin;
   xi = xmin - dx/10;
   xf = xmax + dx/10;
   yc = [];
   for xc = xi : (xf - xi)/99 : xf
      yc=[yc feval(fnctn,xc,varargin{:})];
   end
   xc = linspace(xi,xf,100);
   plot(xc,yc,xpath,ypath,xpath(2),ypath(2),'*', ...
```

```
      x,fnk,'o',[xi xf],[xi,xf],'--')
   axis([xi xf min(yc) max(yc)])
   xlabel('x')
   ylabel('g(x)  [-- : y=x]')
   title('x=g(x)  : The function and path to the root ...
      (* : initial guess ; o : root)')
end

if iter >= itermax
   disp('Warning : Maximum iterations reached.')
end
```

LI.m

```
function x = LI(fnctn,x1,x2,tol,trace,varargin)
%LI Finds a zero of a function by the linear interpolation method.
%
%   LI('F',X1,X2) finds a zero of the function described by the
%   M-file F.M. X1 and X2 are starting points where the function
%   has different signs at these points.
%
%   LI('F',X1,X2,TOL,TRACE) uses tolerance TOL for convergence
%   test. TRACE=1 shows the calculation steps numerically and
%   TRACE=2 shows the calculation steps both numerically and
%   graphically.
%
%   LI('F',X1,X2,TOL,TRACE,P1,P2,...) allows for additional
%   arguments which are passed to the function F(X,P1,P2,...).
%   Pass an empty matrix for TOL or TRACE to use the default
%   value.
%
%   See also FZERO, ROOTS, XGX, NR

% (c) by N. Mostoufi & A. Constantinides
% January 1, 1999

% Initialization
if nargin < 4 | isempty(tol)
   tol = 1e-6;
end
if nargin < 5 | isempty(trace)
   trace = 0;
end
if tol == 0
   tol = 1e-6;
end
if (length(x1) > 1) | (~isfinite(x1)) | (length(x2) > 1) | ...
      (~isfinite(x2))
   error('Second and third arguments must be finite scalars.')
end
if trace
   header = ' Iteration       x                f(x)';
```

```
   disp(' ')
   disp(header)
end
f1 = feval(fnctn,x1,varargin{:});
f2 = feval(fnctn,x2,varargin{:});

iter = 0;
if trace
   % Display initial values
   fprintf('%5.0f    %13.6g %13.6g \n',iter, [x1 f1])
   fprintf('%5.0f    %13.6g %13.6g \n',iter, [x2 f2])
   if trace == 2
      xpath = [x1 x1 x2 x2];
      ypath = [0 f1 f2 0];
   end
end

if f1 < 0
   xm = x1;
   fm = f1;
   xp = x2;
   fp = f2;
else
   xm = x2;
   fm = f2;
   xp = x1;
   fp = f1;
end

iter = iter + 1;
itermax = 100;
x = xp;
x0 = xm;

% Main iteration loop
while abs(x - x0) > tol & iter <= itermax
   x0 = x;
   x = xp - fp * (xm - xp) / (fm - fp);
   fnk = feval(fnctn,x,varargin{:});

   if fnk < 0
      xm = x;
      fm = fnk;
   else
      xp = x;
      fp = fnk;
   end

   % Show the results of calculation
   if trace
      fprintf('%5.0f    %13.6g %13.6g \n',iter, [x fnk])
```

```
        if trace == 2
            xpath = [xpath xm xm xp xp];
            ypath = [ypath 0 fm fp 0];
        end
    end
    iter = iter + 1;
end

if trace == 2
   % Plot the function and path to the root
   xmin = min(xpath);
   xmax = max(xpath);
   dx = xmax - xmin;
   xi = xmin - dx/10;
   xf = xmax + dx/10;
   yc = [];
   for xc = xi : (xf - xi)/99 : xf
      yc=[yc feval(fnctn,xc,varargin{:})];
   end
   xc = linspace(xi,xf,100);
   ax = linspace(0,0,100);
   plot(xc,yc,xpath,ypath,xc,ax,xpath(2:3),ypath(2:3),'*',x,fnk,'o')
   axis([xi xf min(yc) max(yc)])
   xlabel('x')
   ylabel('f(x)')
   title('Linear Interpolation : The function and path to the root
(* : initial guess ; o : root)')
end

if iter >= itermax
   disp('Warning : Maximum iterations reached.')
end
```

NR.m

```
function x = NR(fnctn,x0,tol,trace,varargin)
%NR Finds a zero of a function by the Newton-Raphson method.

%
%   NR('F',X0) finds a zero of the function described by the
%   M-file F.M. X0 is a starting guess.
%
%   NR('F',X0,TOL,TRACE) uses tolerance TOL for convergence
%   test. TRACE=1 shows the calculation steps numerically and
%   TRACE=2 shows the calculation steps both numerically and
%   graphically.
%
%   NR('F',X0,TOL,TRACE,P1,P2,...) allows for additional
%   arguments which are passed to the function F(X,P1,P2,...).
%   Pass an empty matrix for TOL or TRACE to use the default
%   value.
%
%   See also FZERO, ROOTS, XGX, LI
```

```
% (c) by N. Mostoufi & A. Constantinides
% January 1, 1999

% Initialization
if nargin < 3 | isempty(tol)
   tol = 1e-6;
end
if nargin < 4 | isempty(trace)
   trace = 0;
end
if tol == 0
   tol = 1e-6;
end
if (length(x0) > 1) | (~isfinite(x0))
   error('Second argument must be a finite scalar.')
end

iter = 0;
fnk = feval(fnctn,x0,varargin{:});
if trace
   header = ' Iteration        x                f(x)';
   disp(' ')
   disp(header)
   fprintf('%5.0d   %13.6g %13.6g \n',iter, [x0 fnk])
   if trace == 2
      xpath = [x0 x0];
      ypath = [0 fnk];
   end
end

x = x0;
x0 = x + 1;
itermax = 100;

% Main iteration loop
while abs(x - x0) > tol & iter <= itermax
   iter = iter + 1;
   x0 = x;

   % Set dx for differentiation
   if x ~= 0
      dx = x/100;
   else
      dx = 1/100;
   end

   % Differentiation
   a = x - dx;   fa = feval(fnctn,a,varargin{:});
   b = x + dx;   fb = feval(fnctn,b,varargin{:});
   df = (fb - fa)/(b - a);
```

```
   % Next approximation of the root
   if df == 0
      x = x0 + max(abs(dx),1.1*tol);
   else
      x = x0 - fnk/df;
   end

   fnk = feval(fnctn,x,varargin{:});
   % Show the results of calculation
   if trace
      fprintf('%5.0d   %13.6g %13.6g \n',iter, [x fnk])
      if trace == 2
         xpath = [xpath x x];
         ypath = [ypath 0 fnk];
      end
   end
end

if trace == 2
   % Plot the function and path to the root
   xmin = min(xpath);
   xmax = max(xpath);
   dx = xmax - xmin;
   xi = xmin - dx/10;
   xf = xmax + dx/10;
   yc = [];
   for xc = xi : (xf - xi)/99 : xf
      yc = [yc feval(fnctn,xc,varargin{:})];
   end
   xc = linspace(xi,xf,100);
   ax = linspace(0,0,100);
   plot(xc,yc,xpath,ypath,xc,ax,xpath(1),ypath(2),'*',x,fnk,'o')
   axis([xi xf min(yc) max(yc)])
   xlabel('x')
   ylabel('f(x)')
   title('Newton-Raphson : The function and path to the root
     (* : initial guess ; o : root)')
end

if iter >= itermax
   disp('Warning : Maximum iterations reached.')
end
```

Colebrook.m

```
function y = Colebrook(f, Re, e)
% Colebrook.m
% This function evaluates the value of Colebrook equation to be
% solved by the linear interpolation or the Newton-Raphson method.

y = 1/sqrt(f) + 0.86*log(e/3.7 + 2.51/Re/sqrt(f));
```

Colebrookg.m

```
function y = clbrkg(f, Re, e)
% Colebrookg.m
% This function evaluates the value of the rearranged Colebrook
% equation to be solved by x=g(x) method.

y=1/(0.86*log(e/3.7+2.51/Re/sqrt(f)))^2;
```

Input and Results

```
>>Example1_1

Calculating the friction factor from the Colebrook equation

 Reynolds No.        = 1e5
 Relative roughness = 1e-4

  1 ) Successive substitution
  2 ) Linear Interpolation
  3 ) Newton-Raphson
  0 ) Exit

 Choose the method of solution : 1

 Function containing the Colebrook equation : 'Colebrookg'
 Starting value = 0.01

 Iteration          x              g(x)
    1              0.01       0.0201683
    2         0.0201683       0.0187204
    3         0.0187204       0.0188639
    4         0.0188639       0.0188491
    5         0.0188491       0.0188506
    6         0.0188506       0.0188505

f = 0.0189

  1 ) Successive substitution
  2 ) Linear Interpolation
  3 ) Newton-Raphson
  0 ) Exit

 Choose the method of solution : 2

 Function containing the Colebrook equation : 'Colebrook'
 First  starting value = 0.01
 Second starting value = 0.03

 Iteration          x            f(x)
    0              0.01    2.9585
    0              0.03   -1.68128
```

```
    1         0.0227528  -0.723985
    2         0.0202455  -0.282098
    3         0.0193536  -0.105158
    4         0.0190326  -0.0385242
    5         0.0189165  -0.0140217
    6         0.0188744  -0.00509133
    7         0.0188592  -0.00184708
    8         0.0188536  -0.000669888
    9         0.0188516  -0.000242924
   10        0.0188509   -8.80885e-005

f = 0.0189

  1 ) Successive substitution
  2 ) Linear Interpolation
  3 ) Newton-Raphson
  0 ) Exit

 Choose the method of solution : 3

 Function containing the Colebrook equation : 'Colebrook'
 Starting value = 0.01

 Iteration         x             f(x)
    0              0.01      2.9585
    1         0.0154904      0.825216
    2         0.0183977      0.0982029
    3         0.0188425      0.00170492
    4         0.0188505    6.30113e-007
    5         0.0188505    3.79075e-011

f = 0.0189

  1 ) Successive substitution
  2 ) Linear Interpolation
  3 ) Newton-Raphson
  0 ) Exit

 Choose the method of solution : 0
```

Discussion of Results: All three methods are applied to finding the root of the Colebrook equation for the friction factor. Graphs of the step-by-step path to convergence are shown in Figs. E1.1*a*, *b*, and *c* for the three methods. It can be seen that Newton-Raphson converges faster than the other two methods. However, the Newton-Raphson method is very sensitive to the initial guess, and the method may converge to the other roots of the equation, if a different starting point is used. The reader may test other starting points to examine this sensitivity. The convergence criterion in all the above MATLAB functions is $|x_n - x_{n-1}| < 10^{-6}$.

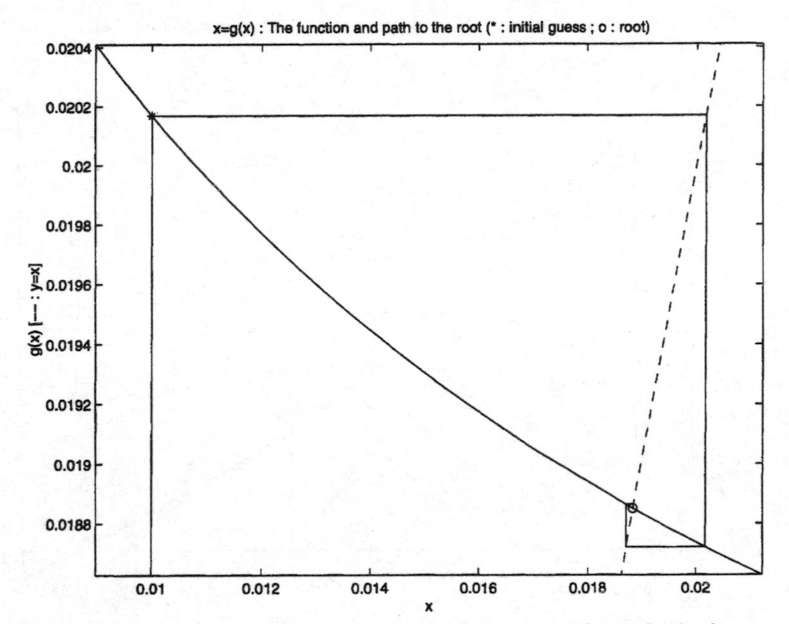

Figure E1.1a Solution using the method of successive substitution.

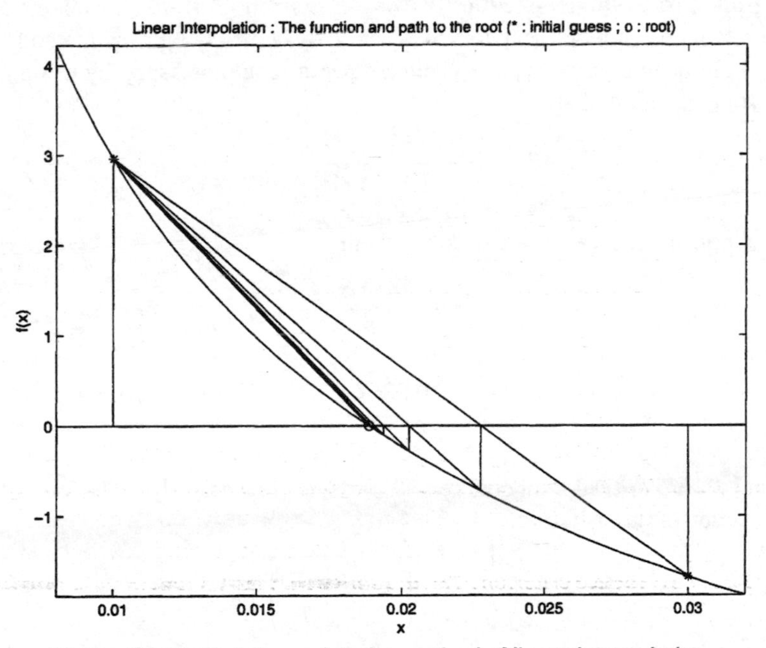

Figure E1.1*b* Solution using the method of linear interpolation.

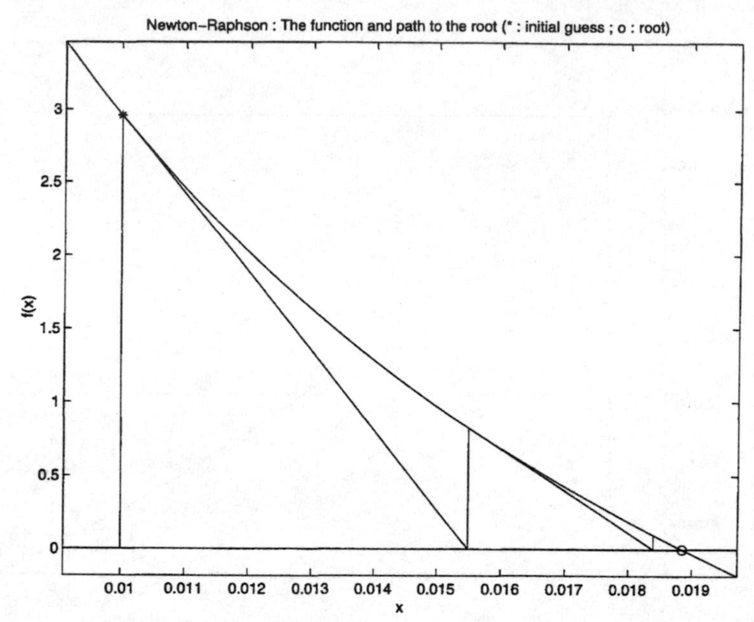

Figure E1.1*c* Solution using the Newton-Raphson method.

Example 1.2: Finding a Root of an *n*th-Degree Polynomial by Newton-Raphson Method Applied to the Soave-Redlich-Kwong Equation of State. Develop a MATLAB function to calculate a root of a polynomial equation by Newton-Raphson method. Calculate the specific volume of a pure gas, at a given temperature and pressure, by using the Soave-Redlich-Kwong equation of state

$$P = \frac{RT}{V - b} - \frac{a\alpha}{V(V + b)}$$

The equation constants, a and b, are obtained from

$$a = \frac{0.4278\, R^2\, T_C^2}{P_C}$$

$$b = \frac{0.0867\, R\, T_C}{P_C}$$

where T_C and P_C are critical temperature and pressure, respectively. The variable α is an empirical function of temperature:

$$\alpha = \left[1 + S\left(1 - \sqrt{\frac{T}{T_C}}\right)\right]^2$$

The value of S is a function of the acentric factor, ω, of the gas:

$$S = 0.48508 + 1.55171\,\omega - 0.15613\,\omega^2$$

The physical properties of n-butane are:

$$T_C = 425.2\ K, \qquad P_C = 3797\ kPa, \qquad \omega = 0.1931$$

and the gas constant is:

$$R = 8314\ J/kmol.K.$$

Calculate the specific volume of n-butane vapor at 500 K and at temperatures from 1 to 40 atm. Compare the results graphically with the ones obtained from using the ideal gas law. What conclusion do you draw from this comparison?

Method of Solution: Eq. (1.42) is used for Newton-Raphson evaluation of the root. For finding the gas specific volume from the Soave-Redlich-Kwong equation of state, Eq. (1.2), which is a third-degree polynomial in compressibility factor, is solved. Starting value for the iterative method is $Z = 1$, which is the compressibility factor of the ideal gas.

Program Description: The MATLAB function *NRpoly.m* calculates a root of a polynomial equation by Newton-Raphson method. The first input argument of this function is the vector of coefficients of the polynomial, and the second argument is an initial guess of the root. The function employs MATLAB functions *polyval* and *polyder* for evaluation of the polynomial and its derivative at each point. The reader can change the convergence criterion by introducing a new value in the third input argument. The default convergence criterion is 10^{-6}. The reader may also see the results of the calculations at each step numerically and graphically by entering the proper value as the fourth argument (1 and 2, respectively). The third and fourth arguments are optional.

MATLAB program *Example1_2.m* solves the Soave-Redlich-Kwong equation of state by utilizing the *NRpoly.m* function. In the beginning of this program, temperature, pressure range and the physical properties of n-butane are entered. The constants of the Soave-Redlich-Kwong equation of state are calculated next. The values of A and B [used in Eq. (1.2)] are also calculated in this section. Evaluation of the root is done in the third part of the program. In this part, the coefficients of Eq. (1.2) are first introduced and the root of the equation, closest to the ideal gas, is determined using the above-mentioned MATLAB function *NRpoly*. The last part of the program, *Example1_2.m,* plots the results of the calculation both for Soave-Redlich-Kwong and ideal gas equations of state. It also shows some of the numerical results.

Program

Example1_2.m

```
% Example1_2.m
% This program solves the problem posed in Example 1.2.
% It calculates the real gas specific volume from the
% SRK equation of state using the Newton-Raphson method
% for calculating the roots of a polynomial.

clear
clc
clf

% Input data
P = input(' Input the vector of pressure range (Pa) = ');
T = input(' Input temperature (K) = ');
R = 8314;      % Gas constant (J/kmol.K)
Tc = input(' Critical temperature (K) = ');
Pc = input(' Critical pressure (Pa) = ');
omega = input(' Acentric factor = ');

% Constants of Soave-Redlich-Kwong equation of state
a = 0.4278 * R^2 * Tc^2 / Pc;
b = 0.0867 * R * Tc / Pc;
sc = [-0.15613, 1.55171, 0.48508];
s = polyval(sc,omega);
alpha = (1 + s * (1 - sqrt(T/Tc)))^2;
A = a * alpha * P / (R^2 * T^2);
B = b * P / (R * T);

for k = 1:length(P)
   % Defining the polynomial coefficients
   coef = [1, -1, A(k)-B(k)-B(k)^2, -A(k)*B(k)];
   v0(k) = R * T / P(k);   % Ideal gas specific volume
   vol(k) = NRpoly(coef , 1) * R * T / P(k);   % Finding the root
end

% Show numerical results
fprintf('\nRESULTS:\n');
fprintf('Pres. =  %5.2f  Ideal gas vol. =%7.4f',P(1),v0(1));
fprintf('  Real gas vol. =%7.4f\n',vol(1));
for k=10:10:length(P)
   fprintf('Pres. = %5.2f  Ideal gas vol. =%7.4f',P(k),v0(k));
   fprintf('  Real gas vol. =%7.4f\n',vol(k));
end

% plotting the results
loglog(P/1000,v0,'.',P/1000,vol)
xlabel('Pressure, kPa')
```

```
ylabel('Specific Volume, m^3/kmol')
legend('Ideal','SRK')
```

NRpoly.m

```
function x = NRpoly(c,x0,tol,trace)
%NRPOLY Finds a root of polynomial by the Newton-Raphson method.
%
%   NRPOLY(C,X0) computes a root of the polynomial whose
%   coefficients are the elements of the vector C.
%   If C has N+1 components, the polynomial is
%   C(1)*X^N + ... + C(N)*X + C(N+1).
%   X0 is a starting point.
%
%   NRPOLY(C,X0,TOL,TRACE) uses tolerance TOL for convergence
%   test. TRACE=1 shows the calculation steps numerically and
%   TRACE=2 shows the calculation steps both numerically and
%   graphically.
%
%   See also ROOTS, NRsdivision, NR.

% (c) by N. Mostoufi & A. Constantinides
% January 1, 1999

% Initialization
if nargin < 3 | isempty(tol)
   tol = 1e-6;
end
if nargin < 4 | isempty(trace)
   trace = 0;
end
if tol == 0
   tol = 1e-6;
end
if (length(x0) > 1) | (~isfinite(x0))
   error('Second argument must be a finite scalar.')
end

iter = 0;
fnk = polyval(c,x0);    % Function
if trace
  header = ' Iteration          x             f(x)';
  disp(header)
  disp([sprintf('%5.0f   %13.6g %13.6g ',iter, [x0 fnk])])
  if trace == 2
     xpath = [x0 x0];
     ypath = [0 fnk];
  end
end
```

```
x = x0;
x0 = x + .1;
maxiter = 100;

% Solving the polynomial by Newton-Raphson method
while abs(x0 - x) > tol & iter < maxiter
   iter = iter + 1;
   x0 = x;
   fnkp = polyval(polyder(c),x0); % Derivative
   if fnkp ~= 0
      x = x0 - fnk / fnkp; % Next approximation
   else
      x = x0 + .01;
   end

   fnk = polyval(c,x);  % Function
   % Show the results of calculation
   if trace
      disp([sprintf('%5.0f   %13.6g %13.6g ',iter, [x fnk])])
      if trace == 2
         xpath = [xpath x x];
         ypath = [ypath 0 fnk];
      end
   end
end

if trace == 2
   % Plot the function and path to the root
   xmin = min(xpath);
   xmax = max(xpath);
   dx = xmax - xmin;
   xi = xmin - dx/10;
   xf = xmax + dx/10;
   yc = [];
   for xc = xi : (xf - xi)/99 : xf
      yc = [yc polyval(c,xc)];
   end
   xc = linspace(xi,xf,100);
   ax = linspace(0,0,100);
   plot(xc,yc,xpath,ypath,xc,ax,xpath(1),ypath(2),'*',x,fnk,'o')
   axis([xi xf min(yc) max(yc)])
   xlabel('x')
   ylabel('f(x)')
   title('Newton-Raphson : The function and path to the root (* :
initial guess ; o : root)')
end

if iter == maxiter
   disp('Warning : Maximum iterations reached.')
end
```

Input and Results

```
>>Example1_2

 Input the vector of pressure range (Pa) : [1:40]*101325
 Input temperature (K) : 500
 Critical temperature (K) : 425.2
 Critical pressure (Pa) : 3797e3
 Acentric factor : 0.1931

Pres. =   101325.00  Ideal gas vol. =41.0264  Real gas vol. =40.8111
Pres. = 1013250.00  Ideal gas vol. = 4.1026  Real gas vol. = 3.8838
Pres. = 2026500.00  Ideal gas vol. = 2.0513  Real gas vol. = 1.8284
Pres. = 3039750.00  Ideal gas vol. = 1.3675  Real gas vol. = 1.1407
Pres. = 4053000.00  Ideal gas vol. = 1.0257  Real gas vol. = 0.7954
```

Discussion of Results: In this example we use the *Example1_2.m* program to calculate the specific volume of a gas using the Soave-Redlich-Kwong equation of state. Because this equation can be arranged in the canonical form of a third-degree polynomial, the function *NRpoly.m* can be used. Additional information such as temperature, pressure, and physical properties are entered by the user through the program.

Above the critical temperature, the Soave-Redlich-Kwong equation of state has only one real root that is of interest, the one located near the value given by the ideal gas law. Therefore, the latter, which corresponds to $Z = 1$, is used as the initial guess of the root.

Direct comparison between the Soave-Redlich-Kwong and ideal gas volumes is made in Fig. E1.2. It can be seen from this figure that the ideal gas equation overestimates gas volumes and, as expected from thermodynamic principles, the deviation from ideality increases as the pressure increases.

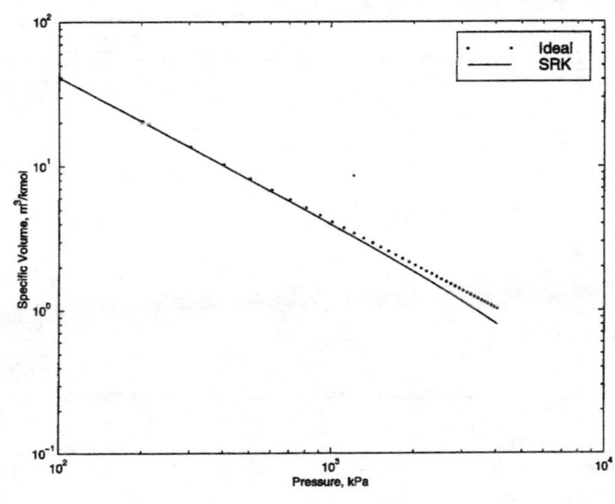

Figure E1.2 Graphical comparison between the Soave-Redlich-Kwong and the ideal gas equations of state.

1.7 SYNTHETIC DIVISION ALGORITHM

If the nonlinear equation being solved is of the polynomial form, each real root (located by one of the methods already discussed) can be removed from the polynomial by synthetic division, thus reducing the degree of the polynomial to $(n - 1)$. Each successive application of the synthetic division algorithm will reduce the degree of the polynomial further, until all real roots have been located.

A simple computational algorithm for synthetic division has been given by Lapidus [4]. Consider the fourth-degree polynomial

$$f(x) = a_4 x^4 + a_3 x^3 + a_2 x^2 + a_1 x + a_0 = 0 \tag{1.48}$$

whose first real root has been determined to be x^*. This root can be factored out as follows:

$$f(x) = (x - x^*)(b_3 x^3 + b_2 x^2 + b_1 x + b_0) = 0 \tag{1.49}$$

In order to determine the coefficients (b_i) of the third-degree polynomial first multiply out Eq. (1.49) and rearrange in descending power of x:

$$f(x) = b_3 x^4 + (b_2 - b_3 x^*) x^3 + (b_1 - b_2 x^*) x^2 + (b_0 - b_1 x^*) x - b_0 x^* \tag{1.50}$$

Equating Eqs. (1.48) and (1.50), the coefficients of like powers of x must be equal to each other, that is,

$$\begin{aligned} a_3 &= b_2 - b_3 x^* \\ a_2 &= b_1 - b_2 x^* \\ a_1 &= b_0 - b_1 x^* \end{aligned} \tag{1.51}$$

Solving Eqs. (1.51) for b_i we obtain:

$$\begin{aligned} b_3 &= a_4 \\ b_2 &= a_3 + b_3 x^* \\ b_1 &= a_2 + b_2 x^* \\ b_0 &= a_1 + b_1 x^* \end{aligned} \tag{1.52}$$

In general notation, for a polynomial of nth-degree, the new coefficients after application of synthetic division are given by

$$b_{n-1} = a_n$$
$$b_{n-1-r} = a_{n-r} + b_{n-r}x^* \tag{1.53}$$

where $r = 1, 2, \ldots, (n - 1)$. The polynomial is then reduced by one degree

$$n_{j+1} = n_j - 1 \tag{1.54}$$

where j is the iteration number, and the newly calculated coefficients are renamed as shown by Eq. (1.55):

$$b_{n-1} = a_n$$
$$b_{n-1-r} = a_{n-r} + b_{n-r}x^* \tag{1.55}$$

This procedure is repeated until all real roots are extracted. When this is accomplished, the remainder polynomial will contain the complex roots. The presence of a pair of complex roots will give a quadratic equation that can be easily solved by quadratic formula. However, two or more pairs of complex roots require the application of more elaborate techniques, such as the eigenvalue method, which is developed in the next section.

1.8 THE EIGENVALUE METHOD

The concept of eigenvalues will be discussed in Chap. 2 of this textbook. As a preview of that topic, we will state that a square matrix has a *characteristic polynomial* whose roots are called the *eigenvalues* of the matrix. However, root-finding methods that have been discussed up to now are not efficient techniques for calculating eigenvalues [5]. There are more efficient eigenvalue methods to find the roots of the characteristic polynomial (see Sec. 2.8).

It can be shown that Eq. (1.11) is the characteristic polynomial of the $(n \times n)$ companion matrix $\boldsymbol{A}$, which contains the coefficients of the original polynomial as shown in Eq. (1.56). Therefore, finding the eigenvalues of $\boldsymbol{A}$ is equivalent to locating the roots of the polynomial in Eq. (1.11).

MATLAB has its own function, *roots.m*, for calculating all the roots of a polynomial equation of the form in Eq. (1.11). This function accomplishes the task of finding the roots of the polynomial equation [Eq. (1.11)] by first converting the polynomial to the companion matrix $\boldsymbol{A}$ shown in Eq. (1.56). It then uses the built-in function *eig.m*, which calculates the eigenvalues of a matrix, to evaluate the eigenvalues of the companion matrix, which are also the roots of the polynomial Eq. (1.11):

$$A = \begin{bmatrix} -\frac{a_{n-1}}{a_n} & -\frac{a_{n-2}}{a_n} & \cdots & -\frac{a_1}{a_n} & -\frac{a_0}{a_n} \\ 1 & 0 & \cdots & 0 & 0 \\ 0 & 1 & \cdots & 0 & 0 \\ \cdot & & & & \cdot \\ \cdot & & & & \cdot \\ \cdot & & & & \cdot \\ 0 & 0 & \cdots & 1 & 0 \end{bmatrix} \tag{1.56}$$

Example 1.3: Solution of *n*th-Degree Polynomials and Transfer Functions Using the Newton-Raphson Method with Synthetic Division and Eigenvalue Method. Consider the isothermal continuous stirred tank reactor (CSTR) shown in Fig. E1.3.

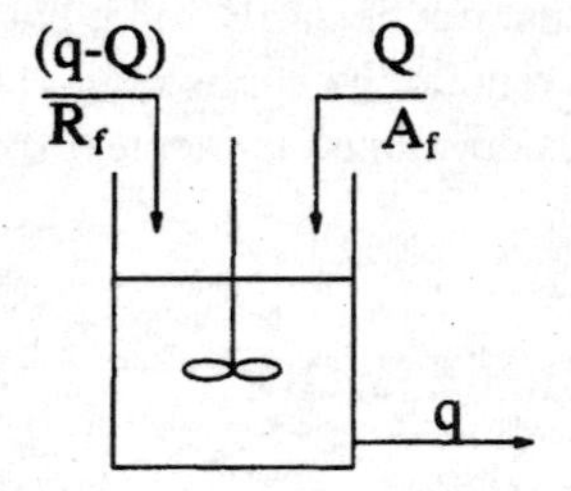

Figure E1.3 The continuous stirred tank reactor.

Components A and R are fed to the reactor at rates of Q and $(q - Q)$, respectively. The following complex reaction scheme develops in the reactor:

$$A + R \rightarrow B$$

$$B + R \rightarrow C$$

$$C + R \rightarrow D$$

$$D + R \rightarrow E$$

This problem was analyzed by Douglas [6] in order to illustrate the various techniques for designing simple feedback control systems. In his analysis of this system, Douglas made the following assumptions:

1. Component R is present in the reactor in sufficiently large excess so that the reaction rates can be approximated by first-order expressions.
2. The feed compositions of components B, C, D, and E are zero.
3. A particular set of values is chosen for feed concentrations, feed rates, kinetic rate constant, and reactor volume.
4. Disturbances are due to changes in the composition of component R in the vessel.

The control objective is to maintain the composition of component C in the reactor as close as possible to the steady-state design value, despite the fact that disturbances enter the system. This objective is accomplished by measuring the actual composition of C and using the difference between the desired and measured values to manipulate the inlet flow rate Q of component A.

Douglas developed the following transfer function for the reactor with a proportional control system:

$$K_C \frac{2.98(s + 2.25)}{(s + 1.45)(s + 2.85)^2(s + 4.35)} = -1$$

where K_C is the gain of the proportional controller. This control system is stable for values of K_C that yield roots of the transfer function having negative real parts.

Using the Newton-Raphson method with synthetic division or eigenvalue method, determine the roots of the transfer function for a range of values of the proportional gain K_C and calculate the critical value of K_C above which the system becomes unstable. Write the program so that it can be used to solve *n*th-degree polynomials or transfer functions of the type shown in the above equation.

Method of Solution: In the Newton-Raphson method with synthetic division, Eq. (1.42) is used for evaluation of each root. Eqs. (1.53)-(1.55) are then applied to perform synthetic division in order to extract each root from the polynomial and reduce the latter by one degree. When the *n*th-degree polynomial has been reduced to a quadratic

$$a_2 x^2 + a_1 x + a_0 = 0$$

the program uses the quadratic solution formula

$$x_{1,2} = \frac{-a_1 \pm \sqrt{a_1^2 - 4a_2 a_0}}{2a_2}$$

to check for the existence of a pair of complex roots. In the eigenvalue method, the MATLAB function *roots* may be used directly.

The numerator and the denominator of the transfer function are multiplied out to yield

$$\frac{(2.98s + 6.705)K_C}{s^4 + 11.50s^3 + 47.49s^2 + 83.0632s + 51.2327} = -1$$

A first-degree polynomial is present in the numerator and a fourth-degree polynomial in the denominator. To convert this to the canonical form of a polynomial, we multiply through by the denominator and rearrange to obtain

$$[s^4 + 11.50\,s^3 + 47.49\,s^2 + 83.0632\,s + 51.2327] + [2.98\,s + 6.705]\,K_C = 0$$

It is obvious that once a value of K_C is chosen, the two bracketed terms of this equation can be added to form a single fourth-degree polynomial whose roots can be evaluated.

When $K_C = 0$, the transfer function has the following four negative real roots, which can be found by inspection of the original transfer function:

$$s_1 = -1.45 \qquad s_2 = -2.85 \qquad s_3 = -2.85 \qquad s_4 = -4.35$$

These are called the poles of the open-loop transfer function.

The value of K_C that causes one or more of the roots of the transfer function to become positive (or have positive real parts) is called the critical value of the proportional gain. This critical value is calculated as follows:

1. A range of search for K_C is established.
2. The bisection method is used to search this range.
3. All the roots of the transfer function are evaluated at each step of the bisection search.
4. The roots are checked for positive real part. The range of K_C, over which the change from negative to positive roots occurs, is retained.
5. Steps 2-4 are repeated until successive values of K_C change by less than a convergence criterion, ϵ.

Program Description: The MATLAB function *NRsdivision.m* calculates all roots of a polynomial by the Newton-Raphson method with synthetic division as described in the Method of Solution. Unlike other functions employing the Newton-Raphson method, this function does not need a starting value as one of the input arguments. Instead, the function generates a starting point at each step according to Eq. (1.22). Only polynomials that have no more than a pair of complex roots can be handled by this function. If the polynomial has more than a pair of complex roots, the function *roots* should be used instead. The function is written in general form and may be used in other programs directly.

The MATLAB program *Example1_3.m* does the search for the desired value of K_C by the bisection method. At the beginning of the program, the user is asked to enter the coefficients of the numerator and the denominator of the transfer function (in descending *s* powers). The numerator and the denominator may be of any degree with the limitation that the numerator cannot have a degree greater than that of the denominator. The user should also enter the range of search and method of root finding. It is good practice to choose zero for the minimum value of the range; thus, poles of the open-loop transfer function are evaluated in the first step of the search. The maximum value must be higher than the critical value, otherwise the search will not arrive at the critical value.

Stability of the system is examined at the minimum, maximum, and midpoints of the range of search of K_C. That half of the interval in which the change from negative to positive (stable to unstable system) occurs is retained by the bisection algorithm. This new interval is bisected again and the evaluation of the system stability is repeated, until the convergence criterion, which is $|\Delta K_C| < 0.001$, is met.

In order to determine whether the system is stable or unstable, the two polynomials are combined, as shown in the Method of Solution, using K_C as the multiplier of the polynomial from the numerator of the transfer function. Function *NRsdivision* (which uses the Newton-Raphson method with synthetic division algorithm) or function *roots* (which uses the eigenvalue algorithm) is called to calculate the roots of the overall polynomial function and the sign of all roots is checked for positive real parts. A flag named stbl indicates that the system is stable (all negative roots; stbl = 1) or unstable (positive root; stbl = 0).

Program

Example1_3.m

```
% Example1_3.m
% Solution to the problem posed in Example 1.3. It calculates the
% critical value of the constant of a proportional controller above
% which the system of chemical reactor becomes unstable. This program
% evaluate all roots of the denominator of the transfer function using
% the Newton-Raphson method with synthetic division or eigenvalue
% methods.

clear
clc

% Input data
num = input(' Vector of coefficients of the numerator polynomial =
');
denom = input (' Vector of coefficients of the denominator
polynomial = ');
disp(' ')
Kc1 = input(' Lower limit of the range of search = ');
Kc2 = input(' Upper limit of the range of search = ');
disp(' ')
disp(' 1 ) Newton-Raphson with synthetic division')
disp(' 2 ) Eigenvalue method')
method = input(' Method of root finding = ');

iter = 0;
n1 = length(num);
n2 = length(denom);
c(1:n2-n1) = denom (1:n2-n1);
```

```
% Main loop
while abs(Kc1 - Kc2) > 0.001
   iter = iter + 1;
   if iter == 1
      Kc = Kc1;                                    % Lower limit
   elseif iter == 2
      Kc = Kc2;                                    % Upper limit
   else
      Kc = (Kc1 + Kc2) / 2;                        % Next approximation
   end

   % Calculation of coefficients of canonical form of polynomial
   for m = n2-n1+1 : n2;
      c(m) = denom(m) + Kc * num(m-n2+n1);
   end

   % Root finding
   switch method
   case 1                          %Newton-Raphson with synthetic division
      root = NRsdivision(c);
   case 2                                          % Eigenvalue method
      root = roots (c);
   end
   realpart = real (root);
   imagpart = imag (root);

   % Show the results of calculations of this step
   fprintf('\n Kc = %6.4f\n Roots = ',Kc)
   for k = 1:length(root)
      if isreal(root(k))
         fprintf('%7.5g    ',root(k))
      else
         fprintf('%6.4g',realpart(k))
         if imagpart(k) >= 0
            fprintf('+%5.4gi    ',imagpart(k))
         else
            fprintf('-%5.4gi    ',abs(imagpart(k)))
         end
      end
   end
   disp(' ')
   % Determining stability or unstability of the system
   stbl = 1;
   for m = 1 : length(root)
      if realpart(m) > 0
         stbl = 0;                                 % System is unstable
         break;
      end
   end
```

```
    if iter == 1
       stbl1 = stbl;
    elseif iter == 2
       stbl2 = stbl;
       if stbl1 == stbl2
          error('Critical value is outside the range of search.')
          break
       end
    else
       if stbl == stbl1
          Kc1 = Kc;
       else
          Kc2 = Kc;
       end
    end
end
```

NRsdivision.m

```
function x = NRsdivision(c,tol)
%NRSDIVISION Finds polynomial roots.
%
%   The function NRSDIVISION(C) evaluates the roots of a
%   polynomial equation whose coefficients are given in the
%   vector C.
%
%   NRSDIVISION(C,TOL) uses tolerance TOL for convergence
%   test. Using the second argument is optional.
%
%   The polynomial may have no more than a pair of complex
%   roots. A root of nth-degree polynomial is determined by
%   Newton-Raphson method. This root is then extracted from
%   the polynomial by synthetic division. This procedure
%   continues until the polynomial reduces to a quadratic.
%
%   See also ROOTS, NRpoly, NR

% (c) by N. Mostoufi & A. Constantinides
% January 1, 1999

% Initialization
if nargin < 2 | isempty(tol)
   tol = 1e-6;
end
if tol == 0
   tol = 1e-6;
end

n = length(c) - 1;                        % Degree of the polynomial
a = c;
```

```
% Main loop
for k = n : -1 : 3
   x0 = -a(2)/a(1);
   x1=x0+0.1;
   iter = 0;
   maxiter = 100;

   % Solving the polynomial by Newton-Raphson method
   while abs(x0 - x1) > tol & iter < maxiter
      iter = iter + 1;
      x0 = x1;
      fnk = polyval(a,x0);                  % Function
      fnkp = polyval(polyder(a),x0);        % Derivative
      if fnkp ~= 0
         x1 = x0 - fnk / fnkp;              % Next approximation
      else
         x1 = x0 + 0.01;
      end
   end

   x(n-k+1) = x1;                           % the root

   % Calculation of new coefficients
   b(1) = a(1);
   for r = 2 : k
      b(r) = a(r) + b(r-1) * x1;
   end

   if iter == maxiter
      disp('Warning : Maximum iteration reached.')
   end

   clear a
   a = b;
   clear b
end

% Roots of the remaining quadratic polynomial
delta = a(2) ^ 2 - 4 * a(1) * a(3);
x(n-1) = (-a(2) - sqrt(delta)) / (2 * a(1));
x(n) = (-a(2) + sqrt(delta)) / (2 * a(1));

x=x';
```

Input and Results

```
>>Example1_3

 Vector of coefficients of the numerator polynomial = [2.98, 6.705]
```

```
 Vector of coefficients of the denominator polynomial = [1, 11.5,
47.49, 83.0632, 51.2327]

 Lower limit of the range of search = 0
 Upper limit of the range of search = 100

 1 ) Newton-Raphson with synthetic division
 2 ) Eigenvalue method
 Method of root finding = 1

 Kc = 0.0000
 Roots =    -4.35   -2.8591   -2.8409      -1.45

 Kc = 100.0000
 Roots =   -9.851    -2.248   0.2995+5.701i    0.2995-5.701i

 Kc = 50.0000
 Roots = -8.4949   -2.2459   -0.3796+4.485i    -0.3796-4.485i

 Kc = 75.0000
 Roots = -9.2487   -2.2473   -0.001993+5.163i    -0.001993-5.163i

 Kc = 87.5000
 Roots = -9.5641   -2.2477   0.1559+5.445i    0.1559-5.445i

 Kc = 81.2500
 Roots = -9.4104   -2.2475   0.07893+5.308i    0.07893-5.308i

 Kc = 78.1250
 Roots = -9.3306   -2.2474    0.039+5.237i     0.039-5.237i

 Kc = 76.5625
 Roots =    -9.29   -2.2473   0.01864+5.2i    0.01864-5.2i

 Kc = 75.7812
 Roots = -9.2694   -2.2473   0.00836+5.182i    0.00836-5.182i

 Kc = 75.3906
 Roots = -9.2591   -2.2473   0.003192+5.173i    0.003192-5.173i

 Kc = 75.1953
 Roots = -9.2539   -2.2473   0.0006016+5.168i    0.0006016-5.168i

 Kc = 75.0977
 Roots = -9.2513   -2.2473   -0.0006953+5.166i    -0.0006953-5.166i

 Kc = 75.1465
 Roots = -9.2526   -2.2473   -4.667e-005+5.167i    -4.667e-005-5.167i
```

```
Kc = 75.1709
Roots = -9.2533    -2.2473    0.0002775+5.167i    0.0002775-5.167i

Kc = 75.1587
Roots = -9.2529    -2.2473    0.0001154+5.167i    0.0001154-5.167i

Kc = 75.1526
Roots = -9.2528    -2.2473    3.438e-005+5.167i    3.438e-005-5.167i

Kc = 75.1495
Roots = -9.2527    -2.2473    -6.147e-006+5.167i    -6.147e-006-5.167i

Kc = 75.1511
Roots = -9.2527    -2.2473    1.412e-005+5.167i    1.412e-005-5.167i

Kc = 75.1503
Roots = -9.2527    -2.2473    3.985e-006+5.167i    3.985e-006-5.167i
```

Discussion of Results: The range of search for the proportional gain (K_c) is chosen to be between 0 and 100. A convergence criterion of 0.001 is used and may be changed by the user if necessary. The bisection method evaluates the roots at the low end of the range ($K_c = 0$) and finds them to have the predicted values of

$$-4.3500 \quad -2.8591 \quad -2.8409 \quad \text{and} \quad -1.4500$$

The small difference between the two middle roots and their actual values is due to rounding off the coefficients of the denominator polynomial. This deviation is very small in comparison with the root itself and it can be ignored. At the upper end of the range ($K_c = 100$) the roots are

$$-9.8510 \quad -2.2480 \quad \text{and} \quad 0.2995 \pm 5.7011i$$

The system is unstable because of the positive real components of the roots. At the midrange ($K_c = 50$) the system is still stable because all the real parts of the roots are negative. The bisection method continues its search in the range 50-100. In a total of 19 evaluations, the algorithm arrives at the critical value of K_c in the range

$$75.1495 < K_c < 75.1503$$

In the event that the critical value of the gain was outside the limits of the original range of search, the program would have detected this early in the search and would have issued a warning and stopped running.

1.9 NEWTON'S METHOD FOR SIMULTANEOUS NONLINEAR EQUATIONS

If the mathematical model involves two (or more) simultaneous nonlinear equations in two (or more) unknowns, the Newton-Raphson method can be extended to solve these equations simultaneously. In what follows, we will first develop the Newton-Raphson method for two equations and then expand the algorithm to a system of k equations.

The model for two unknowns will have the general form

$$\begin{aligned} f_1(x_1, x_2) &= 0 \\ f_2(x_1, x_2) &= 0 \end{aligned} \tag{1.57}$$

where f_1 and f_2 are nonlinear functions of variables x_1 and x_2. Both these functions may be expanded in two-dimensional Taylor series around an initial estimate of $x_1^{(1)}$ and $x_2^{(1)}$:

$$\begin{aligned} f_1(x_1, x_2) &= f_1(x_1^{(1)}, x_2^{(1)}) + \frac{\partial f_1}{\partial x_1}\Big|_{x^{(1)}}(x_1 - x_1^{(1)}) + \frac{\partial f_1}{\partial x_2}\Big|_{x^{(1)}}(x_2 - x_2^{(1)}) + \ldots \\ f_2(x_1, x_2) &= f_2(x_1^{(1)}, x_2^{(1)}) + \frac{\partial f_2}{\partial x_1}\Big|_{x^{(1)}}(x_1 - x_1^{(1)}) + \frac{\partial f_2}{\partial x_2}\Big|_{x^{(1)}}(x_2 - x_2^{(1)}) + \ldots \end{aligned} \tag{1.58}$$

The superscript (1) will be used to designate the iteration number of the estimate.

Setting the left sides of Eqs. (1.58) to zero and truncating the second-order and higher derivatives of the Taylor series, we obtain the following equations:

$$\begin{aligned} \frac{\partial f_1}{\partial x_1}\Big|_{x^{(1)}}(x_1 - x_1^{(1)}) + \frac{\partial f_1}{\partial x_2}\Big|_{x^{(1)}}(x_2 - x_2^{(1)}) &= -f_1(x_1^{(1)}, x_2^{(1)}) \\ \frac{\partial f_2}{\partial x_1}\Big|_{x^{(1)}}(x_1 - x_1^{(1)}) + \frac{\partial f_2}{\partial x_2}\Big|_{x^{(1)}}(x_2 - x_2^{(1)}) &= -f_2(x_1^{(1)}, x_2^{(1)}) \end{aligned} \tag{1.59}$$

If we define the correction variable δ as

$$\begin{aligned} \delta_1^{(1)} &= x_1 - x_1^{(1)} \\ \delta_2^{(1)} &= x_2 - x_2^{(1)} \end{aligned} \tag{1.60}$$

then Eqs. (1.59) simplify to

$$\frac{\partial f_1}{\partial x_1}\Big|_{x^{(1)}} \delta_1^{(1)} + \frac{\partial f_1}{\partial x_2}\Big|_{x^{(1)}} \delta_2^{(1)} = -f_1(x_1^{(1)}, x_2^{(1)})$$
$$\frac{\partial f_2}{\partial x_1}\Big|_{x^{(1)}} \delta_1^{(1)} + \frac{\partial f_2}{\partial x_2}\Big|_{x^{(1)}} \delta_2^{(1)} = -f_2(x_1^{(1)}, x_2^{(1)}) \tag{1.61}$$

Eqs. (1.61) are a set of simultaneous linear algebraic equations, where the unknowns are $\delta_1^{(1)}$ and $\delta_2^{(1)}$. These equations can be written in matrix format as follows:

$$\begin{bmatrix} \frac{\partial f_1}{\partial x_1}\Big|_{x^{(1)}} & \frac{\partial f_1}{\partial x_2}\Big|_{x^{(1)}} \\ \frac{\partial f_2}{\partial x_1}\Big|_{x^{(1)}} & \frac{\partial f_2}{\partial x_2}\Big|_{x^{(1)}} \end{bmatrix} \begin{bmatrix} \delta_1^{(1)} \\ \delta_2^{(1)} \end{bmatrix} = - \begin{bmatrix} f_1^{(1)} \\ f_2^{(1)} \end{bmatrix} \tag{1.62}$$

Because this set contains only two equations in two unknowns, it can be readily solved by the application of Cramer's rule (see Chap. 2) to give the first set of values for the correction vector:

$$\delta_1^{(1)} = -\frac{\left[f_1 \frac{\partial f_2}{\partial x_2} - f_2 \frac{\partial f_1}{\partial x_2}\right]}{\left[\frac{\partial f_1}{\partial x_1}\frac{\partial f_2}{\partial x_2} - \frac{\partial f_1}{\partial x_2}\frac{\partial f_2}{\partial x_1}\right]}$$
$$\delta_2^{(1)} = -\frac{\left[f_2 \frac{\partial f_1}{\partial x_1} - f_1 \frac{\partial f_2}{\partial x_1}\right]}{\left[\frac{\partial f_1}{\partial x_1}\frac{\partial f_2}{\partial x_2} - \frac{\partial f_1}{\partial x_2}\frac{\partial f_2}{\partial x_1}\right]} \tag{1.63}$$

The superscripts, indicating the iteration number of the estimate, have been omitted from the right-hand side of Eqs. (1.63) in order to avoid overcrowding.

The new estimate of the solution may now be obtained from the previous estimate by adding to it the correction vector:

$$x_i^{(n+1)} = x_i^{(n)} + \delta_i^{(n)} \tag{1.64}$$

This equation is merely a rearrangement and generalization to the $(n + 1)$st iteration of Eqs. (1.60).

The method just described for two nonlinear equations is readily expandable to the case of k simultaneous nonlinear equations in k unknowns:

$$\begin{aligned} f_1(x_1, \ldots, x_k) &= 0 \\ &\vdots \\ f_k(x_1, \ldots, x_k) &= 0 \end{aligned} \tag{1.65}$$

The linearization of this set by the application of the Taylor series expansion produces Eq. (1.66).

$$\begin{bmatrix} \dfrac{\partial f_1}{\partial x_1} & \cdots & \dfrac{\partial f_1}{\partial x_k} \\ \cdots & \cdots & \cdots \\ \dfrac{\partial f_k}{\partial x_1} & \cdots & \dfrac{\partial f_k}{\partial x_k} \end{bmatrix} \begin{bmatrix} \delta_1 \\ \cdot \\ \cdot \\ \cdot \\ \delta_k \end{bmatrix} = - \begin{bmatrix} f_1 \\ \cdot \\ \cdot \\ \cdot \\ f_k \end{bmatrix} \tag{1.66}$$

In matrix/vector notation this condenses to

$$\boldsymbol{J}\boldsymbol{\delta} = -\boldsymbol{f} \tag{1.67}$$

where $\boldsymbol{J}$ is the *Jacobian* matrix containing the partial derivatives, $\boldsymbol{\delta}$ is the correction vector, and $\boldsymbol{f}$ is the vector of functions. Eq. (1.67) represents a set of linear algebraic equations whose solution will be discussed in Chap. 2.

Strongly nonlinear equations are likely to diverge rapidly. To prevent this situation, relaxation is used to stabilize the iterative solution process. If $\boldsymbol{\delta}$ is the correction vector without relaxation, then relaxed change is $\rho\boldsymbol{\delta}$ where ρ is the relaxation factor:

$$\boldsymbol{x}^{(n+1)} = \boldsymbol{x}^{(n)} + \rho\,\boldsymbol{\delta} \tag{1.68}$$

A typical value for ρ is 0.5. A value of zero inhibits changes and a value of one is equivalent to no relaxation. Relaxation reduces the correction made to the variable from one iteration to the next and may eliminate the tendency of the solution to diverge.

Example 1.4: Solution of Nonlinear Equations in Chemical Equilibrium Using Newton's Method for Simultaneous Nonlinear Equations. Develop a MATLAB function to solve *n* simultaneous nonlinear equations in *n* unknowns. Apply this function to find the equilibrium conversion of the following reactions:

$$2A + B \rightleftharpoons C \qquad K_1 = \frac{C_C}{C_A^2 C_B} = 5\times10^{-4}$$

$$A + D \rightleftharpoons C \qquad K_2 = \frac{C_C}{C_A C_D} = 4\times10^{-2}$$

Initial concentrations are

$$C_{A,0} = 40 \qquad C_{B,0} = 15 \qquad C_{C,0} = 0 \qquad C_{D,0} = 10$$

All concentrations are in kmol/m^3.

Method of Solution: Eq. (1.67) is applied to calculate the correction vector in each iteration and Eq. (1.68) is used to estimate the new relaxed variables. The built-in MATLAB function *inv* is used to invert the Jacobian matrix.

The variables of this problem are the conversions, x_1 and x_2, of the above reactions. The concentrations of the components can be calculated from these conversions and the initial concentrations

$$C_A = C_{A,0} - 2x_1 C_{B,0} - x_2 C_{D,0} = 40 - 30x_1 - 10x_2$$

$$C_B = (1 - x_1) C_{B,0} = 15 - 15x_1$$

$$C_C = C_{C,0} + x_1 C_{B,0} + x_2 C_{D,0} = 15x_1 + 10x_2$$

$$C_D = (1 - x_2) C_{D,0} = 10 - 10x_2$$

The set of equations that are functions of x_1 and x_2 are

$$f_1(x_1, x_2) = \frac{C_C}{C_A^2 C_B} - 5\times10^{-4} = 0$$

$$f_2(x_1, x_2) = \frac{C_C}{C_A C_D} - 4\times10^{-2} = 0$$

The values of x_1 and x_2 are to be calculated by the program so that $f_1 = f_2 = 0$.

Program Description: The MATLAB function *Newton.m* solves a set of nonlinear equations by Newton's method. The first part of the program is initialization in which the convergence criterion, the relaxation factor, and other initial parameters needed by the program are being set. The main iteration loop comes next. In this part, the components of the Jacobian matrix are calculated by numeric differentiation [Eq. (1.66)], and new estimates of the roots are calculated according to Eq. (1.68). This procedure continues until the convergence criterion is met or a maximum iteration limit is reached. The convergence criterion is $max(|x_i^{(n)} - x_i^{(n-1)}|) < \epsilon$.

The MATLAB program *Example1_4.m* is written to solve the particular problem of this example. This program simply takes the required input data from the user and calls *Newton.m* to solve the set of equations. The program allows the user to repeat the calculation and try new initial values and relaxation factor without changing the problem parameters.

The set of equations of this example are introduced in the MATLAB function *Ex1_4_func.m*. It is important to note that the function *Newton.m* should receive the function values at each point as a column vector. This is considered in the function *Ex1_4_func*.

Program

Example1_4.m

```
% Example1_4.m
% Solution to the problem posed in Example 1.4. It calculates the
% equilibrium concentration of the components of a system of two
% reversible chemical reactions using the Newton's method.

clear
clc

% Input data
c0 = input(' Vector of initial concentration of A, B, C, and D = ');
K1 = input(' 2A + B = C         K1 = ');
K2 = input('  A + D = C         K2 = ');
fname = input(' Name of the file containing the set of equations =
');

repeat = 1;
while repeat
   % Input initial values and relaxation factor
   x0 = input('\n\n Vector of initial guesses = ');
   rho = input(' Relaxation factor = ');

   % Solution of the set of equations
   [x,iter] = Newton(fname,x0,rho,[],c0,K1,K2);

   % Display the results
   fprintf('\n Results :\n x1 = %6.4f   ,   x2 = %6.4f',x)
   fprintf('\n Solution reached after %3d iterations.\n\n',iter)
   repeat = input(' Repeat the calculations (0 / 1) ? ');
end
```

Newton.m

```
function [xnew , iter] = Newton(fnctn , x0 , rho , tol , varargin)
%NEWTON   Solves a set of equations by Newton's method.
%
%   NEWTON('F',X0) finds a zero of the set of equations
%   described by the M-file F.M. X0 is a vector of starting
%   guesses.
%
%   NEWTON('F',X0,RHO,TOL) uses relaxation factor RHO and
%   tolerance TOL for convergence test.
%
%   NEWTON('F',X0,RHO,TOL,P1,P2,...) allows for additional
%   arguments which are passed to the function F(X,P1,P2,...).
%   Pass an empty matrix for TOL or TRACE to use the default
%   value.

% (c) by N. Mostoufi & A. Constantinides
% January 1, 1999

% Initialization
if nargin < 4 | isempty(tol)
   tol = 1e-6;
end
if nargin < 3 | isempty(rho)
   rho = 1;
end

x0 = (x0(:).')'; % Make sure it's a column vector
nx = length(x0);
x = x0 * 1.1;
xnew = x0;
iter = 0;
maxiter = 100;

% Main iteration loop
while max(abs(x - xnew)) > tol & iter < maxiter
    iter = iter + 1;
    x = xnew;
    fnk = feval(fnctn,x,varargin{:});

    % Set dx for derivation
    for k = 1:nx
       if x(k) ~= 0
          dx(k) = x(k) / 100;
       else
          dx(k) = 1 / 100;
       end
    end
```

```
   % Calculation of the Jacobian matrix
   a = x;
   b = x;
   for k = 1 : nx
      a(k) = a(k) - dx(k);   fa = feval(fnctn,a,varargin{:});
      b(k) = b(k) + dx(k);   fb = feval(fnctn,b,varargin{:});
      jacob(:,k) = (fb - fa) / (b(k) - a(k));
      a(k) = a(k) + dx(k);
      b(k) = b(k) - dx(k);
   end

   % Next approximation of the roots
   if det(jacob) == 0
      xnew = x + max([abs(dx), 1.1*tol]);
   else
      xnew = x - rho * inv(jacob) * fnk;
   end
end

if iter >= maxiter
   disp('Warning : Maximum iterations reached.')
end
```

Ex1_4_func.m

```
function f = Ex1_4_func(x,c0,K1,K2)
% Evaluation of set of equations for example 1.4.
% c0(1) = ca0 / c0(2) = cb0 / c0(3) = cc0 / c0(4) = cd0

ca = c0(1) - 2*x(1)*c0(2) - x(2)*c0(4);
cb = (1 - x(1))*c0(2);
cc = c0(3) + x(1)*c0(2) + x(2)*c0(4);
cd = (1 - x(2))*c0(4);

f(1) = cc / ca^2 / cb - K1;
f(2) = cc / ca / cd - K2;
f = f';   % Make it a column vector.
```

Input and Results

```
>>Example1_4

Vector of initial concentration of A, B, C, and D = [40, 15, 0, 10]
 2A + B = C       K1 = 5e-4
  A + D = C       K2 = 4e-2
 Name of the file containing the set of equations = 'Ex1_4_func'

 Vector of initial guesses = [0.1, 0.9]
 Relaxation factor = 1
```

```
Results :
x1 = 0.1203    ,   x2 = 0.4787
Solution reached after   8 iterations.

Repeat the calculations (0 / 1) ? 1


Vector of initial guesses = [0.1, 0.1]
Relaxation factor = 1
Warning : Maximum iterations reached.

Results :
x1 = 3212963175567844000000000.0000   ,   x2 =
-6973207056420978000000000.0000
Solution reached after 100 iterations.

Repeat the calculations (0 / 1) ? 1


Vector of initial guesses = [0.1, 0.1]
Relaxation factor = 0.7

Results :
x1 = 0.1203    ,   x2 = 0.4787
Solution reached after  18 iterations.

Repeat the calculations (0 / 1) ? 0
```

Discussion of Results: Three runs are made to test the sensitivity in the choice of initial guesses and the effectiveness of the relaxation factor. In the first run, initial guesses are set to $x_1^{(0)} = 0.1$ and $x_2^{(0)} = 0.9$. With these guesses, the method converges in 8 iterations. By default, the convergence criterion is $max(|\Delta x_i|) < 10^{-6}$. This value may be changed through the fourth input argument of the function *Newton.*

In the second run, the initial guesses are set to $x_1^{(0)} = 0.1$ and $x_2^{(0)} = 0.1$. The maximum number of iterations defined in the function *Newton* is 100, and the method does not converge in this case, even in 100 iterations. This test shows high sensitivity of Newton's method to the initial guess. Introducing the relaxation factor $\rho = 0.7$ in the next run causes the method to converge in only 18 iterations. This improvement in the speed of convergence was obtained by using a fixed relaxation factor. A more effective way of doing this would be to adjust the relaxation factor from iteration to iteration. This is left as an exercise for the reader.

PROBLEMS

1.1 Evaluate all the roots of the polynomial equations, (a)-(g) given below, by performing the following steps:

(i) Use Descartes' rule to predict how many positive and how many negative roots each polynomial may have.

(ii) Use the Newton-Raphson method with synthetic division to calculate the numerical values of the roots. To do so, first apply the MATLAB function *roots.m* and then the *NRsdivision.m*, which was developed in this chapter. Why does the *NRsdivision.m* program fail to arrive at the answers of some of these polynomials? What is the limitation of this program?

(iii) Classify these polynomials according to the four categories described in Sec. 1.2.

(a) $x^4 - 16x^3 + 96x^2 - 256x + 256 = 0$
(b) $x^4 - 32x^2 + 256 = 0$
(c) $x^4 + 3x^3 + 12x - 16 = 0$
(d) $x^4 + 4x^3 + 18x^2 - 20x + 125 = 0$
(e) $x^5 - 8x^4 + 35x^3 - 106x^2 + 170x - 200 = 0$
(f) $x^4 - 10x^3 + 35x^2 - 5x + 24 = 0$
(g) $x^6 - 8x^5 + 11x^4 + 78x^3 - 382x^2 + 800x - 800 = 0$

1.2 Evaluate roots of the following transcendental equations.

(a) $\sin x - 2\exp(-x^2) = 0$
(b) $ax - a^x = 0$ for $a = 2, e$, or 3
(c) $\ln(1 + x^2) - \sqrt{|x|} = 0$
(d) $e^x/(1 + \cos x) - 1 = 0$

1.3 Repeat Example 1.2 by using the Benedict-Webb-Rubin (BWR) and the Patel-Teja (PT) equations of state. Compare the results with those obtained in Example 1.2.

Benedict-Webb-Rubin equation of state:

$$P = \frac{RT}{V} + \frac{B_0RT - A_0 - (C_0/T^2)}{V^2} + \frac{bRT - a}{V^3} + \frac{a\alpha}{V^6} + \frac{c}{V^3T^2}\left(1 + \frac{\gamma}{V^2}\right)e^{-\gamma/V^2}$$

where A_0, B_0, C_0, a, b, c, α, and γ are constants. When P is in atmosphere, V is in liters per mole, and T is in Kelvin, the values of constants for n-butane are:

$A_0 = 10.0847$ $\quad B_0 = 0.124361$ $\quad C_0 = 0.992830\times10^6$
$a = 1.88231$ $\quad b = 0.0399983$ $\quad c = 0.316400\times10^6$
$\alpha = 1.10132\times10^{-3}$ $\quad \gamma = 3.400\times10^{-2}$ $\quad R = 0.08206$

Patel-Teja equation of state:

$$P = \frac{RT}{V - b} - \frac{a}{V(V + b) + c(V - b)}$$

where a is a function of temperature, and b and c are constants

$$a = \Omega_a (R^2 T_c^2 / P_c)\left[1 + F(1 - \sqrt{T_R})\right]^2$$

$$b = \Omega_b (RT_c / P_c)$$

$$c = \Omega_c (RT_c / P_c)$$

where

$$\Omega_c = 1 - 3\zeta_c$$

$$\Omega_a = 3\zeta_c^2 + 3(1 - 2\zeta_c)\Omega_b + \Omega_b^2 + 1 - 3\zeta_c$$

and Ω_b is the smallest positive root of the cubic

$$\Omega_b^3 + (2 - 3\zeta_c)\Omega_b^2 + 3\zeta_c^2\Omega_b - \zeta_c^3 = 0$$

F and ζ_c are functions of the acentric factor given by the following quadratic correlations

$$F = 0.452413 + 1.30982\,\omega - 0.295937\,\omega^2$$

$$\zeta_c = 0.329032 - 0.076799\,\omega + 0.0211947\,\omega^2$$

Use the data given in Example 1.2 for n-butane to calculate the parameters of PT equation.

1.4 F moles per hour of an n-component natural gas stream is introduced as feed to the flash vaporization tank shown in Fig. P1.4. The resulting vapor and liquid streams are withdrawn at the rate of V and L moles per hour, respectively. The mole fractions of the components in the feed, vapor, and liquid are designated by z_i, y_i, and x_i, respectively ($i = 1, 2, \ldots, n$). Assuming vapor-liquid equilibrium and steady-state operation, we have:

Overall balance	$F = L + V$	
Individual component balances	$z_i F = x_i L + y_i V$	$i = 1, 2, \ldots, n$
Equilibrium relations	$K_i = y_i / x_i$	$i = 1, 2, \ldots, n$

Here, K_i is the equilibrium constant for the ith component at the prevailing temperature and pressure in the tank. From these equations and the fact that

$$\sum_{i=1}^{n} x_i = \sum_{i=1}^{n} y_i = 1$$

derive the following equation:

$$\sum_{i=1}^{n} \frac{z_i K_i F}{V(K_i - 1) + F} = 1$$

Using the data given in Table P1.4, solve the above equation for V. Also calculate the values of L, the x_i, and the y_i by using the first three equations given above. The test data in Table P1.4 relates to flashing of a natural gas stream at 11 MPa and 48 °C. Assume that $F = 100$ mol/h.

What would be a good value V_0 for starting the iteration? Base this answer on your observations of the data given in Table P1.4.

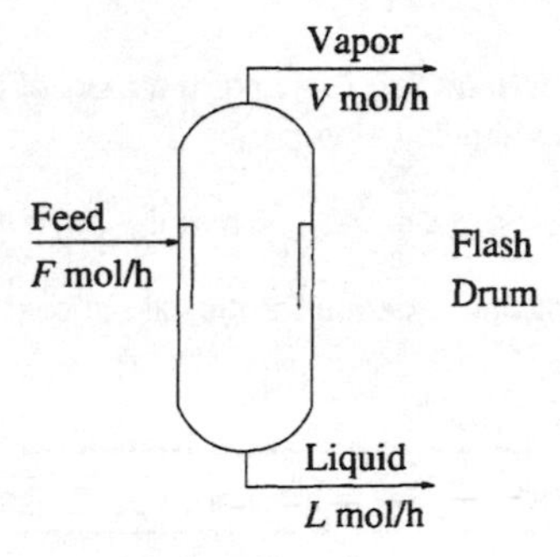

Figure P1.4 Flash drum.

Table P1.4

Component	i	z_i	K_i
Methane	1	0.8345	3.090
Carbon dioxide	2	0.0046	1.650
Ethane	3	0.0381	0.720
Propane	4	0.0163	0.390
i-Butane	5	0.0050	0.210
n-Butane	6	0.0074	0.175
Pentanes	7	0.0287	0.093
Hexanes	8	0.0220	0.065
Heptanes+	9	0.0434	0.036
		1.0000	

1.5 The Underwood equation for multicomponent distillation is given as

$$\left(\sum_{j=1}^{n} \frac{\alpha_j z_{jF} F}{\alpha_j - \phi} \right) - F(1 - q) = 0$$

where
F = molar feed flow rate
n = number of components in the feed
z_{jF} = mole fraction of each component in the feed
q = quality of the feed
α_j = relative volatility of each component at average column conditions
ϕ = root of the equation

It has been shown by Underwood that (n - 1) of the roots of this equation lie between the values of the relative volatilities as shown below:

$$\alpha_n < \phi_{n-1} < \alpha_{n-1} < \phi_{n-2} < \ldots < \alpha_3 < \phi_2 < \alpha_2 < \phi_1 < \alpha_1$$

Evaluate the (n - 1) roots of this equation for the case shown in Table P1.5.

Table P1.5

Component in feed	Mole fraction, z_{jF}	Relative volatility, α_j
C_1	0.05	10.00
C_2	0.05	5.00
C_3	0.10	2.05
C_4	0.30	2.00
C_5	0.05	1.50
C_6	0.30	1.00
C_7	0.10	0.90
C_8	0.05	0.10
	1.00	

F = 100 mol/h q = 1.0 (saturated liquid)

1.6 Carbon monoxide from a water gas plant is burned with air in an adiabatic reactor. Both the carbon monoxide and air are being fed to the reactor at 25°C and atmospheric pressure. For the reaction:

$$CO + \frac{1}{2}O_2 \Rightarrow CO_2$$

the following standard free energy change (at 25°C) has been determined:

$$\Delta G^0_{T_0} = -257 \text{ kJ/(gmol of CO)}$$

The standard enthalpy change at 25°C has been measured as

$$\Delta H^0_{T_0} = -283 \text{ kJ/(gmol of CO)}$$

The standard states for all components are the pure gases at 1 atm.

Calculate the adiabatic flame temperature and the conversion of CO for the following two cases:

(a) 0.4 mole of oxygen per mole of CO is provided for the reaction.

(b) 0.8 mole of oxygen per mole of CO is provided for the reaction.

The constant pressure heat capacities for the various constituents in J/(gmol.K) with T in Kelvin are all of the form

$$C_{p_i} = A_i + B_i T_K + C_i T_K^2$$

For the gases involved here, the constants are as shown in Table P1.6.

Table P1.6

Gas	A	B	C
CO	26.16	8.75×10^{-3}	-1.92×10^{-6}
O_2	25.66	12.52×10^{-3}	-3.37×10^{-6}
CO_2	28.67	35.72×10^{-3}	-10.39×10^{-6}
N_2	26.37	7.61×10^{-3}	-1.44×10^{-6}

Hint: Combine the material balance, enthalpy balance, and equilibrium relationship to form two nonlinear algebraic equations in two unknowns: the temperature and conversion.

1.7 Consider the three-mode feedback control of a stirred-tank heater system (Fig. P1.7).

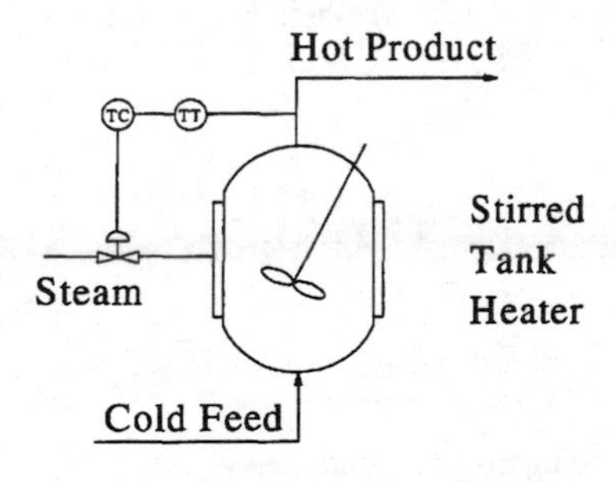

Figure P1.7 Stirred tank heater.

The measured output variable is the feedstream temperature [7]. Using classical methods (i.e., deviation variables, linearization, and Laplace transforms) the overall closed-loop transfer function for the control system is given by

$$\frac{\bar{T}}{\bar{T}_i} = \frac{(\tau_I s)(\tau_v s + 1)(\tau_m s + 1)}{(\tau_I s)(\tau_P s + 1)(\tau_v s + 1)(\tau_m s + 1) + K(\tau_I s + 1 + \tau_D \tau_I s^2)}$$

where τ_I = reset time constant

τ_D = derivative time constant

$K = K_P K_v K_m K_c$

K_P= first-order process static gain

K_v = first-order valve constant

K_m= first-order measurement constant

K_c= proportional gain for the three-mode controller

$\bar{T}$= Laplace transform of the output temperature deviation

$\bar{T}_i$= Laplace transform of the input load temperature deviation

τ_P, τ_m, τ_v = first-order time constants for the process, measurement device, and process valve, respectively.

For a given set of values, the stability of the system can be determined from the roots of the characteristic polynomial (i.e, the polynomial in the denominator of the overall transfer function). Thus:

$$\tau_I \tau_P \tau_m \tau_v s^4 + (\tau_I \tau_P \tau_m + \tau_I \tau_P \tau_v + \tau_I \tau_m \tau_v) s^3 + (K\tau_I \tau_D + \tau_I \tau_P + \tau_I \tau_v + \tau_I \tau_m) s^2$$
$$+ (\tau_I + K\tau_I) s + K = 0$$

For the following set of parameter values, find the four roots to the characteristic polynomial when K_c is equal to its "critical" value:

$\tau_I = 10$ $\tau_D = 1$ $\tau_P = 10$ $\tau_m = 5$ $\tau_v = 5$

$K_P = 10$ $K_v = 2$ $K_m = 0.09$ $K = 1.8K_c$

1.8 In the analytical solution of some parabolic partial differential equations in cylindrical coordinates, it is necessary to calculate roots of the Bessel function first (for example, see Problem 6.11). Find the first N root of the first and the second kind. Use the following approximations for evaluating the initial guesses:

$$J_n(x) \approx \sqrt{\frac{2}{\pi x}} \cos\left(x - \frac{n\pi}{2} - \frac{\pi}{4}\right)$$

$$Y_n(x) \approx \sqrt{\frac{2}{\pi x}} \sin\left(x - \frac{n\pi}{2} - \frac{\pi}{4}\right)$$

Which method of root finding do you recommend?

1.9 A direct-fired tubular reactor is used in the thermal cracking of light hydrocarbons or naphthas for the production of olefins, such as ethylene (see Fig. P1.9). The reactants are preheated in the convection section of the furnace, mixed with steam, and then subjected to high temperatures in the radiant section of the furnace. Heat transfer in the radiant section of the furnace takes place through three mechanisms: radiation, conduction, and convection. Heat is transferred by radiation from the walls of the furnace to the surface of the tubes that carry the reactants, and it is transferred through the walls of the tubes by conduction and finally to the fluid inside the tubes by convection [8].

The three heat-transfer mechanisms are quantified as follows:

1. *Radiation*: The Stefan-Boltzmann law of radiation may be written as

$$\frac{dQ}{dA_o} = \sigma\phi(T_R^4 - T_o^4)$$

where dQ/dA_o is the rate for heat transfer per unit outside surface area of the tubes, T_R is the "effective" furnace radiation temperature, and T_o is the temperature on the outside surface of the tube. In furnaces with tube banks irradiated from both sides, a reasonable approximation is

$$T_R = T_G$$

where T_G is the temperature of the flue gas in the reactor. Therefore, the Stefan-Boltzmann equation is revised to

$$\frac{dQ}{dA_o} = \sigma\phi(T_G^4 - T_o^4)$$

σ is the Stefan-Boltzmann constant and ϕ is the tube geometry emissivity factor, which depends on the tube arrangement and tube surface emissivity. For single rows of tubes irradiated from both sides:

$$\frac{1}{\phi} = \frac{1}{\epsilon} - 1 + \frac{\pi}{2\Omega}$$

$$\Omega = \frac{S}{D_o} + \arctan\sqrt{\left(\frac{S}{D_o}\right)^2 - 1} - \sqrt{\left(\frac{S}{D_o}\right)^2 - 1}$$

where ϵ is the emissivity of the outside surface of the tube and S is the spacing (pitch) of the tubes (center-to-center) and D_o is the outside diameter of the tubes.

2. *Conduction*: Conduction through the tube wall is given by Fourier's equation:

$$\frac{dQ}{dA_o} = \frac{k_t}{t_t}(T_o - T_i)$$

where T_i is the temperature on the inside surface of the tube, k_t is the thermal conductivity of the tube material, and t_t is the thickness of the tube wall.

3. *Convection*: Convection through the fluid film inside the tube is expressed by

$$\frac{dQ}{dA_o} = h_i\left(\frac{D_i}{D_o}\right)(T_i - T_f)$$

where D_i is the inside diameter of the tube, T_f is the temperature of the fluid in the tube, and h_i is the heat-transfer film coefficient on the inside of the tube. The film coefficient may be approximated from the Dittus-Boelter equation [9]:

$$h_i = 0.023\left(\frac{k_f}{D_i}\right) Re_f^{0.8} Pr_f^{0.4}$$

where Re_f is the Reynolds number, Pr_f is the Prandtl number, and k_f is the thermal conductivity of the fluid.

Conditions vary drastically along the length of the tube, as the temperature of the fluid inside the tube rises rapidly. The rate of heat transfer is the highest at the entrance conditions and lowest at the exit conditions of the fluid.

Calculate the rate of heat transfer (dQ/dA_o), the temperature on the outside surface of the tube (T_o), and the temperature on the inside surface of the tube (T_i) at a point along the length of the tube where the following conditions exist:

$T_G = 1200°C$ $\quad T_f = 800°C$ $\quad \epsilon = 0.9$ $\quad \sigma = 5.7\times10^{-8}$ W/m^2.K^4

$S = 0.20$ m $\quad D_i = 0.10$ m $\quad D_o = 0.11$ m $\quad t_t = 0.006$ m

$Re_f = 388{,}000$ $\quad Pr_f = 0.660$ $\quad k_t = 21.6$ W/m.K $\quad k_f = 0.175$ W/m.K

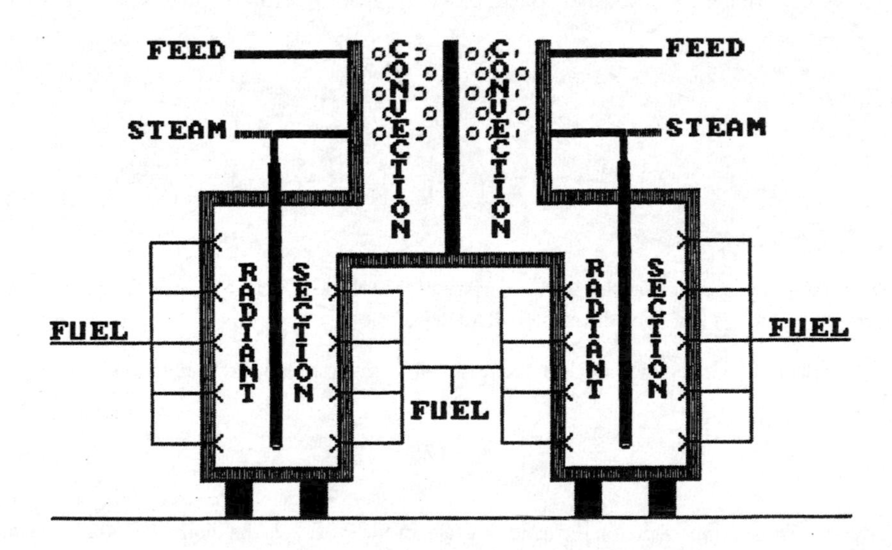

Figure P1.9 Pyrolysis reactor.

1.10 The elementary reaction A → B + C is carried out in a continuous stirred tank reactor (CSTR). Pure A enters the reactor at a flow rate of 12 mol/s and a temperature of 25°C. The reaction is exothermic and cooling water at 50°C is used to absorb the heat generated. The energy balance for this system, assuming constant heat capacity and equal heat capacity of both sides of the reaction, can be written as

$$-F_{A_0} X \Delta H_R = F_{A_0} C_{p_A}(T - T_0) + UA(T - T_a)$$

where F_{A_0} = molar flow rate, mol/s
X = conversion
ΔH_R = heat of reaction, J/mol A
C_{p_A} = heat capacity of A, J/mol.K
T = reactor temperature, °C
T_0 = reference temperature, 25°C
T_a = cooling water temperature, 20°C
U = overall heat transfer coefficient, W/m^2.K
A = heat transfer area, m^2

For a first-order reaction the conversion can be calculated from

$$X = \frac{\tau k}{1 + \tau k}$$

where τ is the residence time of the reactor in seconds and k is the specific reaction rate in s^{-1} defined by the Arrhenius formula:

$$k = 650 \exp[-3800/(T + 273)]$$

Solve the energy balance equation for temperature and find the steady-state operating temperatures of the reactor and the conversions corresponding to these temperatures. Additional data are:

ΔH_R = -1500 kJ/mol τ = 10 s C_{p_A} = 4500 J/mol.K UA/F_{A_0} = 700 W.s/mol.K

REFERENCES

1. Underwood, A. J. V., *Chem. Eng. Prog.*, vol. 44, 1948, p. 603.

2. Treybal, R. E., *Mass Transfer Operations*, 3rd ed., McGraw-Hill, New York, 1980.

3. Salvadori, M. G., and Baron, M. L., *Numerical Methods in Engineering*, Prentice Hall, Englewood Cliffs, NJ, 1961.

4. Lapidus, L., *Digital Computation for Chemical Engineering*, McGraw-Hill, New York, 1962.

5. Press, W. H., Teukolsky, S. A., Vetterling, W. T., and Flannery, B. P., *Numerical Recipes in FORTRAN*, 2nd ed., Cambridge University Press, Cambridge, U.K., 1992.

6. Douglas, J. M., *Process Dynamics and Control*, vol. 2, Prentice Hall, Englewood Cliffs, NJ, 1972.

7. Davidson, B. D., private communication, Department of Chemical and Biochemical Engineering, Rutgers University, Piscataway, NJ, 1984.

8. Constantinides, A., *Applied Numerical Methods with Personal Computers*, McGraw-Hill, New York, 1987.

9. Bennett, C. O., and Meyers, J. E., *Momentum, Heat, and Mass Transfer*, McGraw-Hill, New York, 1973.

CHAPTER 2

Numerical Solution of Simultaneous Linear Algebraic Equations

2.1 INTRODUCTION

The mathematical analysis of linear physicochemical systems often results in models consisting of sets of linear algebraic equations. In addition, methods of solution of nonlinear systems and differential equations use the technique of linearization of the models, thus requiring the repetitive solution of sets of linear algebraic equations. These problems may range in complexity from a set of two simultaneous linear algebraic equations to a set involving 1000 or even 10,000 equations. The solution of a set of two to three linear algebraic equations can be obtained easily by the algebraic elimination of variables or by the application of Cramer's rule. However, for systems involving five or more equations, the algebraic elimination method becomes too complex, and Cramer's rule requires a rapidly escalating number of arithmetic operations, too large even for today's high-speed digital computers.

In the remainder of this section, we give several examples of systems drawing from chemical engineering applications that yield sets of simultaneous linear algebraic equations. In the following sections of this chapter, we discuss several methods for the numerical solution of such problems and demonstrate the application of these methods on the computer.

Material and energy balances are the primary tools of chemical engineers. Such balances applied to multistage or multicomponent processes result in sets of equations that can be either differential or algebraic. Often the systems under analysis are nonlinear, thus resulting in sets of nonlinear equations. However, many procedures have been developed that linearize the equations and apply iterative convergence techniques to arrive at the solution of the nonlinear systems.

A classical example of the use of these techniques is in the analysis of distillation columns, such as the one shown in Fig. 2.1. Steady-state material balances applied to the rectifying section of the column yield the following equations:

Balance around condenser:
$$V_1 y_{1i} = L_0 x_{0i} + D x_{Di} \tag{2.1}$$

Balance above the jth stage:
$$V_j y_{ji} = L_{j-1} x_{j-1,i} + D x_{Di} \tag{2.2}$$

Assuming that the stages are equilibrium stages and that the column uses a total condenser, the following equilibrium relations apply:

$$y_{ji} = K_{ji} x_{ji} \tag{2.3}$$

Substituting Eq. (2.3) in (2.1) and (2.2) and dividing through by Dx_{Di} we get

$$\frac{V_1 y_{1i}}{D x_{Di}} = \frac{L_0}{D x_{Di}} x_{Di} + 1 \tag{2.4}$$

$$\frac{V_j y_{ji}}{D x_{Di}} = \left(\frac{L_{j-1}}{K_{j-1,i} V_{j-1}} \right) \left(\frac{V_{j-1} y_{j-1,i}}{D x_{Di}} \right) + 1 \tag{2.5}$$

The molal flow rates of the individual components are defined as

$$v_{ji} = V_i y_{ji} \tag{2.6}$$

$$d_i = D x_{Di} \tag{2.7}$$

For any stage j, the adsorption ratio is defined as

$$A_{ji} = \frac{L_j}{K_{ji} V_j} \tag{2.8a}$$

and for the total condenser as

$$A_{0i} = \frac{L_0}{D} \tag{2.8b}$$

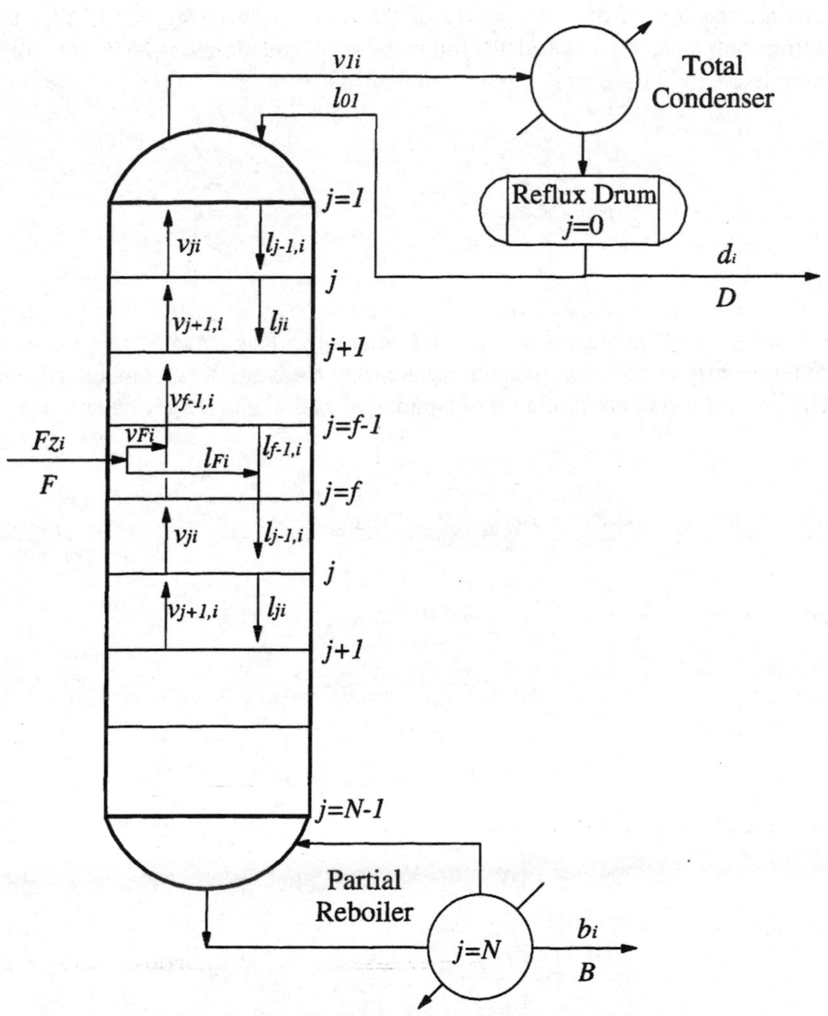

Figure 2.1 Conventional distillation column.

Substitution of Eqs. (2.6)-(2.8) in (2.4) and (2.5) yields

$$\left(\frac{v_{1i}}{d_i}\right) = A_{0i} + 1 \tag{2.9}$$

$$\left(\frac{v_{ji}}{d_i}\right) = A_{j-1,i}\left(\frac{v_{j-1,i}}{d_i}\right) + 1 \qquad 2 \le j \le f - 1 \tag{2.10}$$

For any given trial calculation the As are regarded as constants. The unknowns in the above equations are the groups of terms v_{ji}/d_i. If these are replaced by x_{ji} and the subscript i, designating component i is dropped, the following set of equations can be written for a column containing five equilibrium stages above the feed stage:

$$\begin{aligned}
x_1 \qquad\qquad\qquad\qquad\qquad &= A_0 + 1 \\
-A_1x_1 + x_2 \qquad\qquad\qquad &= 1 \\
-A_2x_2 + x_3 \qquad\qquad &= 1 \\
-A_3x_3 + x_4 \qquad &= 1 \\
-A_4x_4 + x_5 &= 1
\end{aligned} \tag{2.11}$$

This is a set of simultaneous linear algebraic equations. It is actually a special set that has nonzero terms only on the diagonal and one adjacent element. It is a bidiagonal set.

The general formulation of a set of simultaneous linear algebraic equations is

$$\begin{aligned}
a_{11}x_1 + a_{12}x_2 + \dots + a_{1n}x_n &= c_1 \\
a_{21}x_1 + a_{22}x_2 + \dots + a_{2n}x_n &= c_2 \\
\dots\dots\dots\dots\dots\dots\dots & \\
a_{n1}x_1 + a_{n2}x_2 + \dots + a_{nn}x_n &= c_n
\end{aligned} \tag{2.12}$$

where all of the coefficients a_{ij} could be nonzero. This set is usually condensed in vector-matrix notation as

$$\boldsymbol{A}\boldsymbol{x} = \boldsymbol{c} \tag{2.13}$$

where $\boldsymbol{A}$ is the coefficient matrix

$$\boldsymbol{A} = \begin{bmatrix} a_{11} & a_{12} & \dots & a_{1n} \\ a_{21} & a_{22} & \dots & a_{2n} \\ \dots & \dots & \dots & \dots \\ a_{n1} & a_{n2} & \dots & a_{nn} \end{bmatrix} \tag{2.14}$$

$\boldsymbol{x}$ is the vector of unknown variables:

$$\boldsymbol{x} = \begin{bmatrix} x_1 \\ x_2 \\ \cdot \\ \cdot \\ \cdot \\ x_n \end{bmatrix} \tag{2.15}$$

and $\boldsymbol{c}$ is the vector of constants:

$$\boldsymbol{c} = \begin{bmatrix} c_1 \\ c_2 \\ \cdot \\ \cdot \\ \cdot \\ c_n \end{bmatrix} \tag{2.16}$$

When the vector $\boldsymbol{c}$ is the zero vector, the set of equations is called *homogeneous*.

Another example requiring the solution of linear algebraic equations comes from the analysis of complex reaction systems that have monomolecular kinetics. Fig. 2.2 considers a chemical reaction between the three species, whose concentrations are designated by Y_1, Y_2, Y_3, taking place in a batch reactor.

The equations describing the dynamics of this chemical reaction scheme are

$$\begin{aligned} \frac{dY_1}{dt} &= -(k_{21} + k_{31})Y_1 + k_{12}Y_2 + k_{13}Y_3 \\ \frac{dY_2}{dt} &= k_{21}Y_1 - (k_{12} + k_{32})Y_2 + k_{23}Y_3 \\ \frac{dY_3}{dt} &= k_{31}Y_1 + k_{32}Y_2 - (k_{13} + k_{23})Y_3 \end{aligned} \tag{2.17}$$

The above set of linear ordinary differential equations may be condensed into matrix notation

$$\dot{\boldsymbol{y}} = \boldsymbol{K}\boldsymbol{y} \tag{2.18}$$

where $\dot{y}$ is the vector of derivatives:

$$\dot{y} = \begin{bmatrix} \dfrac{dY_1}{dt} \\ \dfrac{dY_2}{dt} \\ \dfrac{dY_3}{dt} \end{bmatrix} \tag{2.19}$$

y is the vector of concentrations of the components

$$y = \begin{bmatrix} Y_1 \\ Y_2 \\ Y_3 \end{bmatrix} \tag{2.20}$$

and K is the matrix of kinetic rate constants:

$$K = \begin{bmatrix} -(k_{21} + k_{31}) & k_{12} & k_{13} \\ k_{21} & -(k_{12} + k_{32}) & k_{23} \\ k_{31} & k_{32} & -(k_{13} + k_{23}) \end{bmatrix} \tag{2.21}$$

The solution of the dynamic problem, which is modeled by Eq. (2.18), would require the evaluation of the characteristic values (eigenvalues) λ_i and characteristic vectors (eigenvectors) x_i of the matrix K. It is shown in Chap. 5 that the solution of a set of linear ordinary differential equations can be obtained by using Eq. (5.53):

$$y = [X e^{\Lambda t} X^{-1}] y_0 \tag{5.53}$$

where X is a matrix whose columns are the eigenvectors x_i of K; $e^{\Lambda t}$ is a matrix with $e^{\lambda_i t}$ on the diagonal and zero elsewhere; X^{-1} is the inverse of X; and y_0 is the vector of initial values of the

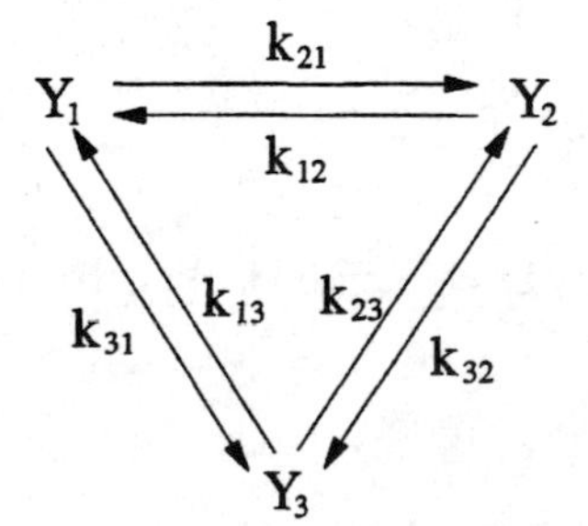

Figure 2.2 System of chemical reactions.

variables Y_i. Methods for calculating eigenvalues and eigenvectors of matrices are developed in Sec. 2.9.

When a chemical reaction reaches steady state, the vector of derivatives in Eq. (2.18) becomes zero and Eq. (2.18) simplifies to

$$\boldsymbol{K}\boldsymbol{y} = 0 \tag{2.22}$$

This is a set of homogeneous linear algebraic equations whose solution describes the steady-state situation of the above chemical reaction problem.

Comparison of Eqs. (2.13) and (2.22) reveals that the difference between nonhomogeneous and homogeneous sets of equations is that, in the latter, the vector of constants $\boldsymbol{c}$ is the zero vector. The steady-state solution of the chemical reaction problem requires finding a unique solution to the set of homogeneous algebraic equations represented by Eq. (2.22).

In the analysis of fermentation processes, the production of the concentration of cell and metabolic products formed in the fermentor can be accomplished by the technique of material balance [1]. For example, the stoichiometry of the fermentation of the microorganism ***Brevibacterium flavum***, which produces glutamic acid, may be represented by the following two chemical reactions:

Biomass formation:

$$C_6H_{12}O_6 + b'O_2 + c'NH_3 \rightarrow d'C_wH_xO_yN_z + e'CO_2 + f'H_2O \tag{2.23}$$

Glutamic acid synthesis:

$$C_6H_{12}O_6 + \frac{3}{2}O_2 + NH_3 \rightarrow C_5H_9O_4N + CO_2 + 3\,H_2O \tag{2.24}$$

$C_6H_{12}O_6$ is glucose, the basic nutrient in this fermentation, and $C_5H_9O_4N$ is glutamic acid. Biomass is represented by $C_wH_xO_yN_z$, where w, x, y, z are the corresponding number of atoms of each element in the cell. This empirical formula can be determined by analyzing the carbon, hydrogen, oxygen, and nitrogen contents of the biomass. The inherent assumption is that C, H, O, and N are the only atoms of significant quantity in the cell biomass. The cellular composition is assumed to remain constant during growth.

Let a be the number of moles of glucose used, g be the fraction of glucose converted to product, and $(1 - g)$ be the fraction of glucose converted to biomass. Then combining Eqs. (2.23) and (2.24) results in the overall reaction

$$aC_6H_{12}O_6 + [b(1-g) + 1.5ag]O_2 + [c(1-g) + ag]NH_3 \rightarrow$$

$$[d(1-g)]C_wH_xO_yN_z + agC_5H_9O_4N + [e(1-g) + ag]CO_2 + [f(1-g) + 3ag]H_2O \tag{2.25}$$

The primed coefficients (b', c', d', e', f') have been replaced by unprimed coefficients (b, c, d, e, f). These are related to each other, as follows:

$$b = ab' \qquad c = ac' \qquad d = ad' \qquad e = ae' \qquad f = af'$$

Eq. (2.25) contains seven unknown quantities: the stoichiometric coefficients (a, b, c, d, e, f) and the fraction of substrate used for product formation (g). Based on Eq. (2.25), four independent equations are derived from material balances on carbon, hydrogen, oxygen, and nitrogen, as shown in Table 2.1.

Table 2.1 Elemental material balances

Elemental material balances from Eq. (2.25)		
Carbon:	$6a = wd(1 - g) + 5ag + e(1 - g) + ag$	(2.26)
Hydrogen:	$12a + 3c(1 - g) + 3ag = xd(1 - g) + 9ag + 2f(1 - g) + 6ag$	(2.27)
Oxygen:	$6a + 2b(1 - g) + 3ag = yd(1 - g) + 4ag + 2e(1 - g) + 2ag + f(1 - g) + 3ag$	(2.28)
Nitrogen:	$c(1 - g) + ag = zd(1 - g) + ag$	(2.29)
Simplified elemental balance		
Carbon:	$(wd + e - 6a)(1 - g) = 0$	(2.30)
Hydrogen:	$(xd + 2f - 3c - 12a)(1 - g) = 0$	(2.31)
Oxygen:	$(yd + 2e + f - 2b - 6a)(1 - g) = 0$	(2.32)
Nitrogen:	$(c - zd)(1 - g) = 0$	(2.33)

In order to determine all seven variables, three more independent relationships are required. This is accomplished by calculating total oxygen consumed (*TOTAL* O_2), total carbon dioxide released (*TOTAL* CO_2), and total glucose consumed (a). Oxygen and carbon dioxide contents of inlet and exit gases are measured by means of a paramagnetic gaseous oxygen analyzer and an IR carbon dioxide analyzer, respectively. These quantities, combined with gas flow rates, complete the oxygen and carbon balances on the gas stream. The trapezoidal rule (Chap. 4) is used to calculate total oxygen consumed and total carbon dioxide released at any time during the fermentation. These quantities are related to the stoichiometric coefficients of Eq. (2.25) as follows:

$$\frac{TOTAL\ O_2}{M_{O_2}} = b(1 - g) + 1.5ag \tag{2.34}$$

$$\frac{TOTAL\ CO_2}{M_{CO_2}} = e(1 - g) + ag \tag{2.35}$$

where M_{O_2} = molecular weight of O_2 (g/mol)
M_{CO_2} = molecular weight of CO_2 (g/mol)

Because this system is a batch process with respect to glucose, the glucose consumption can be evaluated by subtracting the residual glucose concentration from the initial glucose concentration:

$$a = (GLU_0 - GLU_1)\frac{V}{M_G} \tag{2.36}$$

where a = glucose consumption (mol)
GLU_0 = glucose consumption at time 0 (g/L)
GLU_1 = glucose consumption at time t (g/L)
V = liquid volume (L)
M_G = molecular weight of glucose (g/mol)

The residual glucose concentration can be measured by an on-line enzymatic glucose analyzer.

The overall system consists of Eqs. (2.30)- (2.36). This is a set of nonlinear simultaneous algebraic equations that can be solved simultaneously using a combination of Newton's method (Sec. 1.9) and the Gauss-Jordan method to be developed in this chapter (Sec. 2.6).

The solution of some partial differential equations is sometimes carried out by solving sets of simultaneous finite difference equations (see Chap. 6). These equations are often linear in nature and can be solved by the methods to be discussed in this chapter.

Optimization of a complex assembly of unit operations, such as a chemical plant, or of a cluster of interrelated assemblies, such as a group of refineries, can be accomplished by techniques of linear programming that handle large sets of simultaneous linear equations.

The application of linear and nonlinear regression analysis to fit mathematical models to experimental data and to evaluate the unknown parameters of these models (see Chap. 7) requires the repetitive solution of sets of linear algebraic equations. In addition, the ellipse formed by the correlation coefficient matrix in the parameter hyperspace of these systems must be searched in the direction of the major and minor axes. The directions of these axes are defined by the eigenvectors of the correlation coefficient matrix, and the relative lengths of the axes are measured by the eigenvalues of the correlation coefficient matrix.

In developing systematic methods for the solution of linear algebraic equations and the evaluation of eigenvalues and eigenvectors of linear systems, we will make extensive use of matrix-vector notation. For this reason, and for the benefit of the reader, a review of selected matrix and vector operations is given in the next section.

2.2 Review of Selected Matrix and Vector Operations

2.2.1 Matrices and Determinants

A matrix is an array of elements arranged in rows and columns as

$$A = \begin{bmatrix} a_{11} & a_{12} & \dots & a_{1n} \\ a_{21} & a_{22} & \dots & a_{2n} \\ \dots & \dots & \dots & \dots \\ a_{m1} & a_{m2} & \dots & a_{mn} \end{bmatrix} \tag{2.37}$$

The elements a_{ij} of the matrix may be real numbers, complex numbers, or functions of other variables. Matrix $\boldsymbol{A}$ has m rows and n columns and is said to be of order $(m \times n)$. If the number of rows of a matrix is equal to the number of columns, that is, if $m = n$, then the matrix is a square matrix of nth order. A special matrix containing only a single column is called a vector:

$$\boldsymbol{x} = \begin{bmatrix} x_1 \\ x_2 \\ \cdot \\ \cdot \\ \cdot \\ x_n \end{bmatrix} \tag{2.15}$$

Define another matrix $\boldsymbol{B}$ with k rows and l columns:

$$B = \begin{bmatrix} b_{11} & b_{12} & \dots & b_{1l} \\ b_{21} & b_{22} & \dots & b_{2l} \\ \dots & \dots & \dots & \dots \\ b_{k1} & b_{k2} & \dots & b_{kl} \end{bmatrix} \tag{2.38}$$

The two matrices $\boldsymbol{A}$ and $\boldsymbol{B}$ can be added to (or subtracted from) each other if they have the same number of rows $(m = k)$ and the same number of columns $(n = l)$. For example, if both $\boldsymbol{A}$ and $\boldsymbol{B}$ are (3×2) matrices, their sum (or difference) can be written as

$$A \pm B = \begin{bmatrix} a_{11} \pm b_{11} & a_{12} \pm b_{12} \\ a_{21} \pm b_{21} & a_{22} \pm b_{22} \\ a_{31} \pm b_{31} & a_{32} \pm b_{32} \end{bmatrix} = C \tag{2.39}$$

Matrix C is also a (3×2) matrix. The commutative and associative laws for addition and subtraction apply.

Two matrices can multiply each other, if they are conformable. Matrices A and B would be conformable in the order AB if A had the same number of columns as B has rows $(n = k)$. If A is of order (4×2) and B is of order (2×3), then the product AB is

$$AB = \begin{bmatrix} a_{11} & a_{12} \\ a_{21} & a_{22} \\ a_{31} & a_{32} \\ a_{41} & a_{42} \end{bmatrix} \begin{bmatrix} b_{11} & b_{12} & b_{13} \\ b_{21} & b_{22} & b_{23} \end{bmatrix}$$

$$= \begin{bmatrix} a_{11}b_{11} + a_{12}b_{21} & a_{11}b_{12} + a_{12}b_{22} & a_{11}b_{13} + a_{12}b_{23} \\ a_{21}b_{11} + a_{22}b_{21} & a_{21}b_{12} + a_{22}b_{22} & a_{21}b_{13} + a_{22}b_{23} \\ a_{31}b_{11} + a_{32}b_{21} & a_{31}b_{12} + a_{32}b_{22} & a_{31}b_{13} + a_{32}b_{23} \\ a_{41}b_{11} + a_{42}b_{21} & a_{41}b_{12} + a_{42}b_{22} & a_{41}b_{13} + a_{42}b_{23} \end{bmatrix} = E \tag{2.40}$$

The resulting matrix E is of order (4×3). The general equation for performing matrix multiplication is

$$e_{ij} = \sum_{p=1}^{n} a_{ip} b_{pj} \qquad \begin{cases} i = 1, 2, \ldots, m \\ j = 1, 2, \ldots, l \end{cases} \tag{2.41}$$

The resulting matrix would be of order $(m \times l)$.

The commutative law is not usually valid for matrix multiplication, that is,

$$AB \neq BA \tag{2.42}$$

even if the matrices are conformable. The distributive law for multiplication applies to matrices, provided that conformability exists:

$$A(B + C) = AB + AC \tag{2.43}$$

The associative law of multiplication is also valid for matrices:

$$A(BC) = (AB)C \tag{2.44}$$

When working with MATLAB, it should be noted that there is also an element-by-element multiplication for matrices that is completely different from ordinary multiplication of matrices described above. The element-by-element multiplication, whose operator is ".*" (a dot before the ordinary multiplication operator) may be applied only to matrices of the same order, and it simply multiplies corresponding elements of the two matrices. For example, if $\boldsymbol{A}$ and $\boldsymbol{B}$ are of order (3 × 2), then the element-by-element product $\boldsymbol{A}.*\boldsymbol{B}$ is

$$\boldsymbol{A}.*\boldsymbol{B} = \begin{bmatrix} a_{11} & a_{12} \\ a_{21} & a_{22} \\ a_{31} & a_{32} \end{bmatrix} .* \begin{bmatrix} b_{11} & b_{12} \\ b_{21} & b_{22} \\ b_{31} & b_{32} \end{bmatrix} = \begin{bmatrix} a_{11}b_{11} & a_{12}b_{12} \\ a_{21}b_{21} & a_{22}b_{22} \\ a_{31}b_{31} & a_{32}b_{32} \end{bmatrix} \tag{2.45}$$

If the rows of an $(m \times n)$ matrix are written as columns, a new matrix of order $(n \times m)$ is formed. This new matrix is called the *transpose* of the original matrix. For example, if matrix $\boldsymbol{A}$ is

$$\boldsymbol{A} = \begin{bmatrix} 1 & 2 \\ 3 & 4 \\ 5 & 6 \end{bmatrix} \tag{2.46}$$

then the *transpose* $\boldsymbol{A}'$ is

$$\boldsymbol{A}' = \begin{bmatrix} 1 & 3 & 5 \\ 2 & 4 & 6 \end{bmatrix} \tag{2.47}$$

The transpose of the matrix $\boldsymbol{A}$ is sometimes shown as $\boldsymbol{A}^{\mathrm{T}}$. The transpose of the sum of two matrices is given by

$$(\boldsymbol{A} + \boldsymbol{B})' = \boldsymbol{A}' + \boldsymbol{B}' \tag{2.48}$$

The transpose of the product of two matrices is given by

$$(\boldsymbol{AB})' = \boldsymbol{B}'\boldsymbol{A}' \tag{2.49}$$

In MATLAB transpose of a matrix is simply obtained by adding a prime sign (') after the matrix.

The following definitions apply to square matrices only: A *symmetric* matrix is one that obeys the equation

$$\boldsymbol{A} = \boldsymbol{A}' \tag{2.50}$$

If the symmetrically situated elements of a matrix are complex conjugates of each other, the matrix is called *Hermitian.*

A *diagonal* matrix is one with nonzero elements on the principal diagonal and zero elements everywhere else:

$$D = \begin{bmatrix} d_{11} & 0 & \dots & 0 \\ 0 & d_{22} & \dots & 0 \\ \dots & \dots & \dots & \dots \\ 0 & 0 & \dots & d_{nn} \end{bmatrix} \quad (2.51)$$

The built-in MATLAB function *diag*(x) creates a diagonal matrix whose main diagonal elements are the components of the vector x. If x is a matrix, *diag*(x) is a column vector formed from the elements of the diagonal of x.

A *unit* matrix (or *identity* matrix) is a diagonal matrix whose nonzero elements are unity:

$$I = \begin{bmatrix} 1 & 0 & \dots & 0 \\ 0 & 1 & \dots & 0 \\ \dots & \dots & \dots & \dots \\ 0 & 0 & \dots & 1 \end{bmatrix} \quad (2.52)$$

Multiplication of a matrix (or a vector) by the identity matrix does not alter the matrix (or vector):

$$IA = A, \qquad Ix = x \quad (2.53)$$

In MATLAB, the function *eye*(n) returns an $(n \times n)$ unit matrix.

A *tridiagonal* matrix is one which has nonzero elements on the principal diagonal and its two adjacent diagonals, that we will refer to as the subdiagonal (below) and superdiagonal (above), and zero elements everywhere else:

$$T = \begin{bmatrix} t_{11} & t_{12} & 0 & 0 & & & \dots & 0 & 0 \\ t_{21} & t_{22} & t_{23} & 0 & & & \dots & 0 & 0 \\ 0 & t_{32} & t_{33} & t_{34} & & & \dots & 0 & 0 \\ \dots & \dots & \dots & \dots & \dots & \dots & \dots & \dots & \dots \\ 0 & 0 & & \dots & & t_{n-2\,n-3} & t_{n-2\,n-2} & t_{n-2\,n-1} & 0 \\ 0 & 0 & & \dots & & 0 & t_{n-1\,n-2} & t_{n-1\,n-1} & t_{n-1\,n} \\ 0 & 0 & & \dots & & 0 & 0 & t_{n\,n-1} & t_{nn} \end{bmatrix} \quad (2.54)$$

An *upper triangular* matrix is one that has all zero elements below the principal diagonal:

$$U = \begin{bmatrix} u_{11} & u_{12} & u_{13} & \dots & u_{1\,n-1} & u_{1n} \\ 0 & u_{22} & u_{23} & \dots & u_{2\,n-1} & u_{2n} \\ \dots & \dots & \dots & \dots & \dots & \dots \\ 0 & 0 & 0 & \dots & u_{n-1\,n-1} & u_{n-1\,n} \\ 0 & 0 & 0 & \dots & 0 & u_{nn} \end{bmatrix} \tag{2.55}$$

In MATLAB, the function *triu*(A) constructs an upper triangular matrix out of the matrix A, that is, it keeps the elements on the main diagonal and above that unchanged and replaces the elements located under the main diagonal with zero.

A *lower triangular* matrix is one that has all zero elements above the principal diagonal:

$$L = \begin{bmatrix} l_{11} & 0 & 0 & \dots & 0 & 0 \\ l_{21} & l_{22} & 0 & \dots & 0 & 0 \\ \dots & \dots & \dots & \dots & \dots & \dots \\ l_{n-1\,1} & l_{n-1\,2} & l_{n-1\,3} & \dots & l_{n-1\,n-1} & 0 \\ l_{n1} & l_{n2} & l_{n3} & \dots & l_{n\,n-1} & l_{nn} \end{bmatrix} \tag{2.56}$$

In MATLAB, the function *tril*(A) constructs a lower triangular matrix out of the matrix A, that is, it keeps the elements on the main diagonal and below that unchanged and replaces the elements located above the main diagonal with zero.

Outputs of the MATLAB function *lu*(A) are an upper triangular matrix U and a "psychologically lower triangular matrix", that is, a product of lower triangular and permutation matrices, in L so that $LU = A$.

A *supertriangular* matrix, also called a *Hessenberg* matrix, is one that has all zero elements below the subdiagonal, such as the upper Hessenberg matrix of Eq. (2.57):

$$H_U = \begin{bmatrix} h_{11} & h_{12} & h_{13} & & \dots & & h_{1\,n-2} & h_{1\,n-1} & h_{1n} \\ h_{21} & h_{22} & h_{23} & & \dots & & h_{2\,n-2} & h_{2\,n-1} & h_{2n} \\ 0 & h_{32} & h_{33} & & \dots & & h_{3\,n-3} & h_{3\,n-1} & h_{3n} \\ \dots & \dots & \dots & \dots & \dots & \dots & \dots & \dots & \dots \\ 0 & 0 & & & \dots & & h_{n-1\,n-2} & h_{n-1\,n-1} & h_{n-1\,n} \\ 0 & 0 & & & \dots & & 0 & h_{n\,n-1} & h_{nn} \end{bmatrix} \tag{2.57}$$

or above the superdiagonal, such as the lower Hessenberg matrix of Eq. (2.58):

$$H_L = \begin{bmatrix} h_{11} & h_{12} & 0 & & \dots & & 0 & 0 \\ h_{21} & h_{22} & h_{23} & & \dots & & 0 & 0 \\ \dots & \dots & \dots & \dots & \dots & \dots & \dots & \dots \\ h_{n-2\,1} & h_{n-2\,2} & h_{n-2\,3} & & \dots & & h_{n-2\,n-1} & 0 \\ h_{n-1\,1} & h_{n-1\,2} & h_{n-1\,3} & & \dots & & h_{n-1\,n-1} & h_{n-1\,n} \\ h_{n1} & h_{n2} & h_{n3} & & \dots & & h_{n\,n-1} & h_{nn} \end{bmatrix} \tag{2.58}$$

Tridiagonal, triangular, and Hessenberg matrices are called *banded* matrices.

Matrices can be divided into two general categories: *dense* and *sparse* matrices. The dense matrices are usually of low order and may have only few zero elements. The sparse matrices may be of high order with many zero elements. A special subcategory of sparse matrices is the group of banded matrices described above.

The sum of the elements on the main diagonal of a square matrix is called the *trace*:

$$tr\boldsymbol{A} = \sum_{i=1}^{n} a_{ii} \tag{2.59}$$

The sum of the *eigenvalues* of a square matrix is equal to the trace of that matrix:

$$\sum_{i=1}^{n} \lambda_i = tr\boldsymbol{A} \tag{2.60}$$

The MATLAB function *trace*(***A***) calculates the trace of the matrix ***A***.

Matrix division is not defined in the normal algebraic sense. Instead, an inverse operation is defined, which uses multiplication to achieve the same results. If a square matrix ***A*** and another square matrix ***B***, of same order as ***A***, lead to the identity matrix ***I*** when multiplied together:

$$\boldsymbol{AB} = \boldsymbol{I} \tag{2.61}$$

then ***B*** is called the *inverse* of ***A*** and is written as $\boldsymbol{A}^{-1}$. It follows then that

$$\boldsymbol{AA}^{-1} = \boldsymbol{A}^{-1}\boldsymbol{A} = \boldsymbol{I} \tag{2.62}$$

There are several different ways in MATLAB to calculate the inverse of a square matrix. The function *inv*(***A***) gives the inverse of ***A***. Also, the inverse of the matrix may be obtained by

A^(-1). The expression **A/B** in MATLAB is equivalent to $\mathbf{AB}^{-1}$ and the expression **A\B** is equivalent to $\mathbf{A}^{-1}\mathbf{B}$. Note that the expression **A./B** (putting "." before division operator) is element-by-element division of the elements of the two matrices and the expression **A**.^(-1) results in a matrix whose elements are the reciprocals of the elements of the original matrix.

The inverse of the product of two matrices is the product of the inverses of these matrices multiplied in reverse order:

$$(\mathbf{AB})^{-1} = \mathbf{B}^{-1}\mathbf{A}^{-1} \tag{2.63}$$

This can be generalized to products of more than two matrices:

$$(\mathbf{ABC}...\mathbf{KLM})^{-1} = \mathbf{M}^{-1}\mathbf{L}^{-1}\mathbf{K}^{-1}...\mathbf{C}^{-1}\mathbf{B}^{-1}\mathbf{A}^{-1} \tag{2.64}$$

A matrix is singular if the determinant of the matrix is zero. Only *nonsingular* matrices have inverse.

The value of the determinant, which exists for square matrices only, can be calculated from Laplace's expansion theorem, which involves *minors* and *cofactors* of square matrices. If the row and column containing an element a_{ij} in a square matrix **A** are deleted, the determinant of the remaining square array is called the *minor* of a_{ij} and is denoted by M_{ij}. The *cofactor* of a_{ij}, denoted by A_{ij}, is given by

$$A_{ij} = (-1)^{i+j}M_{ij} \tag{2.65}$$

Laplace's expansion theorem states that the determinant of a square matrix **A**, shown as $|\mathbf{A}|$, is equal to the sum of products of the elements of any row (or column) and their respective cofactors:

$$|\mathbf{A}| = \sum_{k=1}^{n} a_{ik}A_{ik} \tag{2.66}$$

for any row i, or

$$|\mathbf{A}| = \sum_{k=1}^{n} a_{kj}A_{kj} \tag{2.67}$$

for any column j.

Determinants have the following properties:

Property 1. If all the elements of any row or column of a matrix are zero, its determinant is equal to zero.

Property 2. If the corresponding rows and columns of a matrix are interchanged, its determinant is unchanged.

Property 3. If two rows or two columns of a matrix are interchanged, the sign of the determinant changes.

Property 4. If the elements of two rows or two columns of a matrix are equal, the determinant of the matrix is zero.

Property 5. If the elements of any row or column of a matrix are multiplied by a scalar, this is equivalent to multiplying the determinant by the scalar.

Property 6. Adding the product of a scalar and any row (or column) to any other row (or column) of a matrix leaves the determinant unchanged.

Property 7. The determinant of a triangular matrix is equal to the product of its diagonal elements:

$$|U| = \prod_{i=1}^{n} a_{ii} \tag{2.68}$$

Calculating determinants by the expansion of cofactors is a very time-intensive task. Each determinant has $n!$ groups of terms and each group is the product of n elements; thus the total number of multiplications is $(n - 1)(n!)$. Evaluating the determinant of a matrix of order (10×10) would require 32,659,200 multiplications. More efficient methods have been developed for evaluating determinants. It will be shown in Sec. 2.5 that the Gauss elimination method can be used to calculate the determinant of a matrix in addition to finding the solution of simultaneous linear algebraic equations. To evaluate the determinant of a square matrix **A** in MATLAB, the built-in function *det*(**A**) may be used.

The inverse of a matrix cannot always be determined accurately. There are many matrices that are ill-conditioned. An ill-conditioned matrix can be identified using the following criterion: When the ratio of the absolute values of the largest and smallest eigenvalues of the matrix is very large, the matrix is ill-conditioned.

The *rank* r of matrix **A** is defined as the order of the largest nonsingular square matrix within **A**. Consider the $(m \times n)$ matrix

$$\mathbf{A} = \begin{bmatrix} a_{11} & a_{12} & \cdots & a_{1n} \\ a_{21} & a_{22} & \cdots & a_{2n} \\ \cdots & \cdots & \cdots & \cdots \\ a_{m1} & a_{m2} & \cdots & a_{mn} \end{bmatrix} \tag{2.37}$$

where $n \geq m$. The largest square submatrix within **A** is of order $(m \times m)$. If the determinant of this $(m \times m)$ submatrix is nonzero, then the rank of **A** is m $(r = m)$. However, if the determinant of the $(m \times m)$ submatrix is equal to zero, then the rank of **A** is less than m $(r < m)$. The order of the next largest nonsingular submatrix that can be located within **A** would determine the value of the rank. In MATLAB, the function *rank*(*A*) gives the value of r, the rank of matrix **A**.

As an example, let us look at the following (3 × 4) matrix:

$$A = \begin{bmatrix} 3 & 1 & 2 & -4 \\ 5 & 2 & 1 & 3 \\ 6 & 2 & 4 & -8 \end{bmatrix} \tag{2.69}$$

There are four submatrices of order (3 × 3), whose determinant are evaluated below using Laplace's expansion theorem:

$$\begin{vmatrix} 3 & 1 & 2 \\ 5 & 2 & 1 \\ 6 & 2 & 4 \end{vmatrix} = (3)(-1)^2 \begin{vmatrix} 2 & 1 \\ 2 & 4 \end{vmatrix} + (1)(-1)^3 \begin{vmatrix} 5 & 1 \\ 6 & 4 \end{vmatrix} + (2)(-1)^4 \begin{vmatrix} 5 & 2 \\ 6 & 2 \end{vmatrix}$$

$$= (3)(8 - 2) - (1)(20 - 6) + (2)(10 - 12) = 0 \tag{2.70}$$

Similarly,

$$\begin{vmatrix} 3 & 1 & -4 \\ 5 & 2 & 3 \\ 6 & 2 & -8 \end{vmatrix} = 0 \qquad \begin{vmatrix} 3 & 2 & -4 \\ 5 & 1 & 3 \\ 6 & 4 & -8 \end{vmatrix} = 0 \qquad \begin{vmatrix} 1 & 2 & -4 \\ 2 & 1 & 3 \\ 2 & 4 & -8 \end{vmatrix} = 0 \tag{2.71}$$

Because all the above (3 × 3) submatrices are singular; the rank of A is less than 3. It is easy to find several (2 × 2) submatrices that are nonsingular; therefore $r = 2$.

The same conclusion, regarding the singularity of the (3 × 3) submatrices, could have been reached by the application of Properties 4 and 5, which were mentioned earlier in this section. Property 4 states, "If the elements of two rows or two columns of a matrix are equal, the determinant of the matrix is zero." Property 5 states, "If the elements of any row or column of a matrix are multiplied by a scalar, this is equivalent to multiplying the determinant by a scalar." Careful inspection of the four (3 × 3) submatrices shows that the first and third rows are multiples of each other. In accordance with Properties 4 and 5, the determinants are zero.

2.2.2 Matrix Transformations

It is often desirable to transform a matrix to a different form which is more amenable to solution. There are several such transformations that convert matrices without significantly changing their properties. We will divide these transformations into two categories: *elementary* transformations and *similarity* transformations.

Elementary transformations usually change the shape of the matrix but preserve the value of its determinant. In addition, if the matrix represents a set of linear algebraic equations, the solution of the set is not affected by the elementary transformation. The following series of matrix multiplications:

$$\boldsymbol{L}_{n-1}\boldsymbol{L}_{n-2}\ldots\boldsymbol{L}_2\boldsymbol{L}_1\boldsymbol{A} = \boldsymbol{U} \tag{2.72}$$

represents an elementary transformation of matrix $\boldsymbol{A}$ to an upper triangular matrix $\boldsymbol{U}$. This operation can be shown in condensed form as

$$\boldsymbol{L}\boldsymbol{A} = \boldsymbol{U} \tag{2.73}$$

where the transformation matrix $\boldsymbol{L}$ is the product of the lower triangular matrices $\boldsymbol{L}_i$. The form of $\boldsymbol{L}_i$ matrices will be define in Sec. 2.5, in conjunction with the development of the Gauss elementary transformation procedure.

Similarity transformations are of the form

$$\boldsymbol{Q}^{-1}\boldsymbol{A}\boldsymbol{Q} = \boldsymbol{B} \tag{2.74}$$

where $\boldsymbol{Q}$ is a nonsingular square matrix. In this operation, matrix $\boldsymbol{A}$ is transformed to matrix $\boldsymbol{B}$, which is said to be *similar* to $\boldsymbol{A}$. Similarity in this case implies that:

1. The determinants of $\boldsymbol{A}$ and $\boldsymbol{B}$ are equal:

$$|\boldsymbol{A}| = |\boldsymbol{B}| \tag{2.75}$$

2. The traces of $\boldsymbol{A}$ and $\boldsymbol{B}$ are the same:

$$tr\,\boldsymbol{A} = tr\,\boldsymbol{B} \tag{2.76}$$

3. The eigenvalues of $\boldsymbol{A}$ and $\boldsymbol{B}$ are identical:

$$|\boldsymbol{A} - \lambda\boldsymbol{I}| = |\boldsymbol{B} - \lambda\boldsymbol{I}| \tag{2.77}$$

If columns of matrix $\boldsymbol{Q}$ are real mutually orthogonal unit vectors, then $\boldsymbol{Q}$ is an orthogonal matrix, and the following relations are true:

$$\boldsymbol{Q}'\boldsymbol{Q} = \boldsymbol{I} \tag{2.78}$$

and

$$\boldsymbol{Q}' = \boldsymbol{Q}^{-1} \tag{2.79}$$

In this case, the similarity transformation, represented by Eq. (2.74), can be written as

$$Q'AQ = B \tag{2.80}$$

and is called an *orthogonal* transformation. Since an orthogonal transformation is a similarity transformation, the three identities [Eqs. (2.75)-(2.77)] pertaining to determinants, traces, and eigenvalues of $\boldsymbol{A}$ and $\boldsymbol{B}$ are equally valid.

In MATLAB, the function *orth*($\boldsymbol{A}$) gives the matrix $\boldsymbol{Q}$ described above.

2.2.3 Matrix Polynomials and Power Series

The definition of a scalar polynomial was given in Chap. 1 as

$$f(x) = a_n x^n + a_{n-1} x^{n-1} + \dots + a_1 x + a_0 \tag{2.81}$$

Similarly, a matrix polynomial can be defined as

$$\boldsymbol{P}(\boldsymbol{A}) = \alpha_n \boldsymbol{A}^n + \alpha_{n-1} \boldsymbol{A}^{n-1} + \dots + \alpha_1 \boldsymbol{A} + \alpha_0 \boldsymbol{I} \tag{2.82}$$

where $\boldsymbol{A}$ is a square matrix, $\boldsymbol{A}^n$ is the product of $\boldsymbol{A}$ by itself n times, and $\boldsymbol{A}^0 = \boldsymbol{I}$.

Matrices can be used in infinite series, such as the exponential, trigonometric, and logarithmic series. For example, the matrix exponential function is defined as

$$e^{\boldsymbol{A}} = \boldsymbol{I} + \boldsymbol{A} + \frac{\boldsymbol{A}^2}{2!} + \frac{\boldsymbol{A}^3}{3!} + \dots \tag{2.83}$$

and the matrix trigonometric functions as

$$\sin \boldsymbol{A} = \boldsymbol{A} - \frac{\boldsymbol{A}^3}{3!} + \frac{\boldsymbol{A}^5}{5!} - \dots \tag{2.84}$$

$$\cos \boldsymbol{A} = \boldsymbol{I} - \frac{\boldsymbol{A}^2}{2!} + \frac{\boldsymbol{A}^4}{4!} - \dots \tag{2.85}$$

Note that in MATLAB, the functions: *exp*($\boldsymbol{A}$), *cos*($\boldsymbol{A}$), *sin*($\boldsymbol{A}$), are element-by-element functions and do not obey the above definitions. The MATLAB functions *expm*($\boldsymbol{A}$), *expm1*($\boldsymbol{A}$), *expm2*($\boldsymbol{A}$), and *expm3*($\boldsymbol{A}$) calculate exponential of the matrix $\boldsymbol{A}$ by different algorithms. The function *expm2*($\boldsymbol{A}$) calculates exponential of the matrix $\boldsymbol{A}$ as in Eq. (2.83).

2.2.4 Vector Operations

Consider two vectors $\boldsymbol{x}$ and $\boldsymbol{y}$:

$$\boldsymbol{x} = \begin{bmatrix} x_1 \\ x_2 \\ \cdot \\ \cdot \\ \cdot \\ x_n \end{bmatrix} \qquad \boldsymbol{y} = \begin{bmatrix} y_1 \\ y_2 \\ \cdot \\ \cdot \\ \cdot \\ y_n \end{bmatrix} \tag{2.86}$$

and their transpose:

$$\boldsymbol{x}' = [x_1 \ \ x_2 \ \ \ldots \ \ x_n] \qquad \boldsymbol{y}' = [y_1 \ \ y_2 \ \ \ldots \ \ y_n] \tag{2.87}$$

The *scalar* product (or *inner* product) of these two vectors is defined as

$$\boldsymbol{x}'\boldsymbol{y} = [x_1 \ \ x_2 \ \ \ldots \ \ x_n] \begin{bmatrix} y_1 \\ y_2 \\ \cdot \\ \cdot \\ \cdot \\ y_n \end{bmatrix} = x_1 y_1 + x_2 y_2 + \ldots + x_n y_n \tag{2.88}$$

As the name implies, this is a scalar quantity. The scalar product is sometimes called the *dot* product. The *dyadic* product of these two vectors is defined as

$$\boldsymbol{x}\boldsymbol{y}' = \begin{bmatrix} x_1 \\ x_2 \\ \cdot \\ \cdot \\ \cdot \\ x_n \end{bmatrix} [y_1 \ y_2 \ \dots \ y_n] = \begin{bmatrix} x_1y_1 & x_1y_2 & \dots & x_1y_n \\ x_2y_1 & x_2y_2 & \dots & x_2y_n \\ \dots & \dots & \dots & \dots \\ x_ny_1 & x_ny_2 & \dots & x_ny_n \end{bmatrix} \tag{2.89}$$

This is a matrix of order $(n \times n)$. The *element-by-element* product of two vectors is a vector:

$$\boldsymbol{x}\boldsymbol{y} = \begin{bmatrix} x_1 \\ x_2 \\ \cdot \\ \cdot \\ \cdot \\ x_n \end{bmatrix} \begin{bmatrix} y_1 \\ y_2 \\ \cdot \\ \cdot \\ \cdot \\ y_n \end{bmatrix} = \begin{bmatrix} x_1y_1 \\ x_2y_2 \\ \cdot \\ \cdot \\ \cdot \\ x_ny_n \end{bmatrix} \tag{2.90}$$

Two nonzero vectors are *orthogonal* if their scalar product is zero:

$$\boldsymbol{x}'\boldsymbol{y} = \boldsymbol{0} \tag{2.91}$$

The length of a vector can be calculated from

$$|\boldsymbol{x}| = \sqrt{\boldsymbol{x}'\boldsymbol{x}} \tag{2.92}$$

A *unit vector* is a vector whose length is unity.
A set of vectors $\boldsymbol{x}, \boldsymbol{y}, \boldsymbol{z}, \dots$ is linearly dependent if there exists a set of scalars $c_1, c_2, c_3, \dots$ so that

$$c_1\boldsymbol{x} + c_2\boldsymbol{y} + c_3\boldsymbol{z} + \dots = 0 \tag{2.93}$$

Otherwise the vectors are *linearly independent*.

2.3 CONSISTENCY OF EQUATIONS AND EXISTENCE OF SOLUTIONS

Consider the set of simultaneous linear algebraic equations represented by

$$\begin{aligned} a_{11}x_1 + a_{12}x_2 + \ldots + a_{1n}x_n &= c_1 \\ a_{21}x_1 + a_{22}x_2 + \ldots + a_{2n}x_n &= c_2 \\ \ldots\ldots\ldots\ldots\ldots\ldots\ldots \\ a_{n1}x_1 + a_{n2}x_2 + \ldots + a_{nn}x_n &= c_n \end{aligned} \tag{2.12}$$

The *coefficient matrix* is $\boldsymbol{A}$, the *vector of unknowns* is $\boldsymbol{x}$, and the *vector of constants* is $\boldsymbol{c}$.

The *augmented* matrix $\boldsymbol{A}_a$ is defined as the matrix resulting from joining the vector $\boldsymbol{c}$ to the columns of matrix $\boldsymbol{A}$ as shown below:

$$\boldsymbol{A}_a = \begin{bmatrix} a_{11} & a_{12} & \ldots & a_{1n} & c_1 \\ a_{21} & a_{22} & \ldots & a_{2n} & c_2 \\ \ldots & \ldots & \ldots & \ldots & \ldots \\ a_{n1} & a_{n2} & \ldots & a_{nn} & c_n \end{bmatrix} \tag{2.94}$$

The set of equations has a solution if, and only if, the rank of the augmented matrix is equal to the rank of the coefficient matrix. If, in addition, the rank is equal to n $(r = n)$, the solution is unique. If the rank is less than n $(r < n)$, there are more unknowns in the set than there are independent equations. In that case, the set of equations can be reduced to r independent equations. The remaining $(n - r)$ unknowns must be assigned arbitrary values. This implies that the system of n equations has an infinite number of possible solutions, because the values of $(n - r)$ unknowns are given arbitrary values, and the rest of unknowns depend on these $(n - r)$ values.

A special subcategory of linear algebraic equations is the set whose vector of constants $\boldsymbol{c}$ is the zero vector:

$$\boldsymbol{A}\boldsymbol{x} = 0 \tag{2.95}$$

This is called the *homogeneous* set of linear algebraic equations. This set always has the solution

$$x_1 = x_2 = \ldots = x_n = 0 \tag{2.96}$$

It is called the *trivial* solution, because it is not of any particular interest. The coefficient matrix and the augmented matrix of a homogeneous set always have the same rank, as the vector $\boldsymbol{c}$ is the zero vector. As stated earlier, if the rank of $\boldsymbol{A}$ is equal to n $(r = n)$, then the set of equations has a unique solution. However, in the case of the homogeneous equations, this unique solution is none other than the trivial one. In order for a homogeneous set to have nontrivial solutions, the determinant of $\boldsymbol{A}$ must be zero, that is, $\boldsymbol{A}$ must be singular.

In summary, the nonhomogeneous set has a *unique nontrivial* solution if the matrix of coefficients $\boldsymbol{A}$ is nonsingular. It has an infinite number of solutions if the matrix $\boldsymbol{A}$ is singular and the ranks of $\boldsymbol{A}$ and $\boldsymbol{A}_a$ are equal to each other. It has no solution at all if the rank of $\boldsymbol{A}$ is lower than the rank of $\boldsymbol{A}_a$.

The homogeneous set has a *unique*, but trivial, solution if the matrix of coefficients $\boldsymbol{A}$ is nonsingular. It has an infinite number of solutions if the matrix $\boldsymbol{A}$ is singular. The rank of $\boldsymbol{A}_a$ is always equal to the rank of $\boldsymbol{A}$ for a homogeneous system, since the vector of constants is the zero vector. (See Table 2.2)

Table 2.2 Existence of Solutions

Condition	**Nonhomogeneous set** $\boldsymbol{Ax} = \boldsymbol{c}$	**Homogeneous set** $\boldsymbol{Ax} = \boldsymbol{0}$
$rank\ \boldsymbol{A} = n$	Unique solution	Unique, but trivial, solution
$rank\ \boldsymbol{A} < n$		Infinite number of solutions
$rank\ \boldsymbol{A} < n$ and $rank\ \boldsymbol{A} = rank\ \boldsymbol{A}_a$	Infinite number of solutions	
$rank\ \boldsymbol{A} < n$ and $rank\ \boldsymbol{A} < rank\ \boldsymbol{A}_a$	No solution	

2.4 CRAMER'S RULE

Cramer's rule calculates the solution of nonhomogeneous linear algebraic equations of the form

$$A x = c \tag{2.13}$$

using the determinants of the coefficient matrix A and the substituted matrix A_j as follows:

$$x_j = \frac{|A_j|}{|A|} \qquad j = 1, 2, \ldots, n \tag{2.97}$$

The substituted matrix A_j is obtained by replacing column j of matrix A with the vector c:

$$A_j = \begin{bmatrix} a_{11} & \cdots & a_{1j-1} & c_1 & a_{1j+1} & \cdots & a_{1n} \\ a_{21} & \cdots & a_{2j-1} & c_2 & a_{2j+1} & \cdots & a_{2n} \\ \cdots & \cdots & \cdots & \cdots & \cdots & \cdots & \cdots \\ a_{nl} & \cdots & a_{nj-1} & c_n & a_{nj+1} & \cdots & a_{nn} \end{bmatrix} \tag{2.98}$$

The set of equations must be nonhomogeneous, as the determinant of A appears in the denominator of Eq. (2.97); the determinant cannot be zero; that is, matrix A must be nonsingular.

For a system of n equations, Cramer's rule evaluates $(n + 1)$ determinants and performs n divisions. The calculation of each determinant requires $(n - 1)(n!)$ multiplications; therefore, the total number of multiplications and divisions is

$$(n + 1)(n - 1)(n!) + n \tag{2.99}$$

Table 2.3 illustrates how the number of operations required by Cramer's rule increases as the value of n increases. For $n = 3$, a total of 51 multiplications and divisions are needed. However, when $n = 10$, this number climbs to 359,251,210. For this reason, Cramer's rule is rarely used for systems with $n > 3$. The Gauss elimination, Gauss-Jordan reduction, and Gauss-Seidel methods, to be described in the next three sections of this chapter, are much more efficient methods of solution of linear equations than Cramer's rule.

Table 2.3 Number of operations needed by Cramer's rule

n	$(n+1)(n-1)(n!)+n$	n	$(n+1)(n-1)(n!)+n$
3	51	7	241,927
4	364	8	2,540,168
5	2,885	9	29,030,409
6	25,206	10	359,251,210

2.5 GAUSS ELIMINATION METHOD

The most widely used method for solution of simultaneous linear algebraic equations is the Gauss elimination method. This is based on the principle of converting the set of n equations in n unknowns:

$$\begin{aligned}
a_{11}x_1 + a_{12}x_2 + \dots + a_{1n}x_n &= c_1 \\
a_{21}x_1 + a_{22}x_2 + \dots + a_{2n}x_n &= c_2 \\
\cdots\cdots\cdots\cdots\cdots\cdots\cdots \\
a_{nl}x_1 + a_{n2}x_2 + \dots + a_{nn}x_n &= c_n
\end{aligned} \tag{2.12}$$

to a triangular set of the form

$$\begin{aligned}
a_{11}x_1 + a_{12}\,x_2 + a_{13}\,x_3 + a_{14}\,x_4 + \dots + a_{1n}\,x_n &= c_1 \\
a'_{22}x_2 + a'_{23}x_3 + a'_{24}x_4 + \dots + a'_{2n}x_n &= c'_2 \\
a'_{33}x_3 + a'_{34}x_4 + \dots + a'_{3n}x_n &= c'_3 \\
&\vdots \\
a'_{n-1,n-1}x_{n-1} + a'_{n-1,n}x_n &= c'_{n-1} \\
a'_{nn}\,x_n &= c'_n
\end{aligned} \tag{2.100}$$

whose solution is the same as that of the original set of equations.

The process is essentially that of converting the set

$$\boldsymbol{Ax} = \boldsymbol{c} \tag{2.13}$$

to the equivalent triangular set

$$\boldsymbol{U x} = \boldsymbol{c}' \tag{2.101}$$

where $\boldsymbol{U}$ is an upper triangular matrix and $\boldsymbol{c}'$ is the modified vector of constants. Once triangularization is achieved, the solution of the set can be obtained easily by back substitution starting with variable n and working backward to variable 1.

2.5.1 Gauss Elimination in Formula Form

The Gauss elimination is accomplished by a series of elementary operations that do not alter the solution of the equation. These operations are:

1. Any equation in the set can be multiplied (or divided) by a nonzero scalar without affecting the solution.
2. Any equation in the set can be added to (or subtracted from) another equation without affecting the solution.
3. Any two equations can interchange positions within the set without affecting the solution.

Two matrices that can be obtained from each other by successive application of the above elementary operations are said to be *equivalent* matrices. The rank and determinant of these matrices are unaltered by the application of elementary operations.

In order to demonstrate the application of the Gauss elimination method, apply the triangularization procedure to obtain the solution of the following set of three equations:

$$\begin{aligned} 3x_1 + 18x_2 + 9x_3 &= 18 \\ 2x_1 + 3x_2 + 3x_3 &= 117 \\ 4x_1 + x_2 + 2x_3 &= 283 \end{aligned} \tag{2.102}$$

First, form the (3×4) augmented matrix of coefficients and constants:

$$\left[\begin{array}{ccc|c} 3 & 18 & 9 & 18 \\ 2 & 3 & 3 & 117 \\ 4 & 1 & 2 & 283 \end{array}\right] \tag{2.103}$$

Each complete row of the augmented matrix represents one of the equations of the linear set (2.102). Therefore, any operations performed on a row of the augmented matrix are automatically performed on the corresponding equation.

To obtain the solution, divide the first row by 3, multiply it by 2, and subtract it from the second row to obtain

$$\begin{bmatrix} 3 & 18 & 9 & | & 18 \\ 0 & -9 & -3 & | & 105 \\ 4 & 1 & 2 & | & 283 \end{bmatrix} \tag{2.104}$$

Divide the first row by 3, multiply it by 4, and subtract it from the third row to obtain

$$\begin{bmatrix} 3 & 18 & 9 & | & 18 \\ 0 & -9 & -3 & | & 105 \\ 0 & -23 & -10 & | & 259 \end{bmatrix} \tag{2.105}$$

Note that the coefficients in the first column below the diagonal have become zero. Continue the elimination by dividing the second row by -9, multiply it by -23, and subtracting it from the third row to obtain

$$\begin{bmatrix} 3 & 18 & 9 & | & 18 \\ 0 & -9 & -3 & | & 105 \\ 0 & 0 & -\frac{7}{3} & | & -\frac{28}{3} \end{bmatrix} \tag{2.106}$$

The triangularization of the coefficient part of the augmented matrix is complete, and matrix (2.106) represents the triangular set of equations

$$3x_1 + 18x_2 + 9x_3 = 18 \tag{2.107a}$$

$$-9x_2 - 3x_3 = 105 \tag{2.107b}$$

$$-\frac{7}{3}x_3 = -\frac{28}{3} \tag{2.107c}$$

whose solution is identical to that of the original set (2.102). The solution is obtained by back substitution. Rearrangement of (2.107*c*) yields

$$x_3 = 4$$

Substitution of the value of x_3 in (2.107*b*) and rearrangement gives

$$x_2 = -13$$

Substitution of the values of x_3 and x_2 in (2.107*a*) and rearrangement yields

$$x_1 = 72$$

The overall Gauss elimination procedure applied on the $(n) \times (n+1)$ augmented matrix is condensed into a three-part mathematical formula for initialization, elimination, and back substitution as shown below:

Initialization formula:

$$\left.\begin{array}{ll} a_{ij}^{(0)} = a_{ij} & j = 1,2,\ldots,n \\ a_{ij}^{(0)} = c_{ij} & j = n+1 \end{array}\right\} \quad \left\{ i = 1,2,\ldots,n \right. \tag{2.108}$$

Elimination formula:

$$\sum a_{ij}^{(k)} = a_{ij}^{(k-1)} - \frac{a_{ik}^{(k-1)}}{a_{kk}^{(k-1)}} a_{kj}^{(k-1)} \quad \begin{cases} j = n+1, n, \ldots, k \\ i = k+1, k+2, \ldots, n \end{cases} \quad \begin{cases} k = 1,2,\ldots,n-1 \\ a_{kk}^{(k-1)} \neq 0 \end{cases} \tag{2.109}$$

where the initialization step places the elements of the coefficient matrix and the vector of constants into the augmented matrix, and the elimination formula reduces to zero the elements below the diagonal. The counter k is the iteration counter of the outside loop in a set of nested loops that perform the elimination.

It should be noted that the element a_{kk} in the denominator of Eq. (2.109) is always the diagonal element. It is called the *pivot* element. This pivot element must not be zero; otherwise, the computer program will result in overflow. The computer program can be written so that it rearranges the equations at each step to attain *diagonal dominance* in the coefficient matrix, that is, the row with the largest absolute value pivot element is chosen. This strategy is called *partial pivoting*, and it serves two purposes in the Gaussian elimination procedure: It reduces the possibility of division by zero, and it increases the accuracy of the Gauss elimination method by using the largest pivot element. If, in addition to rows, the columns are also searched for maximum available pivot element, then the strategy is called *complete pivoting*. If pivoting cannot locate a nonzero element to place on the diagonal, the matrix must be singular. When two columns are interchanged, the corresponding variables must also be interchanged. A program that performs complete pivoting must keep track of the column interchanges in order to interchange the corresponding variables.

When triangularization of the coefficient matrix has been completed, the algorithm transfers the calculation to the back-substitution formula:

$$\begin{aligned} x_n &= \frac{a_{n,n+1}}{a_{nn}} \\ x_i &= \frac{a_{i,n+1} - \sum_{j=i+1}^{n} a_{ij} x_j}{a_{ii}} \qquad i = n-1, n-2, \ldots, 1 \end{aligned} \tag{2.110}$$

The above formulas complete the solution of the equations by the Gauss elimination method by calculating all the unknowns from x_n to x_1. The Gauss elimination algorithm requires $n^3/3$ multiplications to evaluate the vector $\boldsymbol{x}$.

2.5.2 Gauss Elimination in Matrix Form

The Gauss elimination procedure, which was described above in formula form, can also be accomplished by series of matrix multiplications. Two types of special matrices are involved in this operation. Both of these matrices are modifications of the identity matrix. The first type, which we designate as $\boldsymbol{P}_{ij}$, is the identity matrix with the following changes: The unity at position ii switches places with the zero at position ij, and the unity at position jj switches places with the zero at position ji. For example, $\boldsymbol{P}_{23}$ for a fifth-order system is

$$\boldsymbol{P}_{23} = \begin{bmatrix} 1 & 0 & 0 & 0 & 0 \\ 0 & 0 & 1 & 0 & 0 \\ 0 & 1 & 0 & 0 & 0 \\ 0 & 0 & 0 & 1 & 0 \\ 0 & 0 & 0 & 0 & 1 \end{bmatrix} \tag{2.111}$$

Premultiplication of matrix $\boldsymbol{A}$ by $\boldsymbol{P}_{ij}$ has the effect of interchanging rows i and j. Postmultiplication causes interchange of column i and j. By definition, $\boldsymbol{P}_{ii} = \boldsymbol{I}$, and multiplication of $\boldsymbol{A}$ by $\boldsymbol{P}_{ii}$ causes no interchanges. The inverse of $\boldsymbol{P}_{ij}$ is identical to $\boldsymbol{P}_{ij}$.

The second type of matrices used by the Gauss elimination method are unit lower triangular matrices of the form

$$\boldsymbol{L}_1 = \begin{bmatrix} 1 & 0 & 0 & 0 & 0 \\ -\dfrac{a_{21}^{(0)}}{a_{11}^{(0)}} & 1 & 0 & 0 & 0 \\ -\dfrac{a_{31}^{(0)}}{a_{11}^{(0)}} & 0 & 1 & 0 & 0 \\ -\dfrac{a_{41}^{(0)}}{a_{11}^{(0)}} & 0 & 0 & 1 & 0 \\ -\dfrac{a_{51}^{(0)}}{a_{11}^{(0)}} & 0 & 0 & 0 & 1 \end{bmatrix} \tag{2.112}$$

where the superscript (0) indicates that each L_k matrix uses the elements $a_{ik}^{(k-1)}$ of the previous transformation step. Premultiplication of matrix A by L_k has the effect of reducing to zero the elements below the diagonal in column k. The inverse of L_k has the same form as L_k but with the signs of the off-diagonal elements reversed.

Therefore, the entire Gauss elimination method, which reduces a nonsingular matrix A to an upper triangular matrix U, can be represented by the following series of matrix multiplications:

$$L_{n-1}L_{n-2}\ldots P_{ij}\ldots L_2 L_1 P_{ij} A = U \qquad (2.113)$$

where the multiplications by P_{ij} cause pivoting, if and when needed, and the multiplications by L_k cause elimination. If pivoting is not performed, Eq. (2.113) simplifies to

$$L_{n-1}L_{n-2}\ldots L_2 L_1 A = U \qquad (2.72)$$

The matrices L_k are unit lower triangular and their product, defined by matrix L, is also unit lower triangular. With this definition of L, Eq. (2.72) condenses to

$$LA = U \qquad (2.73)$$

Because matrix L is unit lower triangular, it is nonsingular. Its inverse exists and is also a unit lower triangular matrix. If we premultiply both sides of Eq. (2.73) by L^{-1}, we obtain

$$A = L^{-1}U \qquad (2.114)$$

This equation represents the *decomposition* of a nonsingular matrix A into a unit lower triangular matrix and an upper triangular matrix. Furthermore, this decomposition is unique [2]. Therefore, the matrix operation of Eq. (2.73) when applied to the augmented matrix $[A \mid c]$ yields the unique solution:

$$L[A \mid c] \rightarrow [U \mid c'] \qquad (2.115)$$

of the system of linear algebraic equations

$$Ax = c \qquad (2.13)$$

whose matrix of coefficients A is nonsingular.

2.5.3 Calculation of Determinants by the Gauss Method

The Gauss elimination method is also very useful in the calculation of determinants of matrices. The elementary operations used in the Gauss method are consistent with the properties of determinants listed in Sec. 2.2. Therefore, the reduction of a matrix to the equivalent triangular matrix by the Gauss elimination procedure would not alter the value of the determinant of the matrix. The determinant of a triangular matrix is equal to the product of its diagonal elements:

$$|U| = \prod_{i=1}^{n} a_{ii} \tag{2.68}$$

Therefore, a matrix whose determinant is to be evaluated should first be converted to the triangular form using the Gauss method, and then its determinant should be calculated from the product of the diagonal elements of the triangular matrix.

Example 2.1 demonstrates the Gauss elimination method with complete pivoting strategy in solving a set of simultaneous linear algebraic equations and in calculating the determinant of the matrix of coefficients.

Example 2.1: Heat Transfer in a Pipe Using the Gauss Elimination Method for Simultaneous Linear Algebraic Equations. Write a general MATLAB function that implements the Gauss elimination method, with complete pivoting for the solution of nonhomogeneous linear algebraic equations. The function should identify singular matrices and give their rank. Use this function to calculate the interface temperatures in the following problem:

Saturated steam at 130°C is flowing inside a steel pipe having an ID of 20 mm (D_1) and an OD of 25 mm (D_2). The pipe is insulated with 40 mm [$(D_3 - D_2)/2$] of insulation on the outside. The convective heat transfer coefficients for the inside steam and outside of the lagging are estimated as h_i = 1700 W/m^2.K and h_o = 3 W/m^2.K, respectively. The mean thermal conductivity of the metal is k_s = 45 W/m.K and that of the insulation is k_i = 0.064 W/m.K. Ambient air temperature is 25°C (see Fig. E2.1).

There are three interfaces in this problem, and by writing the energy balance at each interface, there will be three linear equations and three unknown temperatures:

Heat transfer from steam to pipe:

$$h_i \pi D_1 (T_s - T_1) = \frac{T_1 - T_2}{\ln(D_2/D_1)/(2\pi k_s)}$$

Heat transfer from pipe to insulation:

$$\frac{T_1 - T_2}{\ln(D_2/D_1)/(2\pi k_s)} = \frac{T_2 - T_3}{\ln(D_3/D_2)/(2\pi k_i)}$$

Heat transfer from insulation to air:

$$\frac{T_2 - T_3}{\ln(D_3/D_2)/(2\pi k_i)} = h_o \pi D_3 (T_3 - T_a)$$

where T_s = temperature of steam = 130°C
T_1 = temperature of inside wall of pipe (unknown)

T_2 = temperature of outside wall of pipe (unknown)
T_3 = temperature of outside of insulation (unknown)
T_a = ambient temperature = 25°C.

Rearranging the above three energy balance equations yields the set of linear algebraic equations, shown below, which can be solved to find the three unknowns T_1, T_2, and T_3:

$$\left[\frac{2k_s}{\ln(D_2/D_1)} + h_i D_1\right] T_1 - \left[\frac{2k_s}{\ln(D_2/D_1)}\right] T_2 = h_i D_1 T_s$$

$$\left[\frac{k_s}{\ln(D_2/D_1)}\right] T_1 - \left[\frac{k_s}{\ln(D_2/D_1)} + \frac{k_i}{\ln(D_3/D_2)}\right] T_2 + \left[\frac{k_i}{\ln(D_3/D_2)}\right] T_3 = 0$$

$$\left[\frac{2k_i}{\ln(D_3/D_2)}\right] T_2 - \left[\frac{2k_i}{\ln(D_3/D_2)} + h_o D_3\right] T_3 = -h_o D_3 T_a$$

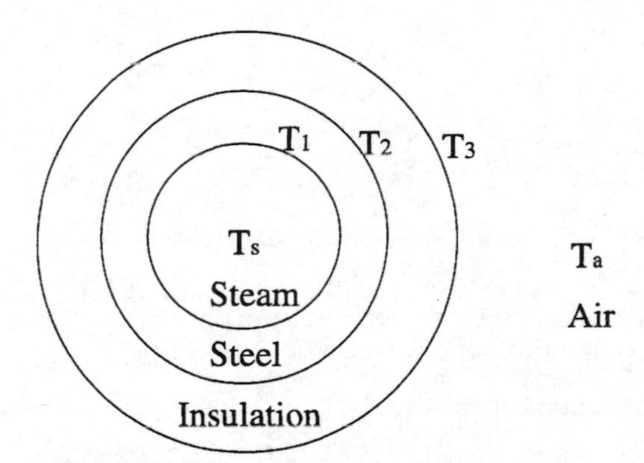

Figure E2.1 Insulated pipe.

Method of Solution: The function is written based on Gauss elimination in matrix form. It applies complete pivoting strategy by searching rows and columns for the maximum pivot element. It keeps track of column interchanges, which affect the positions of the unknown variables. The function applies the back-substitution formula [Eq. (2.110)] to calculate the unknown variables and interchanges their order to correct for column pivoting.

At the beginning, the program checks the determinant of the matrix of coefficients to see if the matrix is singular. If it is singular, the program gives the rank of the matrix and terminates calculations.

Program Description: The MATLAB function *Gauss.m* consists of three main sections. At the beginning, it checks the sizes of input arguments (coefficient matrix and vector of constants) to see if they are consistent. It also checks the coefficient matrix for singularity.

The second part of the function is Gauss elimination. In each iteration, the program finds the location of the maximum pivot element. It interchanges the row and the column of this pivot element to bring it to diagonal and meanwhile keeps track of the interchanged column. At the end of the loop, the program reduces the elements below the diagonal at which the pivot element is placed.

Finally, the function *Gauss.m* applies the back-substitution formula to evaluate all the unknowns. It also interchanges the order of the unknowns at the same time to correct for any column interchanges that took place during complete pivoting.

The program *Example2_1.m* is written to solve the problem of Example 2.1. It mainly acts as an input file, which then builds the coefficient matrix and the vector of constants, and finally calls the function *Gauss.m* to solve the set of equations for the unknown temperatures.

Program

Example2_1.m

```
% Example 2_1.m
% Solution to Example 2.1. This program calculates the
% interface temperatures in an insulated pipe system.
% It calls the function GAUSS.M to solve the heat transfer
% equations for temperature simultaneously.

clc
clear
% Input data
Ts = input(' Temperature of steam (deg C) = ');
Ta = input(' Temperature of air (deg C)   = ');
D1 = 1e-3*input(' Pipe ID (mm) = ');
D2 = 1e-3*input(' Pipe OD (mm) = ');
Ith = 1e-3*input(' Insulation thickness (mm) = ');
D3 = (D2 + 2*Ith);                  % Diameter of pipe with insulation
hi = input(' Inside heat transfer coefficient (W/m2.K)  = ');
ho = input(' Outside heat transfer coefficient (W/m2.K) = ');
ks = input(' Heat conductivity of steel (W/m.K)       = ');
ki = input(' Heat conductivity of insulation (W/m.K) = ');

% Matrix of coefficients
A = ...
[2*ks/log(D2/D1)+hi*D1 , -2*ks/log(D2/D1) , 0
ks/log(D2/D1) , -(ks/log(D2/D1)+ki/log(D3/D2)) , ki/log(D3/D2)
0 , 2*ki/log(D3/D2) , -(2*ki/log(D3/D2)+ho*D3)];

% Matrix of constants
c = [hi*D1*Ts ; 0 ; -ho*D3*Ta];

% Solving the set of equations by Gauss elimination method
T = Gauss (A , c);

% Show the results
disp(' '), disp(' Results :')
fprintf(' T1 = %4.2f\n T2 = %4.2f\n T3 = %4.2f\n',T)
```

Gauss.m

```
function x = Gauss (A , c)
%GAUSS Solves a set of linear algebraic equations by the Gauss
%   elimination method.
%
%   GAUSS(A,C) finds unknowns of a set of linear algebraic
%   equations. A is the matrix of coefficients and C is the
%   vector of constants.
%
%   See also JORDAN, JACOBI

% (c) by N. Mostoufi & A. Constantinides
% January 1, 1999

c = (c(:).')';     % Make sure it's a column vector

n = length(c);
[nr nc] = size(A);

% Check coefficient matrix and vector of constants
if nr ~= nc
   error('Coefficient matrix is not square.')
end
if nr ~= n
   error('Coefficient matrix and vector of constants do not have the
same length.')
end

% Check if the coefficient matrix is singular
if det(A) == 0
   fprintf('\n Rank = %7.3g\n',rank(A))
   error('The coefficient matrix is singular.')
end

unit = eye(n);       % Unit matrix
order = [1 : n];     % Order of unknowns
aug = [A c];     % Augmented matrix

% Gauss elimination
for k = 1 : n - 1
   pivot = abs(aug(k , k));
   prow = k;
   pcol = k;

   % Locating the maximum pivot element
   for row = k : n
      for col = k : n
         if abs(aug(row , col)) > pivot
            pivot = abs(aug(row , col));
            prow = row;
            pcol = col;
         end
      end
   end
```

```
   % Interchanging the rows
   pr = unit;
   tmp = pr(k , :);
   pr(k , : ) = pr(prow , : );
   pr(prow , : ) = tmp;
   aug = pr * aug;

   % Interchanging the columns
   pc = unit;
   tmp = pc(k , : );
   pc(k , : ) = pc(pcol , : );
   pc(pcol , : ) = tmp;
   aug(1 : n , 1 : n) = aug(1 : n , 1 : n) * pc;
   order = order * pc;    % Keep track of the column interchanges

   % Reducing the elements below diagonal to zero in the column k
   lk = unit;
   for m = k + 1 : n
      lk(m , k) = - aug(m , k) / aug(k , k);
   end
   aug = lk * aug;
end

x = zeros(n , 1);

% Back substitution
t(n) = aug(n , n + 1) / aug(n , n);
x(order(n)) = t(n);
for k = n  - 1 : -1 : 1
   t(k) = (aug(k,n+1) - sum(aug(k,k+1:n) .* t(k+1:n))) / aug(k,k);
   x (order(k)) = t(k);
end
```

Input and Results

```
>> Example2_1

 Temperature of steam (deg C) = 130
 Temperature of air (deg C)   = 25
 Pipe ID (mm) = 20
 Pipe OD (mm) = 25
 Insulation thickness (mm) = 40
 Inside heat transfer coefficient (W/m2.K)   = 1700
 Outside heat transfer coefficient (W/m2.K) = 3
 Heat conductivity of steel (W/m.K)      = 45
 Heat conductivity of insulation (W/m.K) = 0.064

 Results :
 T1 = 129.79
 T2 = 129.77
 T3 = 48.12
```

Discussion of Results: The Gauss elimination method finds the interface temperatures as T_1 = 129.79°C, T_2 = 129.77°C, and T_3 = 48.12°C. These values are quite predictable, because the heat transfer coefficient of steam and the heat conductivity of steel are very high. Therefore, the temperatures at steam-pipe interface and pipe-insulation interface are very close to the steam temperature. The main resistance to heat transfer is due to insulation.

The values obtained from the function *Gauss.m* may be verified easily in MATLAB by using the original method of solution of the set of linear equations in matrix form, that is, *T = inv(A)*c*.

2.6 Gauss-Jordan Reduction Method

The Gauss-Jordan reduction method is an extension of the Gauss elimination method. It reduces a set of *n* equations from its canonical form of

$$\boldsymbol{A}\boldsymbol{x} = \boldsymbol{c} \tag{2.13}$$

to the diagonal set of the form

$$\boldsymbol{I}\boldsymbol{x} = \boldsymbol{c}' \tag{2.116}$$

where $\boldsymbol{I}$ is the unit matrix. Eq.(2.116) is identical to

$$\boldsymbol{x} = \boldsymbol{c}' \tag{2.117}$$

that is, the solution vector is given by the $\boldsymbol{c}'$ vector.

The Gauss-Jordan reduction method applies the same series of elementary operations that are used by the Gauss elimination method. It applies these operations both below and above the diagonal in order to reduce all the off-diagonal elements of the matrix to zero. In addition, it converts the elements on the diagonal to unity.

2.6.1 Gauss-Jordan Reduction in Formula Form

We will apply the Gauss-Jordan procedure, without pivoting, to the set of Eqs. (2.102) shown in Sec. 2.5.1 in order to observe the difference between the Gauss-Jordan and the Gauss method. Starting with the augmented matrix

$$\left[\begin{array}{ccc|c} 3 & 18 & 9 & 18 \\ 2 & 3 & 3 & 117 \\ 4 & 1 & 2 & 283 \end{array}\right] \tag{2.103}$$

normalize the first row by dividing it by 3:

$$\begin{bmatrix} 1 & 6 & 3 & | & 6 \\ 2 & 3 & 3 & | & 117 \\ 4 & 1 & 2 & | & 283 \end{bmatrix} \tag{2.118}$$

Multiply the normalized first row by 2 and subtract it from the second row:

$$\begin{bmatrix} 1 & 6 & 3 & | & 6 \\ 0 & -9 & -3 & | & 105 \\ 4 & 1 & 2 & | & 283 \end{bmatrix} \tag{2.119}$$

Multiply the normalized first row by 4 and subtract it from the third row:

$$\begin{bmatrix} 1 & 6 & 3 & | & 6 \\ 0 & -9 & -3 & | & 105 \\ 0 & -23 & -10 & | & 259 \end{bmatrix} \tag{2.120}$$

Normalize the second row by dividing it by -9:

$$\begin{bmatrix} 1 & 6 & 3 & | & 6 \\ 0 & 1 & \frac{1}{3} & | & -\frac{35}{3} \\ 0 & -23 & -10 & | & 259 \end{bmatrix} \tag{2.121}$$

Multiply the normalized second row by 6 and subtract it from the first row:

$$\begin{bmatrix} 1 & 0 & 1 & | & 76 \\ 0 & 1 & \frac{1}{3} & | & -\frac{35}{3} \\ 0 & -23 & -10 & | & 259 \end{bmatrix} \tag{2.122}$$

Multiply the normalized second row by -23 and subtract it from the third row:

$$\begin{bmatrix} 1 & 0 & 1 & | & 76 \\ 0 & 1 & \frac{1}{3} & | & -\frac{35}{3} \\ 0 & 0 & -\frac{7}{3} & | & -\frac{28}{3} \end{bmatrix} \tag{2.123}$$

Normalize the third row by dividing by -7/3:

$$\begin{bmatrix} 1 & 0 & 1 & | & 76 \\ 0 & 1 & \frac{1}{3} & | & -\frac{35}{3} \\ 0 & 0 & 1 & | & 4 \end{bmatrix} \tag{2.124}$$

Multiply the third row by 1 and subtract it from the first row:

$$\begin{bmatrix} 1 & 0 & 0 & | & 72 \\ 0 & 1 & \frac{1}{3} & | & -\frac{35}{3} \\ 0 & 0 & 1 & | & 4 \end{bmatrix} \tag{2.125}$$

Finally, multiply the third row by 1/3 and subtract it from the second row:

$$\begin{bmatrix} 1 & 0 & 0 & | & 72 \\ 0 & 1 & 0 & | & -13 \\ 0 & 0 & 1 & | & 4 \end{bmatrix} \tag{2.126}$$

This reduced matrix [Eq. (2.126)] is equivalent to the set of equations

$$\boldsymbol{Ix} = \boldsymbol{c}' \tag{2.116}$$

The vector $\boldsymbol{c}'$, which is the last column of the reduced matrix, is the solution of the original set of equations (2.102). There is no need for back substitution because the solution is obtained in its final from in vector $\boldsymbol{c}'$.

The Gauss-Jordan reduction procedure applied to the $(n) \times (n+1)$ augmented matrix can be given in a three-part mathematical formula for the initialization, normalization, and reduction steps as shown below:

Initialization formula:

$$\begin{aligned} a_{ij}^{(0)} &= a_{ij} \quad \{ j = 1, 2, \ldots, n \\ a_{ij}^{(0)} &= c_i \quad \{ j = n+1 \end{aligned} \quad \Big\{ i = 1, 2, \ldots, n \tag{2.127}$$

Normalization formula:

$$a_{kj}^{(k)} = \frac{a_{kj}^{(k-1)}}{a_{kk}^{(k-1)}} \quad \{ j = n+1, n, \ldots, k \quad \begin{cases} k = 1, 2, \ldots, n \\ a_{kk}^{(k-1)} \neq 0 \end{cases} \tag{2.128}$$

Reduction formula:

$$a_{ij}^{(k)} = a_{ij}^{(k-1)} - a_{ik}^{(k-1)} a_{kj}^{(k)} \; \{ j = n+1, n, \ldots, k \quad \begin{cases} i = 1, 2, \ldots, n \\ i \neq k \end{cases} \begin{cases} k = 1, 2, \ldots, n \\ a_{kk}^{(k-1)} \neq 0 \end{cases} \tag{2.129}$$

The initialization formula places the elements of the coefficient matrix in columns 1 to n and the vector of constants in column $(n+1)$ of the augmented matrix. The normalization formula divides each row of the augmented matrix by its pivot element and makes this change permanent, thus causing the diagonal elements of the coefficient segment of the augmented matrix to become unity. Finally, the reduction formula reduces to zero the off-diagonal elements in each row and column in the coefficient segment of the augmented matrix, and converts column $(n+1)$ to the solution vector.

2.6.2 Gauss-Jordan Reduction in Matrix Form

The Gauss-Jordan reduction procedure can also be accomplished by a series of matrix multiplications, similar to those performed in the Gauss elimination method (Sec. 2.5.2). The matrix $\boldsymbol{P}_{ij}$, which causes pivoting, is identical to that defined by Eq. (2.111). The matrix $\boldsymbol{L}_k$ must have additional terms above the diagonal to cause the reduction to zero of elements above, as well as below, the diagonal, and a term on the diagonal in order to normalize the element on the diagonal of the original matrix. We will designate this matrix as $\overline{\boldsymbol{L}}_k$ and give an example for a fourth-order system with $k = 2$, where the superscript (1) indicates that each $\overline{\boldsymbol{L}}_k$ matrix uses the elements $a_{ik}^{(k-1)}$ of the previous transformation step [Eq. (2.130)]:

$$\overline{\boldsymbol{L}}_2 = \begin{bmatrix} 1 & -\dfrac{a_{12}^{(1)}}{a_{22}^{(1)}} & 0 & 0 \\ 0 & \dfrac{1}{a_{22}^{(1)}} & 0 & 0 \\ 0 & -\dfrac{a_{32}^{(1)}}{a_{22}^{(1)}} & 1 & 0 \\ 0 & -\dfrac{a_{42}^{(1)}}{a_{22}^{(1)}} & 0 & 1 \end{bmatrix} \tag{2.130}$$

The Gauss-Jordan algorithm reduces a nonsingular matrix $\boldsymbol{A}$ to the identity matrix $\boldsymbol{I}$ by the following series of matrix multiplications:

$$\bar{L}_n \bar{L}_{n-1} \dots P_{ij} \dots \bar{L}_2 \bar{L}_1 P_{ij} A = I \tag{2.131}$$

where the multiplications by P_{ij} cause pivoting, if and when needed, and the multiplications by $\bar{L}_k$ cause normalization and reduction. If pivoting is not performed, Eq. (2.131) simplifies to

$$\bar{L}_n \bar{L}_{n-1} \dots \bar{L}_2 \bar{L}_1 A = I \tag{2.132}$$

By defining the product of all the $\bar{L}_k$ matrices as $\bar{L}$, we can condense Eq. (2.132) to

$$\bar{L} A = I \tag{2.133}$$

The matrix operation of Eq. (2.133), when applied to the augmented matrix $[A \mid c]$, yields the unique solution

$$\bar{L}[A \mid c] \Rightarrow [I \mid c'] \tag{2.134}$$

of the system of linear algebraic equations

$$A x = c \tag{2.13}$$

whose matrix of coefficients A is nonsingular.

2.6.3 Gauss-Jordan Reduction with Matrix Inversion

Matrix $\bar{L}$, in Eq. (2.133), is a nonsingular matrix; therefore, its inverse exists. Premultiplying both sides of Eq. (2.133) by $\bar{L}^{-1}$, we obtain

$$A = \bar{L}^{-1} I \tag{2.135}$$

Taking the inverse of both sides of Eq. (2.135) results in

$$A^{-1} = \bar{L} I \tag{2.136}$$

This simply states that the inverse of A is equal to $\bar{L}$. This has very important implications in numerical methods because it shows that the Gauss-Jordan reduction method is essentially a matrix inversion algorithm. Eq. (2.136), when rearranged, clearly shows that the application of the reduction operation $\bar{L}$ on the identity matrix yields the inverse of A:

$$\bar{L} I = A^{-1} \tag{2.136}$$

This observation can be used to extend the formula form of the Gauss-Jordan algorithm to give the inverse of matrix A every time it calculates the solution to the set of equations

$$\mathbf{A}\,\mathbf{x} = \mathbf{c} \tag{2.13}$$

This is done by forming the augmented matrix of order $(n) \times (2n+1)$:

$$[\,\mathbf{A} \mid \mathbf{c} \mid \mathbf{I}\,] \tag{2.137}$$

and applying the Gauss-Jordan reduction to the augmented matrix. In this case, the three-part mathematical formula for the initialization, normalization, and reduction steps is the following:

Initialization formula:

$$\left.\begin{array}{lll} a_{ij}^{(0)} = a_{ij} & & \{ j = 1,2,\ldots,n \\ a_{ij}^{(0)} = c_i & & \{ j = n+1 \\ a_{ij}^{(0)} = 0 & \{ i \neq j & \multirow{2}{*}{$\{ j = n+2,\ldots,2n+1$} \\ a_{ij}^{(0)} = 1 & \{ i = j & \end{array}\right\} \quad \{ i = 1,2,\ldots,n \tag{2.138}$$

Normalization formula:

$$a_{kj}^{(k)} = \frac{a_{kj}^{(k-1)}}{a_{kk}^{(k-1)}} \quad \{ j = 2n+1, 2n, \ldots, k \quad \begin{cases} k = 1,2,\ldots,n \\ a_{kk}^{(k-1)} \neq 0 \end{cases} \tag{2.139}$$

Reduction formula:

$$a_{ij}^{(k)} = a_{ij}^{(k-1)} - a_{ik}^{(k-1)} a_{kj}^{(k)} \quad \{ j = 2n+1, 2n, \ldots, k \quad \begin{cases} i = 1,2,\ldots,n \\ i \neq k \end{cases} \quad \begin{cases} k = 1,2,\ldots,n \\ a_{kk}^{(k-1)} \neq 0 \end{cases} \tag{2.140}$$

The first two parts of the initialization formula place the elements of the coefficient matrix in columns 1 to n and the vector of constants in column $(n + 1)$ of the augmented matrix. The last two parts of the initialization step expands the augmented matrix to include the identity matrix in columns $(n + 2)$ to $(2n + 1)$. The normalization formula divides each row of the entire matrix by its pivot element, thus causing the diagonal elements of the coefficient segment of the augmented matrix to become unity. Finally, the reduction formula reduces to zero the off-diagonal elements in each row and column in the coefficient segment of the augmented matrix, converts column $(n + 1)$ to the solution vector, and converts the identity matrix in columns $(n + 2)$ to $(2n + 1)$ to the inverse of $\mathbf{A}$.

Example 2.2 demonstrates the use of the Gauss-Jordan reduction method for the solution of simultaneous linear algebraic equations.

Example 2.2: Solution of a Steam Distribution System Using the Gauss-jordan Reduction Method for Simultaneous Linear Algebraic Equations. Figure E2.2 represents the steam distribution system of a chemical plant.[1] The material and energy balances of this system are given below:

$$181.60 - x_3 - 132.57 - x_4 - x_5 = -y_1 - y_2 + y_5 + y_4 = 5.1 \quad (1)$$

$$1.17x_3 - x_6 = 0 \quad (2)$$

$$132.57 - 0.745x_7 = 61.2 \quad (3)$$

$$x_5 + x_7 - x_8 - x_9 - x_{10} + x_{15} = y_7 + y_8 - y_3 = 99.1 \quad (4)$$

$$x_8 + x_9 + x_{10} + x_{11} - x_{12} - x_{13} = -y_7 = -8.4 \quad (5)$$

$$x_6 - x_{15} = y_6 - y_5 = 24.2 \quad (6)$$

$$-1.15(181.60) + x_3 - x_6 + x_{12} + x_{16} = 1.15y_1 - y_9 + 0.4 = -19.7 \quad (7)$$

$$181.60 - 4.594x_{12} - 0.11x_{16} = -y_1 + 1.0235y_9 + 2.45 = 35.05 \quad (8)$$

$$-0.0423(181.60) + x_{11} = 0.0423y_1 = 2.88 \quad (9)$$

$$-0.016(181.60) + x_4 = 0 \quad (10)$$

$$x_8 - 0.147x_{16} = 0 \quad (11)$$

$$x_5 - 0.07x_{14} = 0 \quad (12)$$

$$-0.0805(181.60) + x_9 = 0 \quad (13)$$

$$x_{12} - x_{14} + x_{16} = 0.4 - y_9 = -97.9 \quad (14)$$

There are four levels of steam in this plant: 680, 215, 170, and 37 psia. The 14 x_i, $i = 3, \ldots, 16$ are the unknowns, and the y_i are given parameters for the system. Both x_i and y_i have the units of 1000 lb/h.

Use the Gauss-Jordan method to determine the values of the 14 unknown quantities x_i, $i = 3, \ldots, 16$.

Method of Solution: The 14 equations of this problem represent balances around the following 14 units, respectively:

Eq. (1) 680 psia header
Eq. (2) Desuperheater
Eq. (3) Alternator turbine
Eq. (4) 170 psia header
Eq. (5) 37 psia header
Eq. (6) 215 psia steam
Eq. (7) BFW balance
Eq. (8) Condensate quench drum
Eq. (9) Blow down flash drum
Eq. (10) Boiler atomizing
Eq. (11) Treated feedwater pump
Eq. (12) Boiler feedwater pump
Eq. (13) Boiler fan
Eq. (14) Deaerator-quench.

[1] This problem was adopted from Himmelblau [Ref. 3] by permission of the author.

The set of equations given in the statement of the problem simplifies to a set containing 14 unknowns (x_3 - x_{16}):

$$x_3 + x_4 + x_5 = 43.93 \qquad (1)$$
$$1.17x_3 - x_6 = 0 \qquad (2)$$
$$x_7 = 95.798 \qquad (3)$$
$$x_5 + x_7 - x_8 - x_9 - x_{10} + x_{15} = 99.1 \qquad (4)$$
$$x_8 + x_9 + x_{10} + x_{11} - x_{12} - x_{13} = -8.4 \qquad (5)$$
$$x_6 - x_{15} = 24.2 \qquad (6)$$
$$x_3 - x_6 + x_{12} + x_{16} = 189.14 \qquad (7)$$
$$4.594x_{12} + 0.11x_{16} = 146.55 \qquad (8)$$
$$x_{11} = 10.56 \qquad (9)$$
$$x_4 = 2.9056 \qquad (10)$$
$$x_8 - 0.0147x_{16} = 0 \qquad (11)$$
$$x_5 - 0.07x_{14} = 0 \qquad (12)$$
$$x_9 = 14.6188 \qquad (13)$$
$$x_{12} - x_{14} + x_{16} = -97.9 \qquad (14)$$

For convenience in programming, we change the numbering order of the variables so that x_3 becomes x_1, x_4 becomes x_2, and so forth.

$$x_1 + x_2 + x_3 = 43.93 \qquad (1)$$
$$1.17x_1 - x_4 = 0 \qquad (2)$$
$$x_5 = 95.798 \qquad (3)$$
$$x_3 + x_5 - x_6 - x_7 - x_8 + x_{13} = 99.1 \qquad (4)$$
$$x_6 + x_7 + x_8 + x_9 - x_{10} - x_{11} = -8.4 \qquad (5)$$
$$x_4 - x_{13} = 24.2 \qquad (6)$$
$$x_1 - x_4 + x_{10} + x_{14} = 189.14 \qquad (7)$$
$$4.594x_{10} + 0.11x_{14} = 146.55 \qquad (8)$$
$$x_9 = 10.56 \qquad (9)$$
$$x_2 = 2.9056 \qquad (10)$$
$$x_6 - 0.0147x_{14} = 0 \qquad (11)$$
$$x_3 - 0.07x_{12} = 0 \qquad (12)$$
$$x_7 = 14.6188 \qquad (13)$$
$$x_{10} - x_{12} + x_{14} = -97.9 \qquad (14)$$

The above set of simultaneous linear algebraic equations can be represented by

$$\boldsymbol{A}\boldsymbol{x} = \boldsymbol{c}$$

where A is a sparse matrix containing many zeros. It is not a banded set or a predominantly diagonal set. The Gauss or Gauss-Jordan methods may be used for the solution of this problem. The computer program, which is described in the next section, implements the Gauss-Jordan algorithm in matrix form. The program uses a complete pivoting strategy.

Program Description: The MATLAB function *Jordan.m* consists of three main sections. At the beginning, it checks the sizes of input arguments (coefficient matrix and vector of constants) to see if they are consistent. It also checks the coefficient matrix for singularity.

The second part of the function is the Gauss-Jordan algorithm with application of complete pivoting strategy. In each iteration, the program finds the location of the maximum pivot element. It interchanges the row and the column of this pivot element to bring it to the diagonal position; meanwhile the program keeps track of the interchanged columns. At the end of the loop, the program reduces the elements below and above the diagonal position at which the pivot element is placed.

Finally, the function *Jordan.m* sets the unknowns equal to the elements of the last column of the modified augmented matrix. It also interchanges the order of the unknowns at the same time to correct for any column interchanges that took place during complete pivoting.

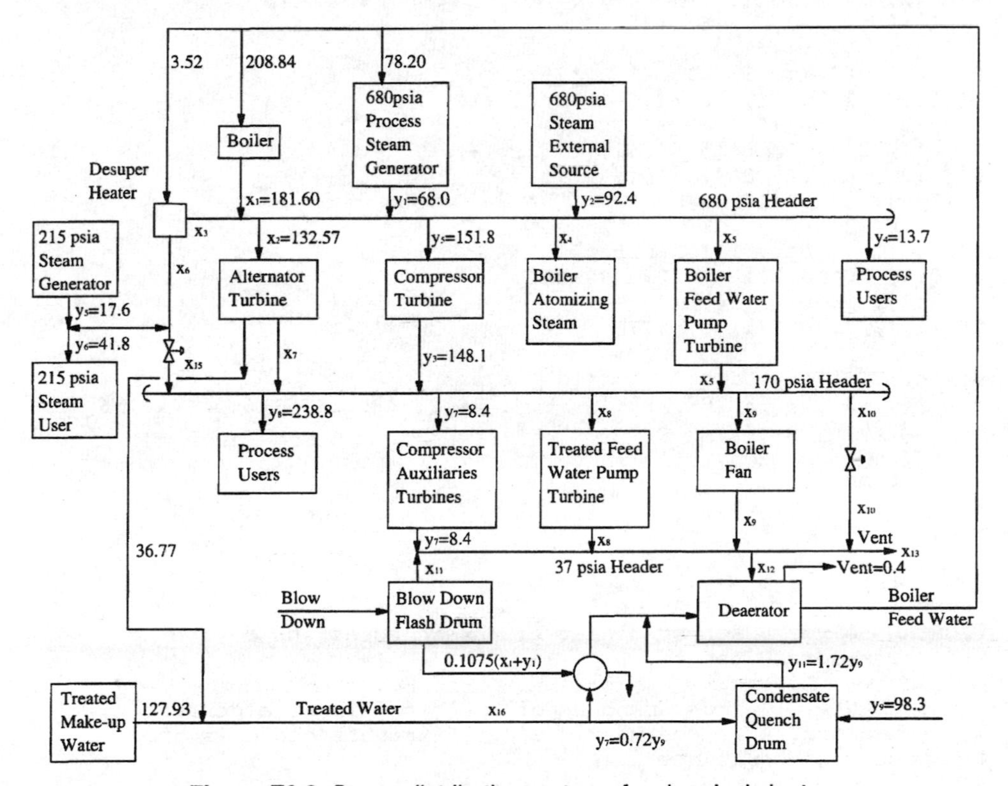

Figure E2.2 Steam distribution system of a chemical plant.

The program *Example2_2.m* is written to solve the problem of Example 2.2. It builds the coefficient matrix and the vector of constants and then calls the function *Jordan.m* to solve the set of equations for the unknown flow rates. Although it is easy to write a few statements at the beginning of the program to ask the user to input the matrix of coefficients and the vector of constants, it is not convenient to do so in this problem because the number of equations is relatively high, and it is not easy to input the data from the screen.

Program

Example2_2.m
```
% Example2_2.m
% Solution to Example 2.2. This program solves the material
% and energy balances equations of a steam distribution
% system using the function JORDAN.M.

clc
clear

% Matrix of coefficients
A = [1, 1, 1, 0*[4:14]
1.17, 0, 0, -1, 0*[5:14]
0*[1:4], 1, 0*[6:14]
0, 0, 1, 0, 1, -1, -1, -1, 0*[9:12], 1, 0
0*[1:5], 1, 1, 1, 1, -1, -1, 0*[12:14]
0*[1:3], 1, 0*[5:12], -1, 0
1, 0*[2:3], -1, 0*[5:9], 1, 0*[11:13], 1
0*[1:9], 4.594, 0*[11:13], 0.11
0*[1:8], 1, 0*[10:14]
0, 1, 0*[3:14]
0*[1:5], 1, 0*[7:13], -0.0147
0, 0, 1, 0*[4:11], -0.07, 0, 0
0*[1:6], 1, 0*[8:14]
0*[1:9], 1, 0, -1, 0, 1];

% Vector of constants
c = [43.93, 0, 95.798, 99.1, -8.4, 24.2, 189.14, 146.55, 10.56, ...
     2.9056, 0, 0, 14.6188, -97.9];

% Solution
X = Jordan (A , c)
```

Jordan.m
```
function x = Jordan (A , c)
%JORDAN Solves a set of linear algebraic equations by the
%   Gauss-Jordan method.
%
%   JORDAN(A,C) finds unknowns of a set of linear algebraic
%   equations. A is the matrix of coefficients and C is the
%   vector of constants.
%
```

```
%   See also GAUSS, JACOBI
% (c) by N. Mostoufi & A. Constantinides
% January 1, 1999

c = (c(:).')';      % Make sure it's a column vector
n = length(c);
[nr nc] = size(A);

% Check coefficient matrix and vector of constants
if nr ~= nc
   error('Coefficient matrix is not square.')
end
if nr ~= n
   error('Coefficient matrix and vector of constants do not have the
same length.')
end

% Check if the coefficient matrix is singular
if det(A) == 0
   fprintf('\n Rank = %7.3g\n',rank(A))
   error('The coefficient matrix is singular.')
end

unit = eye(n);         % Unit matrix
order = [1 : n];       % Order of unknowns
aug = [A c];       % Augmented matrix

% Gauss - Jordan algorithm
for k = 1 : n
   pivot = abs(aug(k , k));
   prow = k;
   pcol = k;

   % Locating the maximum pivot element
   for row = k : n
      for col = k : n
         if abs(aug(row , col)) > pivot
            pivot = abs(aug(row , col));
            prow = row;
            pcol = col;
         end
      end
   end

   % Interchanging the rows
   pr = unit;
   tmp = pr(k , :);
   pr(k , : ) = pr(prow , : );
   pr(prow , : ) = tmp;
   aug = pr * aug;

   % Interchanging the columns
   pc = unit;
```

```
    tmp = pc(k , : );
    pc(k , : ) = pc(pcol , : );
    pc(pcol , : ) = tmp;
    aug(1 : n , 1 : n) = aug(1 : n , 1 : n) * pc;
    order = order * pc;      % Keep track of the column interchanges

    % Reducing the elements above and below diagonal to zero
    lk = unit;
    for m = 1 : n
       if m == k
          lk(m , k) = 1 / aug(k , k);
       else
          lk(m , k) = - aug(m , k) / aug(k , k);
       end
    end
    aug = lk * aug;
end

x = zeros(n , 1);

% Solution
for k = 1 : n
   x(order(k)) = aug(k , n + 1);
end
```

Input and Results

```
>>Example2_2

X =
    20.6854
     2.9056
    20.3390
    24.2020
    95.7980
     2.4211
    14.6188
    -0.0010
    10.5600
    27.9567
     8.0422
   290.5565
     0.0020
   164.6998
```

Discussion of Results: The results are listed in Table E2.2 and show the correspondence between the program-variable numbering sequence and the problem-variable numbering sequence.

The units of the above quantities are 1000 lb/h. The values of variables x_{10} and x_{15} are zero, as may be expected from the flow diagram of Fig. E2.2.

Table E2.2

Program variable	Value	Problem variable
X(1)	20.7	x_3
X(2)	2.9	x_4
X(3)	20.3	x_5
X(4)	24.2	x_6
X(5)	95.8	x_7
X(6)	2.4	x_8
X(7)	14.6	x_9
X(8)	0.0	x_{10}
X(9)	10.6	x_{11}
X(10)	28.0	x_{12}
X(11)	8.0	x_{13}
X(12)	290.6	x_{14}
X(13)	0.0	x_{15}
X(14)	164.7	x_{16}

2.7 GAUSS-SEIDEL SUBSTITUTION METHOD

Certain engineering problems yield sets of simultaneous linear algebraic equations that are *predominantly diagonal* systems. A *predominantly diagonal* system of linear equations has coefficients on the diagonal that are larger in absolute value than the sum of the absolute values of the other coefficients. For example, the set of equations:

$$\begin{aligned} -10x_1 + 2x_2 + 3x_3 &= 6 \\ x_1 + 8x_2 - 2x_3 &= 9 \\ -3x_1 - x_2 - 7x_3 &= -33 \end{aligned} \tag{2.141}$$

is a predominantly diagonal set because

$$|-10| > |2| + |3|$$

and

$$|8| > |1| + |2|$$

and

$$|-7| > |-3| + |-1|$$

Each equation in set (2.141) can be solved for the unknown on its diagonal:

$$x_1 = \frac{6 - (2x_2 + 3x_3)}{-10} \tag{2.142a}$$

$$x_2 = \frac{9 - (x_1 - 2x_3)}{8} \tag{2.142b}$$

$$x_3 = \frac{-33 - (-3x_1 - x_2)}{-7} \tag{2.142c}$$

The Gauss-Seidel substitution method requires an initial guess of the values of the unknowns x_2 and x_3. The initial guesses are used in Eq. (2.142*a*) to calculate a new estimate of x_1. This estimate of x_1 and the guessed value of x_3 are replaced in Eq. (2.142*b*) to evaluate the new estimate of x_2. The new estimate of x_3 is then calculated from Eq. (2.142*c*). The iteration continues until all the newly calculated x values converge to within a convergence criterion ϵ of their previous values.

The Gauss-Seidel method converges to the correct solution, no matter what the initial estimate is, provided that the system of equations is predominantly diagonal. On the other hand, if the system is not predominantly diagonal, the correct solution may still be obtained if the initial estimate of the values of x_2 to x_n is close with the correct set. The Gauss-Seidel method is a very simple algorithm to program, and it is computationally very efficient, in comparison with the other methods described in this chapter, provided that the system is predominantly diagonal. These advantages account for this method's wide use in the solution of engineering problems.

For a general set of n equations in n unknowns:

$$\boldsymbol{A}\boldsymbol{x} = \boldsymbol{c} \tag{2.13}$$

the Gauss-Seidel substitution method corresponds to the formula

$$x_i = \frac{1}{a_{ii}} \left[c_i - \sum_{\substack{j=1 \\ j \neq i}}^{n} a_{ij} x_j \right] \qquad i = 1, 2, \ldots, n \tag{2.143}$$

Eq. (2.143) is the Gauss-Seidel method in formula form. Calculation starts with an initial guess of the values x_2 to x_n. Each newly calculated x_i from Eq. (2.143) replaces its previous value in subsequent calculations. Substitution continues until the convergence criterion is met.

2.8 JACOBI METHOD

The Jacobi iterative method is similar to the Gauss-Seidel method with the exception that the newly calculated variables are not replaced until the end of each iteration is reached. In this section, we develop the Jacobi method in matrix form.

The matrix of coefficients $\boldsymbol{A}$ can be written as

$$\boldsymbol{A} = (\boldsymbol{A} - \boldsymbol{D}) + \boldsymbol{D} \tag{2.144}$$

where $\boldsymbol{D}$ is a diagonal matrix whose elements are those of the main diagonal of matrix $\boldsymbol{A}$. Therefore, the matrix $(\boldsymbol{A} - \boldsymbol{D})$ is similar to $\boldsymbol{A}$, with the difference that its main diagonal elements are equal to zero. Replace Eq. (2.144) into Eq. (2.13) and rearrange results in

$$\boldsymbol{D}\boldsymbol{x} = \boldsymbol{c} - (\boldsymbol{A} - \boldsymbol{D})\boldsymbol{x} \tag{2.145}$$

from where the vector $\boldsymbol{x}$ can be evaluated:

$$\boldsymbol{x} = \boldsymbol{D}^{-1}\boldsymbol{c} - \boldsymbol{D}^{-1}(\boldsymbol{A} - \boldsymbol{D})\boldsymbol{x}$$

$$= \boldsymbol{D}^{-1}\boldsymbol{c} - (\boldsymbol{D}^{-1}\boldsymbol{A} - \boldsymbol{I})\boldsymbol{x} \tag{2.145}$$

In an iterative procedure, Eq. (2.145) should be written as

$$\boldsymbol{x}^{(k)} = \boldsymbol{D}^{-1}\boldsymbol{c} - (\boldsymbol{D}^{-1}\boldsymbol{A} - \boldsymbol{I})\boldsymbol{x}^{(k-1)} \tag{2.146}$$

where superscript (k) represents the iteration number. The Jacobi method requires an initial guess of all unknowns (rather than one less in the Gauss-Seidel method) and the newly calculated values of the vector $\boldsymbol{x}$ replace the old ones only at the end of each iteration. The substitution procedure continues until convergence is achieved.

It is worth mentioning that the solution of a set of equations by the Gauss-Seidel method in formula form needs fewer iterations to converge than using the Jacobi method in matrix form. This is because the unknowns change during each iteration in the Gauss-Seidel method, whereas in the Jacobi method they are not changed until the very end of each iteration (see Problem 2.1).

Example 2.3: Solution of Chemical Reaction and Material Balance Equations Using the Jacobi Iteration for Predominantly Diagonal Systems of Linear Algebraic Equations. A chemical reaction takes place in a series of four continuous stirred tank reactors arranged as shown in Fig. E2.3.

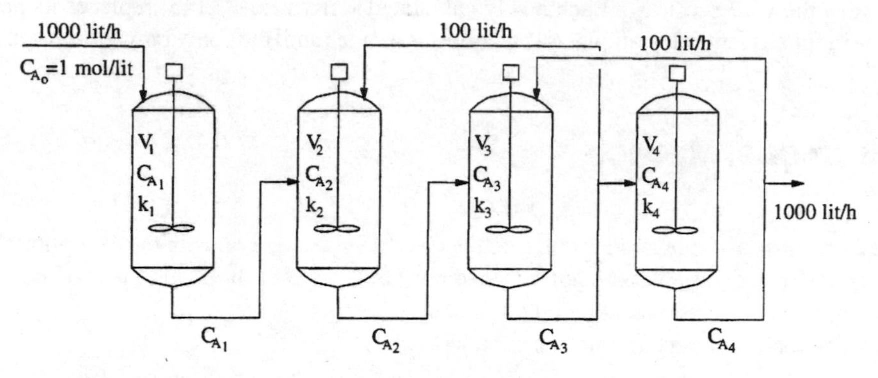

Figure E2.3 Series of continuous stirred tank reactors.

The chemical reaction is a first-order irreversible reaction of the type

$$A \xrightarrow{k_i} B$$

The conditions of temperature in each reactor are such that the value of the rate constant k_i is different in each reactor. Also, the volume of each reactor V_i is different. The values of k_i and V_i are given in Table E2.3. The following assumptions can be made regarding this system:

1. The system is at steady state.
2. The reactions are in the liquid phase.
3. There is no change in volume or density of the liquid.
4. The rate of disappearance of component A in each reactor is given by

$$R_i = V_i k_i c_{A_i} \qquad \text{mol/h}$$

Table E2.3

Reactor	V_i (L)	k_i (h^{-1})	Reactor	V_i (L)	k_i (h^{-1})
1	1000	0.1	3	100	0.4
2	1500	0.2	4	500	0.3

Respond to the following questions:

a. Set up the material balance equation for each of the four reactors. What type of equations do you have in this set of material balances?

b. What method do you recommend as the best one to use to solve for the exit concentration (c_{A_i}) from each reactor?

c. Write a MATLAB script to solve this set of equations and find the exit concentration from each reactor.

Method of Solution: *Part* (*a*): The general unsteady-state material balance for each reactor is

Input = output + disappearance by reaction + accumulation

Because the system is at steady state, the accumulation term is zero; therefore, the material balance simplifies to

Input = output + disappearance by reaction

This balance applied to each of the four reactors yields the following set of equations:

$$(1000)(1) = 1000\,c_{A_1} + V_1 k_1 c_{A_1}$$

$$1000\,c_{A_1} + 100\,c_{A_3} = 1100\,c_{A_2} + V_2 k_2 c_{A_2}$$

$$1100\,c_{A_2} + 100\,c_{A_4} = 1200\,c_{A_3} + V_3 k_3 c_{A_3}$$

$$1100\,c_{A_3} = 1100\,c_{A_4} + V_4 k_4 c_{A_4}$$

Substituting the values of V_i and k_i and rearranging:

$$1100\,c_{A_1} = 1000$$

$$1000\,c_{A_1} - 1400\,c_{A_2} + 100\,c_{A_3} = 0$$

$$1100\,c_{A_2} - 1240\,c_{A_3} + 100\,c_{A_4} = 0$$

$$1100\,c_{A_3} - 1250\,c_{A_4} = 0$$

The above is a set of four simultaneous linear algebraic equations. It appears to be a predominantly diagonal system of equations, as the coefficients on the diagonal are larger in absolute value than the sum of the absolute values of the other coefficients.

Part (*b*): From the discussion of the Jacobi method (Sec. 2.8), it would seem that Gauss-Seidel would be the best method of solution for a predominantly diagonal set. However, because calculations in MATLAB are based on matrices, the Jacobi method in matrix form is considerably faster than the Gauss-Seidel method in formula form in the MATLAB workspace.

Part (*c*): The general program, which uses the Jacobi iterative method in matrix form, is described in the next section. An initial guess of unknowns c_{A_1} to c_{A_4} is needed to start the Jacobi algorithm. This system of equations is a predominantly diagonal set; therefore,any initial guess for unknowns will yield convergence. However, the initial guess of 0.6 for all four unknowns seems to be an appropriate choice based on the fact that $c_{A_0} = 1.0$. Two cases will be run in order to test the ability of the Jacobi method to converge. The first case will use 0.6 as the initial values and the second will use 100 as the starting values.

Program Description: The MATLAB function *Jacobi.m* is written to solve a set of linear algebraic equations by the Jacobi iterative method. Inputs to the function are the coefficient matrix, the vector of constants, and the vector of initial guesses for all the unknowns. The default convergence criterion is $|x_i^{(k)} - x_i^{(k-1)}| < 10^{-6}$. However, the user may change this convergence criterion by introducing another value as the fourth input argument into the function.

The next step in the program is to build the modified coefficient matrix $(\mathbf{D}^{-1}\mathbf{A} - \mathbf{I})$ and the modified vector of constants $(\mathbf{D}^{-1}\mathbf{c})$. The function then starts the substitution procedure according to Eq. (2.146), which continues until the convergence criterion is reached for all of the unknowns.

In the program *Example2_3.m*, the coefficient matrix and the vector of constants of the set of equations developed in this example are introduced as input data. The program also asks the user to input the convergence criterion and if the user wants to see the results of the calculations at the end of each step. The vector of initial guesses are introduced to the program in a loop so that the user can redo the calculations with different initial guesses. Then it calls the function *Jacobi.m* to solve the set of equations. Finally, the program shows the final results of calculation.

Program

Example2_3.m

```
% Example2_3.m
% Solution to Example 2.3. This program solves a set of
% linear algebraic equations by the Jacobi iterative
% method, using the function JACOBI.M, to find the
% concentrations of a seres of CSTRs.

clc
clear

% Input data
fprintf(' Solution of set of linear algebraic equations by the
Jacobi method\n\n')
n = input(' Number of equations = ');
for k = 1 : n
   fprintf('\n Coefficients of eq. %2d =',k)
   A(k,1:n) = input(' ');
   fprintf(' Constant of eq. %2d =',k)
   c(k) = input(' ');
end
disp(' ')
tol = input(' Convergence criterion = ');
trace = input(' Show step-by-step path to results (0/1) ? ');

redo = 1;
while redo
   disp(' ')
   guess = input(' Vector of initial guess = ');
```

```
    % Solution
    ca = Jacobi(A, c, guess, tol, trace);
    fprintf('\n\n Results :\n')

    for k = 1 : n
       fprintf(' CA(%2d) = %6.4g\n',k,ca(k))
    end

   disp(' ')
   redo=input(' Repeat the calculations with another guess (0/1)? ');
   disp(' ')
end
```

Jacobi.m

```
function x = Jacobi(A, c, x0, tol, trace)
%JACOBI Solves a set of linear algebraic equations by the
%   Jacobi iterative method.
%
%   JACOBI(A,C,X0) finds unknowns of a set of linear algebraic
%   equations. A is the matrix of coefficients, C is the vector
%   of constants and X0 is the vector of initial guesses.
%
%   JACOBI(A,C,X0,TOL,TRACE) finds unknowns of a set of linear
%   algebraic equations and uses TOL as the convergence test.
%   A nonzero value for TRACE results in showing calculated
%   unknowns at the end of each iteration.
%
%   See also GAUSS, JORDAN

% by N. Mostoufi & A. Constantinides
% January 1, 1999

% Initialization
if nargin < 4 | isempty(tol)
   tol = 1e-6;
end
if nargin >= 4 & tol == 0
   tol = 1e-6;
end
if nargin < 5 | isempty(trace)
   trace = 0;
end
if trace
   fprintf('\n Initial guess :\n')
   fprintf('%8.6g  ',x0)
end

c = (c(:).')';        % Make sure it's a column vector
x0 = (x0(:).')';      % Make sure it's a column vector

n = length(c);
[nr nc] = size(A);
```

```
% Check coefficient matrix, vector of constants and
% vector of unknowns
if nr ~= nc
   error('Coefficient matrix is not square.')
end
if nr ~= n
   error('Coefficient matrix and vector of constants do not have the
same length.')
end
if length(x0) ~= n
   error('Vector of unknowns and vector of constants do not have the
same length.')
end

% Check if the coefficient matrix is singular
if det(A) == 0
   fprintf('\n Rank = %7.3g\n',rank(A))
   error('The coefficient matrix is singular.')
end

% Building modified coefficient matrix and modified
% vector of coefficients
D = diag(diag(A));            % The diagonal matrix
a0 = inv(D)*A - eye(n);       % Modified matrix of coefficients
c0 = inv(D)*c;                % Modified vector of constants

x = x0;
x0 = x + 2 * tol;
iter = 0;

% Substitution procedure
while max(abs(x - x0)) >= tol
   x0 = x;
   x = c0 - a0 * x0;
   if trace
      iter = iter + 1;
      fprintf('\n Iteration no. %3d\n',iter)
      fprintf('%8.6g  ',x)
   end
end
```

Input and Results

```
>> Example2_3

 Solution of set of linear algebraic equations by the Jacobi method

 Number of equations = 4

 Coefficients of eq.  1 = [1100, 0, 0, 0]
 Constant of eq.  1 = 1000

 Coefficients of eq.  2 = [1000, -1400, 100, 0]
```

```
Constant of eq.  2 = 0

Coefficients of eq.  3 = [0, 1100, -1240, 100]
Constant of eq.  3 = 0

Coefficients of eq.  4 = [0, 0, 1100, -1250]
Constant of eq.  4 = 0

Convergence criterion = 1e-5
Show step-by-step path to results (0/1) ? 1

Vector of initial guess = 0.6*ones(1,4)

Initial guess :
    0.6         0.6         0.6         0.6
Iteration no.    1
  0.909091  0.471429  0.580645      0.528
Iteration no.    2
  0.909091  0.690825  0.460783  0.510968
Iteration no.    3
  0.909091  0.682264  0.654036  0.405489
Iteration no.    4
  0.909091  0.696068  0.637935  0.575552
Iteration no.    5
  0.909091  0.694917  0.663895  0.561383
Iteration no.    6
  0.909091  0.696772  0.661732  0.584227
Iteration no.    7
  0.909091  0.696617  0.665219  0.582324
Iteration no.    8
  0.909091  0.696866  0.664928  0.585393
Iteration no.    9
  0.909091  0.696846  0.665397  0.585137
Iteration no.   10
  0.909091  0.696879  0.665358  0.585549
Iteration no.   11
  0.909091  0.696876  0.665421  0.585515
Iteration no.   12
  0.909091  0.696881  0.665416   0.58557
Iteration no.   13
  0.909091   0.69688  0.665424  0.585566

Results :
CA( 1) = 0.9091
CA( 2) = 0.6969
CA( 3) = 0.6654
CA( 4) = 0.5856

Repeat the calculations with another guess (0/1) ? 1

Vector of initial guess = 100*ones(1,4)

Initial guess :
    100         100         100         100
```

```
Iteration no.    1
 0.909091    78.5714    96.7742         88
Iteration no.    2
 0.909091    7.56179    76.7972    85.1613
Iteration no.    3
 0.909091    6.13487    13.5759    67.5816
Iteration no.    4
 0.909091    1.61906    10.8923    11.9468
Iteration no.    5
 0.909091    1.42738    2.39971    9.58527
Iteration no.    6
 0.909091   0.820759    2.03923    2.11175
Iteration no.    7
 0.909091    0.79501   0.898394    1.79452
Iteration no.    8
 0.909091   0.713522    0.84997   0.790587
Iteration no.    9
 0.909091   0.710063    0.69672   0.747973
Iteration no.   10
 0.909091   0.699116   0.690215   0.613113
Iteration no.   11
 0.909091   0.698652   0.669628   0.607389
Iteration no.   12
 0.909091   0.697181   0.668755   0.589273
Iteration no.   13
 0.909091   0.697119   0.665989   0.588504
Iteration no.   14
 0.909091   0.696921   0.665872   0.586071
Iteration no.   15
 0.909091   0.696913     0.6655   0.585967
Iteration no.   16
 0.909091   0.696886   0.665485    0.58564
Iteration no.   17
 0.909091   0.696885   0.665435   0.585626
Iteration no.   18
 0.909091   0.696882   0.665433   0.585583
Iteration no.   19
 0.909091   0.696882   0.665426   0.585581

Results :
CA( 1) = 0.9091
CA( 2) = 0.6969
CA( 3) = 0.6654
CA( 4) = 0.5856

Repeat the calculations with another guess (0/1) ? 0
```

Discussion of Results: The first case uses the value of 0.6 as the initial guess for the values of the unknowns c_{A_1} to c_{A_4}. The Jacobi method converges to the solution in 13 iterations. The convergence criterion, which is satisfied by all the unknowns, is 0.000001.

In the second case, the value of 100 is used as the initial guess for each of the unknowns c_{A_1} to c_{A_4}. Convergence to exactly the same answer as in the first case is accomplished in 19 iterations.

2.9 HOMOGENEOUS ALGEBRAIC EQUATIONS AND THE CHARACTERISTIC-VALUE PROBLEM

We mentioned earlier that a homogeneous set of equations

$$\boldsymbol{A}\boldsymbol{x} = 0 \qquad (2.95)$$

has a nontrivial solution, if and only if the matrix $\boldsymbol{A}$ is singular; that is, if the rank r of $\boldsymbol{A}$ is less than n. The system of equations would consist of r independent equations, r unknowns that can be evaluated independently, and $(n - r)$ unknowns that must be chosen arbitrarily in order to complete the solution. Choosing nonzero values for the $(n - r)$ unknowns transforms the homogeneous set to a nonhomogeneous set of order r. The Gauss and Gauss-Jordan methods, which are applicable to nonhomogeneous systems, can then be used to obtain the complete solution of the problem. In fact, these methods can be used first on the homogeneous system to determine the number of independent equations (or the rank of $\boldsymbol{A}$) and then applied to the set of r nonhomogeneous independent equations to evaluate the r unknowns. This concept will be demonstrated later in this section in conjunction with the calculation of eigenvectors.

A special class of homogeneous linear algebraic equations arises in the study of vibrating systems, structure analysis, and electric circuit system analysis, and in the solution and stability analysis of linear ordinary differential equations (Chap. 5). This system of equations has the form

$$\boldsymbol{A}\boldsymbol{x} = \lambda\boldsymbol{x} \qquad (2.147)$$

which can be alternatively expressed as

$$(\boldsymbol{A} - \lambda\boldsymbol{I})\boldsymbol{x} = 0 \qquad (2.148)$$

where the scalar λ is called an *eigenvalue* (or a *characteristic value*) of matrix $\boldsymbol{A}$. The vector $\boldsymbol{x}$ is called the *eigenvector* (or *characteristic vector*) corresponding to λ. The matrix $\boldsymbol{I}$ is the identity matrix. The problem often requires the solution of the homogeneous set of equations, represented by Eq. (2.148), to determine the values of λ and $\boldsymbol{x}$ that satisfy this set. In MATLAB, *eig(A)* is a vector containing the eigenvalues of $\boldsymbol{A}$. The statement $[\boldsymbol{V}, \boldsymbol{D}] = eig(\boldsymbol{A})$ produces a diagonal matrix $\boldsymbol{D}$ of eigenvalues and a full matrix $\boldsymbol{V}$ whose columns are the corresponding eigenvectors, so that $\boldsymbol{A}\boldsymbol{V} = \boldsymbol{V}\boldsymbol{D}$.

Before we proceed with developing methods of solution, we examine Eq. (2.147) from a geometric perspective. The multiplication of a vector by a matrix is a linear transformation of the original vector to a new vector of different direction and length. For example, matrix $\boldsymbol{A}$ transforms the vector $\boldsymbol{y}$ to the vector $\boldsymbol{z}$ in the operation

$$\boldsymbol{A y} = \boldsymbol{z} \tag{2.149}$$

In contrast to this, if $\boldsymbol{x}$ is the eigenvector of $\boldsymbol{A}$, then the multiplication of the eigenvector $\boldsymbol{x}$ by matrix $\boldsymbol{A}$ yields the same vector $\boldsymbol{x}$ multiplied by a scalar λ, that is, the same vector but of different length:

$$\boldsymbol{A x} = \lambda \boldsymbol{x} \tag{2.147}$$

It can be stated that for a nonsingular matrix $\boldsymbol{A}$ of order n, there are n characteristic directions in which the operation by $\boldsymbol{A}$ does not change the direction of the vector, but only changes its length. More simply stated, matrix $\boldsymbol{A}$ has n eigenvectors and n eigenvalues. The types of eigenvalues that exist for a set of special matrices are listed in Table 2.4.

The homogeneous problem

$$(\boldsymbol{A} - \lambda \boldsymbol{I})\boldsymbol{x} = 0 \tag{2.148}$$

possesses nontrivial solutions if the determinant of matrix $(\boldsymbol{A} - \lambda \boldsymbol{x})$, called the *characteristic matrix* of $\boldsymbol{A}$, vanishes:

$$|\boldsymbol{A} - \lambda \boldsymbol{I}| = \begin{vmatrix} a_{11} - \lambda & a_{12} & \dots & a_{1n} \\ a_{21} & a_{22} - \lambda & \dots & a_{2n} \\ \dots & \dots & \dots & \dots \\ a_{n1} & a_{n2} & \dots & a_{nn} - \lambda \end{vmatrix} = 0 \tag{2.150}$$

The determinant can be expanded by minors to yield a polynomial of nth degree

$$\lambda^n - \alpha_1 \lambda^{n-1} - \alpha_2 \lambda^{n-2} - \dots - \alpha_n = 0 \tag{2.151}$$

This polynomial, which is called the *characteristic equation* of matrix $\boldsymbol{A}$, has n roots, which are the eigenvalues of $\boldsymbol{A}$. These roots may be real distinct, real repeated, or complex, depending on matrix $\boldsymbol{A}$ (see Table 2.4). A nonsingular real symmetric matrix of order n has n real nonzero eigenvalues and n linearly independent eigenvectors. The eigenvectors of a real symmetric matrix are orthogonal to each other. The coefficients α_i of the characteristic polynomial are functions of the matrix elements a_{ij} and must be determined before the polynomial can be used.

The well-known Cayley-Hamilton theorem states that a square matrix satisfies its own characteristic equation, that is,

$$\boldsymbol{A}^n - \alpha_1 \boldsymbol{A}^{n-1} - \alpha_2 \boldsymbol{A}^{n-2} - \dots - \alpha_n = 0 \tag{2.152}$$

Table 2.4

Matrix	Eigenvalue
Singular, $\|A\| = 0$	At least one zero eigenvalue
Nonsingular, $\|A\| \neq 0$	No zero eigenvalues
Symmetric, $A = A'$	All real eigenvalues
Hermitian	All real eigenvalues
Zero matrix, $A = 0$	All zero eigenvalues
Identity, $A = I$	All unity eigenvalues
Diagonal, $A = D$	Equal to diagonal elements of A
Inverse, A^{-1}	Inverse eigenvalues of A
Transformed, $B = Q^{-1}AQ$	Eigenvalues of B = eigenvalues of A

The problem of evaluating the eigenvalues and eigenvectors of matrices is a complex multistep procedure. Several methods have been developed for this purpose. Some of these apply to symmetric matrices, others to tridiagonal matrices, and a few can be used for general matrices. We can classify these methods into two categories:

a. The methods in this category work with the original matrix A and its characteristic polynomial [Eq. (2.151)] to evaluate the coefficients α_i of the polynomial. One such method is the Faddeev-Leverrier procedure, which will be described later. Once the coefficients of the polynomial are known, the methods use root-finding techniques, such as the Newton-Raphson method, to determine the eigenvalues. Finally, the algorithms employ a reduction method, such as Gauss elimination, to calculate the eigenvectors.

b. The methods in this category reduce the original matrix A to tridiagonal form (when A is symmetric) or to Hessenberg form (when A is nonsymmetric) by orthogonal transformations or elementary similarity transformations. They apply successive factorization procedures, such as LR or QR algorithms, to extract the eigenvalues, and, finally, they use a reduction method to calculate the eigenvectors.

In the remaining part of this chapter we will discuss the following methods: (*a*) the Faddeev-Leverrier procedure for calculating the coefficients of the characteristic polynomial, (*b*) the elementary similarity transformation for converting a matrix to Hessenberg form, (*c*) the QR algorithm of successive factorization for the determination of the eigenvalues, and finally, (*d*) the Gauss elimination method applied for the evaluation of the eigenvectors. These methods were chosen for their general applicability to both symmetric and nonsymmetric matrices. For a complete discussion of these and other methods, the reader is referred to Ralston and Rabinowitz [2].

2.9.1 The Faddeev-Leverrier Method

The Faddeev-Leverrier method [4] calculates the coefficients α_1 to α_n of the characteristic polynomial [Eq. (2.151)] by generating a series of matrices A_k whose traces are equal to the coefficients of the polynomial. The starting matrix and first coefficient are

$$A_1 = A \qquad \alpha_1 = tr A_1 \tag{2.153}$$

and the subsequent matrices are evaluated from the recursive equations:

$$\left.\begin{aligned} A_k &= A(A_{k-1} - \alpha_{k-1}I) \\ \alpha_k &= \frac{1}{k} tr A_k \end{aligned}\right\} \quad k = 2, 3, \ldots, n \tag{2.154}$$

In addition to this, the Faddeev-Leverrier method yields the inverse of the matrix A by

$$A^{-1} = \frac{1}{\alpha_n}(A_{n-1} - \alpha_{n-1}I) \tag{2.155}$$

To elucidate this method, we will determine the coefficients of the characteristic polynomial of the following set of homogeneous equations:

$$\begin{aligned} (1 - \lambda)x_1 + 2x_2 + x_3 &= 0 \\ 3x_1 + (1 - \lambda)x_2 + 2x_3 &= 0 \\ 4x_1 + 2x_2 + (3 - \lambda)x_3 &= 0 \end{aligned} \tag{2.156}$$

The characteristic polynomial for this third-order system is

$$\lambda^3 - \alpha_1\lambda^2 - \alpha_2\lambda - \alpha_3 = 0 \tag{2.157}$$

The matrix A is

$$A = \begin{bmatrix} 1 & 2 & 1 \\ 3 & 1 & 2 \\ 4 & 2 & 3 \end{bmatrix} \tag{2.158}$$

Application of Eq. (2.153) gives

$$A_1 = A \quad \text{and} \quad \alpha_1 = tr A_1 = 5 \tag{2.159}$$

Application of Eq. (2.154), with $k = 2$, yields

$$A_2 = A(A_1 - \alpha_1 I)$$

$$= \begin{bmatrix} 1 & 2 & 1 \\ 3 & 1 & 2 \\ 4 & 2 & 3 \end{bmatrix} \left\{ \begin{bmatrix} 1 & 2 & 1 \\ 3 & 1 & 2 \\ 4 & 2 & 3 \end{bmatrix} - \begin{bmatrix} 5 & 0 & 0 \\ 0 & 5 & 0 \\ 0 & 0 & 5 \end{bmatrix} \right\} \tag{2.160}$$

$$= \begin{bmatrix} 6 & -4 & 3 \\ -1 & 6 & 1 \\ 2 & 6 & 2 \end{bmatrix}$$

$$\alpha_2 = \frac{1}{2} tr A_2 = 7 \tag{2.161}$$

Repetition of Eq. (2.154), with $k = 3$, results in

$$A_3 = A(A_2 - \alpha_2 I)$$

$$= \begin{bmatrix} 1 & 2 & 1 \\ 3 & 1 & 2 \\ 4 & 2 & 3 \end{bmatrix} \left\{ \begin{bmatrix} 6 & -4 & 3 \\ -1 & 6 & 1 \\ 2 & 6 & 2 \end{bmatrix} - \begin{bmatrix} 7 & 0 & 0 \\ 0 & 7 & 0 \\ 0 & 0 & 7 \end{bmatrix} \right\} \tag{2.162}$$

$$= \begin{bmatrix} -1 & 0 & 0 \\ 0 & -1 & 0 \\ 0 & 0 & -1 \end{bmatrix}$$

$$\alpha_3 = \frac{1}{3} tr A_3 = -1 \tag{2.163}$$

Therefore, the characteristic polynomial is

$$\lambda^3 - 5\lambda^2 - 7\lambda + 1 = 0 \tag{2.164}$$

The root-finding techniques described in Chap. 1 may be used to determine the λ values of this polynomial. The eigenvectors corresponding to each eigenvalue may be calculated using the Gauss elimination method. The Faddeev-Leverrier method, the Newton-Raphson method with synthetic division, and the Gauss elimination method constitute a complete algorithm for the evaluation of all the eigenvalues and eigenvectors of this characteristic-value problem. This combination of methods, however, is "fraught with peril," because it is too sensitive to small changes in the coefficients. Use of the QR algorithm, discussed in Sec. 2.9.3, is preferable.

2.9.2 Elementary Similarity Transformations

In Sec. 2.5.2, we showed that the Gauss elimination method can be represented in matrix form as

$$\boldsymbol{LA} = \boldsymbol{U} \tag{2.73}$$

Matrix $\boldsymbol{A}$ is nonsingular, matrix $\boldsymbol{L}$ is unit lower triangular, and matrix $\boldsymbol{U}$ is upper triangular. The inverse of $\boldsymbol{L}$ is also a unit lower triangular matrix. Postmultiplying both sides of Eq. (2.73) by $\boldsymbol{L}^{-1}$ we obtain

$$\boldsymbol{LAL}^{-1} = \boldsymbol{UL}^{-1} = \boldsymbol{B} \tag{2.165}$$

This is a similarity transformation of the type described in Sec. 2.2.2. The transformation coverts matrix $\boldsymbol{A}$ to a *similar* matrix $\boldsymbol{B}$. The two matrices, $\boldsymbol{A}$ and $\boldsymbol{B}$, have identical eigenvalues, determinants, and traces.

We, therefore, conclude that if the Gauss elimination method is extended so that matrix $\boldsymbol{A}$ is postmultiplied by $\boldsymbol{L}^{-1}$, at each step of the operation, in addition to being premultiplied by $\boldsymbol{L}$, the resulting matrix $\boldsymbol{B}$ is similar to $\boldsymbol{A}$. This operation is called the *elementary similarity transformation.*

In the determination of eigenvalues, it is desirable to reduce matrix $\boldsymbol{A}$ to a super-triangular matrix of upper Hessenberg form:

$$\boldsymbol{H}_U = \begin{bmatrix} h_{11} & h_{12} & h_{13} & & \cdots & & h_{1\,n-2} & h_{1\,n-1} & h_{1n} \\ h_{21} & h_{22} & h_{23} & & \cdots & & h_{2\,n-2} & h_{2\,n-1} & h_{2n} \\ 0 & h_{32} & h_{33} & & \cdots & & h_{3\,n-3} & h_{3\,n-1} & h_{3n} \\ \cdots & \cdots & \cdots & \cdots & \cdots & \cdots & \cdots & \cdots & \cdots \\ 0 & 0 & & & \cdots & & h_{n-1\,n-2} & h_{n-1\,n-1} & h_{n-1\,n} \\ 0 & 0 & & & \cdots & & 0 & h_{n\,n-1} & h_{nn} \end{bmatrix} \tag{2.57}$$

This can be done by using the $(k + 1)$st row to eliminate the elements $(k + 2)$ to n of column k. Consequently, the elements of the subdiagonal do not vanish. The transformation matrices that perform this elimination are unit lower triangular of the form shown in Eq. (2.166). The elimination matrix that would eliminate the elements of column 1 below the subdiagonal is

$$\bar{L}_1 = \begin{bmatrix} 1 & 0 & 0 & 0 & \dots & 0 \\ 0 & 1 & 0 & 0 & \dots & 0 \\ 0 & -\dfrac{h_{31}^{(0)}}{h_{21}^{(0)}} & 1 & 0 & \dots & 0 \\ 0 & -\dfrac{h_{41}^{(0)}}{h_{21}^{(0)}} & 0 & 1 & \dots & 0 \\ \dots & \dots & \dots & \dots & \dots & \dots \\ 0 & -\dfrac{h_{n1}^{(0)}}{h_{21}^{(0)}} & 0 & 0 & \dots & 1 \end{bmatrix} \tag{2.166}$$

where the superscript (0) indicates that each matrix uses the elements of the previous transformation step. The reader is encouraged to compare with L_1 of Eq. (2.112).

The inverse of is given by Eq. (2.167):

$$\bar{L}_1^{-1} = \begin{bmatrix} 1 & 0 & 0 & 0 & \dots & 0 \\ 0 & 1 & 0 & 0 & \dots & 0 \\ 0 & \dfrac{h_{31}^{(0)}}{h_{21}^{(0)}} & 1 & 0 & \dots & 0 \\ 0 & \dfrac{h_{41}^{(0)}}{h_{21}^{(0)}} & 0 & 1 & \dots & 0 \\ \dots & \dots & \dots & \dots & \dots & \dots \\ 0 & \dfrac{h_{n1}^{(0)}}{h_{21}^{(0)}} & 0 & 0 & \dots & 1 \end{bmatrix} \tag{2.167}$$

The complete elementary similarity transformation that converts matrix A to the upper Hessenberg matrix H is shown by

$$\bar{L}_{n-1}\bar{L}_{n-2}\dots\bar{L}_2\bar{L}_1 A \bar{L}_1^{-1}\bar{L}_2^{-1}\dots\bar{L}_{n-2}^{-1}\bar{L}_{n-1}^{-1} = H \tag{2.168}$$

Each postmultiplication step by the inverse $\overline{L}_i^{-1}$ preserves the zeros previously obtained in the premultiplication step by L_i [5].

For simplicity in the above discussion, the partial pivoting matrices P_{ij} were not applied. However, use of partial pivoting is strongly recommended in order to reduce roundoff errors. Premultiplication by P_{ij} interchanges two rows and causes the sign of the determinant to change. Postmultiplication by P_{ij}^{-1} (which is identical to P_{ij}) interchanges the corresponding two columns and causes the sign of the determinant to change again. The premultiplication step must be followed immediately by the postmultiplication step in order to balance the symmetry of the transformation and to preserve the form of the transformed matrix.

The elementary similarity transformation to produce an upper Hessenberg matrix in formula form is as follows:

Initialization step:

$$h_{ij}^{(0)} = a_{ij} \qquad \begin{cases} i = 1,2,\ldots,n \\ j = 1,2,\ldots,n \end{cases} \tag{2.169a}$$

Transformation formula:

$$m_{i,k+1}^{(k)} = \frac{h_{ij}^{(k-1)}}{h_{k+1,k}^{(k-1)}} \qquad \begin{cases} j = n,n-1,\ldots,k \\ i = k+2,\ldots,n \end{cases} \tag{2.169b}$$

Premultiplication step:

$$h_{ij}^{(k-1/2)} = h_{ij}^{(k-1)} - m_{i,k+1}^{(k)} h_{k+1,j}^{(k-1)} \qquad \begin{cases} j = n,n-1,\ldots,k \\ i = k+2,\ldots,n \end{cases} \qquad \begin{array}{|l} k = 1,2,\ldots,n-2 \\ h_{k+1,k} \neq 0 \end{array} \tag{2.170}$$

Postmultiplication step:

$$h_{i,k+1}^{(k)} = h_{i,k+1}^{(k-1/2)} + h_{ij}^{(k-1/2)} m_{j,k+1}^{(k)} \qquad \begin{cases} j = k+2,\ldots,n \\ i = n,n-1,\ldots,1 \end{cases} \tag{2.171}$$

where the superscript (k - ½) means that only half the complete transformation (that is, only premultiplication) has been completed at the point.

The QR algorithm, which will be discussed next, utilizes the upper Hessenberg matrix H to determine its eigenvalues, which are equivalent to the eigenvalues of matrix A.

2.9.3 The QR Algorithm of Successive Factorization

The QR algorithm is based on the possible decomposition of a matrix A into a product of two matrices

$$A = QR \tag{2.172}$$

where Q is orthogonal and R is upper triangular with nonnegative diagonal elements. The decomposition always exists, and when A is nonsingular, the decomposition is unique [2].

The above decomposition can be used to form a series of successive matrices A_k that are similar to the original matrix A; therefore, their eigenvalues are the same. To do this, let us first define $A_1 = A$ and convert Eq. (2.172) to

$$A_1 = Q_1 R_1 \tag{2.173}$$

Premultiply each side by Q_1^{-1} and rearrange to obtain

$$R_1 = Q_1^{-1} A_1 \tag{2.174}$$

Form a second matrix A_2 from the product of R_1 with Q_1:

$$A_2 = R_1 Q_1 \tag{2.175}$$

and use Eq. (2.174) to eliminate R_1 from Eq. (2.175)

$$A_2 = Q_1^{-1} A_1 Q_1 \tag{2.176}$$

Because Q_1 is an orthogonal matrix, this is an orthogonal transformation of A_1 to A_2; therefore, these two matrices are similar. They have the same eigenvalues. The inverse of an orthogonal matrix is equal to its transpose; thus Eq. (2.176) can also be written as

$$A_2 = Q_1{}' A_1 Q_1 \tag{2.177}$$

In the particular case where matrix A is symmetric, an orthogonal transformation of A can be found that yields a diagonal matrix D:

$$D = Q' A Q \tag{2.178}$$

whose diagonal elements are the eigenvalues of A. Our discussion, however, will focus on nonsymmetric matrices that transform to triangular matrices.

The orthogonal matrix Q_1 is determined by finding a series of $S_{ij}{}'$ orthogonal transformation matrices, each of which eliminates one element, in position ij, below the diagonal of the matrix it is postmultiplying. The complete set of transformations converts matrix A_1 to upper triangular form with nonnegative diagonal elements:

$$S_{n,n-1}{}' \ldots S_{ij}{}' \ldots S_{nl}{}' \ldots S_{31}{}' S_{21}{}' A_1 = R_1 \tag{2.179}$$

where the counter i increases from $j + 1$ to n, and the counter j increases from 1 to $(n - 1)$.

Each of the S_{ij}' matrices is orthogonal, and the product of orthogonal matrices is also orthogonal. Direct comparison of Eq. (2.179) with Eq. (2.174) reveals that Q_1^{-1} is equal to the product of the S_{ij}' matrices:

$$Q_1^{-1} = S_{n,n-1}' \ldots S_{ij}' \ldots S_{31}' S_{21}' \tag{2.180}$$

The transpose of an orthogonal matrix is equal to its inverse, so it follows that

$$Q_1' = S_{n,n-1}' \ldots S_{ij}' \ldots S_{31}' S_{21}' \tag{2.181}$$

and

$$Q_1 = S_{21} S_{31} \ldots S_{ij} \ldots S_{n,n-1} \tag{2.182}$$

Therefore, Eq. (2.176) can be rewritten in terms of the S_{ij} matrices:

$$A_2 = S_{n,n-1}' \ldots S_{ij}' \ldots S_{31}' S_{21}' A_1 S_{21} S_{31} \ldots S_{ij} \ldots S_{n,n-1} \tag{2.183}$$

As an example of the orthogonal transformation matrices S_{ij} we give the S_{pq} matrix for a (6×6)-order system, with $p = 6$ and $q = 3$:

$$S_{6,3} = \begin{bmatrix} 1 & 0 & 0 & 0 & 0 & 0 \\ 0 & 1 & 0 & 0 & 0 & 0 \\ 0 & 0 & s_{33} & 0 & 0 & s_{36} \\ 0 & 0 & 0 & 1 & 0 & 0 \\ 0 & 0 & 0 & 0 & 1 & 0 \\ 0 & 0 & s_{63} & 0 & 0 & s_{66} \end{bmatrix} \tag{2.184}$$

where the diagonal elements of this matrix are specified as

$$s_{pp} = s_{qq} = \cos\theta \tag{2.185}$$

$$s_{ii} = 1 \quad \text{for } i \neq p \text{ or } q \tag{2.186}$$

and the off-diagonal elements as

$$s_{pq} = -s_{qp} = \sin\theta \tag{2.187}$$

$$s_{ij} = 0 \quad \text{everywhere else} \tag{2.188}$$

Premultiplication of matrix $\boldsymbol{A}$ by $\boldsymbol{S}_{pq}'$ eliminates the element pq and causes a rotation of axes in the (p, q) plane. The $\boldsymbol{S}_{ij}$ matrices clearly satisfy the orthogonality requirement that This is left as an exercise for the reader to verify.

The angle of axis rotation θ, in Eqs. (2.185) and (2.187), is chosen so that by element pq:

$$\boldsymbol{S}_{ij}'\boldsymbol{S}_{ij} = \boldsymbol{I} \tag{2.189}$$

of the matrix being transformed, vanishes. It has been shown by Givens [6] that it is not necessary to actually calculate the value of θ itself. The trigonometric terms $\cos\theta$ and $\sin\theta$ can be obtained from the values of the elements of the matrix being transformed. Givens has determined that the elements of the matrix $\boldsymbol{S}_{pq}$ are calculated as follows:

Diagonal elements:

$$s_{pp}^{(k)} = s_{qq}^{(k)} = \frac{a_{qq}^{(k-1)}}{\sqrt{(a_{qq}^{(k-1)})^2 + (a_{pq}^{(k-1)})^2}} \tag{2.190}$$

$$s_{ii}^{(k)} = 1 \quad \text{for } i \neq p \text{ or } q \tag{2.191}$$

Off-diagonal elements:

$$s_{pq}^{(k)} = -s_{qp}^{(k)} = \frac{a_{pq}^{(k-1)}}{\sqrt{(a_{qq}^{(k-1)})^2 + (a_{pq}^{(k-1)})^2}} \tag{2.192}$$

$$s_{ij}^{(k)} = 0 \quad \text{everywhere else} \tag{2.193}$$

The superscripts $(k-1)$ have been used in the above equations to remind the reader that the elements $a_{pq}^{(k-1)}$ and $a_{qq}^{(k-1)}$ are those of the matrix from the previous transformation step and not those of the original matrix.

Givens' method of plane rotations can reduce a nonsymmetric matrix to upper triangular form and a symmetric matrix to tridiagonal form. However, a large number of computations is required. It is computationally more efficient to apply first the elementary similarity transformation to reduce the matrix to upper Hessenberg form, as we described in Sec. 2.9.2, and then to use plane rotations to reduce it to triangular form. In the rest of this section we will assume that the matrix $\boldsymbol{A}$ has been already reduced to upper Hessenberg form, $\boldsymbol{H}_1$, and we will show how QR algorithm further reduces the matrix to obtain its eigenvalues.

If the eigenvalues of matrix $\boldsymbol{H}_1$ are λ, then the eigenvalues of matrix $(\boldsymbol{H}_1 - \gamma_1\boldsymbol{I})$ are $(\lambda - \gamma_1)$, where γ_1 is called the *shift factor*. The orthogonal transformation applied to $\boldsymbol{A}_1$ above can also be applied to the shifted matrix $(\boldsymbol{H}_1 - \gamma_1\boldsymbol{I})$ as follows:

Decompose the matrix $(\boldsymbol{H}_1 - \gamma_1\boldsymbol{I})$ into $\boldsymbol{Q}_1$ and $\boldsymbol{R}_1$ matrices:

$$\boldsymbol{H}_1 - \gamma_1\boldsymbol{I} = \boldsymbol{Q}_1\boldsymbol{R}_1 \tag{2.194}$$

Rearrange the above equation to obtain $\boldsymbol{R}_1$:

$$\boldsymbol{R}_1 = \boldsymbol{Q}_1^{-1}(\boldsymbol{H}_1 - \gamma_1 \boldsymbol{I}) \tag{2.195}$$

Form a new matrix $(\boldsymbol{H}_2 - \gamma_1\boldsymbol{I})$ from the product of $\boldsymbol{R}_1$ and $\boldsymbol{Q}_1$:

$$\boldsymbol{H}_2 - \gamma_1 \boldsymbol{I} = \boldsymbol{R}_1 \boldsymbol{Q}_1 \tag{2.196}$$

Eliminate $\boldsymbol{R}_1$ using Eq. (2.195):

$$\boldsymbol{H}_2 - \gamma_1 \boldsymbol{I} = \boldsymbol{Q}_1^{-1}(\boldsymbol{H}_1 - \gamma_1 \boldsymbol{I})\boldsymbol{Q}_1 \tag{2.197}$$

Solve for $\boldsymbol{H}_2$:

$$\boldsymbol{H}_2 = \boldsymbol{Q}_1^{-1}(\boldsymbol{H}_1 - \gamma_1 \boldsymbol{I})\boldsymbol{Q}_1 + \gamma_1 \boldsymbol{I} \tag{2.198}$$

It has been shown [2] that if the shift factor γ_1 is chosen to be a good estimate of one of the eigenvalues and that if the magnitudes of the eigenvalues are

$$|\lambda_1| > |\lambda_2| > \ldots > |\lambda_n| \tag{2.199}$$

then the matrix $\boldsymbol{H}_k$ will converge to a triangular form with the elements $h_{n,n-1} \rightarrow 0$ and $h_{nn} \rightarrow \lambda_n$.

Estimation of the shift factor γ_1 is relatively easy when the matrix has been reduced to upper Hessenberg form:

$$\boldsymbol{H}_1 = \begin{bmatrix} h_{11} & h_{12} & h_{13} & & \cdots & & h_{1\,n-2} & h_{1\,n-1} & h_{1n} \\ h_{21} & h_{22} & h_{23} & & \cdots & & h_{2\,n-2} & h_{2\,n-1} & h_{2n} \\ 0 & h_{32} & h_{33} & & \cdots & & h_{3\,n-3} & h_{3\,n-1} & h_{3n} \\ \cdots & \cdots & \cdots & \cdots & \cdots & \cdots & \cdots & \cdots & \cdots \\ 0 & 0 & & & \cdots & & h_{n-1\,n-2} & h_{n-1\,n-1} & h_{n-1\,n} \\ 0 & 0 & & & \cdots & & 0 & h_{n\,n-1} & h_{nn} \end{bmatrix} \tag{2.57}$$

The eigenvalues of the lower (2×2) submatrix:

$$\begin{bmatrix} h_{n-1,n-1} & h_{n-1,n} \\ h_{n,n-1} & h_{nn} \end{bmatrix} \tag{2.200}$$

can be used to determine the shift factor. The two eigenvalues of this matrix are obtained from the quadratic characteristic equation

$$\gamma^2 - (h_{n-1,n-1} + h_{nn})\gamma + (h_{n-1,n-1}h_{nn} - h_{n,n-1}h_{n-1,n}) = 0 \tag{2.201}$$

whose solution is given by the quadratic formula

$$\gamma_{+,-} = \frac{1}{2}\left[(h_{n-1,n-1} + h_{nn}) \pm \sqrt{(h_{n-1,n-1} + h_{nn})^2 - 4(h_{n-1,n-1}h_{nn} - h_{n,n-1}h_{n-1,n})}\right] \tag{2.202}$$

The value of γ closest to h_{nn} is chosen from the two roots. In the case where the roots are complex conjugates, the real part of the root is chosen as the shift factor.

In the QR iteration procedure the subsequent values of the shift factor, $\gamma_2 \ldots \gamma_k$, are similarly chosen from matrices $\boldsymbol{H}_2 \ldots \boldsymbol{H}_k$.

The steps of the QR algorithm for calculating the eigenvalues and eigenvectors of a nonsingular nonsymmetric matrix $\boldsymbol{A}$ with real eigenvalues are the following:

1. Use the elementary similarity transformations [Eqs. (2.168)] to transform matrix $\boldsymbol{A}$ to the upper Hessenberg matrix $\boldsymbol{H}_1$.
2. Utilize the lower (2×2) submatrix of $\boldsymbol{H}_1$ [Eq. (2.200)] to estimate the shift factor γ_1 from Eq. (2.201).
3. Construct the shifted matrix $(\boldsymbol{H}_1 - \gamma_1 \boldsymbol{I})$.
4. Calculate the elements of the transformation matrix $\boldsymbol{S}_{21}$ from the elements of the shifted matrix $(\boldsymbol{H}_1 - \gamma_1 \boldsymbol{I})$ using Eqs. (2.190)-(2.193).
5. Perform the premultiplication $\boldsymbol{S}_{21}'(\boldsymbol{H}_1 - \gamma_1 \boldsymbol{I})$, which eliminates the elements in position (2, 1) of the matrix $(\boldsymbol{H}_1 - \gamma_1 \boldsymbol{I})$.
6. Repeat steps 4 and 5, calculating the transformation matrix $\boldsymbol{S}_{pq}$ and eliminating one element on subdiagonal in each set of steps. The application of steps 4 and 5 for $(n - 1)$ times, with the counter q increasing from 1 to $(n - 1)$ and the counter p set at $(q + 1)$, will convert the Hessenberg matrix $\boldsymbol{H}_1$ to a triangular matrix $\boldsymbol{R}_1$:
$$\boldsymbol{S}_{n,n-1}' \ldots \boldsymbol{S}_{32}' \boldsymbol{S}_{21}' \boldsymbol{H}_1 = \boldsymbol{R}_1 \tag{2.203}$$
7. Perform the postmultiplication of $\boldsymbol{R}_1$ by $\boldsymbol{S}_{pq}$ to obtain the transformed shifted matrix$(\boldsymbol{H}_2 - \gamma_1 \boldsymbol{I})$:
$$\boldsymbol{H}_2 - \gamma_1 \boldsymbol{I} = \boldsymbol{R}_1 \boldsymbol{S}_{21} \boldsymbol{S}_{32} \ldots \boldsymbol{S}_{n,n-1} \tag{2.204}$$
8. Solve Eq. (2.204) for the transformed Hessenberg matrix $\boldsymbol{H}_2$:
$$\boldsymbol{H}_2 = \boldsymbol{R}_1 \boldsymbol{S}_{21} \boldsymbol{S}_{32} \ldots \boldsymbol{S}_{n,n-1} + \gamma_1 \boldsymbol{I} \tag{2.205}$$
9. Use $\boldsymbol{H}_2$ as the new Hessenberg matrix and repeat steps 2-8 until $|h_{n,n-1}| \le \epsilon$, where ϵ is a small convergence criterion. At this point, the element h_{nn} will give one eigenvalue λ_n.
10. Deflate the $\boldsymbol{H}_k$ matrix to order $(n - 1)$ by eliminating the nth row and nth column, and repeat steps 2 to 10 until all the eigenvalues are calculated.
11. Apply the Gauss elimination method with complete pivoting to the matrix $(\boldsymbol{A} - \lambda \boldsymbol{I})$ to evaluate the eigenvectors corresponding to each eigenvalue. Several different possibilities exist when the eigenvalues are real:

a. *Distinct nonzero eigenvalues*: Matrix A is nonsingular, and matrix $(A - \lambda I)$ is singular of rank $(n - 1)$. Application of the Gauss elimination method with complete pivoting on the matrix $(A - \lambda I)$ triangularizes the matrix and causes the last row to contain all zero values, because the rank is $(n - 1)$. Assume the value of the nth element of the eigenvector to be equal to unity and reduce the problem to finding the remaining $(n - 1)$ elements.

b. *One zero eigenvalue*: Matrix A is singular of rank $(n - 1)$, and matrix $(A - \lambda I)$ is singular of rank $(n - 1)$. Application of the Gauss elimination method proceeds as in *a*. One element of each eigenvector will be found to be a zero element.

c. *One pair of repeated eigenvalues*: Matrix A is nonsingular, and matrix $(A - \lambda I)$ is of rank $(n - 2)$. Application of the Gauss elimination method with complete pivoting on the matrix $(A - \lambda I)$ triangularizes the matrix and causes the last two rows to contain all zero values, because the rank is $(n - 2)$. Assume the values of the last two elements in the eigenvector to be equal to unity and reduce the problem to finding the remaining $(n - 2)$ elements.

The QR algorithm described in this section applies well to both symmetric and nonsymmetric matrices with real eigenvalues. A more general method, called the double QR algorithm, which can evaluate complex eigenvalues, is described by Ralston and Rabinowitz [2].

PROBLEMS

2.1 Solve the following set of equations by both Gauss-Seidel method and Jacobi method.

$$y + x = 3$$

$$-y + 2x = 0$$

In both cases, start the iteration from (0, 0). Compare the results of both methods at the end of each iteration.

2.2 When a pure sample of gas is bombarded by low-energy electrons in a mass spectrometer, the galvanometer shows peak heights that correspond to individual m/e (mass-to-charge) ratios for the resulting mixture of ions. For the ith peak produced by a pure sample j, one can then assign a sensitivity S_{ij} [peak height per micron (μm) of Hg sample pressure]. These coefficients are unique for each type of gas.

A distribution of peak heights may also be obtained for an n-component gas mixture to be analyzed for the partial pressures $p_1, p_2, \ldots, p_n$ of each of its constituents. The height h_i of a certain peak is a linear combination of the products of the individual sensitivities and partial pressures:

$$\sum_{j=1}^{n} S_{ij} p_j = h_i$$

In general, more than n peaks may be available. However, if the n most distinct ones are chosen, we have $i = 1, 2, \ldots, n$, so that the individual partial pressures are given by solution of n simultaneous linear equations.

Table P2.2

Peak index		Component index (j)						
		1	2	3	4	5	6	7
i	m/e	Hydrogen	Methane	Ethylene	Ethane	Propylene	Propane	n-Pentane
1	2	16.87	0.165	0.2019	0.317	0.234	0.182	0.110
2	16	0.0	27.70	0.862	0.062	0.073	0.131	0.120
3	26	0.0	0.0	22.35	13.05	4.42	6.001	3.043
4	30	0.0	0.0	0.0	11.28	0.0	1.110	0.371
5	40	0.0	0.0	0.0	0.0	9.85	1.168	2.108
6	44	0.0	0.0	0.0	0.0	0.299	15.98	2.107
7	72	0.0	0.0	0.0	0.0	0.0	0.0	4.670

The sensitivities given in Table P2.2 were reported by Carnahan et al. [7] in connection with the analysis of a hydrogen gas mixture. Write a program that will accept values for the sensitivities, $S_{11}, \ldots, S_{nn}$ and the peak heights $h_1, \ldots, h_n$ and compute values for the individual partial pressures $p_1, \ldots, p_n$.

A particular gas mixture produced the following peak heights: $h_1 = 17.1$, $h_2 = 65.1$, $h_3 = 186.0$, $h_4 = 82.7$, $h_5 = 84.2$, $h_6 = 63.7$, and $h_7 = 119.7$. The measured total pressure of the mixture was 38.78 μm of Hg, which can be compared with the sum of the computed partial pressures.

2.3 Aniline is being removed from water by solvent extraction using toluene [8]. The unit is a 10-stage countercurrent tower, shown in Fig. P2.3. The equilibrium relationship valid at each stage is, to a first approximation:

$$m = \frac{Y_i}{X_i} = 9$$

where Y_i = (lb of aniline in the toluene phase) / (lb of toluene in the toluene phase)
X_i = (lb of aniline in the water phase) / (lb of water in the water phase).

(a) The solution to this problem is a set of 10 simultaneous equations. Derive these equations from material balances around each stage. Present these equations using compact notation.

(b) Solve the above set of equations to find the concentration in both the aqueous and organic phases leaving each stage of the system (X_i and Y_i).

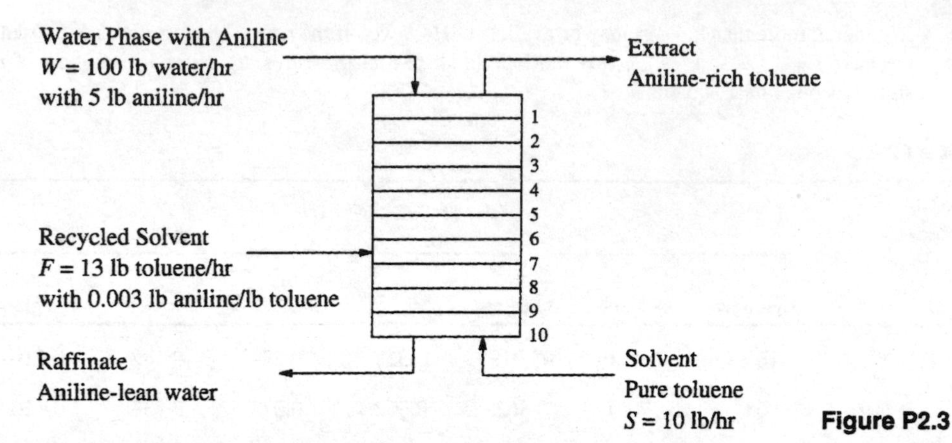

Figure P2.3

(c) If the slope of the equilibrium relationship is replaced by the expression $m = 9 + 20X_i$, the solution becomes a set of simultaneous nonlinear equations. Describe a procedure that would solve this problem.

(d) Solve the problem described in (*c*) above.

2.4 In the study of chemical reaction, Aris [9] developed a technique of writing simultaneous chemical reactions in the form of linear algebraic equations. For example, the following two simultaneous chemical equations

$$C_2H_6 \rightleftharpoons C_2H_4 + H_2$$

$$2\,C_2H_6 \rightleftharpoons C_2H_4 + 2\,CH_4$$

can be rearranged in the form

$$C_2H_4 + H_2 - C_2H_6 = 0$$

$$C_2H_4 + 2\,CH_4 - 2\,C_2H_6 = 0$$

If we identify A_1 with C_2H_4, A_2 with H_2, A_3 with CH_4, and A_4 with C_2H_6, the set of equations becomes

$$A_1 + A_2 - A_4 = 0$$

$$A_1 + 2A_3 - 2A_4 = 0$$

This can be generalized to a system of R reactions between S chemical species by the set of equations represented by

$$\sum_{j=1}^{S} \alpha_{ij} A_j = 0 \qquad i = 1, 2, \ldots, R$$

where α_{ij} are the stoichiometric coefficients of each species A_j in each reaction i.

Aris demonstrated that the number of independent chemical reactions in a set of R reactions is equal to the rank of the matrix of stoichiometric coefficients α_{ij}. Using Aris' method and the techniques developed in this chapter, determine the number of independent chemical reactions in the following reaction system:

$$4NH_3 + 5O_2 \rightleftharpoons 4NO + 6H_2O$$

$$4NH_3 + 3O_2 \rightleftharpoons 2N_2 + 6H_2O$$

$$4NH_3 + 6NO \rightleftharpoons 5N_2 + 6H_2O$$

$$2NO + O_2 \rightleftharpoons 2NO_2$$

$$2NO \rightleftharpoons N_2 + O_2$$

$$N_2 + 2O_2 \rightleftharpoons 2NO_2$$

2.5 The multistage distillation tower shown in Fig. P2.5 is equipped with a total condenser and a partial boiler. This tower will be used for the separation of a multicomponent mixture. Assume that for this particular mixture, the tower contains the equivalent of 10 equilibrium stages, including the reboiler; that is, $N = 10$ and $j = 11$.

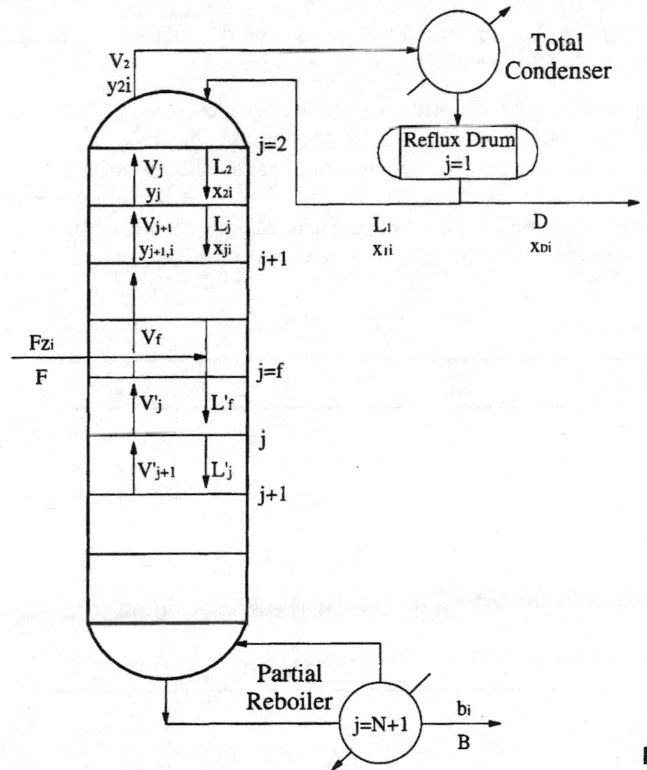

Figure P2.5

The feed to the column has a flow rate $F = 1000$ mol/h. It is a saturated liquid, and it enters the column on equilibrium stage 5 ($j = 6$). It contains five components ($n = 5$) whose mole fractions are

$z_1 = 0.06$ $z_2 = 0.17$ $z_3 = 0.22$ $z_4 = 0.20$ $z_5 = 0.35$

It is desired to recover a distillate product at a rate of 500 mol/h.

Develop all the material balances for component i for all 10 equilibrium stages and for the condenser. For this problem make the following assumptions:

1. The external reflux ratio is

$$\frac{L_1}{D} = 2.5$$

2. Constant molal overflow occurs in each section of the tower.
3. The initial guesses of the temperatures corresponding to the equilibrium stages are $T_2 = 140°F$, $T_3 = 150°F$, $T_4 = 160°F$, $T_5 = 170°F$, $T_6 = 180°F$, $T_7 = 190°F$, $T_8 = 200°F$, $T_9 = 210°F$, $T_{10} = 220°F$, and $T_{11} = 230°F$.
4. The equilibrium constant K_{ji} can be approximated by the following equation:

$$K_{ji} = \alpha_i + \beta_i T_j + \gamma_i T_j^2$$

where the temperatures are in degrees Fahrenheit and the coefficients for each individual component are listed in Table P2.5 [10].

Solve the resulting set of equations in order to determine the following:

(a) The molal flow rates of all vapor and liquid streams in the tower

(b) The mole fraction of each component in the vapor and liquid streams.

Note that the mole fractions in each stage do not add up to unity, because the above solution is only a single step in the solution of multicomponent distillation problem. Assumptions 2 and 3 are only initial guesses that must be subsequently corrected from energy balances and bubble point calculations.

Table P2.5

Component i	α_i	β_i	γ_i
1	0.70	0.30×10^{-2}	0.65×10^{-4}
2	2.21	1.95×10^{-2}	0.90×10^{-4}
3	1.50	-1.60×10^{-2}	0.80×10^{-4}
4	0.86	-0.97×10^{-2}	0.46×10^{-4}
5	0.71	-0.87×10^{-2}	0.42×10^{-4}

2.6 The following equations can be shown to relate the temperatures and pressures on either side of a detonation wave that is moving into a zone of unburned gas [7]:

$$\frac{\gamma_2 m_2 T_1}{m_1 T_2}\left(\frac{P_2}{P_1}\right)^2 - (\gamma_2 + 1)\frac{P_2}{P_1} + 1 = 0$$

$$\frac{\Delta H_{R1}}{c_{p2}T_1} + \frac{T_2}{T_1} - 1 = \frac{(\gamma_2 - 1)m_2}{2\gamma_2 m_1}\left(\frac{P_2}{P_1} - 1\right)\left(1 + \frac{m_1 T_2 P_1}{m_2 T_1 P_2}\right)$$

Here, T = absolute temperature, P = absolute pressure, γ_2 = ratio of specific heat at constant pressure to that at constant volume, m = mean molecular weight, ΔH_{R1} = heat of reaction, c_{p2} = specific heat, and the subscripts 1 and 2 refer to the unburned and burned gas, respectively.

Write a program that accepts values for m_1, m_2, γ_2, ΔH_{R1}, c_{p2}, T_1, and P_1 as data and that will proceed to compute and print values for T_2 and P_2. Run the program with the following data, which apply to the detonation of a mixture of hydrogen and oxygen:

$m_1 = 12$ g/g mol	$m_2 = 18$ g/g mol	$T_1 = 300$ K	$\gamma_2 = 1.31$
$\Delta H_{R1} = -58{,}300$ cal/g mol	$c_{p2} = 9.86$ cal/(g mol.K)	$P_1 = 1$ atm	

2.7 The system of highly coupled chemical reactions shown in Fig. P2.7 takes place in a batch reactor. The conditions of temperature and pressure in the reactor are such that the kinetic rate constants attain the following values:

$k_{21} = 0.2$	$k_{31} = 0.1$	$k_{32} = 0.1$	$k_{34} = 0.1$	$k_{54} = 0.05$	$k_{64} = 0.2$	$k_{65} = 0.1$
$k_{12} = 0.1$	$k_{13} = 0.05$	$k_{23} = 0.05$	$k_{43} = 0.2$	$k_{45} = 0.1$	$k_{46} = 0.2$	$k_{56} = 0.1$

If the chemical reaction starts with the following initial concentrations:

$A_0 = 1.0$ mol/L	$D_0 = 0$
$B_0 = 0$	$E_0 = 1.0$ mol/L
$C_0 = 0$	$F_0 = 0$

calculate the steady-state concentration of all components. Assume that all reactions are of first order.

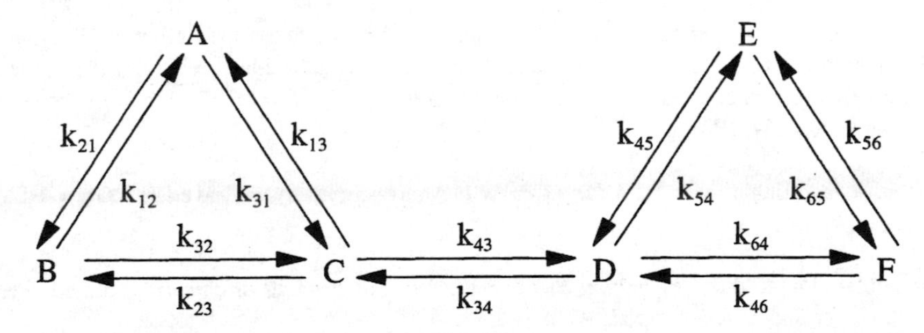

Figure P2.7 System of coupled chemical reactions.

2.8 A linear mathematical model that has three independent variables, X_1, X_2, and X_3, may be written as

$$Y = b_1 X_1 + b_2 X_2 + b_3 X_3$$

where b_1, b_2, and b_3 are parameter constants to be determined from experimental observations. It can be shown that the vector of parameters $\boldsymbol{b}$ may be calculated from

$$\boldsymbol{b} = (\boldsymbol{X}'\boldsymbol{X})^{-1}\boldsymbol{X}'\boldsymbol{Y}$$

where $\boldsymbol{X}$ is the matrix that contains the vectors of independent variable observations, $\boldsymbol{X}_1$, $\boldsymbol{X}_2$, $\boldsymbol{X}_3$, as columns:

$$\boldsymbol{X} = [\boldsymbol{X}_1 \ \ \boldsymbol{X}_2 \ \ \boldsymbol{X}_3]$$

and $\boldsymbol{Y}$ is the vector of dependent variable observations (see Chap. 7).

Using the experimental observations shown in Table P2.8 determine the values of the parameters b_1, b_2, and b_3 for this linear model.

Table P2.8

X_1	X_2	X_3	Y
1	0.2	5.0	1.0
2	0.6	4.1	5.0
3	0.7	3.0	7.0
4	1.0	2.0	10.0
5	1.5	1.2	12.5
6	2.0	0.5	15.0

REFERENCES

1. Constantinides, A., and Shao, P., "Material Balancing Applied to the Prediction of Glutamic Acid Production and Cell Mass Formation," in A. Constantinides, W. R. Vieth, and K. Venkatasubramanian (eds.), *Biochemical Engineering II*, New York Academy of Sciences Annuals, vol. 369, 1981, p. 167.

2. Ralston, A., and Rabinowitz, P., *A First Course in Numerical Analysis*, 2nd ed., McGraw-Hill, New York, 1978.

3. Himmelblau, D. M., *Basic Principles and Calculations in Chemical Engineering*, 6th ed., Prentice Hall, Upper Saddle River, NJ, 1996, p. 602.

4. Faddeev, D. K., and Faddeeva, U. N., *Computational Methods in Linear Algebra*, trans. by R. C. Williams, W. H. Freeman, New York, 1963.

5. Johnson, L. W., and Riess, R. D., *Numerical Analysis*, 2nd ed., Addison-Wesley, Inc., Reading, MA, 1982.

6. Givens, M., "Computation of Plane Unitary Rotation Transforming a General Matrix to Triangular Form," *J. Soc. Ind. Appl. Math.*, vol. 6, 1958, pp. 26-50.

7. Carnahan, B., Luther, H. A., and Wilkes, J. O., *Applied Numerical Methods*, Wiley, New York, 1969, p. 331.

8. Freeman, R., private communication, Department of Chemical and Biochemical Engineering, Rutgers University, Piscataway, N.J., 1984.

9. Aris, R., *Introduction to the Analysis of Chemical Reactors*, Prentice Hall, Englewood Cliffs, NJ, 1965.

10. Chang, H. Y., and Over, I. E., *Selected Numerical Methods and Computer Programs for Chemical Engineers*, Sterling Swift, Manchaca, TX, 1981.

CHAPTER 3

Finite Difference Methods and Interpolation

3.1 INTRODUCTION

The most commonly encountered mathematical models in engineering and science are in the form of differential equations. The dynamics of physical systems that have one independent variable can be modeled by ordinary differential equations, whereas systems with two, or more, independent variables require the use of partial differential equations. Several types of ordinary differential equations, and a few partial differential equations, render themselves to analytical (closed-form) solutions. These methods have been developed thoroughly in differential calculus. However, the great majority of differential equations, especially the nonlinear ones and those that involve large sets of simultaneous differential equations, do not have analytical solutions but require the application of numerical techniques for their solution.

Several numerical methods for differentiation, integration, and the solution of ordinary and partial differential equations are discussed in Chaps. 4-6 of this book. These methods are based on the concept of *finite differences*. Therefore, the purpose of this chapter is to develop the systematic terminology used in the *calculus of finite differences* and to derive the relationships between finite differences and differential operators, which are needed in the numerical solution of ordinary and partial differential equations.

The calculus of finite differences may be characterized as a "two-way street" that enables the user to take a differential equation and integrate it numerically by calculating the values of the function at a discrete (finite) number of points. Or, conversely, if a set of finite values is available, such as experimental data, these may be differentiated, or integrated, using the calculus of finite differences. It should be pointed out, however, that numerical differentiation is inherently less accurate than numerical integration.

Another very useful application of the calculus of finite differences is in the derivation of interpolation/extrapolation formulas, the so-called *interpolating polynomials*, which can be used to represent experimental data when the actual functionality of these data is not known. A very common example of the application of interpolation is in the extraction of physical properties of water from the steam tables. Interpolating polynomials are also used to estimate numerical derivative and integral of the tabulated data (see Chap. 4). The discussion of several interpolating polynomials is given in Secs. 3.7-3.10.

3.2 SYMBOLIC OPERATORS

In differential calculus, the definition of the derivative is given as

$$\frac{df(x)}{dx}\bigg|_{x_0} = f'(x_0) = \lim_{x \to x_0} \frac{f(x) - f(x_0)}{x - x_0} \tag{3.1}$$

In the calculus of finite differences, the value of $x - x_0$ does not approach zero but remains a finite quantity. If we represent this quantity by h:

$$h = x - x_0 \tag{3.2}$$

then the derivative may be *approximated* by

$$f'(x_0) \simeq \frac{f(x_0 + h) - f(x_0)}{h} \tag{3.3}$$

Under certain circumstances, there is a point, ξ, in the interval (a, b) for which the derivative can be calculated *exactly* from Eq. (3.3). This is confirmed by the mean-value theorem of differential calculus:

Mean-value theorem: Let $f(x)$ be continuous in the range $a \le x \le b$ and differentiable in the range $a < x < b$; then there exists at least one ξ, $a < \xi < b$, for which

$$f'(\xi) = \frac{f(b) - f(a)}{b - a} \tag{3.4}$$

This theorem forms the basis for both the differential calculus and the finite difference calculus.

A function $f(x)$, which is continuous and differentiable in the interval $[x_0, x]$, can be represented by a Taylor series

$$f(x) = f(x_0) + (x - x_0)f'(x_0) + \frac{(x - x_0)^2 f''(x_0)}{2!} + \frac{(x - x_0)^3 f'''(x_0)}{3!} + \ldots + \frac{(x - x_0)^n f^{(n)}(x_0)}{n!} + R_n(x) \tag{3.5}$$

where $R_n(x)$ is called the remainder. This term lumps together the remaining terms in the infinite series from $(n + 1)$ to infinity; it, therefore, represents the *truncation* error, when the function is evaluated using the terms up to, and including, the nth-order term of the infinite series.

The mean-value theorem can be used to show that there exists a point ξ in the interval (x_0, x) so that the remainder term given by

$$R_n(x) = \frac{(x - x_0)^{n+1} f^{(n+1)}(\xi)}{(n + 1)!} \tag{3.6}$$

The value of ξ is an unknown function of x; therefore, it is impossible to evaluate the remainder, or truncation error, term exactly. The remainder is a term of order $(n + 1)$, because it is a function of $(x - x_0)^{n+1}$ and of the $(n + 1)$th derivative. For this reason, in our discussion of truncation errors, we will always specify the order of the remainder term and will usually abbreviate it using the notation $O(h^{n+1})$.

The calculus of finite differences is used in conjunction with a series of discrete values, which can be either experimental data, such as

$$y_{i-3} \qquad y_{i-2} \qquad y_{i-1} \qquad y_i \qquad y_{i+1} \qquad y_{i+2} \qquad y_{i+3}$$

or discrete values of a continuous function $y(x)$:

$$y(x - 3h) \qquad y(x - 2h) \qquad y(x - h) \qquad y(x) \qquad y(x + h) \qquad y(x + 2h) \qquad y(x + 3h)$$

or, equivalently, values of a function $f(x)$:

$$f(x-3h) \quad f(x-2h) \quad f(x-h) \quad f(x) \quad f(x+h) \quad f(x+2h) \quad f(x+3h)$$

In all the above cases, the values of the dependent variable, y or f, are those corresponding to *equally spaced* values of the independent variable x. This concept is demonstrated in Fig. 3.1 for a smooth function $y(x)$.

A set of *linear symbolic operators* drawn from differential calculus and from finite

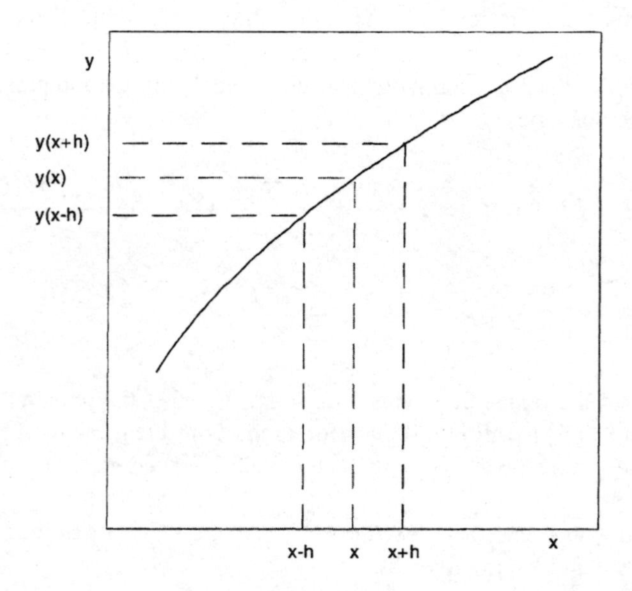

Figure 3.1 Values of function $y(x)$ at equally spaced points of the independent variable x.

difference calculus will be defined in conjunction with the above series of discrete values. These definitions will then be used to derive the interrelationships between the operators. The linear symbolic operators are

D = differential operator
I = integral operator
E = shift operator
Δ = forward difference operator
∇ = backward difference operator
δ = central difference operator
μ = averager operator.

All these operators may be treated as algebraic variables because they satisfy the distributive, commutative, and associative laws of algebra.

The first two operators are well known from differential calculus. The *differential operator D* has the following effect when applied to the function $y(x)$:

$$D\,y(x) = \frac{dy(x)}{dx} = y'(x) \tag{3.7}$$

and the *integral operator* is

$$I\,y(x) = \int_x^{x+h} y(x)\,dx \tag{3.8}$$

The integral operator is equivalent to the inverse of the differential operator

$$I = D^{-1} \tag{3.9}$$

The *shift operator* causes the function to shift to the next successive value of the independent variable:

$$E\,y(x) = y(x+h) \tag{3.10}$$

The inverse of the shift operator, E^{-1} causes the function to shift in the negative direction of the independent variable:

$$E^{-1}y(x) = y(x-h) \tag{3.11}$$

Higher powers of the shift operator are defined as

$$E^{n}y(x) = y(x+nh) \tag{3.12}$$

The shift operator can be expressed in terms of the differential operator by expanding the function $y(x+h)$ into a Taylor series about x:

$$y(x+h) = y(x) + \frac{h}{1!}y'(x) + \frac{h^2}{2!}y''(x) + \frac{h^3}{3!}y'''(x) + \ldots \tag{3.13}$$

Using the differential operator D to indicate the derivatives of y, we obtain

$$y(x+h) = y(x) + \frac{h}{1!}Dy(x) + \frac{h^2}{2!}D^2y(x) + \frac{h^3}{3!}Dy(x) + \ldots \tag{3.14}$$

Factoring out the term $y(x)$ from the right-hand side of Eq. (3.14):

$$y(x+h) = \left(1 + \frac{h}{1!}D + \frac{h^2}{2!}D^2 + \frac{h^3}{3!}D^3 + \dots\right)y(x) \tag{3.15}$$

The terms in the parentheses are equivalent to the series expansion

$$e^{hD} = 1 + \frac{h}{1!}D + \frac{h^2}{2!}D^2 + \frac{h^3}{3!}D^3 + \dots \tag{3.16}$$

Therefore, Eq. (3.15) can be written as

$$y(x+h) = e^{hD}y(x) \tag{3.17}$$

Comparing Eq. (3.10) with (3.17), we conclude that the shift operator can be expressed in terms of the differential operator by the relation

$$E = e^{hD} \tag{3.18}$$

Similarly, the inverse of the shift operator can be related to the differential operator by expanding the function $y(x - h)$ into a Taylor series about x:

$$y(x-h) = y(x) - \frac{h}{1!}y'(x) + \frac{h^2}{2!}y''(x) - \frac{h^3}{3!}y'''(x) + \dots \tag{3.19}$$

Replacing the derivatives with the differential operators and rearranging, we obtain

$$y(x-h) = \left(1 - \frac{h}{1!}D + \frac{h^2}{2!}D^2 - \frac{h^3}{3!}D^3 + \dots\right)y(x) \tag{3.20}$$

The terms in the parenthesis are equivalent to the series expansion

$$e^{-hD} = 1 - \frac{h}{1!}D + \frac{h^2}{2!}D^2 - \frac{h^3}{3!}D^3 + \dots \tag{3.21}$$

Therefore, Eq. (3.19) can be written as

$$y(x-h) = e^{-hD}y(x) \tag{3.22}$$

It follows from a comparison of Eq. (3.11) with Eq. (3.22) that

$$E^{-1} = e^{-hD} \tag{3.23}$$

With these introductory concepts in mind, let us proceed to develop the backward, forward, and central difference operators and the relationships between these and the differential operators.

3.3 BACKWARD FINITE DIFFERENCES

Consider the set of values

$$y_{i-3} \qquad y_{i-2} \qquad y_{i-1} \qquad y_i \qquad y_{i+1} \qquad y_{i+2} \qquad y_{i+3}$$

or the equivalent set

$$y(x-3h) \quad y(x-2h) \quad y(x-h) \qquad y(x) \qquad y(x+h) \quad y(x+2h) \quad y(x+3h)$$

The *first backward difference* of y at *i* (or *x*) is defined as

$$\nabla y_i = y_i - y_{i-1}$$

or

$$\nabla y(x) = y(x) - y(x-h) \tag{3.24}$$

The *second backward difference* of y at *i* (or *x*) is defined as

$$\nabla^2 y_i = \nabla(\nabla y_i) = \nabla(y_i - y_{i-1}) = \nabla y_i - \nabla y_{i-1}$$

$$= (y_i - y_{i-1}) - (y_{i-1} - y_{i-2})$$

$$\nabla^2 y_i = y_i - 2y_{i-1} + y_{i-2}$$

or

$$\nabla^2 y(x) = y(x) - 2y(x-h) + y(x-2h) \tag{3.25}$$

The *third backward difference* of y at *i* is defined as

$$\nabla^3 y_i = \nabla(\nabla^2 y_i) = \nabla(y_i - 2y_{i-1} + y_{i-2})$$

$$= \nabla y_i - 2\nabla y_{i-1} + \nabla y_{i\;2} \tag{3.26}$$

$$= (y_i - y_{i-1}) - 2(y_{i-1} - y_{i-2}) + (y_{i-2} - y_{i-3})$$

$$\nabla^3 y_i = y_i - 3y_{i-1} + 3y_{i-2} - y_{i-3}$$

Higher-order backward differences are similarly derived:

$$\nabla^4 y_i = y_i - 4y_{i-1} + 6y_{i-2} - 4y_{i-3} + y_{i-4} \tag{3.27}$$

$$\nabla^5 y_i = y_i - 5y_{i-1} + 10y_{i-2} - 10y_{i-3} + 5y_{i-4} - y_{i-5} \tag{3.28}$$

The coefficients of the terms in each of the above finite differences correspond to those of the binomial expansion $(a - b)^n$, where n is the order of the finite difference. Therefore, the general formula of the nth-order backward finite difference can be expressed as

$$\nabla^n y_i = \sum_{m=0}^{n} (-1)^m \frac{n!}{(n-m)!\,m!} y_{i-m} \tag{3.29}$$

It should also be noted that the sum of the coefficients of the binomial expansion is always equal to zero. This can be used as a check to ensure that higher-order differences have been expanded correctly.

The relationship between backward difference operators and differential operators can now be established. Combine Eqs. (3.22) and (3.24) to obtain

$$\nabla y(x) = y(x) - y(x-h) = y(x) - e^{-hD} y(x)$$
$$= (1 - e^{-hD}) y(x) \tag{3.30}$$

which shows that the backward difference operator is given by

$$\nabla = 1 - e^{-hD} \tag{3.31}$$

Using the infinite series expression of e^{-hD} [Eq. (3.21)], Eq. (3.31) becomes

$$\nabla = hD - \frac{h^2D^2}{2} + \frac{h^3D^3}{6} - \dots \tag{3.32}$$

The higher-order backward difference operator, ∇^2, ∇^3, . . ., can be obtained by raising the first backward difference operator to higher powers[1]:

$$\nabla^2 = (1 - e^{-hD})^2 = (1 - 2e^{-hD} + e^{-2hD}) \tag{3.33}$$

$$\nabla^3 = (1 - e^{-hD})^3 = (1 - 3e^{-hD} + 3e^{-2hD} - e^{-3hD}) \tag{3.34}$$

$$\vdots$$

$$\nabla^n = (1 - e^{-hD})^n \tag{3.35}$$

Expansion of the exponential terms and rearrangement yields the following equations for the second and third backward difference operators:

$$\nabla^2 = h^2D^2 - h^3D^3 + \frac{7}{12}h^4D^4 - \dots \tag{3.36}$$

---

[1] These relationships can also be obtained by combining the definitions of the backward differences [Eqs. (3.25) and (3.26)] with the definition of the inverse shift operator [Eqs. (3.11) and (3.23)].

$$\nabla^3 = h^3D^3 - \frac{3}{2}h^4D^4 + \frac{5}{4}h^5D^5 - \ldots \tag{3.37}$$

Equations (3.32), (3.36), and (3.37) express the backward difference operators in terms of infinite series of differential operators. In order to complete the set of relationships, equations that express the differential operators in terms of backward difference operators will also be derived. To do so, first rearrange Eq. (3.31) to solve for e^{-hD}:

$$e^{-hD} = 1 - \nabla \tag{3.38}$$

Take the natural logarithm of both sides of this equation:

$$\ln e^{-hD} = -hD = \ln(1 - \nabla) \tag{3.39}$$

Utilize the infinite series expansion:

$$\ln(1 - \nabla) = -\nabla - \frac{\nabla^2}{2} - \frac{\nabla^3}{3} - \frac{\nabla^4}{4} - \frac{\nabla^5}{5} - \ldots \tag{3.40}$$

Combine Eq. (3.39) with Eq. (3.40) to obtain:

$$hD = \nabla + \frac{\nabla^2}{2} + \frac{\nabla^3}{3} + \frac{\nabla^4}{4} + \frac{\nabla^5}{5} + \ldots \tag{3.41}$$

The higher-order differential operators can be obtained by simply raising both sides of Eq. (3.41) to higher powers:

$$h^2D^2 = \nabla^2 + \nabla^3 + \frac{11}{12}\nabla^4 + \frac{5}{6}\nabla^5 + \ldots \tag{3.42}$$

$$h^3D^3 = \nabla^3 + \frac{3}{2}\nabla^4 + \frac{7}{4}\nabla^5 + \ldots \tag{3.43}$$

$$\vdots$$

$$h^nD^n = \left(\nabla + \frac{\nabla^2}{2} + \frac{\nabla^3}{3} + \frac{\nabla^4}{4} + \frac{\nabla^5}{5} + \ldots\right)^n \tag{3.44}$$

The complete set of relationships between backward difference operators and differential operators is summarized in Table 3.1.

Table 3.1 Backward finite differences

Backward difference operators	Differential operators
$\nabla = hD - \frac{h^2D^2}{2} + \frac{h^3D^3}{6} - \ldots$	$hD = \nabla + \frac{\nabla^2}{2} + \frac{\nabla^3}{3} + \frac{\nabla^4}{4} + \ldots$
$\nabla^2 = h^2D^2 - h^3D^3 + \frac{7}{12}h^4D^4 - \ldots$	$h^2D^2 = \nabla^2 + \nabla^3 + \frac{11}{12}\nabla^4 + \frac{5}{6}\nabla^5 + \ldots$
$\nabla^3 = h^3D^3 - \frac{3}{2}h^4D^4 + \frac{5}{4}h^5D^5 - \ldots$	$h^3D^3 = \nabla^3 + \frac{3}{2}\nabla^4 + \frac{7}{4}\nabla^5 + \ldots$
$\nabla^n = (1 - e^{-hD})^n$	$h^nD^n = \left(\nabla + \frac{\nabla^2}{2} + \frac{\nabla^3}{3} + \frac{\nabla^4}{4} + \ldots\right)^n$

3.4 Forward Finite Differences

The development of forward finite differences follows a course parallel to that used in the development of backward differences.

Consider the set of values

$$y_{i-3} \quad y_{i-2} \quad y_{i-1} \quad y_i \quad y_{i+1} \quad y_{i+2} \quad y_{i+3}$$

or the equivalent set

$$y(x-3h) \quad y(x-2h) \quad y(x-h) \quad y(x) \quad y(x+h) \quad y(x+2h) \quad y(x+3h)$$

The *first forward difference* of y at i (or x) is defined as

$$\Delta y_i = y_{i+1} - y_i$$

or

$$\Delta y(x) = y(x + h) - y(x) \tag{3.45}$$

The *second forward difference* of y at i (or x) is defined as

$$\begin{aligned}\Delta^2 y_i &= \Delta(\Delta y_i) = \Delta(y_{i+1} - y_i) = \Delta y_{i+1} - \Delta y_i \\ &= (y_{i+2} - y_{i+1}) - (y_{i+1} - y_i)\end{aligned} \tag{3.46}$$

$$\Delta^2 y_i = y_{i+2} - 2y_{i+1} + y_i$$

or (3.46)

$$\Delta^2 y(x) = y(x + 2h) - 2y(x + h) + y(x)$$

The *third forward difference* of y at i is defined as

$$\Delta^3 y_i = \Delta(\Delta^2 y_i) = \Delta(y_{i+2} - 2y_{i+1} + y_i)$$

$$= \Delta y_{i+2} - 2\Delta y_{i+1} + \Delta y_i \tag{3.47}$$

$$= (y_{i+3} - y_{i+2}) - 2(y_{i+2} - y_{i+1}) + (y_{i+1} - y_i)$$

$$\Delta^3 y_i = y_{i+3} - 3y_{i+2} + 3y_{i+1} - y_i$$

Higher-order forward differences are similarly derived:

$$\Delta^4 y_i = y_{i+4} - 4y_{i+3} + 6y_{i+2} - 4y_{i+1} + y_i \tag{3.48}$$

$$\Delta^5 y_i = y_{i+5} - 5y_{i+4} + 10y_{i+3} - 10y_{i+2} + 5y_{i+1} - y_i \tag{3.49}$$

In similarity to the backward finite differences, the forward finite differences also have coefficients which correspond to those of the binomial expansion $(a - b)^n$. Therefore, the general formula of the nth-order forward finite difference can be expressed as

$$\Delta^n y_i = \sum_{m=0}^{n} (-1)^m \frac{n!}{(n-m)!\,m!} y_{i+n-m} \tag{3.50}$$

In MATLAB, the function *diff*(*y*) returns forward finite differences of y. Values of nth-order forward finite difference may be obtained from *diff*(y, n).

The relationship between forward difference operators and differential operators can now be developed. Combine Eqs. (3.45) and (3.17) to obtain

$$\Delta y(x) = y(x + h) - y(x)$$

$$= e^{hD} y(x) - y(x)$$

$$= (e^{hD} - 1) y(x) \tag{3.51}$$

which shows that the forward difference operator is given by

$$\Delta = e^{hD} - 1 \tag{3.52}$$

Using the infinite series expression of e^{hD} [Eq. (3.16)], Eq. (3.52) becomes

$$\Delta = hD + \frac{h^2D^2}{2} + \frac{h^3D^3}{6} + \dots \tag{3.53}$$

The higher-order forward difference operator, $\Delta^2, \Delta^3, \dots$, can be obtained by raising the first forward difference operator to higher powers:

$$\Delta^2 = (e^{hD} - 1)^2 = (e^{2hD} - 2e^{hD} + 1) \tag{3.54}$$

$$\Delta^3 = (e^{hD} - 1)^3 = (e^{3hD} - 3e^{2hD} + 3e^{hD} - 1) \tag{3.55}$$

$$\vdots$$

$$\Delta^n = (e^{hD} - 1)^n \tag{3.56}$$

Expansion of the exponential terms and rearrangement yields the following equations for the second and third forward difference operators:

$$\Delta^2 = h^2D^2 + h^3D^3 + \frac{7}{12}h^4D^4 + \dots \tag{3.57}$$

$$\Delta^3 = h^3D^3 + \frac{3}{2}h^4D^4 + \frac{5}{4}h^5D^5 + \dots \tag{3.58}$$

Eqs. (3.53), (3.57), and (3.58) express the forward difference operators in terms of infinite series of differential operators. In order to complete the set of relationships, equations that express the differential operators in terms of forward difference operators will also be derived. To do this, first rearrange Eq. (3.52) to solve for e^{hD}:

$$e^{hD} = 1 + \Delta \tag{3.59}$$

Take the natural logarithm of both sides of this equation:

$$\ln e^{hD} = hD = \ln(1 + \Delta) \tag{3.60}$$

Utilize the infinite series expansion:

$$\ln(1 + \Delta) = \Delta - \frac{\Delta^2}{2} + \frac{\Delta^3}{3} - \frac{\Delta^4}{4} + \frac{\Delta^5}{5} - \dots \tag{3.61}$$

Combine Eq. (3.60) with Eq. (3.61) to obtain

$$hD = \Delta - \frac{\Delta^2}{2} + \frac{\Delta^3}{3} - \frac{\Delta^4}{4} + \frac{\Delta^5}{5} - \ldots \tag{3.62}$$

The higher-order differential operators can be obtained by simply raising both sides of Eq. (3.62) to higher powers:

$$h^2D^2 = \Delta^2 - \Delta^3 + \frac{11}{12}\Delta^4 - \frac{5}{6}\Delta^5 + \ldots \tag{3.63}$$

$$h^3D^3 = \Delta^3 - \frac{3}{2}\Delta^4 + \frac{7}{4}\Delta^5 - \ldots \tag{3.64}$$

$$\vdots$$

$$h^nD^n = \left(\Delta - \frac{\Delta^2}{2} + \frac{\Delta^3}{3} - \frac{\Delta^4}{4} + \frac{\Delta^5}{5} - \ldots\right)^n \tag{3.65}$$

The complete set of relationships between forward difference operators and differential operators is summarized in Table 3.2.

Table 3.2 Forward finite differences

Forward difference operators	Differential operators
$\Delta = hD + \frac{h^2D^2}{2} + \frac{h^3D^3}{6} + \ldots$	$hD = \Delta - \frac{\Delta^2}{2} + \frac{\Delta^3}{3} - \frac{\Delta^4}{4} + \ldots$
$\Delta^2 = h^2D^2 + h^3D^3 + \frac{7}{12}h^4D^4 + \ldots$	$h^2D^2 = \Delta^2 - \Delta^3 + \frac{11}{12}\Delta^4 - \frac{5}{6}\Delta^5 + \ldots$
$\Delta^3 = h^3D^3 + \frac{3}{2}h^4D^4 + \frac{5}{4}h^5D^5 + \ldots$	$h^3D^3 = \Delta^3 - \frac{3}{2}\Delta^4 + \frac{7}{4}\Delta^5 - \ldots$
$\Delta^n = (e^{hD} - 1)^n$	$h^nD^n = \left(\Delta - \frac{\Delta^2}{2} + \frac{\Delta^3}{3} - \frac{\Delta^4}{4} + \ldots\right)^n$

3.5 CENTRAL FINITE DIFFERENCES

As their name implies, central finite differences are *centered* at the pivot position and are evaluated utilizing the values of the function to the right and to the left of the pivot position, but located only $h/2$ distance from it.

Consider the series of values used in the previous two sections, but with the additional values at the midpoints of the intervals

$$y_{i-2} \quad y_{i-1\frac{1}{2}} \quad y_{i-1} \quad y_{i-\frac{1}{2}} \quad y_i \quad y_{i+\frac{1}{2}} \quad y_{i+1} \quad y_{i+1\frac{1}{2}} \quad y_{i+2}$$

or the equivalent set

$$y(x-2h) \quad y(x-1\tfrac{1}{2}h) \quad y(x-h) \quad y(x-\tfrac{1}{2}h) \quad y(x) \quad y(x+\tfrac{1}{2}h) \quad y(x+h) \quad y(x+1\tfrac{1}{2}h) \quad y(x+2h)$$

The *first central difference* of y at i (or x) is defined as

$$\delta y_i = y_{i+1/2} - y_{i-1/2}$$

or

$$\delta y(x) = y(x + \tfrac{1}{2}h) - y(x - \tfrac{1}{2}h) \tag{3.66}$$

The *second central difference* of y at i (or x) is defined as

$$\delta^2 y_i = \delta(\delta y_i) = \delta(y_{i+\frac{1}{2}} - y_{i-\frac{1}{2}}) = \delta y_{i+\frac{1}{2}} - \delta y_{i-\frac{1}{2}}$$

$$= (y_{i+1} - y_i) - (y_i - y_{i-1})$$

$$\delta^2 y_i = y_{i+1} - 2y_i + y_{i-1}$$

or

$$\delta^2 y(x) = y(x + h) - 2y(x) + y(x - h) \tag{3.67}$$

The *third central difference* of y at i is defined as

$$\delta^3 y_i = \delta(\delta^2 y_i) = \delta(y_{i+1} - 2y_i + y_{i-1})$$

$$= \delta_{i+1} - 2\delta y_i + \delta y_{i-1}$$

$$= (y_{i+1\frac{1}{2}} - y_{i+\frac{1}{2}}) - 2(y_{i+\frac{1}{2}} - y_{i-\frac{1}{2}}) + (y_{i-\frac{1}{2}} - y_{i-1\frac{1}{2}})$$

$$= y_{i+1\frac{1}{2}} - 3y_{i+\frac{1}{2}} + 3y_{i-\frac{1}{2}} - y_{i-1\frac{1}{2}} \tag{3.68}$$

Higher-order central differences are similarly derived:

$$\delta^4 y_i = y_{i+2} - 4y_{i+1} + 6y_i - 4y_{i-1} + y_{i-2} \tag{3.69}$$

$$\delta^5 y_i = y_{i+2½} - 5y_{i+1½} + 10y_{i+½} - y_{i-½} + 5y_{i-1½} - y_{i-2½} \tag{3.70}$$

Consistent with the other finite differences, the central finite differences also have coefficients that correspond to those of the binomial expansion $(a - b)^n$. Therefore, the general formula of the nth-order central finite difference can be expressed as

$$\delta^n y_i = \sum_{m=0}^{n} (-1)^m \frac{n!}{(n-m)!\,m!} y_{i-m+n/2} \tag{3.71}$$

It should be noted that the *odd*-order central differences involve values of the function at the midpoint of the intervals, whereas the even-order central differences involve values at the full intervals. To fully utilize odd- and even-order central differences, we need a set of values of the function y that includes twice as many points as that used in either backward or forward differences. This situation is rather uneconomical, especially in the case where these values must be obtained experimentally. To alleviate this difficulty, we make use of the *averager operator* μ, which is defined as

$$\mu = \frac{1}{2}[E^{1/2} + E^{-1/2}] \tag{3.72}$$

The averager operator shifts its operand by a half interval to the right of the pivot and by a half interval to the left of the pivot, evaluates it at these two positions, and averages the two values.

Application of the averager on the odd central differences gives the *first averaged central difference* as follows:

$$\begin{aligned} \mu\delta y_i &= \frac{1}{2}(E^{1/2}\delta y_i + E^{-1/2}\delta y_i) \\ &= \frac{1}{2}(\delta y_{i+½} + \delta y_{i-½}) \\ &= \frac{1}{2}[(y_{i+1} - y_i) + (y_i - y_{i-1})] \\ &= \frac{1}{2}(y_{i+1} - y_{i-1}) \end{aligned} \tag{3.73}$$

The *third averaged central difference* is given by

$$\mu\delta^3 y_i = \frac{1}{2}(E^{1/2}\delta^3 y_i + E^{-1/2}\delta^3 y_i)$$

$$= \frac{1}{2}(\delta^3 y_{i+1/2} + \delta^3 y_{i-1/2})$$

$$= \frac{1}{2}[(y_{i+2} - 3y_{i+1} + 3y_i - y_{i-1}) + (y_{i+1} - 3y_i + 3y_{i-1} - y_{i-2})]$$

$$= \frac{1}{2}(y_{i+2} - 2y_{i+1} + 2y_{i-1} - y_{i-2}) \tag{3.74}$$

As expected, the effect of the averager is to remove the midpoint values of the function y from the odd central differences.

It will be shown in Chap. 4 that central differences are more accurate than either backward or forward differences when used to evaluate the derivatives of functions.

The relationships between central difference operators and differential operators can now be developed. Eq. (3.73), representing the first averaged central difference, is combined with Eqs. (3.17) and (3.22) to yield

$$\mu\delta y(x) = \frac{1}{2}[y(x + h) - y(x - h)]$$

$$= \frac{1}{2}[e^{hD}y(x) - e^{-hD}y(x)]$$

$$= \frac{1}{2}(e^{hD} - e^{-hD})y(x) \tag{3.75}$$

which shows that the first averaged central difference operator is given by

$$\mu\delta = \frac{1}{2}(e^{hD} - e^{-hD}) = \sinh hD \tag{3.76}$$

Using the infinite series expansions of e^{hD} and e^{-hD}, or equivalently the infinite series expansion of the hyperbolic sine:

$$\sinh hD = hD + \frac{(hD)^3}{3!} + \frac{(hD)^5}{5!} + \frac{(hD)^7}{7!} + \dots \tag{3.77}$$

Eq. (3.76) becomes

$$\mu\delta = hD + \frac{h^3D^3}{6} + \frac{h^5D^5}{120} + \frac{h^7D^7}{5040} + \dots \tag{3.78}$$

Similarly, using Eq. (3.67) for the second central difference, and combining it with Eqs. (3.17) and (3.22), we obtain

$$\begin{aligned}\delta^2 y(x) &= y(x+h) - 2y(x) + y(x-h) \\ &= e^{hD}y(x) - 2y(x) + e^{-hD}y(x) \\ &= (e^{hD} - 2 + e^{-hD})\,y(x)\end{aligned} \tag{3.79}$$

which shows that the second central difference operator is equivalent to

$$\delta^2 = e^{hD} + e^{-hD} - 2 = 2(\cosh hD - 1) = E + E^{-1} - 2 \tag{3.80}$$

Expanding the exponentials into their infinite series, or equivalently the infinite series expansion of the hyperbolic cosine in Eq. (3.80), we obtain

$$\delta^2 = h^2D^2 + \frac{h^4D^4}{12} + \frac{h^6D^6}{360} + \frac{h^8D^8}{20160} + \ldots \tag{3.81}$$

The higher-order averaged odd central difference operators are obtained by taking products of Eqs. (3.78) and (3.81). The higher-order even central differences are formulated by taking powers of Eq. (3.81). The third and fourth central operators, thus obtained, are listed below:

$$\mu\delta^3 = h^3D^3 + \frac{h^5D^5}{4} + \frac{h^7D^7}{40} + \ldots \tag{3.82}$$

$$\delta^4 = h^4D^4 + \frac{h^6D^6}{6} + \frac{h^8D^8}{80} + \ldots \tag{3.83}$$

In order to develop the inverse relationships, i.e., equations for the differential operators in terms of the central difference operators, we must first derive an algebraic relationship between μ and δ. To do this, we start with Eqs. (3.72) and (3.80). Squaring both sides of Eq. (3.72), we obtain

$$\mu^2 = \frac{1}{4}(E + E^{-1} + 2) \tag{3.84}$$

Rearranging Eq. (3.80), we get

$$\delta^2 + 2 = E + E^{-1} \tag{3.85}$$

Combining Eqs. (3.84) and (3.85), and rearranging, we arrive at the desired relationship

$$\mu^2 = \frac{\delta^2}{4} + 1 \tag{3.86}$$

Now taking the inverse of Eq. (3.76):

$$hD = \sinh^{-1}\mu\delta \tag{3.87}$$

The infinite series expansion of the inverse hyperbolic sine is

$$\sinh^{-1}\mu\delta = \mu\delta - \frac{(\mu\delta)^3}{6} + \frac{3(\mu\delta)^5}{40} - \dots \tag{3.88}$$

Therefore, Eq. (3.87) expands to

$$hD = \mu\delta - \frac{\mu^3\delta^3}{6} + \frac{3\mu^5\delta^5}{40} - \dots \tag{3.89}$$

The even powers of μ are eliminated from Eq. (3.89) by using Eq. (3.86) to obtain the first differential operator in terms of central difference operators:

$$hD = \mu(\delta - \frac{\delta^3}{6} + \frac{\delta^5}{30} - \dots) \tag{3.90}$$

Higher-order differential operators are obtained by raising Eq. (3.90) to the appropriate power and using Eq. (3.86) to eliminate the even powers of μ. The second, third, and fourth differential operators obtained by this way are

$$h^2D^2 = \delta^2 - \frac{\delta^4}{12} + \frac{\delta^6}{90} - \dots \tag{3.91}$$

$$h^3D^3 = \mu(\delta^3 - \frac{\delta^5}{4} + \frac{7\delta^7}{120} - \dots) \tag{3.92}$$

$$h^4D^4 = \delta^4 - \frac{\delta^6}{6} + \frac{7\delta^8}{240} - \dots \tag{3.93}$$

The complete set of relationships between central difference operators and differential operators is summarized in Table 3.3. These relationships will be used in Chap. 4 to develop a set of formulas expressing the derivatives in terms of central finite differences. These formulas will have higher accuracy than those developed using backward and forward finite differences.

Table 3.3 Central finite differences

Central difference operators	Differential operators
$\mu\delta = hD + \frac{h^3D^3}{6} + \frac{h^5D^5}{120} + \frac{h^7D^7}{5040} + \dots$	$hD = \mu\left(\delta - \frac{\delta^3}{6} + \frac{\delta^5}{30} - \frac{\delta^7}{140} \dots\right)$
$\delta^2 = h^2D^2 + \frac{h^4D^4}{12} + \frac{h^6D^6}{360} + \frac{h^8D^8}{20160} + \dots$	$h^2D^2 = \delta^2 - \frac{\delta^4}{12} + \frac{\delta^6}{90} - \dots$
$\mu\delta^3 = h^3D^3 + \frac{h^5D^5}{4} + \frac{h^7D^7}{40} + \dots$	$h^3D^3 = \mu\left(\delta^3 - \frac{\delta^5}{4} + \frac{7\delta^7}{120} - \dots\right)$
$\delta^4 = h^4D^4 + \frac{h^6D^6}{6} + \frac{h^8D^8}{80} + \dots$	$h^4D^4 = \delta^4 - \frac{\delta^6}{6} + \frac{7\delta^8}{240} - \dots$

3.6 Difference Equations and Their Solutions

The application of forward, backward, or central finite differences in the solution of differential equations transforms these equations to *difference equations* of the form

$$f(y_k, y_{k+1}, \dots, y_{k+n}) = 0 \tag{3.94}$$

In addition, difference equations are obtained from the application of material balances on multistage operations, such as distillation and extraction.

Depending on their origin, difference equations may be linear or nonlinear, homogeneous or nonhomogeneous, with constant or variable coefficients. For the purposes of this book, it will be necessary to discuss only the methods of solution of *homogeneous linear difference equations with constant coefficients*.

The *order* of a difference equation is the difference between the highest and lowest subscript of the dependent variable in the equation, that is, it is the number of finite steps spanned by the equation. The order of Eq. (3.94) is given by

$$Order = (k + n) - k = n \tag{3.95}$$

The process of obtaining y_k is called *solving the difference equation*. The methods of obtaining such solutions are analogous to those used in finding analytical solutions of differential equations. As a matter of fact, the theory of difference equations is parallel to the corresponding theory of differential equations. Difference equations resemble ordinary differential equations. For example, Eq. (3.96) is a second-order homogeneous linear ordinary *differential* equation:

$$y'' + 3y' - 4y = 0 \tag{3.96}$$

whereas Eq. (3.97) is a second-order homogeneous linear *difference* equation:

$$y_{k+2} + 3y_{k+1} - 4y_k = 0 \tag{3.97}$$

The solution of the differential equation (3.96) can be obtained from the methods of differential calculus applied as follows:

1. Replace the derivatives in (3.96) with the differential operators:

$$D^2 y + 3Dy - 4y = 0$$

2. Factor out the y:

$$(D^2 + 3D - 4)y = 0$$

3. Find the roots of the *characteristic equation*:

$$D^2 + 3D - 4 = 0$$

These roots are called the *eigenvalues* of the differential equation. In this case they are

$$\lambda_1 = 1 \quad \text{and} \quad \lambda_2 = -4$$

4. Construct the solution of the homogeneous differential equation as follows:

$$\begin{aligned} y &= C_1 e^{\lambda_1 x} + C_2 e^{\lambda_2 x} \\ &= C_1 e^{(1)x} + C_2 e^{(-4)x} \end{aligned} \tag{3.98}$$

where C_1 and C_2 are constants that must be evaluated from the boundary conditions of the differential equation.

Similarly, the solution of the difference equation (3.97) can be obtained by using the shift operator E:

1. Replace each term of Eq. (3.97) with its equivalent using the shift operator:

$$E^2 y_k + 3Ey_k - 4y_k = 0$$

2. Factor out the y_k:

$$(E^2 + 3E - 4)y_k = 0$$

3. Find the roots of the characteristic equation:

$$E^2 + 3E - 4 = 0$$

These roots are $\lambda_1 = 1$ and $\lambda_2 = -4$.

4. Construct the solution of the homogeneous difference equation as follows:

$$\begin{aligned} y_k &= C_1\lambda_1^k + C_2\lambda_2^k \\ &= C_1(1)^k + C_2(-4)^k \end{aligned} \tag{3.99}$$

where C_1 and C_2 are constants that must be evaluated from the boundary conditions of the difference equation.

In the above case, both eigenvalues were real and distinct. When the eigenvalues are real and repeated, the solution for a second-order equation with both roots identical is formed as follows:

$$y_k = (C_1 + C_2 k)\lambda^k \tag{3.100}$$

For an nth-order equation, which has m repeated roots (λ_m) and one distinct root (λ_n), the general formulation of the solution is obtained by superposition:

$$y_k = (C_1 + C_2 k + C_3 k^2 + \ldots + C_m k^{m-1})\lambda_m^k + C_n\lambda_n^k \tag{3.101}$$

In the case where the characteristic equation contains two complex roots

$$\lambda_1 = \alpha + \beta i \qquad \text{and} \qquad \lambda_2 = \alpha - \beta i \tag{3.102}$$

the solution is

$$y_k = C_1(\alpha + \beta i)^k + C_2(\alpha - \beta i)^k \tag{3.103}$$

This solution may be also expressed in terms of trigonometric quantities by utilizing the trigonometric (polar) form of complex numbers:

$$\alpha \pm \beta i = r(\cos\theta \pm i\sin\theta) \tag{3.104}$$

This is obtained by showing the complex number as a vector in the complex plane represented in Fig. 3.2. The *modulus* r of the complex number is obtained from the Pythagorean theorem

$$r = \sqrt{\alpha^2 + \beta^2} \tag{3.105}$$

The values of α and β are expressed in terms of the *phase angle* θ:

$$\alpha = r\cos\theta \tag{3.106}$$

$$\beta = r\sin\theta \tag{3.107}$$

and the phase angle is given by

$$\theta = \tan^{-1}\frac{\beta}{\alpha} \tag{3.108}$$

Substituting Eq. (3.104) in Eq. (3.103) and utilizing de Moivre's theorem

$$(\cos\theta \pm i\sin\theta)^k = \cos k\theta \pm i\sin k\theta \tag{3.109}$$

we obtain the solution of the difference equation as

$$y_k = r^k[C_1'\cos k\theta + C_2'\sin k\theta] \tag{3.110}$$

where $C_1' = C_1 + C_2$ and $C_2' = (C_1 - C_2)i$.

It can be concluded from the above discussion that the solution of homogeneous linear difference equations with constant coefficients is of the form

$$y_k = f(k,\lambda) \tag{3.111}$$

where k is the forward-marching counter and λ is the vector of eigenvalues of the characteristic equation. The stability and convergence of these solutions depend on the values of the eigenvalues. The following stability cases apply to the solutions of difference equations:

1. The solution is stable, converging without oscillations, when
   a. All the eigenvalues are real distinct and have absolute values less than, or equal to, unity:

$$\lambda = \text{real distinct}$$
$$|\lambda| \leq 1.0$$

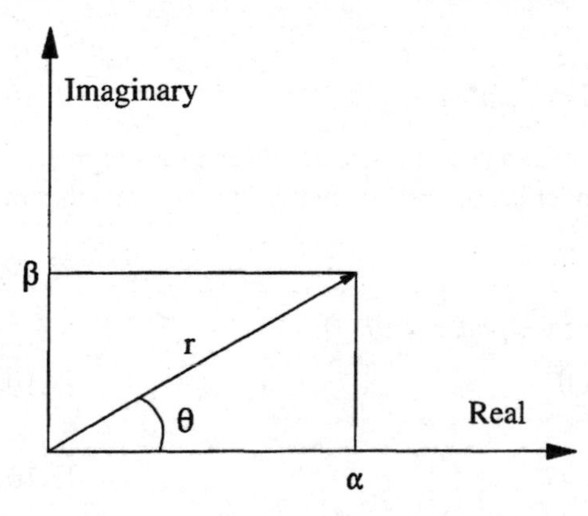

Figure 3.2 Representation of a complex number in a plane.

b. The eigenvalues are real, but repeated, and have absolute values less than unity:

$$\lambda = \text{real repeated}$$
$$|\lambda| < 1.0$$

2. The solution is stable, converging with damped oscillations, when

a. Complex distinct eigenvalues are present, and the moduli of the eigenvalues are less than, or equal to, unity:

$$\lambda = \text{complex distinct}$$
$$|r| \leq 1.0$$

b. Complex repeated eigenvalues are present, and the moduli of the eigenvalues are less than unity:

$$\lambda = \text{complex repeated}$$
$$|r| < 1.0$$

3. The solution is unstable and nonoscillatory, when

a. All the eigenvalues are real distinct, and one or more of these have absolute values greater than unity:

$$\lambda = \text{real distinct}$$
$$|\lambda| > 1.0$$

b. The eigenvalues are real, but repeated, and one or more of these have absolute values equal to, or greater than, unity:

$$\lambda = \text{real repeated}$$
$$|\lambda| \geq 1.0$$

4. The solution is unstable and oscillatory, when

a. Complex distinct eigenvalues are present, and the moduli of one or more of these are greater than unity:

$$\lambda = \text{complex distinct}$$
$$|r| > 1.0$$

b. Complex repeated eigenvalues are present, and the moduli of one or more of these are equal to, or greater than, unity:

$$\lambda = \text{complex repeated}$$
$$|r| \geq 1.0$$

The numerical solutions of ordinary and partial differential equations are based on the finite difference formulation of these differential equations. Therefore, the stability and convergence considerations of finite difference solutions have important implications on the numerical solutions of differential equations. This topic will be discussed in more detail in Chaps. 5 and 6.

3.7 INTERPOLATING POLYNOMIALS

Engineers and scientists often face the task of interpreting and correlating experimental observations, which are usually in the form of discrete data, and are called upon to either integrate or differentiate these data numerically or graphically. This task is facilitated by the use of interpolation/extrapolation formulas. The calculus of finite differences enables us to develop *interpolating polynomials* that can represent experimental data when the actual functionality of these data is not well known. But, even more significantly, these polynomials can be used to approximate functions that are difficult to integrate or differentiate, thus making the task somewhat easier, albeit approximate.

Let us assume that values of functions $f(x)$ are known at a set of $(n + 1)$ values of the independent variables x:

$$\begin{array}{cc} x_0 & f(x_0) \\ x_1 & f(x_1) \\ x_2 & f(x_2) \\ x_3 & f(x_3) \\ \vdots & \vdots \\ x_n & f(x_n) \end{array}$$

These values are called the *base points* of the function. They are shown graphically in Fig. 3.3*a*.

The general objective in developing interpolating polynomials is to choose a polynomial of the form

$$P_n(x) = a_0 + a_1 x + a_2 x^2 + a_3 x^3 + \dots + a_n x^n \qquad (3.112)$$

so that this equation fits exactly the base points of the function and connects these points with a smooth curve, as shown in Fig. 3.3*b*. This polynomial can then be used to approximate the function at any value of the independent variable x between the base points.

For the given set of $(n + 1)$ known base points, the polynomial must satisfy the equation

$$P_n(x_i) = f(x_i) \qquad i = 0, 1, 2, \dots, n \qquad (3.113)$$

Substitution of the known values of $(x_i, f(x_i))$ in Eq. (3.112) yields a set of $(n + 1)$ simultaneous linear algebraic equations whose unknowns are the coefficients $a_0, \dots, a_n$ of the polynomial equation. The solution of this set of linear algebraic equations may be obtained using one of the algorithms discussed in Chap. 2. However, this solution results in an ill-conditioned linear system; therefore, other methods have been favored in the development of interpolating polynomials.

MATLAB has several functions for interpolation. The function y_i = *interp1*(x, y, x_i) takes the values of the independent variable x and the dependent variable y (base points) and does the one-dimensional interpolation based on x_i to find y_i. The default method of interpolation is linear. However, the user can choose the method of interpolation in the fourth input argument from '*nearest*' (nearest neighbor interpolation), '*linear*' (linear interpolation), '*spline*' (cubic spline interpolation), and '*cubic*' (cubic interpolation). If the vector of independent variable is not equally spaced, the function *interp1q* may be used instead. It is faster than *interp1* because it does not check the input arguments. MATLAB also has the function *spline* to perform one-dimensional interpolation by cubic splines, using *not-a-knot* method. It can also return coefficients of piecewise polynomials, if required. The functions *interp2*, *interp3*, and *interpn* perform two-, three-, and n-dimensional interpolation, respectively.

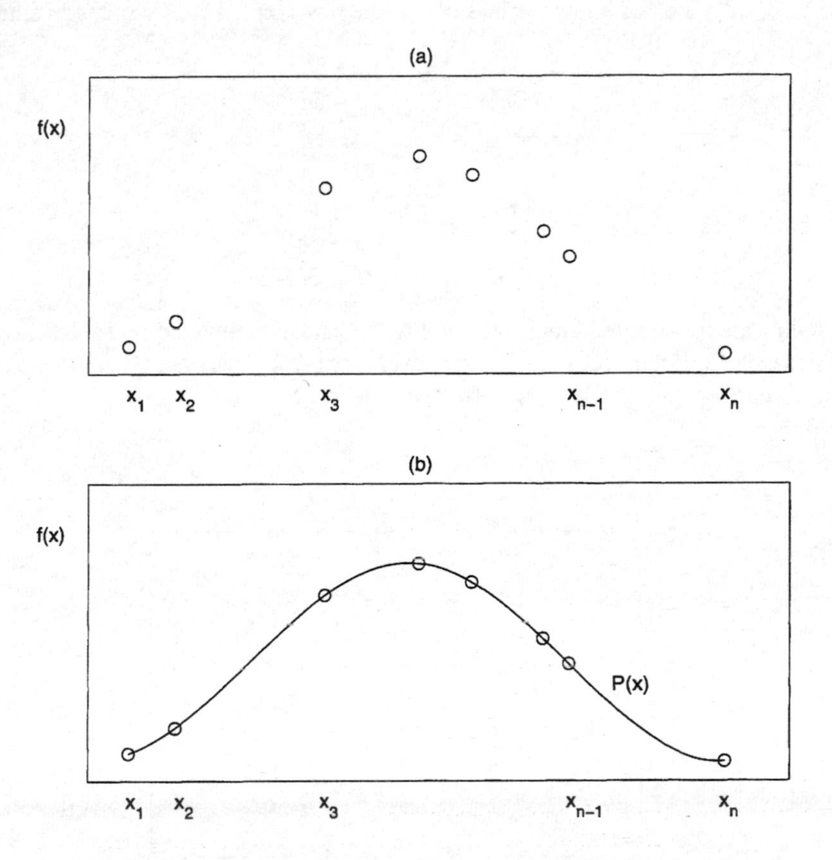

Figure 3.3 (*a*) Unequally spaced base points of the function $f(x)$. (*b*) Unequally spaced base points with interpolating polynomial.

3.8 Interpolation of Equally Spaced Points

In this section, we will develop two interpolation methods for equally spaced data: (1) the *Gregory-Newton formulas,* which are based on forward and backward differences, and (2) *Stirling's interpolation formula,* based on central differences.

3.8.1 Gregory-Newton Interpolation

First, we consider a set of known values of the function $f(x)$ at *equally spaced* values of x:

$x - 3h$	$f(x - 3h)$
$x - 2h$	$f(x - 2h)$
$x - h$	$f(x - h)$
x	$f(x)$
$x + h$	$f(x + h)$
$x + 2h$	$f(x + 2h)$
$x + 3h$	$f(x + 3h)$

These points are represented graphically in Fig. 3.4 and are tabulated in Tables 3.4 and 3.5. The first, second, and third forward differences of these base points are also tabulated in Table 3.4 and the corresponding backward differences in Table 3.5.

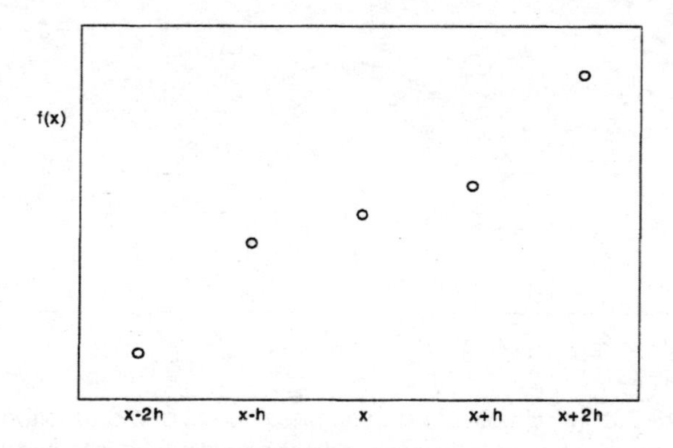

Figure 3.4 Equally spaced base points for interpolating polynomials.

Table 3.4 Forward difference table

i	x_i	$f(x_i)$	$\Delta f(x_i)$	$\Delta^2 f(x_i)$	$\Delta^3 f(x_i)$
0	x	$f(x)$	$f(x+h)-f(x)$	$f(x+2h)-2f(x+h)+f(x)$	$f(x+3h)-3f(x+2h)+3f(x+h)-f(x)$
1	$x+h$	$f(x+h)$	$f(x+2h)-f(x+h)$	$f(x+3h)-2f(x+2h)+f(x+h)$	
2	$x+2h$	$f(x+2h)$	$f(x+3h)-f(x+2h)$		
3	$x+3h$	$f(x+3h)$			

Table 3.5 Backward difference table

i	x_i	$f(x_i)$	$\nabla f(x_i)$	$\nabla^2 f(x_i)$	$\nabla^3 f(x_i)$
-3	$x-3h$	$f(x-3h)$			
-2	$x-2h$	$f(x-2h)$	$f(x-2h)-f(x-3h)$		
-1	$x-h$	$f(x-h)$	$f(x-h)-f(x-2h)$	$f(x-h)-2f(x-2h)+f(x-3h)$	
0	x	$f(x)$	$f(x)-f(x-h)$	$f(x)-2f(x-h)+f(x-2h)$	$f(x)-3f(x-h)+3f(x-2h)-f(x-3h)$

The *Gregory-Newton forward interpolation formula* can be derived using the forward finite difference relations derived in Secs. 3.2 and 3.4. Eq. (3.17), written for the function f,

$$f(x + h) = e^{hD} f(x) \tag{3.114}$$

relates the value of the function at one interval forward of the pivot point x to the value of the function at the pivot point. Applying this equation for n intervals forward, that is, replacing h with nh, we obtain

$$f(x + nh) = e^{nhD} f(x) \tag{3.115}$$

or equivalently

$$f(x + nh) = (e^{hD})^n f(x) \tag{3.116}$$

We note from Eq. (3.59) that

$$e^{hD} = 1 + \Delta \tag{3.59}$$

Combining Eqs. (3.116) and (3.59) we obtain

$$f(x + nh) = (1 + \Delta)^n f(x) \tag{3.117}$$

The term $(1 + \Delta)^n$ can be expanded using the *binomial series*

$$(1 + \Delta)^n = 1 + n\Delta + \frac{n(n-1)}{2!}\Delta^2 + \frac{n(n-1)(n-2)}{3!}\Delta^3 + \frac{n(n-1)(n-2)(n-3)}{4!}\Delta^4 + \dots \tag{3.118}$$

Therefore, Eq. (3.117) becomes

$$f(x + nh) = f(x) + n\Delta f(x) + \frac{n(n-1)}{2!}\Delta^2 f(x) + \frac{n(n-1)(n-2)}{3!}\Delta^3 f(x) + \frac{n(n-1)(n-2)(n-3)}{4!}\Delta^4 f(x) + \dots \tag{3.119}$$

When n is a positive integer, the binomial series has $(n + 1)$ terms; therefore, Eq. (3.119) is a polynomial of degree n. If $(n + 1)$ base-point values of the function f are known, this polynomial fits all $(n + 1)$ points exactly. Assume that these $(n + 1)$ base-points are $(x_0, f(x_0))$, $(x_1, f(x_1)), \dots, (x_n, f(x_n))$, where $(x_0, f(x_0))$ is the pivot point and x_i is defined as

$$x_i = x_0 + ih \tag{3.120}$$

We can now designate the distance of the point of interest from the pivot point as $(x - x_0)$. The value of n is no longer an integer and is replaced by

$$n = \frac{x - x_0}{h} \tag{3.121}$$

These substitutions convert Eq. (3.119) to

$$f(x) = f(x_0) + \frac{(x - x_0)}{h}\Delta f(x_0) + \frac{(x - x_0)(x - x_1)}{2!\,h^2}\Delta^2 f(x_0)$$
$$+ \frac{(x - x_0)(x - x_1)(x - x_2)}{3!\,h^3}\Delta^3 f(x_0)$$
$$+ \frac{(x - x_0)(x - x_1)(x - x_2)(x - x_3)}{4!\,h^4}\Delta^4 f(x_0) + \dots \tag{3.122}$$

This is the *Gregory-Newton forward interpolation formula*. The general formula of the above series is

$$f(x) = f(x_0) + \sum_{k=1}^{n}\left(\prod_{m=0}^{k-1}(x - x_m)\right)\frac{\Delta^k f(x_0)}{k!\,h^k} \tag{3.123}$$

In a similar derivation, using backward differences, the *Gregory-Newton backward interpolation formula* is derived as

$$f(x) = f(x_0) + \frac{(x - x_0)}{h}\nabla f(x_0) + \frac{(x - x_0)(x - x_{-1})}{2!\,h^2}\nabla^2 f(x_0)$$
$$+ \frac{(x - x_0)(x - x_{-1})(x - x_{-2})}{3!\,h^3}\nabla^3 f(x_0) \tag{3.124}$$
$$+ \frac{(x - x_0)(x - x_{-1})(x - x_{-2})(x - x_{-3})}{4!\,h^4}\nabla^4 f(x_0) + \dots$$

The general formula of the above series is

$$f(x) = f(x_0) + \sum_{k=1}^{n}\left(\prod_{m=0}^{k-1}(x - x_{-m})\right)\frac{\nabla^k f(x_0)}{k!\,h^k} \tag{3.125}$$

It was stated earlier that the binomial series [Eq. (3.118)] has a finite number of terms, $(n + 1)$, when n is a positive integer. However, in the Gregory-Newton interpolation formulas,

n is not usually an integer; therefore, these polynomials have an infinite number of terms. It is known from algebra that if $|\Delta| \leq 1$, then the binomial series for $(1 + \Delta)^n$ converges to the value of $(1 + \Delta)^n$ as the number of terms become larger and larger. This implies that the finite differences must be small. This is true for a flat, smooth function, or, alternatively, if the known base points are close together; that is, if h is small. Of course, the number of terms that can be used in each formula depends on the highest order of finite differences that can be evaluated from the available known data. It is common sense that for evenly spaced data, the accuracy of interpolation is higher for a large number of data points that are closely spaced together.

For a given set of data points, the accuracy of interpolation can be further enhanced by choosing the pivot point as close to the point of interest as possible, so that $x < h$. If this is satisfied, then the series should utilize as many term as possible; that is, the number of finite differences in the equation should be maximized. The order of error of the formula applied in each case is equivalent to the order of the finite difference contained in the first truncated term of the series. Examination of Table 3.4 reveals that points at the top of the table have the largest available number of forward differences, whereas Table 3.5 reveals that points at the bottom of the table have the largest number of backward differences. Therefore, the forward formula should be used for interpolating between points near the top of the table, and the backward formula should be used for interpolation near the bottom of the table.

Example 3.1: Gregory-Newton Method for Interpolation of Equally Spaced Data. An exothermic, relatively slow reaction takes place in a reactor under your supervision. Yesterday, after you left the plant, the temperature of the reactor went out of control, for a yet unknown reason, until the operator put it under control by changing the cooling water flow rate. Your supervisor has asked you to prepare a report regarding this incident. As the first step, you must know when the reactor reached its maximum temperature and what was the value of this maximum temperature. A computer was recording the temperature of the reactor at one-hour intervals. These time-temperature data are given in Table E3.1. Write a general MATLAB function for n-order one-dimensional interpolation by Gregory-Newton forward interpolation formula to solve this problem.

Method of Solution: The function uses the general formula of the Gregory-Newton forward interpolation [Eq. (3.123)] to perform the n-order interpolation. The input to the function specifying the number of base points must be at least $(n + 1)$.

Program Description: The MATLAB function *GregoryNewton.m* is developed to perform the Gregory-Newton forward interpolation. The first and second input arguments are the coordinates of the base points. The third input argument is the vector of independent variable at which the interpolation of the dependent variable is required. The fourth input, n, is the order of interpolation. If no value is introduced to the function through the fourth argument, the function does linear interpolation. For obtaining the results of the higher-order interpolation, this value should be entered as the fourth input argument.

At the beginning, the function checks the inputs. The vectors of coordinates of base points have to be of the same size. The function also checks to see if the vector of independent variable is monotonic; otherwise, the function terminates calculations. The order of interpolation cannot be more than the intervals (number of base points minus one). In this case, the function displays a warning and continues with the maximum possible order of interpolation. The function then performs the interpolation according to Eq. (3.123).

The main program *Example3_1.m* is written to solve the problem of Example 3.1. It asks the user to input the vector of time (independent variable), vector of temperature of the reactor (dependent variable), and the order of interpolation. The program applies the function *GregoryNewton.m* to interpolate the temperature between the recorded temperatures and finds its maximum. The user can repeat the calculations with another order of interpolation.

Table E3.1

Time (p.m.)	Temperature (°C)	Time (p.m.)	Temperature (°C)
4	70	9	93
5	71	10	81
6	75	11	68
7	83	12	70
8	92		

Program

Example3_1.m

```
% Example3_1.m
% Solution to Example 3.1. It interpolates the time-temperature data
% given in Table E3.1 by Gregory-Newton forward interpolation
% formula and finds the maximum temperature and the time this
% maximum happened.

clc
clear
clf

% Input data
time = input(' Vector of time = ');
temp = input(' Vector of temperature = ');
ti=linspace(min(time),max(time)); % Vector of time for interpolation
```

```
redo = 1;
while redo
   disp(' ')
   n = input(' Order of interpolation = ');
   te = GregoryNewton(time,temp,ti,n); % Interpolation
   [max_temp,k] = max(te);
   max_time = ti(k);
   % Show the results
   fprintf('\n Maximum temperature of %4.1f C reached at %5.2f.\n',
max_temp,max_time)
   % Show the results graphically
   plot(time,temp,'o',ti,te)
   xlabel('Time (hr)')
   ylabel('Temperature (deg C)')
   disp(' ')
   redo = input(' Repeat the calculation (1/0) : ');
end
```

GregoryNewton.m

```
function yi = GregoryNewton(x,y,xi,n)
%GregoryNewton One dimensional interpolation.
%
%   YI = GregoryNewton(X,Y,XI,N) applies the Nth-order
%   Gregory-Newton forward interpolation to find YI, the
%   values of the underlying function Y at the points in
%   the vector XI.  The vector X specifies the points at
%   which the data Y is given.
%
%   YI = GregoryNewton(X,Y,XI) is equivalent to the
%   linear interpolation.
%
%   See also INTERP1, NATURALSPLINE, Lagrange, SPLINE, INTERP1Q

% (c) by N. Mostoufi & A. Constantinides
% January 1, 1999

% Initialization
if nargin < 3
   error('Invalid number of inputs.')
end

% Check x for equal spacing and determining h
if min(diff(x)) ~= max(diff(x))
   error('Independent variable is not monotonic.')
else
   h = x(2) - x(1);
end

x = (x(:).')';      % Make sure it's a column vector
y = (y(:).')';      % Make sure it's a column vector
```

```
nx = length(x);
ny = length(y);
if nx ~= ny
   error('X and Y vectors are not the same size.');
end

% Check the order of interpolation
if nargin == 3 | n < 1
   n = 1;
end
n = floor(n);
if n >= nx
 fprintf('\nNot enough data points for %2d-order interpolation.', n)
 fprintf('\n%2d-order interpolation will be performed instead.\n',
nx-1)
 n = nx - 1;
end

deltax(1,1:length(xi)) = ones(1,length(xi));
% Locating the required number of base points
for m = 1:length(xi)
   dx = xi(m) - x;
   % Locating xi
   [dxm , loc(m)] = min(abs(dx));
   % locating the first base point
   if dx(loc(m)) < 0
      loc(m) = loc(m) - 1;
   end
   if loc(m) < 1
      loc(m) = 1;
   end
   if loc(m)+n > nx
      loc(m) = nx - n;
   end
   deltax(2:n+1,m) = dx(loc(m):loc(m)+n-1);
   ytemp(1:n+1,m) = y(loc(m):loc(m)+n);
end

% Interpolation
yi = y(loc)';
for k = 1 : n
   yi = yi + prod(deltax(1:k+1,:)) .* diff(ytemp(1:k+1,:),k) /...
   (gamma(k+1) * h^k);
end
```

Input and Results

```
>>Example3_1

 Vector of time = [4, 5, 6, 7, 8, 9, 10, 11, 12]
```

```
Vector of temperature = [70, 71, 75, 83, 92, 93, 81, 68, 70]
Order of interpolation = 2

Maximum temperature of 94.2 C reached at  8.61.

Repeat the calculation (1/0) : 0
```

Discussion of Results: Graphical results are shown in Fig. E3.1. As can be seen from this plot and also from the numerical results, the reactor has reached the maximum temperature of 94.2°C at 8:37 p.m. The reader can repeat the calculations with other values for order of interpolation.

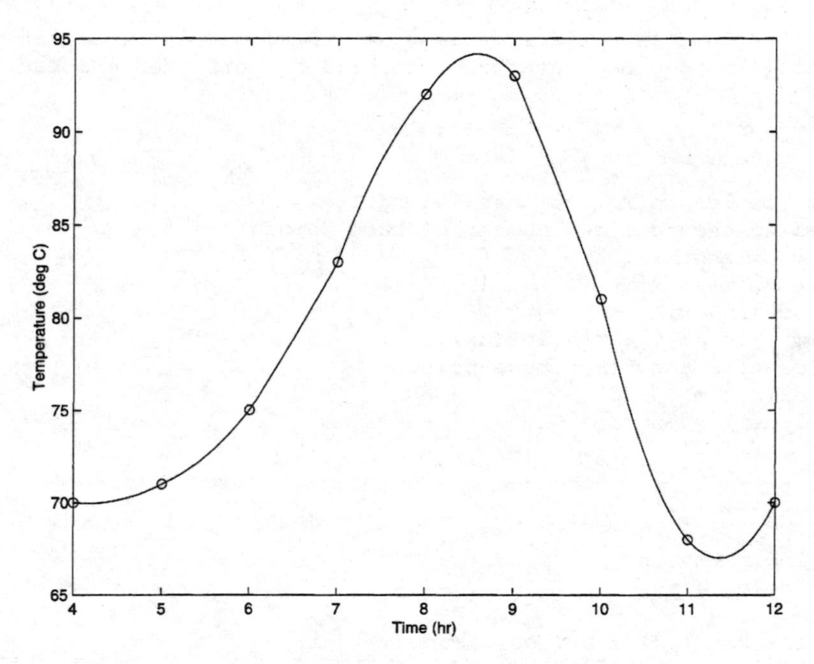

Figure E3.1 Interpolation of equally spaced points.

3.8.2 Stirling's Interpolation

Stirling's interpolation formula is based on central differences. Its derivation is similar to that of the Gregory-Newton formulas and can be arrived at by using either the symbolic operator relations or the Taylor series expansion of the function. We will use the latter and expand the function $f(x + nh)$ in a Taylor series around x:

$$f(x + nh) = f(x) + \frac{nh}{1!}f'(x) + \frac{n^2h^2}{2!}f''(x) + \frac{n^3h^3}{3!}f'''(x) + \dots \tag{3.126}$$

We replace the derivatives of $f(x)$ with the differential operators to obtain

$$f(x + nh) = f(x) + \frac{nh}{1!}Df(x) + \frac{n^2h^2}{2!}D^2f(x) + \frac{n^3h^3}{3!}D^3f(x) + \dots \tag{3.127}$$

The odd-order differential operators in Eq. (3.127) are replaced by averaged central differences and the even-order differential operators by central differences, all taken from Table 3.3. Substituting these into Eq. (3.127) and regrouping of terms yield the formula

$$f(x + nh) = f(x) + n\mu\delta f(x) + \frac{n^2}{2!}\delta^2 f(x) + \frac{n(n^2 - 1)}{3!}\mu\delta^3 f(x) + \frac{n^2(n^2 - 1)}{4!}\delta^4 f(x) + \dots \tag{3.128}$$

By applying Eq. (3.121) into Eq. (3.128), we obtain the final form of *Stirling's interpolation formula*

$$\begin{aligned} f(x) = f(x_0) &+ \frac{(x - x_0)}{h}\mu\delta f(x_0) + \frac{(x - x_0)^2}{2!\,h^2}\delta^2 f(x_0) \\ &+ \frac{(x - x_{-1})(x - x_0)(x - x_1)}{3!\,h^3}\mu\delta^3 f(x_0) \\ &+ \frac{(x - x_{-1})(x - x_0)^2(x - x_1)}{4!\,h^4}\delta^4 f(x_0) + \dots \end{aligned} \tag{3.129}$$

The general formula for determining the higher-order terms containing odd differences in the above series is

$$\left[\frac{1}{k!\,h^k}\prod_{m=-(k-1)/2}^{(k-1)/2}(x - x_m)\right]\mu\delta^k f(x_0) \tag{3.130}$$

where $k = 1, 3, \dots$, and the formula for terms with even differences is

$$\left[\frac{(x - x_0)}{k!\,h^k}\prod_{m=-(k-2)/2}^{(k-2)/2}(x - x_m)\right]\delta^k f(x_0) \tag{3.131}$$

where $k = 2, 4, \dots$

Other forms of Stirling's interpolation formula exist, which make use of base points spaced at half intervals (i.e., at $h/2$). Our choice of using *averaged* central differences to replace the odd differential operators eliminated the need for having base points located at the midpoints. The central differences for Eq. (3.129) are tabulated in Table 3.6.

Table 3.6 Central difference table*

i	x_i	$f(x_i)$	$\mu\delta f(x_i)$	$\delta^2 f(x_i)$
-3	$x-3h$	$f(x-3h)$		
-2	$x-2h$	$f(x-2h)$	$\frac{1}{2}[f(x-h)-f(x-3h)]$	$f(x-h)-2f(x-2h)+f(x-3h)$
-1	$x-h$	$f(x-h)$	$\frac{1}{2}[f(x)-f(x-2h)]$	$f(x)-2f(x-h)+f(x-2h)$
0	x	$f(x)$	$\frac{1}{2}[f(x+h)-f(x-h)]$	$f(x+h)-2f(x)+f(x-h)$
1	$x+h$	$f(x+h)$	$\frac{1}{2}[f(x+2h)-f(x)]$	$f(x+2h)-2f(x+h)+f(x)$
2	$x+2h$	$f(x+2h)$	$\frac{1}{2}[f(x+3h)-f(x+h)]$	$f(x+3h)-2f(x+2h)+f(x+h)$
3	$x+3h$	$f(x+3h)$		

i	$\mu\delta^3 f(x_i)$	$\delta^4 f(x_i)$
-2		
-1	$\frac{1}{2}[f(x+h)-2f(x)+2f(x-2h)-f(x-3h)]$	$f(x+h)-4f(x)+6f(x-h)-4f(x-2h)+f(x-3h)$
0	$\frac{1}{2}[f(x+2h)-2f(x+h)+2f(x-h)-f(x-2h)]$	$f(x+2h)-4f(x+h)+6f(x)-4f(x-h)+f(x-2h)$
1	$\frac{1}{2}[f(x+3h)-2f(x+2h)+2f(x)-f(x-h)]$	$f(x+3h)-4f(x+2h)+6f(x+h)-4f(x)+f(x-h)$
2		

i	$\mu\delta^5 f(x_i)$	$\delta^6 f(x_i)$
-1		
0	$\frac{1}{2}[f(x+3h)-4f(x+2h)+5f(x+h)-5f(x-h)$ $+4f(x-2h)-f(x-3h)]$	$f(x+3h)-6f(x+2h)+15f(x+h)-20f(x)+15f(x-h)$ $-6f(x-2h)+f(x-3h)$
1		

* Read this table from left to right, starting with top section and continuing with middle and bottom sections.

3.9 INTERPOLATION OF UNEQUALLY SPACED POINTS

In this section, we will develop two interpolation methods for unequally spaced data: the *Lagrange polynomials* and *spline interpolation.*

3.9.1 Lagrange Polynomials

Consider a set of unequally spaced base points, such as those shown in Fig. 3.3*a*. Define the polynomial

$$P_n(x) = \sum_{k=0}^{n} p_k(x) f(x_k) \tag{3.132}$$

which is the sum of the *weighted* values of the function at all $(n + 1)$ base points. The weights $p_k(x)$ are nth-degree polynomial functions corresponding to each base point. Eq. (3.132) is actually a linear combination of nth-degree polynomials; therefore, $P_n(x)$ is also an nth-degree polynomial.

In order for the interpolating polynomial to fit the function exactly at all the base points, each particular weighting polynomial $p_k(x)$ must be chosen so that it has the value of unity when $x = x_k$, and the value of zero at all other base points, that is,

$$p_k(x) = \begin{cases} 0 & i \neq k \\ 1 & i = k \end{cases} \tag{3.133}$$

The *Lagrange polynomials*, which have the form

$$p_k(x) = C_k \prod_{\substack{i=0 \\ i \neq k}}^{n} (x - x_i) \tag{3.134}$$

satisfy the first part of condition (3.133), because there will be a term $(x_i - x_i)$ in the product series of Eq. (3.134) whenever $x = x_i$. The constant C_k is evaluated to make the Lagrange polynomial satisfy the second part of condition (3.133):

$$C_k = \frac{1}{\prod\limits_{\substack{i=0 \\ i \neq k}}^{n} (x_k - x_i)} \tag{3.135}$$

Combination of Eqs. (3.134) and (3.135) gives the Lagrange polynomials

$$p_k(x) = \prod_{\substack{i=0 \\ i \neq k}}^{n} \left(\frac{x - x_i}{x_k - x_i} \right) \tag{3.136}$$

The interpolating polynomial $P_n(x)$ has a remainder term, which can be obtained from Eq. (3.6):

$$R_n(x) = \prod_{i=0}^{n} (x - x_i) \frac{f^{(n+1)}(\xi)}{(n + 1)!} \qquad x_0 < \xi < x_n \tag{3.137}$$

3.9.2 Spline Interpolation

When we deal with a large number of data points, high-degree interpolating polynomials are likely to fluctuate between base points instead of passing smoothly through them. This situation is illustrated in Fig. 3.5*a*. Although the interpolating polynomial passes through all the base points, it is not able to predict the value of the function satisfactorily in between these points. In order to avoid such an undesired behavior of the high-degree interpolating polynomial, a series of lower-degree interpolating polynomials may be used to connect smaller number of base points. These sets of interpolating polynomials are called *spline functions.* Fig. 3.5*b* shows the result of such interpolation using third-degree (or cubic) splines. Compared with the higher-order interpolation illustrated in Fig. 3.5*a*, third-degree splines shown in Fig. 3.5*b* provide a much more acceptable approximation.

(a)

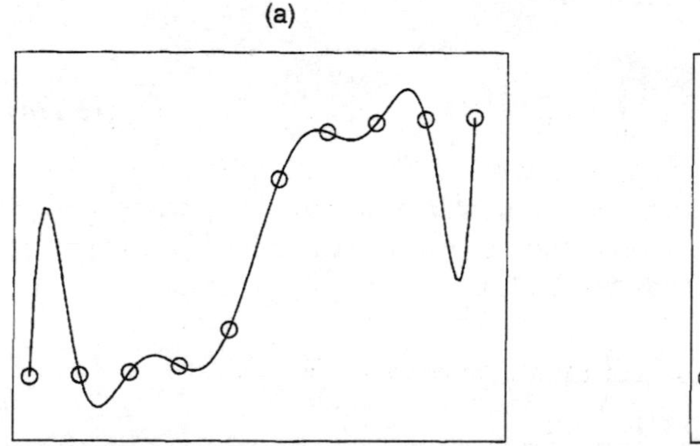

(b)

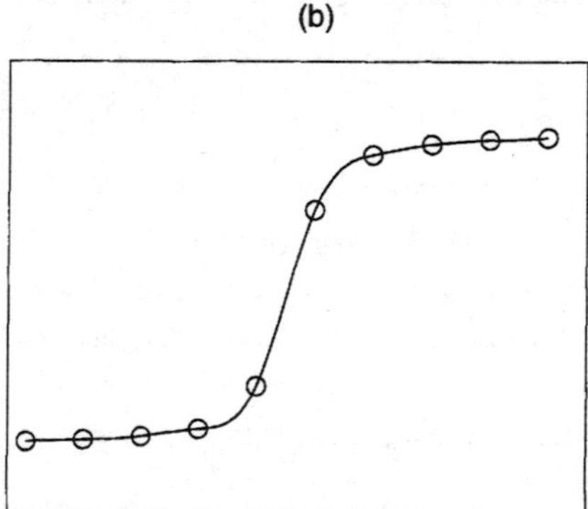

Figure 3.5 (*a*) Fluctuation of high-degree interpolating polynomials between base points. (*b*) Cubic spline interpolation.

The most common spline used in engineering problems is the *cubic spline*. In this method, a cubic polynomial is used to approximate the curve between each two adjacent base points. Because there would be an infinite number of third-degree polynomials passing through each pair of points, additional constraints are necessary to make the spline unique. Therefore, it is set that all the polynomials should have equal first and second derivatives at the base points. These conditions imply that the slope and the curvature of the spline polynomials are continuous across the base points.

The cubic spline of the interval $[x_{i-1}, x_i]$ has the following general form

$$P_i(x) = a_i x^3 + b_i x^2 + c_i x + d_i \tag{3.138}$$

There are four unknown coefficients in Eq. (3.138) and n such polynomials for the whole range of data points $[x_0, x_n]$. Therefore, there are $4n$ unknown coefficients and we need $4n$ equations to evaluate these coefficients. The required equations come from the following conditions:

a. Each spline passes from the base points of the edge of its interval ($2n$ equations).

b. The first derivative of the splines are continuous across the interior base points (n - 1 equations).

c. The second derivative of the splines are continuous across the interior base points (n - 1 equations).

d. The second derivative of the end splines are zero at the end base points (2 equations). This is called the *natural* condition. Another commonly used condition is to set the third derivative of the end splines equal to the third derivative of the neighboring splines. The latter is called *not-a-knot* condition.

Simultaneous solution of the above $4n$ linear algebraic equations results in the determination of all cubic interpolating polynomials. However, for programming purposes, there is an alternative method of determination of the coefficients that needs simultaneous solution of only (n - 1) algebraic equations. This method is described in detail in this section.

The second derivative of the Eq. (3.138) is a line

$$P_i''(x) = y'' = 6a_i x + 2b_i \tag{3.139}$$

From Eq. (3.139) it can be concluded that the second derivative of the interpolating polynomial at any point in the interval $[x_{i-1}, x_i]$ can be given by the first-order Lagrange interpolation formula

$$y'' = \frac{x - x_i}{x_{i-1} - x_i} y_{i-1}'' + \frac{x - x_{i-1}}{x_i - x_{i-1}} y_i'' \tag{3.140}$$

The expression for the spline can be obtained by twice integrating Eq. (3.140):

$$y = \frac{(x - x_i)^3}{6(x_{i-1} - x_i)} y_{i-1}'' + \frac{(x - x_{i-1})^3}{6(x_i - x_{i-1})} y_i'' + C_1 x + C_2 \tag{3.141}$$

where the constants C_1 and C_2 in Eq. (3.141) are evaluated from the following boundary conditions:

$$y(x_{i-1}) = y_{i-1}$$
$$y(x_i) = y_i \tag{3.142}$$

By evaluating the constants C_1 and C_2 from the conditions (3.142), substituting them into Eq. (3.141), and further rearrangement, we find the following cubic equation:

$$P_i(x) = y = \frac{1}{6}\left[\frac{(x - x_i)^3}{x_{i-1} - x_i} - (x_{i-1} - x_i)(x - x_i)\right]y''_{i-1}$$
$$+ \frac{1}{6}\left[\frac{(x - x_{i-1})^3}{x_i - x_{i-1}} - (x_i - x_{i-1})(x - x_{i-1})\right]y''_i$$
$$+ \left(\frac{x - x_i}{x_{i-1} - x_i}\right)y_{i-1} + \left(\frac{x - x_{i-1}}{x_i - x_{i-1}}\right)y_i \tag{3.143}$$

Note that Eqs. (3.138) and (3.143) are equivalent and the relations between their coefficients are given by Eq. (3.144):

$$a_i \equiv \frac{1}{6}\left(\frac{y''_{i-1} - y''_i}{x_{i-1} - x_i}\right) \qquad b_i \equiv \frac{1}{2}\left(\frac{x_{i-1}y''_i - x_i y''_{i-1}}{x_{i-1} - x_i}\right)$$

$$c_i \equiv \frac{1}{2}\left(\frac{x_i^2 y''_{i-1} - x_{i-1}^2 y''_i}{x_{i-1} - x_i}\right) + \frac{1}{6}(x_{i-1} - x_i)(y''_i - y''_{i-1}) + \frac{y_{i-1} - y_i}{x_{i-1} - x_i}$$

$$d_i \equiv \frac{1}{6}\left(\frac{x_{i-1}^3 y''_i - x_i^3 y''_{i-1}}{x_{i-1} - x_i}\right) + \frac{1}{6}(x_{i-1} - x_i)(x_i y''_{i-1} - x_{i-1}y''_i)$$
$$+ \frac{x_{i-1}y_i - x_i y_{i-1}}{x_{i-1} - x_i} \tag{3.144}$$

Although Eq. (3.143) is a more complicated expression than Eq. (3.138), it contains only two unknowns, namely y''_{i-1} and y''_i. In order to determine the y″ values, we apply the condition of continuity of the first derivative of splines at the interior base points; that is,

$$y'_{i-1} = y'_i \tag{3.145}$$

Differentiating Eq. (3.143) and applying the resulting expression in the condition (3.145), followed by rearranging of the terms, results in

$$(x_i - x_{i-1})y''_{i-1} + 2(x_{i+1} - x_{i-1})y''_i + (x_{i+1} - x_i)y''_{i+1} = 6\frac{y_{i+1} - y_i}{x_{i+1} - x_i} - 6\frac{y_i - y_{i-1}}{x_i - x_{i-1}} \tag{3.146}$$

where $i = 1, 2, \ldots, n - 1$ and $y''_0 = y''_n = 0$ (natural spline).

Eq. (3.146) represents an $(n - 1)$-order tridiagonal set of simultaneous equations, which in matrix form becomes Eq. (3.147):

$$\begin{bmatrix} 2(x_2-x_0) & (x_2-x_1) & 0 & 0 & \cdots & 0 & 0 \\ (x_2-x_1) & 2(x_3-x_1) & (x_3-x_2) & 0 & \cdots & 0 & 0 \\ 0 & (x_3-x_2) & 2(x_4-x_2) & (x_4-x_3) & \cdots & 0 & 0 \\ \cdots & \cdots & \cdots & \cdots & \cdots & \cdots & \cdots \\ 0 & 0 & \cdots & 0 & (x_{n-2}-x_{n-3}) & 2(x_{n-1}-x_{n-3}) & (x_{n-1}-x_{n-2}) \\ 0 & 0 & \cdots & 0 & 0 & (x_{n-1}-x_{n-2}) & 2(x_n-x_{n-2}) \end{bmatrix}$$

$$\times \begin{bmatrix} y''_1 \\ y''_2 \\ y''_3 \\ \cdot \\ \cdot \\ y''_{n-2} \\ y''_{n-1} \end{bmatrix} = 6 \begin{bmatrix} \dfrac{y_2 - y_1}{x_2 - x_1} - \dfrac{y_1 - y_0}{x_1 - x_0} \\ \dfrac{y_3 - y_2}{x_3 - x_2} - \dfrac{y_2 - y_1}{x_2 - x_1} \\ \dfrac{y_4 - y_3}{x_4 - x_3} - \dfrac{y_3 - y_2}{x_3 - x_2} \\ \cdot \\ \dfrac{y_{n-1} - y_{n-2}}{x_{n-1} - x_{n-2}} - \dfrac{y_{n-2} - y_{n-3}}{x_{n-2} - x_{n-3}} \\ \dfrac{y_n - y_{n-1}}{x_n - x_{n-1}} - \dfrac{y_{n-1} - y_{n-2}}{x_{n-1} - x_{n-2}} \end{bmatrix} \tag{3.147}$$

After calculating the values of the second derivatives at each base point, Eq. (3.143) can be used for interpolating the value of the function in every interval.

Example 3.2: The Lagrange Polynomials and Cubic Splines for Interpolation of Unequally Spaced Data. The pressure drop of a basket-type filter is measured at different flow rates as shown in Table E3.2. Write a program to estimate pressure drop of the filter at any flow rate within the experimental range. This program should call general MATLAB functions for interpolating unequally spaced data using Lagrange polynomials and cubic splines.

Method of Solution: The Lagrange interpolation is done based on Eqs. (3.132) and (3.136). The order of interpolation is an input to the function. The cubic spline interpolation is done based on Eq. (3.143). The values of the second derivatives at base points, assuming a natural spline, are calculated from Eq. (3.147).

Program Description: The general MATLAB function *Lagrange.m* performs the nth-order Lagrange interpolation. This function consists of the following three parts:

At the beginning, it checks the inputs and sets the order of interpolation if necessary. If not introduced to the function, the interpolation is done by the first-order Lagrange polynomial (linear interpolation).

In the second part of the function, locations of all the points at which the values of the function are to be evaluated are found in between the base points. Because matrix operations are much faster than element-by-element operations in MATLAB, the required number of independent and dependent variables are arranged in two interim matrices at each location. These matrices are used at the interpolation section for doing the interpolation in vector form.

The last part of the function is interpolation itself. In this section, $p_k(x)$ subpolynomials are calculated according to Eq. (3.136). The terms of summation (3.132) are then calculated, and, finally, the function value is determined based on Eq. (3.132). In order to be time efficient, all these calculations are done in vector form and at all the required points simultaneously.

The MATLAB function *NaturalSPLINE.m* also consists of three parts. The first and second parts are more or less similar to those of *Lagrange.m*. However, instead of forming the interim matrices, the interpolation locations are kept in a vector.

Table E3.2 Pressure drop of a basket-type filter

Flow Rate (L/s)	Pressure drop (kPa)	Flow rate (L/s)	Pressure drop (kPa)
0.00	0.000	32.56	1.781
10.80	0.299	36.76	2.432
16.03	0.576	39.88	2.846
22.91	1.036	43.68	3.304

In the last section of the function, the matrix of coefficients and the vector of constants are built according to Eq. (3.147), and the values of the second derivatives at the base points are evaluated. The interpolation is then performed, in the vector form, based on Eq. (3.143).

Program

Example3_2.m

```
% Example3_2.m
% Solution to Example 3.2. It uses Lagrange and cubic spline
% interpolations to find the pressure drop of a filter at any
% point in between the experimental data.

clc
clear
clf

% Input data
Q = input(' Vector of flow rates = ');
dP = input (' Vector of pressure drops = ');
disp(' ')
n = input(' Order of the Lagrange interpolation = ');

q = linspace(min(Q) , max(Q));
% Interpolation
dP1 = Lagrange(Q , dP , q , n);
dP2 = NaturalSPLINE(Q , dP , q);

% Plotting the results
plot(Q,dP,'o',q,dP1,q,dP2,'.')
xlabel('Flow Rate (lit./s)')
ylabel('Pressure Drop (kPa)')
legend('Experimental Data','Lagrange Interpolation','Natural Spline
Interpolation',2)
```

Lagrange.m

```
function yi = Lagrange(x,y,xi,n)
%Lagrange One dimensional interpolation.
%
%   YI = Lagrange(X,Y,XI,N) applies the Nth-order Lagrange
%   interpolation to find YI, the values of the underlying
%   function Y at the points in the vector XI.  The vector
%   X specifies the points at which the data Y is given.
%
%   YI = Lagrange(X,Y,XI) is equivalent to the linear
%   interpolation.
%
%   See also NATURALSPLINE, GregoryNewton, SPLINE, INTERP1, INTERP1Q
```

```
% (c) by N. Mostoufi & A. Constantinides
% January 1, 1999

% Initialization
if nargin < 3
   error('Invalid number of inputs.')
end

x = (x(:).')';      % Make sure it's a column vector
y = (y(:).')';      % Make sure it's a column vector
nx = length(x);
ny = length(y);
if nx ~= ny
   error('X and Y vectors are not the same size.');
end

% Check the order of interpolation
if nargin == 3 | n < 1
   n = 1;
end
n = floor(n);
if n >= nx
 fprintf ('\nNot enough data points for %2d-oredr interpolation.',n)
 fprintf('\n%2d-order interpolation will be performed instead.\n',
nx-1)
   n = nx - 1;
end

lxi = length(xi);
deltax(1,:) = ones(1,lxi);
% Locating the required number of base points
for m = 1:lxi
   dx = xi(m) - x;
   % Locating xi
   [dxm , loc] = min(abs(dx));
   % locating the first base point
   if dx(loc) < 0
      loc = loc - 1;
   end
   if loc < 1
      loc = 1;
   end
   if loc+n > nx
      loc = nx - n;
   end
   deltax(2:n+2,m) = dx(loc:loc+n);
   xtemp(1:n+1,m) = x(loc:loc+n);
   ytemp(1:n+1,m) = y(loc:loc+n);
```

```
end
% Interpolation
for k = 1 : n+1
   for m = 1 : n+1
      if k ~= m
         den(m,:) = xtemp(k,:) - xtemp(m,:);
      else
         den(m,:) = ones(1,lxi);
      end
   end
   p(k,:) = prod([deltax(1:k,:) ; deltax(k+2:n+2,:)]) ./ prod(den);
   s(k,:) = p(k,:) .* ytemp(k,:);
end
yi = sum(s);
```

NaturalSPLINE.m

```
function yi = NaturalSPLINE(x,y,xi)
%NATURALSPLINE One dimensional interpolation.
%
%   YI = NATURALSPLINE(X,Y,XI) applies the natural spline
%   interpolation to find YI, the values of the underlying
%   function Y at the points in the vector XI.  The vector
%   X specifies the points at which the data Y is given.
%
%   See also Lagrange, GregoryNewton, INTERP1, INTERPQ, SPLINE

% (c) by N. Mostoufi & A. Constantinides
% January 1, 1999

% Initialization
if nargin < 3
   error('Invalid number of inputs.')
end

x = (x(:).')';     % Make sure it's a column vector
y = (y(:).')';     % Make sure it's a column vector
xi = (xi(:).')';   % Make sure it's a column vector
nx = length(x);
ny = length(y);
if nx ~= ny
   error('X and Y vectors are not the same size.');
end

lxi = length(xi);
% Locating the required number of base points
for m = 1:lxi
   d = xi(m) - x;
```

```
   % Locating xi
   [dm , loc(m)] = min(abs(d));
   % locating the first base point
   if d(loc(m)) < 0 | loc(m) == nx
      loc(m) = loc(m) - 1;
   end
   if loc(m) < 1
      loc(m) = 1;
   end
end

dx = diff(x);
dy = diff(y);
yox = dy ./ dx;
% Matrix of coefficients
A = 2 * diag(x(3:nx)-x(1:nx-2)) + ...
   [zeros(nx-2,1) [diag(dx(2:nx-2)) ; zeros(1,nx-3)]] + ...
   [zeros(1,nx-2) ; [diag(dx(2:nx-2)) zeros(nx-3,1)]];
% Vector of constants
c = 6 * (yox(2:nx-1) - yox(1:nx-2));
% Solution of the set of linear equations
y2 = [0; inv(A) * c; 0];% Interpolation
yi = (1/6) * ((xi - x(loc+1)).^3 ./ (x(loc) - x(loc+1)) ...
   - (x(loc) - x(loc+1)) .* (xi - x(loc+1))) .* y2(loc) ...
   + (1/6) * ((xi - x(loc)).^3 ./ (x(loc+1) - x(loc)) ...
   - (x(loc+1) - x(loc)) .* (xi - x(loc))) .* y2(loc+1) ...
   + (xi - x(loc+1)) ./ (x(loc) - x(loc+1)) .* y(loc) ...
   + (xi - x(loc)) ./ (x(loc+1) - x(loc)) .* y(loc+1);
```

Input and Results

```
>>Example3_2

 Vector of flow rates = [0, 10.80, 16.03, 22.91, 28.24, 32.56,
36.76, 39.88, 43.68]
 Vector of pressure drops = [0, 0.299, 0.576, 1.036, 1.383, 1.781,
2.432, 2.846, 3.304]

 Order of the Lagrange interpolation = 3
```

Discussion of Results: Order of the Lagrange interpolation is chosen to be three for comparison of the results with that of the cubic spline, which is also third-order interpolation. Fig. E3.2 shows the results of calculations. There is no essential difference between the two methods. The cubic spline, however, passes smoothly through the base points, as expected. Because the Lagrange interpolation is performed in the subsets of four base points with no restriction related to their neighboring base points, it can be seen that the slope of the resulting curve is not continuous through most of the base points.

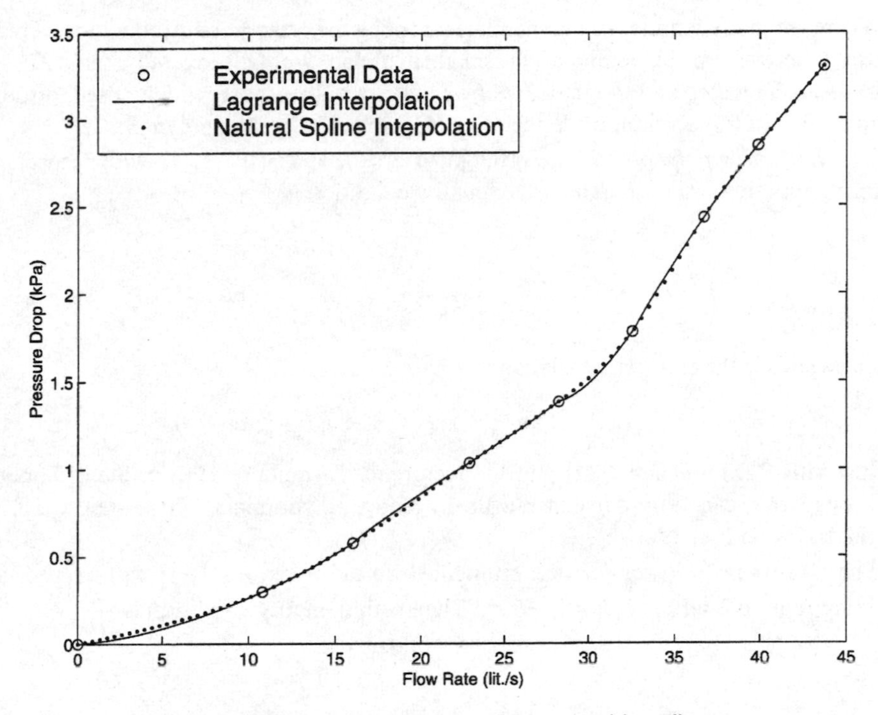

Figure E3.2 Lagrange polynomials and cubic splines.

3.10 ORTHOGONAL POLYNOMIALS

Orthogonal polynomials are a special category of functions that satisfy the following orthogonality condition with respect to a weighting function $w(x) \geq 0$, on the interval $[a, b]$:

$$\int_a^b w(x)\, g_n(x)\, g_m(x)\, dx = \begin{cases} 0 & \text{if } n \neq m \\ c(n) > 0 & \text{if } n = m \end{cases} \tag{3.148}$$

This orthogonality condition can be viewed as the continuous analog of the orthogonality property of two vectors (see Chap. 2)

$$\mathbf{x}'\mathbf{y} = 0 \tag{2.91}$$

in n-dimensional space, where n becomes very large, and the elements of the vectors are represented as continuous functions of some independent variable.

There are many families of polynomials that obey the orthogonality condition. These are generally known by the name of the mathematician who discovered them: *Legendre, Chebyshev, Hermite,* and *Laguerre polynomials* are the most widely used orthogonal polynomials. In this section, we list the Legendre and Chebyshev polynomials.

The *Legendre polynomials* are orthogonal on the interval [-1, 1] with respect to the weighting function $w(x) = 1$. The orthogonality condition is

$$\int_{-1}^{1} P_n(x)\,P_m(x)\,dx = \begin{cases} 0 & \text{if } n \neq m \\ \dfrac{2}{2n+1} & \text{if } n = m \end{cases} \tag{3.149}$$

They also satisfy the recurrence relation

$$(n + 1)P_{n+1}(x) - (2n + 1)xP_n(x) + nP_{n-1}(x) = 0 \tag{3.150}$$

Starting with $P_0(x) = 1$ and $P_1(x) = x$, the recurrence formula (3.150) or the orthogonality condition (3.149) can be used to generate the Legendre polynomials. These are listed in Table 3.7 and drawn on Fig. 3.6.

The *Chebyshev polynomials* are orthogonal on the interval [-1, 1] with respect to the weighting function $w(x) = 1/\sqrt{1 - x^2}$. Their orthogonality condition is

$$\int_{-1}^{1} \frac{1}{\sqrt{1-x^2}}\,T_n(x)\,T_m(x)\,dx = \begin{cases} 0 & \text{if } n \neq m \\ \pi & \text{if } n = m = 0 \\ \dfrac{\pi}{2} & \text{if } n = m > 0 \end{cases} \tag{3.151}$$

and their recurrence relation is

$$T_{n+1} - 2xT_n + T_{n-1} = 0 \tag{3.152}$$

Starting with $T_0(x) = 1$ and $T_1(x) = x$, the recurrence formula (3.152) or orthogonality condition (3.151) can be used to generate the Chebyshev polynomials listed in Table 3.8 and drawn on Fig. 3.7.

It should be noticed from Figs. 3.6 and 3.7 that these orthogonal polynomials have their zeros (roots) more closely packed near the ends of the interval of integration. This property can be used to advantage in order to improve the accuracy of interpolation of unequally spaced points. This can be done in the case where the choice of base points is completely free. The interpolation can be performed using Lagrange interpolation method described in Sec. 3.9.1, but the base points are chosen at the roots of the appropriate orthogonal polynomial. This concept is demonstrated in Chap. 4 in connection with the development of *Gauss quadrature*.

Table 3.7 Legendre polynomials

n	$P_n(x)$
0	$P_0(x) = 1$
1	$P_1(x) = x$
2	$P_2(x) = \dfrac{3x^2 - 1}{2}$
3	$P_3(x) = \dfrac{5x^3 - 3x}{2}$
4	$P_4(x) = \dfrac{35x^4 - 30x^2 + 3}{8}$
⋮	⋮
n	$P_n(x) = \sum_{m=0}^{[n/2]^*} (-1)^m \dfrac{(2n - 2m)!}{2^n m!(n - m)!(n - 2m)!} x^{n-2m}$

* The notation [n/2] represents the integer part of n/2.

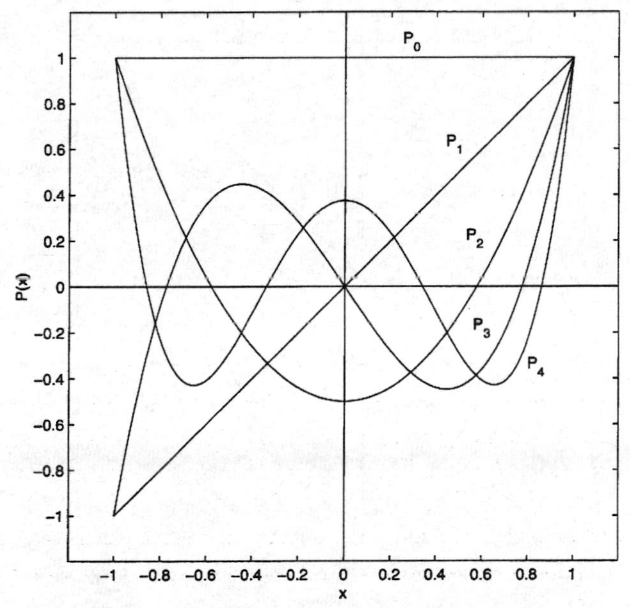

Figure 3.6 The Legendre orthogonal polynomials.

Table 3.8 Chebyshev polynomials

n	$T_n(x)$
0	$T_0(x) = 1$
1	$T_1(x) = x$
2	$T_2(x) = 2x^2 - 1$
3	$T_3(x) = 4x^3 - 3x$
4	$T_4(x) = 8x^4 - 8x^2 + 1$
⋮	⋮
n	$T_n(x) = \sum_{m=0}^{[n/2]^*} \frac{n!\,x^{n-2m}(x^2-1)^m}{(2m)!(n-2m)!}$

* The notation [n/2] represents the integer part of n/2.

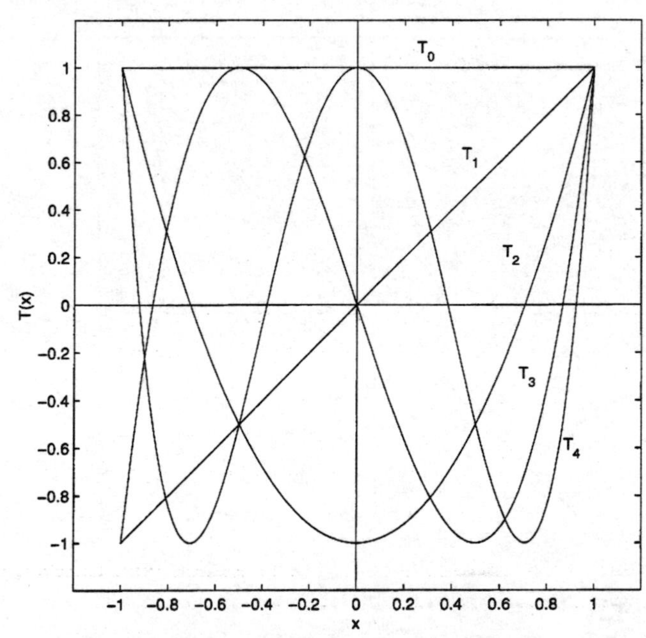

Figure 3.7 The Chebyshev orthogonal polynomials.

PROBLEMS

3.1 Show that all the interpolating formulas discussed here reduce to the same formula when a first-order interpolation is used.

3.2 Derive the Gregory-Newton backward interpolation formula.

3.3 Using the experimental data in Table P3.3:
(a) Develop the forward difference table.
(b) Develop the backward difference table.
(c) Apply the Gregory-Newton interpolation formulas to evaluate the function at $x = 10, 50, 90, 130, 170,$ and 190.

Table P3.3 Data of penicillin fermentation

Time (h)	Penicillin concentration (units/mL)	Time (h)	Penicillin concentration (units/mL)
0	0	120	9430
20	106	140	10950
40	1600	160	10280
60	3000	180	9620
80	5810	200	9400
100	8600		

3.4 Write a MATLAB function that uses the Gregory-Newton backward interpolation formula to evaluate the function $f(x)$ from a set of $(n + 1)$ equally spaced input values. Write the function in a general fashion so that n can be any positive integer. Also write a MATLAB script that reads the data and shows how this MATLAB function fits the data. Use the experimental data of Table 3.3 to verify the program, and evaluate the function at $x = 10, 50, 90, 130, 170,$ and 190.

3.5 Using the experimental data of Prob. 3.3,
(a) Develop the central difference table.
(b) Apply Stirling's interpolation formula to evaluate the function at $x = 10, 50, 90, 130, 170,$ and 190.

3.6 Write a MATLAB function which uses the Stirling's interpolation formula to evaluate the function $f(x)$ from a set of $(n + 1)$ equally spaced input values. Write the function in a general fashion so that n can be any positive integer. Also write a MATLAB script that reads the data and shows how this

MATLAB function fits the data. Use the experimental data of Table 3.3 to verify the program, and evaluate the function at $x = 10, 50, 90, 130, 170$, and 190.

3.7 With the set of unequally spaced data points in Table P3.7 use Lagrange polynomials and spline interpolation to evaluate the function at $x = 2, 4, 5, 8, 9$, and 11.

Table P3.7

x	$f(x)$	x	$f(x)$
1	7.0	10	8.2
3	3.5	12	9.0
6	3.2	13	9.2
7	3.9		

3.8 Vapor pressure of lithium chloride is given in Table P3.8 [4]. Use these data to present the vapor pressure of lithium chloride in the following tables:
(a) From 800°C to 1350°C at 50°C increment.
(b) From 10 kPa to 100 kPa at 10 kPa increment.

3.9 The zeta-potential of particles in a suspension is an indication of the sign and the density of the surface charge of the particles. The iso-electric point (i.e.p.) refers to the pH where zeta-potential is zero. Use data from Rashchi et al. [5] [Table P3.9] to determine the iso-electric points of silica in the presence of 10^{-4}M $Pb(NO_3)_2$.

3.10 Obtain the solution of the difference equation (3.97) directly from the solution of the differential equation (3.96) by utilizing the relationship $E = e^{hD}$.

Table P3.8 Vapor pressure of lithium chloride

Pressure (mm Hg)	Temperature (°C)	Pressure (mm Hg)	Temperature (°C)
1	783	60	1081
5	883	100	1129
10	932	200	1203
20	987	400	1290
40	1045	760	1382

Table P3.9 Zeta-potential of silica in the presence of 10^{-4}M $Pb(NO_3)_2$ as a function of pH

pH	zeta-potential (mV)	pH	zeta-potential (mV)
1.74	-5.3	6.00	-33.2
2.72	-10.8	6.53	-15.7
3.72	-21.8	6.70	-10.0
4.09	-32.0	7.29	13.7
4.32	-35.8	8.06	32.2
4.70	-36.9	10.02	24.0
5.00	-36.7	11.12	6.9
5.55	-37.7	12.15	-30.0

REFERENCES

1. Salvadori, M. G., and Baron, M. L., *Numerical Methods in Engineering*, Prentice Hall, Englewood Cliffs, NJ, 1961.

2. Chorlton, F., *Ordinary Differential and Difference Equations*, Van Nostrand, London, U.K., 1965.

3. Gel'fond, A. O., *Calculus of Finite Differences*, English trans. of the third Russian edition, Hindustan Publishing Corp., Delhi, India, 1971.

4. Green, D. W., and Maloney, J. O., *Perry's Chemical Engineers' Handbook*, 7th ed., McGraw-Hill, New York, 1997.

5. Rashchi, F., Xu, Z., and Finch, J. A., "Adsorption of Silica in Pb- and Ca-SO_4-CO_3 Systems," *Colloids and Surfaces A: Physicochemical and Engineering Aspects*, vol. 132, 1998, p. 159.

CHAPTER 4

Numerical Differentiation and Integration

4.1 INTRODUCTION

The solution of many engineering problems requires calculation of the derivative of a function at a known point or integration of the derivative over a known range of the independent variable. The simplest example of such problems is root-finding by the Newton-Raphson method, which needs calculation of the derivative of the function in each iteration (see Sec. 1.6). Although, in some cases, the analytical derivative of the function may be derived, it is more convenient to obtain it numerically if the function is complicated and/or the calculation is done by a computer program. In the following examples, there is no algebraic expression for the experimental data, or analytical integration does not exist for the function, therefore, numerical differentiation or integration is inevitable.

In chemical reaction kinetics, one of the methods for determination of the order of a chemical reaction is the method of initial rates. In this method, the reaction starts with different initial concentrations of the reactant A, and changes in the concentration of A with time are measured. For each initial concentration, the initial reaction rate can be calculated from differentiation of concentration with respect to time at the beginning of the reaction:

$$-r_{A_0} = -\frac{dC_A}{dt}\Big|_{t=0} \tag{4.1}$$

If the reaction rate could be expressed by

$$-r_A = kC_A^n \tag{4.2}$$

then taking the logarithm of both sides of this equation at $t = 0$ results in

$$\ln(-r_{A_0}) = \ln k + n \ln C_{A_0} \tag{4.3}$$

The reaction order can be obtained by calculation of the slope of the line $\ln(-r_{A_0})$ versus $\ln(C_{A_0})$.

Experimental determination of the rate of drying of a given material can be done by placing the moist material in a tray that is exposed to the drying air stream. A balance indicates the weight of the moist material, which is being recorded at different time intervals, during drying. The drying rate is calculated for each point by

$$R = -\frac{1}{A}\frac{dW}{dt} \tag{4.4}$$

where R is the drying rate, A is the exposed surface area for drying, W is the mass of the moist material, and t is time.

In the study of hydrodynamics of multiphase reactors, the velocity profiles of solids may be determined experimentally by Radioactive Particle Tracking (RPT) velocimetry technique [1]. In this technique, a radioactive tracer is being followed for several time intervals, and coordinates of this tracer are evaluated at each time interval. Instantaneous velocity of the tracer can be calculated then from

$$V_i = \frac{dx_i}{dt} \tag{4.5}$$

where V_i is the velocity of the tracer in direction i, x_i is the ith component of the coordinate of the tracer, and dt is the time increment used at the time of data acquisition. The steady-state velocity profile of solid particles in the reactor is calculated by averaging the instantaneous velocities in small compartments inside the reactor. Once the velocity profile is determined, solids velocity fluctuation is calculated by

$$V_i' = V_i - \langle V_i \rangle \tag{4.6}$$

where V_i' is velocity fluctuation (a function of time) and $<V_i>$ is the average velocity (a function of position), both in i-direction. Having the above information, turbulent eddy diffusivity of solids (D_i) may be obtained from the Lagrangian autocorrelation integral of velocity fluctuations:

$$D_i(t) = \int_0^t <V_i'(t)\, V_i'(\tau)> d\tau \tag{4.7}$$

The height of a cooling tower is calculated from the following equation:

$$z = \frac{G}{MK_G aP} \int_{H_1}^{H_2} \frac{dH}{H^* - H} \tag{4.8}$$

where z is height of tower, G is dry air mass flow, M is molecular weight of air, $K_G a$ is overall mass transfer coefficient, P is pressure, H is enthalpy of moist air, and H^* is enthalpy of moist air at saturation. The integral in Eq. (4.8) should be calculated from H_1 at inlet of the tower to H_2 at its outlet. In order to calculate this integral, enthalpies between H_1 and H_2 may be read from the psychrometric chart.

Calculation of the volume of a nonisothermal chemical reactor usually needs the use of numerical integration. For example, consider the first order reaction $A \rightarrow B$ in liquid phase, taking place in an adiabatic plug flow reactor. Pure A enters the reactor, and it is desired to have the conversion X_1 at the outlet. The volume of this reactor is given by

$$V = \frac{v_0}{k_1} \int_0^{X_1} \frac{dX}{(1 - X) \exp\left[\frac{E_a}{R}\left(\frac{1}{T_1} - \frac{1}{T}\right)\right]} \tag{4.9}$$

where V is the volume of the reactor, v_0 is the inlet volumetric flow rate of A, k_1 is the rate constant at the temperature T_1, E_a is the activation energy of the reaction, R is the ideal gas constant, T is the temperature of the reactor where the conversion is X, and T_1 is a reference temperature.

We must relate X and T through the energy balance to carry out this integration. For an adiabatic plug flow reactor, assuming constant heat capacities for both A and B, T is given by

$$T = T_0 + \frac{X(-\Delta H_R)}{C_{p_A} + X(C_{p_B} - C_{p_A})} \tag{4.10}$$

In this equation, ΔH_R is the heat of the reaction, C_{p_A} and C_{p_B} are heat capacities of A and B, respectively, and T_0 is a reference temperature.

In order to calculate the volume of the reactor from Eq. (4.9), one has to divide the interval [0, X] into small ΔXs first, and from Eq. (4.10) the temperature in each increment can be evaluated. Knowing both X and T, the function in the denominator of the integral in Eq. (4.9) is calculated. Finally, using a numerical technique for integration, the volume of the reactor can be calculated from Eq. (4.9).

In addition to calculating definite integrals, numerical integration can also be used to solve simple differential equations of the form

$$y' = \frac{dy}{dx} = f(x) \tag{4.11}$$

Solution to the differential equation (4.11), after rearrangement, is given as

$$y = y(x_0) + \int_{x_0}^{x} f(x)\,dx \tag{4.12}$$

In this chapter we deal with numerical differentiation in Secs. 4.2-4.5 and integration in Secs. 4.6-4.10.

4.2 Differentiation by Backward Finite Differences

The relationships between backward difference operators and differential operators, which are summarized in Table 3.1, enable us to develop a variety of formulas expressing derivatives of functions in terms of backward finite differences, and vice versa. In addition, these formulas may have any degree of accuracy desired, provided that a sufficient number of terms is retained in the manipulation of these infinite series. This concept will be demonstrated in the remainder of this section.

4.2.1 First-Order Derivative in Terms of Backward Finite Differences with Error of Order *h*

Rearrange Eq. (3.32) to solve for the differential operator D:

$$D = \frac{1}{h}\nabla + \frac{hD^2}{2} - \frac{h^2D^3}{6} + \dots \tag{4.13}$$

Apply this operator to the function y at i:

$$Dy_i = \frac{1}{h}\nabla y_i + \frac{hD^2y_i}{2} - \frac{h^2D^3y_i}{6} + \dots \tag{4.14}$$

Truncate the series, retaining only the first term, and show the order of the truncation error:

$$Dy_i = \frac{1}{h}\nabla y_i + O(h) \tag{4.15}$$

Express the differential and backward operators in terms of their respective definitions:

$$\frac{dy_i}{dx} = \frac{1}{h}(y_i - y_{i-1}) + O(h) \tag{4.16}$$

Eq. (4.16), therefore, enables us to evaluate the first-order derivative of y at position i in terms of backward finite differences.

The term $O(h)$ is used to represent the order of the first term in the truncated portion of the series. When $h < 1.0$ and the function is smooth and continuous, the first term in the truncated portion of the series is the predominant term. It should be emphasized that for $h < 1.0$:

$$h > h^2 > h^3 > h^4 > \ldots > h^n$$

Therefore, when $h < 1.0$, formulas with higher-order error term, $O(h^n)$, have smaller truncation errors, i.e., they are more accurate approximations of derivatives.

On the other hand, when $h > 1.0$:

$$h < h^2 < h^3 < h^4 < \ldots < h^n$$

Therefore, formulas with higher-order error terms have larger truncation errors and are less-accurate approximations of derivatives.

It is obvious then, that the choice of step size h is very important in determining the accuracy and stability of numerical integration and differentiation. This concept will be discussed in detail in Chaps. 5 and 6.

4.2.2 Second-Order Derivative in Terms of Backward Finite Differences with Error of Order *h*

Rearrange Eq. (3.36) to solve for D^2:

$$D^2 = \frac{1}{h^2}\nabla^2 + hD^3 - \frac{7}{12}h^2D^4 + \ldots \tag{4.17}$$

Apply this operator to the function y at i:

$$D^2y_i = \frac{1}{h^2}\nabla^2 y_i + hD^3y_i - \frac{7}{12}h^2D^4y_i + \ldots \tag{4.18}$$

Truncate the series, retaining only the first term, and express the operators in terms of their respective definition:

$$\frac{d^2y_i}{dx^2} = \frac{1}{h^2}(y_i - 2y_{i-1} + y_{i-2}) + O(h) \tag{4.19}$$

This equation evaluates the second-order derivative of y at position i, in terms of backward finite differences, with error of order h.

4.2.3 First-Order Derivative in Terms of Backward Finite Differences with Error of Order h^2

Rearrange Eq. (3.32) to solve for hD:

$$hD = \nabla + \frac{h^2D^2}{2} - \frac{h^3D^3}{6} + \dots \tag{4.20}$$

Rearrange Eq. (3.36) to solve for h^2D^2:

$$h^2D^2 = \nabla^2 + h^3D^3 - \frac{7}{12}h^4D^4 + \dots \tag{4.21}$$

Combine these two equations to eliminate h^2D^2:

$$hD = \nabla + \frac{1}{2}\left(\nabla^2 + h^3D^3 - \frac{7}{12}h^4D^4 + \dots\right) - \frac{h^3D^3}{6} + \dots$$

$$= \nabla + \frac{1}{2}\nabla^2 + \frac{h^3D^3}{3} - \dots \tag{4.22}$$

Divide through by h, and apply this operator to the function y at i:

$$Dy_i = \frac{1}{h}\nabla y_i + \frac{1}{2h}\nabla^2 y_i + \frac{h^2D^3y_i}{3} - \dots \tag{4.23}$$

Truncate the series, retaining only the first *two* terms, and express the operators in terms of their respective definitions:

$$\frac{dy_i}{dx} = \frac{1}{h}(y_i - y_{i-1}) + \frac{1}{2h}(y_i - 2y_{i-1} + y_{i-2}) + O(h^2)$$

$$= \frac{1}{2h}(3y_i - 4y_{i-1} + y_{i-2}) + O(h^2) \tag{4.24}$$

In this section, the first derivative of y is obtained with error of order h^2. For the case where $h < 1.0$, Eq. (4.24) is a more accurate approximation of the first derivative than Eq. (4.16). To obtain the higher accuracy, however, a larger number of terms is involved in the calculation.

4.2.4 Second-Order Derivative in Terms of Backward Finite Differences with Error of Order h^2

Rearrange Eq. (3.36) to solve for h^2D^2:

$$h^2D^2 = \nabla^2 + h^3D^3 - \frac{7}{12}h^4D^4 + \ldots \tag{4.25}$$

Rearrange Eq. (3.37) to solve for h^3D^3:

$$h^3D^3 = \nabla^3 + \frac{3}{2}h^4D^4 - \frac{5}{4}h^5D^5 + \ldots \tag{4.26}$$

Combine these two equations to eliminate h^3D^3:

$$h^2D^2 = \nabla^2 + \left(\nabla^3 + \frac{3}{2}h^4D^4 - \frac{5}{4}h^5D^5 + \ldots\right) - \frac{7}{12}h^4D^4 + \ldots$$

$$= \nabla^2 + \nabla^3 + \frac{11}{12}h^4D^4 - \ldots \tag{4.27}$$

Divide through by h^2 and apply the operator to the function y at i:

$$D^2y_i = \frac{1}{h^2}\nabla^2y_i + \frac{1}{h^2}\nabla^3y_i + \frac{11}{12}h^2D^4y_i - \ldots \tag{4.28}$$

Truncate the series, retaining only the first two terms, and express the operators in terms of their respective definitions:

$$\frac{d^2y_i}{dx^2} = \frac{1}{h^2}(y_i - 2y_{i-1} + y_{i-2}) + \frac{1}{h^2}(y_i - 3y_{i-1} + 3y_{i-2} - y_{i-3}) + O(h^2)$$

$$= \frac{1}{h^2}(2y_i - 5y_{i-1} + 4y_{i-2} - y_{i-3}) + O(h^2) \tag{4.29}$$

It should be noted that this same equation could have been derived using Eq. (3.42) and an equation for ∇^4 (not shown here). This statement applies to all these examples, which can be solved utilizing both sets of equations shown in Table 3.1.

The formulas for the first- and second-order derivatives, developed in the preceding four sections, together with those of the third- and fourth-order derivative, are summarized in Table 4.1. It can be concluded from these examples that any derivative can be expressed in terms of finite differences with any degree of accuracy desired. These formulas may be used to differentiate the function $y(x)$ given a set of values of this function at equally spaced intervals of x, such as a set of experiment data. Conversely, these same formulas may be used in the numerical integration of differential equations, as shown in Chaps. 5 and 6.

Table 4.1 Derivatives in terms of backward finite differences

Error of order h

$$\frac{dy_i}{dx} = \frac{1}{h}(y_i - y_{i-1}) + O(h)$$

$$\frac{d^2y_i}{dx^2} = \frac{1}{h^2}(y_i - 2y_{i-1} + y_{i-2}) + O(h)$$

$$\frac{d^3y_i}{dx^3} = \frac{1}{h^3}(y_i - 3y_{i-1} + 3y_{i-2} - y_{i-3}) + O(h)$$

$$\frac{d^4y_i}{dx^4} = \frac{1}{h^4}(y_i - 4y_{i-1} + 6y_{i-2} - 4y_{i-3} + y_{i-4}) + O(h)$$

Error of order h^2

$$\frac{dy_i}{dx} = \frac{1}{2h}(3y_i - 4y_{i-1} + y_{i-2}) + O(h^2)$$

$$\frac{d^2y_i}{dx^2} = \frac{1}{h^2}(2y_i - 5y_{i-1} + 4y_{i-2} - y_{i-3}) + O(h^2)$$

$$\frac{d^3y_i}{dx^3} = \frac{1}{2h^3}(5y_i - 18y_{i-1} + 24y_{i-2} - 14y_{i-3} + 3y_{i-4}) + O(h^2)$$

$$\frac{d^4y_i}{dx^4} = \frac{1}{h^4}(3y_i - 14y_{i-1} + 26y_{i-2} - 24y_{i-3} + 11y_{i-4} - 2y_{i-5}) + O(h^2)$$

4.3 Differentiation by Forward Finite Differences

The relationships between forward difference operators and differential operators, which are summarized in Table 3.2, enable us to develop a variety of formulas expressing derivatives of functions in terms of forward finite differences and vice versa. As was demonstrated in Sec. 4.2, these formulas may have any degree of accuracy desired, provided that a sufficient number of terms are retained in the manipulation of these infinite series. A set of expressions, parallel to those of Sec. 4.2, will be derived using the forward finite differences.

4.3.1 First-Order Derivative in Terms of Forward Finite Differences with Error of Order *h*

Rearrange Eq. (3.53) to solve for the differential operator *D*:

$$D = \frac{1}{h}\Delta - \frac{hD^2}{2} - \frac{h^2D^3}{6} - \ldots \tag{4.30}$$

Apply this operator to the function *y* at *i*:

$$Dy_i = \frac{1}{h}\Delta y_i - \frac{hD^2y_i}{2} - \frac{h^2D^3y_i}{6} - \ldots \tag{4.31}$$

Truncate the series, retaining only the first term:

$$Dy_i = \frac{1}{h}\Delta y_i + O(h) \tag{4.32}$$

Express the differential and forward operators in terms of their respective definitions:

$$\frac{dy_i}{dx} = \frac{1}{h}(y_{i+1} - y_i) + O(h) \tag{4.33}$$

Eq. (4.33) enables us to evaluate the first-order derivative of *y* at position *i* in terms of forward finite differences with error of order *h*.

4.3.2 Second-Order Derivative in Terms of Forward Finite Differences with Error of Order h

Rearrange Eq. (3.57) to solve for D^2:

$$D^2 = \frac{1}{h^2}\Delta^2 - hD^3 - \frac{7}{12}h^2D^4 - \dots \tag{4.34}$$

Apply this operator to the function y at i:

$$D^2y_i = \frac{1}{h^2}\Delta^2y_i - hD^3y_i - \frac{7}{12}h^2D^4y_i - \dots \tag{4.35}$$

Truncate the series, retaining only the first term, and express the operators in terms of their respective definitions:

$$\frac{d^2y_i}{dx^2} = \frac{1}{h^2}(y_{i+2} - 2y_{i+1} + y_i) + O(h) \tag{4.36}$$

This equation evaluates the second-order derivative of y at position i, in terms of forward finite differences, with error of order h.

4.3.3 First-Order Derivative in Terms of Forward Finite Differences with Error of Order h^2

Rearrange Eq. (3.53) to solve for hD:

$$hD = \Delta - \frac{h^2D^2}{2} - \frac{h^3D^3}{6} - \dots \tag{4.37}$$

Rearranging Eq. (3.57) to solve for h^2D^2:

$$h^2D^2 = \Delta^2 - h^3D^3 - \frac{7}{12}h^4D^4 - \dots \tag{4.38}$$

Combine these two equations to eliminate h^2D^2:

$$hD = \Delta - \frac{1}{2}\left(\Delta^2 - h^3D^3 - \frac{7}{12}h^4D^4 - \dots\right) - \frac{h^3D^3}{6} - \dots$$

$$= \Delta - \frac{1}{2}\Delta^2 + \frac{h^3D^3}{3} + \dots \tag{4.39}$$

Divide through by h, and apply this operator to the function y at i:

$$Dy_i = \frac{1}{h}\Delta y_i - \frac{1}{2h}\Delta^2 y_i + \frac{h^2D^3y_i}{3} + \dots \tag{4.40}$$

Truncate the series, retaining only the first *two* terms, and express the operators in terms of their respective definitions:

$$\frac{dy_i}{dx} = \frac{1}{h}(y_{i+1} - y_i) - \frac{1}{2h}(y_{i+2} - 2y_{i+1} + y_i) + O(h^2)$$

$$= \frac{1}{2h}(-y_{i+2} + 4y_{i+1} - 3y_i) + O(h^2) \tag{4.41}$$

4.3.4 Second-Order Derivative in Terms of Forward Finite Differences with Error of Order h^2

Rearrange Eq. (3.57) to solve for h^2D^2:

$$h^2D^2 = \Delta^2 - h^3D^3 - \frac{7}{12}h^4D^4 - \dots \tag{4.42}$$

Rearrange Eq. (3.58) to solve for h^3D^3:

$$h^3D^3 = \Delta^3 - \frac{3}{2}h^4D^4 - \frac{5}{4}h^5D^5 - \dots \tag{4.43}$$

Combine these two equations to eliminate h^3D^3:

$$h^2D^2 = \Delta^2 - \left(\Delta^3 - \frac{3}{2}h^4D^4 - \frac{5}{4}h^5D^5 - \dots\right) - \frac{7}{12}h^4D^4 - \dots$$

$$= \Delta^2 - \Delta^3 + \frac{11}{12}h^4D^4 - \dots \tag{4.44}$$

Divide through by h^2 and apply the operator to the function y at i:

$$D^2 y_i = \frac{1}{h^2}\Delta^2 y_i - \frac{1}{h^2}\nabla^3 y_i + \frac{11}{12}h^2 D^4 y_i - \dots \tag{4.45}$$

Truncate the series, retaining only the first two terms, and express the operators in terms of their respective definitions:

$$\frac{d^2 y_i}{dx^2} = \frac{1}{h^2}(-y_{i+3} + 4y_{i+2} - 5y_{i+1} + 2y_i) + O(h^2) \tag{4.46}$$

The formulas developed in these sections for the first- and second-order derivatives are summarized in Table 4.2, together with those of the third- and fourth-order derivatives.

It should be pointed out that all the finite difference approximations of derivatives obtained in this section and the previous section have coefficients that add up to zero. This is a rule of thumb that applies to all such combinations of finite differences.

From a comparison between Tables 4.1 and 4.2, we conclude that derivatives can be expressed in their backward or forward differences, with formulas that are very similar to each other in the number of terms involved and in the order of truncation error. The choice between using forward or backward differences will depend on the geometry of the problem and its boundary conditions. This will be discussed further in Chaps. 5 and 6.

4.4 Differentiation by Central Finite Differences

The relationships between central difference operators and differential operators, which are summarized in Table 3.3, will be used in the following sections to develop a set of formulas expressing the derivatives in terms of central finite differences. These formulas will have higher accuracy than those developed in the previous two sections using backward and forward finite differences.

4.4.1 First-Order Derivative in Terms of Central Finite Differences with Error of Order h^2

Rearrange Eq. (3.78) to solve for D:

$$D = \frac{1}{h}\mu\delta - \frac{h^2 D^3}{6} - \frac{h^4 D^5}{120} - \dots \tag{4.47}$$

Table 4.2 Derivatives in terms of forward finite differences

Error of order h

$$\frac{dy_i}{dx} = \frac{1}{h}(y_{i+1} - y_i) + O(h)$$

$$\frac{d^2y_i}{dx^2} = \frac{1}{h^2}(y_{i+2} - 2y_{i+1} + y_i) + O(h)$$

$$\frac{d^3y_i}{dx^3} = \frac{1}{h^3}(y_{i+3} - 3y_{i+2} + 3y_{i+1} - y_i) + O(h)$$

$$\frac{d^4y_i}{dx^4} = \frac{1}{h^4}(y_{i+4} - 4y_{i+3} + 6y_{i+2} - 4y_{i+1} + y_i) + O(h)$$

Error of order h^2

$$\frac{dy_i}{dx} = \frac{1}{2h}(-y_{i+2} + 4y_{i+1} - 3y_i) + O(h^2)$$

$$\frac{d^2y_i}{dx^2} = \frac{1}{h^2}(-y_{i+3} + 4y_{i+2} - 5y_{i+1} + 2y_i) + O(h^2)$$

$$\frac{d^3y_i}{dx^3} = \frac{1}{2h^3}(-3y_{i+4} + 14y_{i+3} - 24y_{i+2} + 18y_{i+1} - 5y_i) + O(h^2)$$

$$\frac{d^4y_i}{dx^4} = \frac{1}{h^4}(-2y_{i+5} + 11y_{i+4} - 24y_{i+3} + 26y_{i+2} - 14y_{i+1} + 3y_i) + O(h^2)$$

Apply this operator to the function y at i:

$$Dy_i = \frac{1}{h}\mu\delta y_i - \frac{h^2D^3y_i}{6} - \frac{h^4D^5y_i}{120} - \dots \tag{4.48}$$

Truncate the series, retaining only the first term:

$$Dy_i = \frac{1}{h}\mu\delta y_i + O(h^2) \tag{4.49}$$

Express the differential and averaged central difference operators in terms of their respective definitions:

$$\frac{dy_i}{dx} = \frac{1}{2h}(y_{i+1} - y_{i-1}) + O(h^2) \tag{4.50}$$

Eq. (4.50) enables us to evaluate the first-order derivative of y at position i in terms of central finite differences. Comparing this equation with Eq. (4.16) and Eq. (4.33) reveals that use of central differences increases the accuracy of the formulas, for the same number of terms retained.

4.4.2 Second-Order Derivative in Terms of Central Finite Differences with Error of Order h^2

Rearrange Eq. (3.81) to solve for D^2:

$$D^2 = \frac{1}{h^2}\delta^2 - \frac{h^2 D^4}{12} - \frac{h^4 D^6}{360} - \ldots \tag{4.51}$$

Apply this operator to the function y at i:

$$D^2 y_i = \frac{1}{h^2}\delta^2 y_i - \frac{h^2 D^4 y_i}{12} - \frac{h^4 D^6 y_i}{360} - \ldots \tag{4.52}$$

Truncate the series, retaining only the first term:

$$D^2 y_i = \frac{1}{h^2}\delta^2 y_i + O(h^2) \tag{4.53}$$

Express the differential and central difference operators in terms of their respective definitions:

$$\frac{d^2 y_i}{dx^2} = \frac{1}{h^2}(y_{i+1} - 2y_i + y_{i-1}) + O(h^2) \tag{4.54}$$

4.4.3 First-Order Derivative in Terms of Central Finite Differences with Error of Order h^4

Rearrange Eq. (3.78) to solve for hD:

$$hD = \mu\delta - \frac{h^3 D^3}{6} - \frac{h^5 D^5}{120} - \ldots \tag{4.55}$$

Rearrange Eq. (3.82) to solve for h^3D^3:

$$h^3D^3 = \mu\delta^3 - \frac{h^5D^5}{4} - \frac{h^7D^7}{40} - \dots \tag{4.56}$$

Combine these two equations to eliminate h^3D^3:

$$hD = \mu\delta - \frac{1}{6}\left(\mu\delta^3 - \frac{h^5D^5}{4} - \frac{h^7D^7}{40} - \dots\right) - \frac{h^5D^5}{120} - \dots$$

$$= \mu\delta - \frac{1}{6}\mu\delta^3 + \frac{h^5D^5}{30} + \dots \tag{4.57}$$

Divide through by h and apply this operator to the function y at i:

$$Dy_i = \frac{1}{h}\mu\delta y_i - \frac{1}{6h}\mu\delta^3 y_i + \frac{h^4D^5y_i}{30} + \dots \tag{4.58}$$

Truncate the series, retaining only the first *two* terms, and express the operators in terms of their respective definitions:

$$\begin{aligned}\frac{dy_i}{dx} &= \frac{1}{2h}(y_{i+1} - y_{i-1}) - \frac{1}{12h}(y_{i+2} - 2y_{i+1} + 2y_{i-1} - y_{i-2}) + O(h^4)\\ &= \frac{1}{12h}(-y_{i+2} + 8y_{i+1} - 8y_{i-1} + y_{i-2}) + O(h^4)\end{aligned} \tag{4.59}$$

4.4.4 Second-Order Derivative in Terms of Central Finite Differences with Error of Order h^4

Rearrange Eq. (3.81) to solve for h^2D^2:

$$h^2D^2 = \delta^2 - \frac{h^4D^4}{12} - \frac{h^6D^6}{360} - \dots \tag{4.60}$$

Rearrange Eq. (3.83) to solve for h^4D^4:

$$h^4D^4 = \delta^4 - \frac{h^6D^6}{6} - \frac{h^8D^8}{80} - \dots \tag{4.61}$$

Combine these two equations to eliminate h^4D^4:

$$h^2D^2 = \delta^2 - \frac{1}{12}\left(\delta^4 - \frac{h^6D^6}{6} - \frac{h^8D^8}{80} - \dots\right) - \frac{h^6D^6}{360} - \dots$$

$$= \delta^2 - \frac{1}{12}\delta^4 + \frac{h^6D^6}{90} - \dots \tag{4.62}$$

Divide through by h^2 and apply this operator to function y at i:

$$D^2y_i = \frac{1}{h^2}\delta^2 y_i - \frac{1}{12h^2}\delta^4 y_i + \frac{h^4D^6}{90} - \dots \tag{4.63}$$

Truncate the series, retaining only the first *two* terms, and express the operators in terms of their respective definitions:

$$\frac{d^2y_i}{dx^2} = \frac{1}{h^2}(y_{i+1} - 2y_i + y_{i-1}) - \frac{1}{12h^2}(y_{i+2} - 4y_{i+1} + 6y_i - 4y_{i-1} + y_{i-2}) + O(h^4)$$

$$= \frac{1}{12h^2}(-y_{i+2} + 16y_{i+1} - 30y_i + 16y_{i-1} - y_{i-2}) + O(h^4) \tag{4.64}$$

The formulas derived in Sec. 4.4.1-4.4.4 for the first- and second-order derivatives are summarized in Table 4.3, along with those for the third- and fourth-order derivatives. Development of formulas with higher accuracy and for the higher-order derivatives are left as exercises for the reader (see Problems).

Example 4.1: Mass Transfer Flux from an Open Vessel. Develop a MATLAB function for numerical differentiation of a function $f(x)$ over the range $[x_0, x_n]$ with truncation error of order h^2. Apply this function to evaluate the unsteady-state flux of water evaporated into air at 1 atm and 25°C from the top of an open vessel. Consider the distance between water level and the open top of the vessel to be 0.1, 0.2, and 0.3 m. The flux of water vapor at a level z above the level of water is given by Bird et al. [2]:

$$N_z = -cD\frac{\partial x}{\partial z} - x\left(\frac{cD}{1 - x_0}\right)\frac{\partial x}{\partial z}\bigg|_{z=0} \tag{1}$$

where N_z is the flux of water vapor at level z, c is the total concentration of the gas phase, D is the diffusion coefficient of water vapor in air, x is the mole fraction of water vapor, and x_0 indicates the value of x at $z = 0$.

The unsteady-state concentration profile for this problem, assuming no air flow at the top of the vessel, is obtained from

$$X = \frac{1 - erf(Z - \varphi)}{1 + erf\,\varphi} \tag{2}$$

where

$$X = \frac{x}{x_0} \qquad Z = \frac{z}{\sqrt{4Dt}} \tag{3}$$

where t is time and φ is obtained from the solution of the following nonlinear equation:

$$x_0 = \frac{1}{1 + \left[\sqrt{\pi}(1 + erf\,\varphi)\varphi \exp \varphi^2\right]^{-1}} \tag{4}$$

For the air-water system,

$$D = 2.2\times10^{-5}\ \text{m}^2/\text{s}\,, \quad x_0 = P^{sat}(25°\text{C})/P_t = 0.0312\,, \quad c = P_t/RT = 0.0409\ \text{Kmol/m}^3$$

Table 4.3 Derivatives in terms of central finite differences

Error of order h^2

$$\frac{dy_i}{dx} = \frac{1}{2h}(y_{i+1} - y_{i-1}) + O(h^2)$$

$$\frac{d^2y_i}{dx^2} = \frac{1}{h^2}(y_{i+1} - 2y_i + y_{i-1}) + O(h^2)$$

$$\frac{d^3y_i}{dx^3} = \frac{1}{2h^3}(y_{i+2} - 2y_{i+1} + 2y_{i-1} - y_{i-2}) + O(h^2)$$

$$\frac{d^4y_i}{dx^4} = \frac{1}{h^4}(y_{i+2} - 4y_{i+1} + 6y_i - 4y_{i-1} + y_{i-2}) + O(h^2)$$

Error of order h^4

$$\frac{dy_i}{dx} = \frac{1}{12h}(-y_{i+2} + 8y_{i+1} - 8y_{i-1} + y_{i-2}) + O(h^4)$$

$$\frac{d^2y_i}{dx^2} = \frac{1}{12h^2}(-y_{i+2} + 16y_{i+1} - 30y_i + 16y_{i-1} - y_{i-2}) + O(h^4)$$

$$\frac{d^3y_i}{dx^3} = \frac{1}{8h^3}(-y_{i+3} + 8y_{i+2} - 13y_{i+1} + 13y_{i-1} - 8y_{i-2} + y_{i-3}) + O(h^4)$$

$$\frac{d^4y_i}{dx^4} = \frac{1}{6h^4}(-y_{i+3} + 12y_{i+2} - 39y_{i+1} + 56y_i - 39y_{i-1} + 12y_{i-2} - y_{i-3}) + O(h^4)$$

Method of Solution: In order to calculate the flux of water vapor, first we have to determine the value of φ from the solution of the nonlinear Eq. (4). The concentration of water vapor is then obtained from Eqs. (2) and (3). Having the concentration profile, its derivatives at the water surface level ($z = 0$) and any desired level above the surface of water can be evaluated. Finally, the mass transfer flux is calculated from Eq. (1).

Differentiation of the function $y = f(x)$ is done based on equations shown in Tables 4.1-4.3. According to the chosen method of finite difference, the equation for derivation with truncation error of order h^2 will be employed. If the method of finite difference is not determined by the user, central finite difference will be used for differentiation.

Program Description: The MATLAB function *fder.m* evaluates the derivative of a function. The first part of the program is initialization, where inputs to the function are examined and default values for differentiation increment (h) and method of finite difference are applied, if required. Introducing these two inputs to the function is optional. The program then switches to a different part of the program, according to the choice of method of finite difference, and then it switches to the proper section according to the order of differentiation.

The first input argument should be a string variable giving the name of the m-file that contains the function whose derivative is to be evaluated. The second input argument is the order of derivation, which has to be less than or equal to 4. The third input argument may be a scalar or a vector at which the derivative is to be evaluated. The fourth and fifth input arguments are optional and represent the increment of the independent variable and the method of finite differences, respectively. The default value of h is 1/100 of the minimum value of the independent variable or 0.001 if this minimum is zero. If the assigned value for h is smaller than the floating point relative accuracy, *eps*, the function assumes the relative accuracy as the differentiation increment. The user may specify the value of -1, 0, or +1 for method of finite difference if it is required to evaluate the derivative based on backward, central, or forward finite difference, respectively. The default method is the central difference. Any additional argument will be carried directly to the m-file that represents the function and may contain parameters (such as constants) needed for the function.

To solve the problem posed in this example, three more MATLAB programs are written. The main program, named *Example4_1.m*, does the necessary calculations and plots the results. The function *Ex4_1_profile.m* represents the concentration profile of this problem [Eqs. (2) and (3)]. The independent variable of this function z and other variables necessary to evaluate the function are entered as parameters. The function *Ex4_1_phi.m* is the nonlinear function from which the value of φ is calculated [Eq. (4)].

Program

Example4_1.m

```
% Example4_1.m
% This program solves Example 4.1. It calculates and plots the unsteady
% flux of water vapor from the open top of a vessel. The program uses
```

```
% the function FDER to obtain the concentration gradient.

clear
clc
clf

% Input data
z0 = 0;
t = input(' Vector of time (s) = ');
z = input(' Vector of axial positions (m) = ');
D = input(' Diffusion coefficient of the vapor in air (m2/s) = ');
T = 273.15 + input(' System Temperature (deg C) = ');
P = input(' System pressure (Pa) = ');
Psat = input(' Vapor pressure at the system temperature (Pa) = ');
x0 = Psat/P;                        % Mole fraction
R = 8314;                           % Gas constant
c = P/(R*T);                        % Gas concentration
phifile = input(' Name of the m-file containing the equation for phi
= ');
profile = input(' Name of the m-file containing the concentration
profile = ');

% Solving the nonlinear equation for phi
phi = fzero(phifile,1,1e-6,0,x0);
% Concentration gradient at z=z0
dxdz0 = fder(profile,1,z0,[],[],t,x0,D,phi);
for k = 1 : length(z)
   % Concentration gradient
   dxdz(k,:) = fder(profile,1,z(k),[],[],t,x0,D,phi);
   % Mole fraction profile
   x(k,:) = feval(profile,z(k),t,x0,D,phi);
   % Molar flux
   Nz(k,:) = -c*D*dxdz(k,:)-x(k,:)*c*D/(1-x0).*dxdz0;
end

% Plotting the results
figure(1)
plot(t/60,Nz*3600*18*1000)
xlabel('t (min.)')
ylabel('N_z (gr/m2.hr)')
legend('z_1','z_2','z_3',1)
figure(2)
plot(t/60,x)
xlabel('t (min.)')
ylabel('Mole fraction of the Vapor')
legend('z_1','z_2','z_3',2)
```

Ex4_1_profile.m

```
function x = Ex4_1_profile(z,t,x0,D,phi)
% Function Ex4_1_profile.m
% Concentration profile evaluation in Example 4.1.

Z = z./sqrt(4*D*t);                    % Dimensionless axial position
X = (1-erf(Z-phi))/(1+erf(phi));       % Dimensionless concentration
x = x0*X;                              % Mole fraction
```

Ex_4_phi.m

```
function phif = Ex4_1_phi(x , x0)
% Function Ex4_1_phi.m
% Nonlinear equation for calculation of phi in Example 4.1.

phif=1/(1+1/(sqrt(pi)*(1+erf(x))*x*exp(x^2)))-x0;
```

fder.m

```
function df = fder(fnctn, order, x, h, method, varargin)
%FDER Evaluates nth-order derivative (n<=4) of a function
%     with truncation error of the order h^2.
%
%     FDER('F',N,X) evaluates Nth order derivative of the
%     function described by the M-file F.M at X. X may be a
%     scalar or a vector.
%
%     FDER('F',N,X,H,METHOD,PARAMETER) evaluates Nth order
%     derivative of the function using H as increment of X used
%     in differentiation.
%     METHOD is the finite difference method used
%       Use METHOD = -1 for backward finite difference
%       Use METHOD =  0 for central finite difference
%       Use METHOD =  1 for forward finite difference
%     PARAMETER is a scalar or a vector of parameters that are
%     passed to the function F.M. Pass an empty matrix for H
%     or METHOD to use the default values.
%
%     See also DERIV

% (c) N. Mostoufi & A. Constantinides
% January 1, 1999

% Initialization
if nargin == 3 | isempty(h)
   method = 0;
   h = min(abs(x))/100;
   if h == 0
      h = 0.001;
   end
   if h < eps
```

```
      h = eps;
   end
end

if nargin<5 | isempty(method)
   method = 0;
end

if (order < 1) | (order > 4) | (h <= 0)
   error('Invalid input.')
   break
end

switch method
case -1                          % Backward finite difference
   yi = feval(fnctn , x , varargin{:});
   yim1 = feval(fnctn , x-h , varargin{:});
   yim2 = feval(fnctn , x-2*h , varargin{:});
   switch order
   case 1                        % 1st order derivative
      df = (3*yi-4*yim1+yim2)/(2*h);
   case 2                        % 2nd order derivative
      yim3 = feval(fnctn , x-3*h , varargin{:});
      df = (2*yi-5*yim1+4*yim2-yim3)/h^2;
   case 3                        % 3rd order derivative
      yim3 = feval(fnctn , x-3*h , varargin{:});
      yim4 = feval(fnctn , x-4*h , varargin{:});
      df = (5*yi-18*yim1+24*yim2-14*yim3+3*yim4);
   case 4                        % 4th order derivative
      yim3 = feval(fnctn , x-3*h , varargin{:});
      yim4 = feval(fnctn , x-4*h , varargin{:});
      yim5 = feval(fnctn , x-5*h , varargin{:});
      df = (3*yi-14*yim1+26*yim2-24*yim3+11*yim4-2*yim5)/h^4;
   end
case 0                           % Central finite difference
   yim1 = feval(fnctn , x-h , varargin{:});
   yip1 = feval(fnctn , x+h , varargin{:});
   switch order
   case 1                        % 1st order derivative
      df = (yip1-yim1)/(2*h);
   case 2                        % 2nd order derivative
      yi = feval(fnctn , x , varargin{:});
      df = (yip1-2*yi+yim1)/h^2;
   case 3                        % 3rd order derivative
      yim2 = feval(fnctn , x-2*h , varargin{:});
      yip2 = feval(fnctn , x+2*h , varargin{:});
      df = (yip2-2*yip1+2*yim1-yim2)/(2*h^3);
   case 4                        % 4th order derivative
      yim3 = feval(fnctn , x-3*h , varargin{:});
```

```
      yim2 = feval(fnctn , x-2*h , varargin{:});
      yi = feval(fnctn , x , varargin{:});
      yip2 = feval(fnctn , x+2*h , varargin{:});
      yip3 = feval(fnctn , x+3*h , varargin{:});
     df = (-yip3+12*yip2-39*yip1+56*yi-39*yim1+12*yim2-yim3)/(6*h^4);
   end
case 1                          % Forward finite difference
   yi = feval(fnctn , x , varargin{:});
   yip1 = feval(fnctn , x+h , varargin{:});
   yip2 = feval(fnctn , x+2*h , varargin{:});
   switch order
   case 1                       % 1st order derivative
      df = (-yip2+4*yip1-3*yi)/(2*h);
   case 2                       % 2nd order derivative
      yip3 = feval(fnctn , x+3*h , varargin{:});
      df = (-yip3+4*yip2-5*yip1+2*yi)/h^2;
   case 3                       % 3rd order derivative
      yip3 = feval(fnctn , x+3*h , varargin{:});
      yip4 = feval(fnctn , x+4*h , varargin{:});
      df = (-3*yip4+14*yip3-24*yip2+18*yip1-5*yi)/(2*h^3);
   case 4                       % 4th order derivative
      yip3 = feval(fnctn , x+3*h , varargin{:});
      yip4 = feval(fnctn , x+4*h , varargin{:});
      yip5 = feval(fnctn , x+5*h , varargin{:});
      df = (-2*yip5+11*yip4-24*yip3+26*yip2-14*yip1+3*yi)/h^4;
   end
end
```

Input and Results

```
>>Example4_1

 Vector of time (s) = eps:3600
 Vector of axial positions (m) = [0.1, 0.2, 0.3]
 Diffusion coefficient of the vapor in air (m2/s) = 2.2e-5
 System Temperature (deg C) = 25
 System pressure (Pa) = 101325
 Vapor pressure at the system temperature (Pa) = 3161
 Name of the m-file containing the equation for phi = 'Ex4_1_phi'
 Name of the m-file containing the concentration profile =
'Ex4_1_profile'
```

Discussion of Results: Fig. E4.1*a* shows the unsteady diffusive mass transfer flux from the open top of the vessel.[1] The concentration profiles with respect to time are also plotted in

[1] When running *Example4_1.m*, solution results will be shown on the screen by solid lines of different color. However, results for three different levels chosen here are illustrated by different signs in Figs. E4.1*a* and *b* in order to be discriminated.

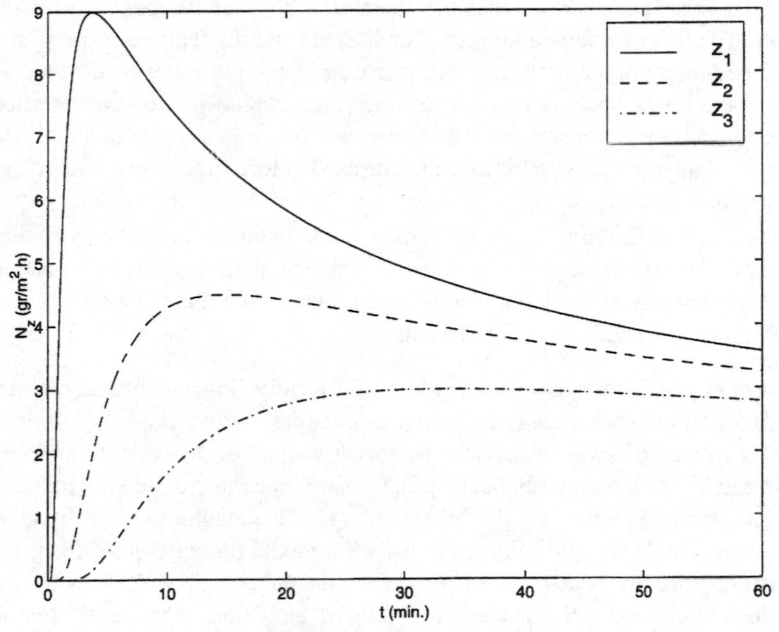

Figure E4.1*a* Flux versus time.

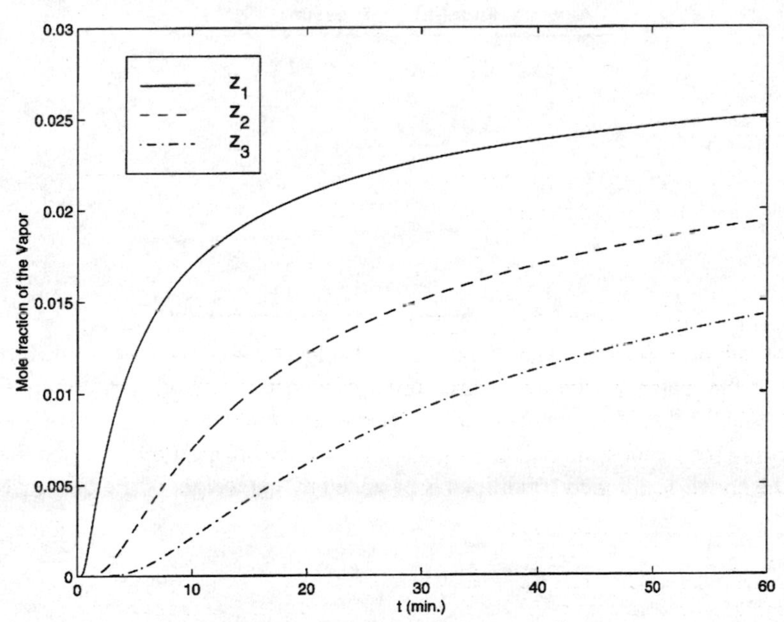

Figure E4.1*b* Mole fraction versus time.

Fig. E4.1*b*. Fig. E4.1*a* shows that the mass flux rises at the beginning and reaches a maximum. The flux of water vapor then decreases constantly. This behavior is due to the fact that at the beginning of the process, there is dry air above the water level, and mass transfer is taking place faster because of high driving force, which is the concentration gradient. However, the mass transfer rate decreases, after enough time has passed and the water vapor concentration moves closer to saturation and the driving force decreases. Vapor concentration decreases with increasing height.

In the program *Example4_1.m*, defaults are used for choice of method of finite difference and differentiation increment. The defaults are central finite difference and $h = 0.001$, respectively. The reader is encouraged to repeat the calculation using the other methods of finite differences and with different increments.

Example 4.2: Derivative of Vectors of Equally Spaced Points. Write a general MATLAB function to calculate first- to fourth-order derivatives of a series of data presented numerically in a matrix whose columns represent vectors of dependent variable. The user should be able to choose between backward, forward, or central differentiation as well as the order of the truncation error. Apply this function to calculate the solids volume fraction in a riser of a bench-scale gas-solid fluidized bed whose axial pressure profile is given in Table E4.2. Assume fully developed solids flow in the riser and neglect wall shear and solids stress. The densities of gas and solids phases are 1.2 kg/m^3 and 2650 kg/m^3, respectively.

Table E4.2

Axial position (m)	Pressure (kPa g)
0.0	1.80
0.5	1.38
1.0	1.09
1.5	0.63
2.0	0.18

Method of Solution: The equations in Tables 4.1-4.3 are used to differentiate the columns of the matrix ***y*** with the desired order of truncation error. Differentiation is done based on equally spaced segments of the independent variable.

Writing the momentum balance equation for the two-phase flow, we find that pressure drop in the above-mentioned conditions is balanced by the weight of the bed, that is:

$$-\frac{dP}{dz} = [\rho_g(1 - \epsilon_s) + \rho_s\epsilon_s]g \tag{1}$$

where P is the pressure, z is the axial position, ρ_g and ρ_s are the densities of gas and solids, respectively, ϵ_s is the volume fraction of the solids, and g is the gravitational acceleration. Eq. (1) can be solved for ϵ_s:

$$\epsilon_s = \frac{(-dP/dz) - \rho_g g}{(\rho_s - \rho_g)g} \qquad (2)$$

The solids volume fraction profile can be calculated from Eq. (2) once the pressure gradient is extracted from the data tabulated in Table E4.2.

Program Description: The MATLAB function *deriv.m* is written to calculate first- to fourth-order derivatives of a matrix of input data. The first part of the program is initialization in which the values of h, order of derivative, method of finite difference used, and order of truncation error are assigned if not entered as input to function. If only **y** is given as input to the function, the program calculates the central finite differences of **y** as the output. The second input argument is the increment of the independent variable. The third one is the order of derivative. A value of -1, 0, or 1 as the fourth input argument results in calculation of the derivative based on backward, central, or forward finite differences, respectively. The fifth argument is the value of the order of truncation error (1 or 2 for backward and forward differences and 2 or 4 for central differences).

The derivative matrix returned by the function *deriv.m* has the same number of elements as the vector of input data itself. However, it is important to note that, depending on the method of finite difference used, some elements at one or both ends of the derivative vector are evaluated by a different method of differentiation. For example, in first-order differentiation with the forward finite difference method with truncation error $O(h)$, the last element of the returned derivative vector is calculated by backward differences. Another example is the calculation of the second-order derivative of a vector by the central finite difference method with truncation error $O(h^2)$, where the function evaluates the first two elements of the vector of derivatives by forward differences and the last two elements of the vector of derivatives by backward differences. The reader should pay special attention to the fact that when the function calculates the derivative by the central finite difference method with the truncation error of the order $O(h^4)$, the starting and ending rows of derivative values are calculated by forward and backward finite differences, with truncation error of the order $O(h^2)$.

The main program *Example4_2.m* asks the reader to input the data from the keyboard. It then applies the function *deriv.m* to evaluate the pressure gradient and calculates the solids volume fraction from Eq. (2). At the end, the program plots the result of the above calculations.

Program

Example4_2.m

```
% Example4_2.m
% This program solves Example 4.2. It calculates and plots
```

```
% the volume fraction profile in a gas-solid fluidized bed.
% The program uses the function DERIV to obtain the pressure
% gradient.

clear
clc
clf

% Input data
P = input(' Vector of pressures (kPa) = ')*1e3;
dz = input(' Axial distance between pressure probes (m) = ');
rhog = input(' Density of the gas (kg/m3) = ');
rhos = input(' Density of solids (kg/m3) = ');
g = 9.81;
% Pressure gradient
dP = deriv(P,dz);
% Solids concentration
epsilon_s = (-dP-0*rhog*g) / (g*(rhos-rhog));
fprintf('\n Average solids concentration = %4.2f%% \n',...
   100*mean(epsilon_s))

% Plotting the results
z = [0:length(P)-1]*dz;
plot(z,100*epsilon_s)
xlabel('z (m)')
ylabel('Solids volume fraction (%)')
```

deriv.m

```
function dy = deriv(y, h, order, method, err)
%DERIV Differentiates a matrix of data numerically
%
%   DERIV(Y) calculates the central differences of each vector
%   of matrix Y.
%
%   DERIV(Y,H) calculated the first-order derivative of Y by
%   central finite differences using H as the independent
%   variable interval.
%
%   DY = DERIV(Y,H,ORDER,METHOD,ERR) returns the derivative
%   of columns of the matrix Y where
%   H       is the independent variable interval
%   ORDER   is the order of differentiation (up to 4th order)
%   METHOD is the finite difference method used
%     Use METHOD = -1 for backward finite difference
%     Use METHOD =  0 for central finite difference
%     Use METHOD =  1 for forward finite difference
%   ERR     is the order of error of calculation. ERR may be
%     1 or 2 for backward and forward finite difference
%     and 2 or 4 for central finite difference.
%
```

```
%   See also FDER, DIFF

% (c) N. Mostoufi & A. Constantinides
% January 1, 1999

% Initialization
if nargin == 1 | isempty(h)
   h = 1;
end

if nargin < 3 | isempty(order)
   order = 1;
end

if nargin < 4 | isempty(method)
   method = 0;
   err = 2;
end

if nargin == 4
   if method == 0
      err = 2;
   else
      err = 1;
   end
end
if abs(method) == 1 & (err < 1 | err > 2)
   err = 1;
   warning(' Order of truncation error is set to 1.')
end
if method ==0 & ~(err ~= 2 | err ~= 4)
   err = 2;
   warning(' Order of truncation error is set to 2.')
end

[r , c] = size(y);
if r == 1 % If y is a row vector
   y = y';    % Make it a column vector
   r = c;
   c = 1;
end
n = r; % Number of points
dy = zeros(r , c);

% Differentiation
switch method
case -1   % Backward finite differences
   switch err
   case 1     % O(h)
    switch order
```

```
    case 1    % 1st order derivative
      dy(2:n , :) = (y(2:n , :) - y(1:n-1 , :))/h;
      dy(1 , :) = (y(2 , :) - y(1 , :))/h;
    case 2    % 2nd order derivative
      dy(3:n , :)=(y(3:n , :) - 2*y(2:n-1 , :) + y(1:n-2 , :))/h^2;
      dy(1:2 , :)=(y(3:4 , :) - 2*y(2:3 , :) + y(1:2 , :))/h^2;
    case 3    % 3rd order derivative
      dy(4:n , :)=(y(4:n , :) - 3*y(3:n-1 , :) + 3*y(2:n-2 , :) ...
           - y(1:n-3 , :))/h^3;
      dy(1:3 , :)=(y(4:6 , :) - 3*y(3:5 , :) + 3*y(2:4 , :) ...
           - y(1:3 , :))/h^3;
    case 4    % 4th order derivative
      dy(5:n , :)=(y(5:n , :) - 4*y(4:n-1 , :) + 6*y(3:n-2 , :) ...
          - 4*y(2:n-3 , :) + y(1:n-4 , :))/h^4;
      dy(1:4 , :)=(y(5:8 , :) - 4*y(4:7 , :) + 6*y(3:6 , :) ...
          - 4*y(2:5 , :)  + y(1:4 , :))/h^4;
      end
   case 2     % O(h^2)
    switch order
    case 1    % 1st order derivative
     dy(3:n , :)=(3*y(3:n , :) - 4*y(2:n-1 , :)+y(1:n-2, :))/(2*h);
     dy(1:2 , :)=(-y(3:4 , :) + 4*y(2:3 , :) - 3*y(1:2 , :))/(2*h);
    case 2    % 2nd order derivative
     dy(4:n , :)=(2*y(4:n , :)-5*y(3:n-1 , :) + 4*y(2:n-2 , :) ...
          - y(1:n-3 , :))/h^2;
     dy(1:3 , :) = (-y(4:6 , :) + 4*y(3:5 , :) - 5*y(2:4 , :) ...
          + 2*y(1:3 , :))/h^2;
    case 3    % 3rd order derivative
     dy(5:n , :)=(5*y(5:n , :)-18*y(4:n-1 , :)+24*y(3:n-2 , :) ...
          - 14*y(2:n-3 , :) + 3*y(1:n-4 , :))/(2*h^3);
     dy(1:4 , :)=(-3*y(5:8 , :) + 14*y(4:7 , :) - 24*y(3:6 , :) ...
         + 18*y(2:5 , :) - 5*y(1:4 , :))/(2*h^3);
    case 4    % 4th order derivative
     dy(6:n , :)=(3*y(6:n , :)-14*y(5:n-1 , :)+26*y(4:n-2 , :) ...
         - 24*y(3:n-3 , :) + 11*y(2:n-4 , :) - 2*y(1:n-5 , :))/h^4;
     dy(1:5 , :)=(-2*y(6:10 , :)+11*y(5:9 , :) - 24*y(4:8 , :) ...
         + 26*y(3:7 , :) - 14*y(2:6 , :) + 3*y(1:5 , :))/h^4;
    end
   end
case 0 % Central finite differences
   switch err
   case 2 % O(h^2)
    switch order
    case 1    % 1st order derivative
     dy(1 , :) = (-y(3 , :) + 4*y(2 , :) - 3*y(1 , :))/(2*h);
     dy(2:n-1 , :) = (y(3:n , :) - y(1:n-2 , :))/(2*h);
     dy(n , :) = (3*y(n , :) - 4*y(n-1 , :) + y(n-2 , :))/(2*h);
    case 2    % 2nd order derivative
     dy(1 , :)=(-y(4 , :)+4*y(3 , :)-5*y(2 , :)+2*y(1 , :))/h^2;
     dy(2:n-1 , :)=(y(3:n , :)-2*y(2:n-1 , :)+y(1:n-2 , :))/h^2;
```

```
      dy(n , :) = (2*y(n , :) - 5*y(n-1 , :) + 4*y(n-2 , :) ...
          - y(n-3 , :))/h^2;
       case 3 % 3rd order derivative
      dy(1:2 , :)=(-3*y(5:6 , :) + 14*y(4:5 , :) - 24*y(3:4 , :) ...
              + 18*y(2:3 , :) - 5*y(1:2 , :))/(2*h^3);
      dy(3:n-2 , :)=(y(5:n , :) - 2*y(4:n-1 , :) + 2*y(2:n-3 , :) ...
           - y(1:n-4 , :))/(2*h^3);
      dy(n-1:n , :) = (5*y(n-1:n , :) - 18*y(n-2:n-1 , :) ...
           + 24*y(n-3:n-2 , :) - 14*y(n-4:n-3 , :) ...
           + 3*y(n-5:n-4 , :))/(2*h^3);
       case 4 % 4th order derivative
      dy(1:2 , :)=(-2*y(6:7 , :) + 11*y(5:6 , :) - 24*y(4:5 , :) ...
           + 26*y(3:4 , :) - 14*y(2:3 , :) + 3*y(1:2 , :))/h^4;
      dy(3:n-2 , :)=(y(5:n , :) - 4*y(4:n-1 , :) + 6*y(3:n-2 , :) ...
           - 4*y(2:n-3 , :) + y(1:n-4 , :))/h^4;
      dy(n-1:n , :) = (3*y(n-1:n , :) - 14*y(n-2:n-1 , :) ...
           + 26*y(n-3:n-2 , :) - 24*y(n-4:n-3 , :) ...
           + 11*y(n-5:n-4 , :) - 2*y(n-6:n-5 , :))/h^4;
       end
   case 4 % O(h^4)
    switch order
    case 1
     dy(1:2 , :)=(-y(3:4 , :) + 4*y(2:3 , :) - 3*y(1:2 , :))/(2*h);
     dy(3:n-2 , :)=(-y(5:n , :)+8*y(4:n-1 , :) - 8*y(2:n-3 , :) ...
            + y(1:n-4 , :))/(12*h);
     dy(n-1:n , :)=(3*y(n-1:n , :)-4*y(n-2:n-1 , :) ...
            +y(n-3:n-2 , :))/(2*h);
      case 2
     dy(1:2 , :) = (-y(4:5 , :) + 4*y(3:4 , :) - 5*y(2:3 , :) ...
            + 2*y(1:2 , :))/h^2;
     dy(3:n-2 , :)=(-y(5:n , :)+16*y(4:n-1 , :)-30*y(3:n-2 , :) ...
            + 16*y(2:n-3 , :) - y(1:n-4 , :))/(12*h^2);
     dy(n-1:n , :) = (2*y(n-1:n , :) - 5*y(n-2:n-1 , :) ...
            + 4*y(n-3:n-2 , :) - y(n-4:n-3 , :))/h^2;
    case 3
     dy(1:3 , :)=(-3*y(5:7 , :) + 14*y(4:6 , :) - 24*y(3:5 , :) ...
            + 18*y(2:4 , :) - 5*y(1:3 , :))/(2*h^3);
     dy(4:n-3 , :)=(-y(7:n , :)+8*y(6:n-1 , :) - 13*y(5:n-2 , :) ...
         + 13*y(3:n-4 , :) - 8*y(2:n-5 , :) + y(1:n-6 , :))/(8*h^3);
     dy(n-2:n , :) = (5*y(n-2:n , :) - 18*y(n-3:n-1 , :) ...
            + 24*y(n-4:n-2 , :) - 14*y(n-5:n-3 , :) ...
            + 3*y(n-6:n-4 , :))/ (2*h^3);
    case 4
     dy(1:3 , :) = (-2*y(6:8 , :) + 11*y(5:7 , :) ...
          - 24*y(4:6 , :) + 26*y(3:5 , :) ...
          - 14*y(2:4 , :) + 3*y(1:3 , :))/h^4;
     dy(4:n-3 , :) = (-y(7:n , :) + 12*y(6:n-1 , :) ...
         - 39*y(5:n-2 , :) + 56*y(4:n-3 , :) - 39*y(3:n-4 , :) ...
         + 12*y(2:n-5 , :) - y(1:n-6 , :))/(6*h^4);
     dy(n-2:n , :) = (3*y(n-2:n , :) - 14*y(n-3:n-1 , :) ...
```

```
            + 26*y(n-4:n-2 , :) - 24*y(n-5:n-3 , :) ...
            + 11*y(n-6:n-4 , :) - 2*y(n-7:n-5 , :))/h^4;
      end
   end
case 1 % Forward finite differences
   switch err
   case 1 % O(h)
    switch order
    case 1   % 1st order derivative
     dy(1:n-1 , :) = (y(2:n , :) - y(1:n-1 , :))/h;
     dy(n , :) = (y(n , :) - y(n-1 , :))/h;
    case 2   % 2nd order derivative
     dy(1:n-2 , :)=(y(3:n , :) - 2*y(2:n-1 , :) + y(1:n-2 , :))/h^2;
     dy(n-1:n , :) = (y(n-1:n , :) - 2*y(n-2:n-1 , :) ...
          + y(n-3:n-2 , :))/h^2;
    case 3   % 3rd order derivative
     dy(1:n-3 , :) = (y(4:n , :) - 3*y(3:n-1 , :) ...
          + 3*y(2:n-2 , :) - y(1:n-3 , :))/h^3;
     dy(n-2:n , :) = (y(n-2:n , :) - 3*y(n-3:n-1 , :) ...
          + 3*y(n-4:n-2 , :) - y(n-5:n-3 , :))/h^3;
   case 4 % 4th order derivative
     dy(1:n-4 , :)=(y(5:n , :) - 4*y(4:n-1 , :) + 6*y(3:n-2 , :) ...
          - 4*y(2:n-3 , :) + y(1:n-4 , :))/h^4;
     dy(n-3:n , :) = (y(n-3:n , :) - 4*y(n-4:n-1 , :) ...
       + 6*y(n-5:n-2 , :) - 4*y(n-6:n-3 , :) + y(n-7:n-4 , :))/h^4;
   end
   case 2 % O(h^2)
    switch order
   case 1 % 1st order derivative
    dy(1:n-2 , :) = (-y(3:n , :) + 4*y(2:n-1 , :) ...
         - 3*y(1:n-2 , :))/(2*h);
    dy(n-1:n , :) = (3*y(n-1:n , :) - 4*y(n-2:n-1 , :) ...
          + y(n-3:n-2 , :))/(2*h);
   case 2 % 2nd order derivative
    dy(1:n-3 , :)=(-y(4:n , :)+4*y(3:n-1 , :)-5*y(2:n-2 , :) ...
          + 2*y(1:n-3 , :))/h^2;
    dy(n-2:n , :) = (2*y(n-2:n , :) - 5*y(n-3:n-1 , :) ...
          + 4*y(n-4:n-2 , :) - y(n-5:n-3 , :))/h^2;
   case 3 % 3rd order derivative
    dy(1:n-4 , :) = (-3*y(5:n , :) + 14*y(4:n-1 , :) ...
        - 24*y(3:n-2 , :) + 18*y(2:n-3 , :) ...
        - 5*y(1:n-4 , :))/(2*h^3);
    dy(n-3:n , :) = (5*y(n-3:n , :) - 18*y(n-4:n-1 , :) ...
          + 24*y(n-5:n-2 , :) - 14*y(n-6:n-3 , :) ...
          + 3*y(n-7:n-4 , :))/(2*h^3);
   case 4 % 4th order derivative
    dy(1:n-5 , :) = (-2*y(6:n , :) + 11*y(5:n-1 , :) ...
         - 24*y(4:n-2 , :) + 26*y(3:n-3 , :) - 14*y(2:n-4 , :) ...
          + 3*y(1:n-5 , :))/h^4;
    dy(n-4:n , :) = (3*y(n-4:n , :) - 14*y(n-5:n-1 , :) ...
```

```
            + 26*y(n-6:n-2 , :) - 24*y(n-7:n-3 , :) ...
            + 11*y(n-8:n-4 , :) - 2*y(n-9:n-5 , :))/h^4;
      end
   end
end
```

Input and Results

```
>>Example4_2

 Vector of pressures (kPa) = [1.80, 1.38, 1.09, 0.63, 0.18]
 Axial distance between pressure probes (m) = 0.5
 Density of the gas (Kg/m3) = 1.2
 Density of solids (Kg/m3) = 2650

 Average solids concentration = 3.26%
```

Discussion of Results: Fig. E4.2 shows the results graphically. It can be seen from this figure that the solids fraction does not change appreciably with height. The value of ϵ_s varies between 2.7% and 3.8% (approximately) with its mean value at 3.26%. This confirms the assumption made at the beginning that the measurements are done in the fully developed zone where the solids move with a constant velocity.

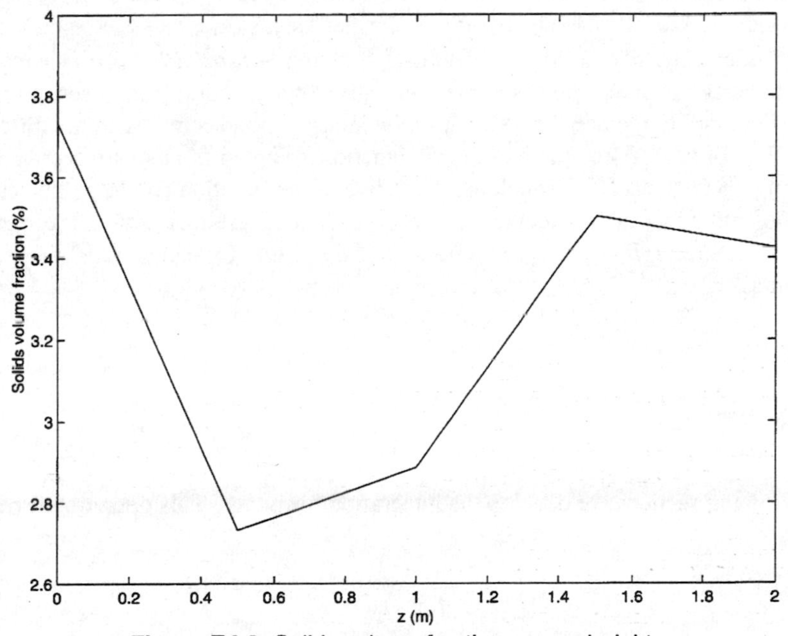

Figure E4.2 Solids volume fraction versus height.

4.5 Spline Differentiation

In some situations, tabulated function values are available, instead of their algebraic expression, and it is desired to evaluate the derivative of the function at a point (or points) between the tabulated values. A practical method in this situation is to interpolate the base points first and calculate the value of the derivative from differentiating the interpolating polynomial. Among different interpolating polynomials, cubic splines have the advantage of continuity of the first derivative through all base points.

By cubic spline interpolation of the function, its derivative at any point x in the interval $[x_{i-1}, x_i]$ can be calculated from differentiating Eq. (3.143):

$$\frac{dy}{dx} = \frac{1}{2}\frac{(x - x_i)^2}{x_{i-1} - x_i}y''_{i-1} + \frac{1}{2}\frac{(x - x_{i-1})^2}{x_i - x_{i-1}}y''_i$$

$$- \frac{1}{6}(x_{i-1} - x_i)(y''_{i-1} - y''_i) + \frac{y_{i-1} - y_i}{x_{i-1} - x_i} \tag{4.65}$$

Prior to calculating the derivative from Eq. (4.65), the values of the second derivative at the base points should be calculated from Eq. (3.147). Note that if a natural spline interpolation is employed, the second derivatives for the first and the last intervals are equal to zero.

The reader can easily modify the MATLAB function *NaturalSPLINE.m* (see Example 3.2) in order to calculate at any point the first derivative of a function from a series of tabulated data. It is enough to replace the formula of the interpolation section with the differentiation formula, Eq. (4.65). Also, the MATLAB function *spline.m* is able to give the piecewise polynomial coefficients from which the derivative of the function can be evaluated. A good example of applying such a method can be found in Hanselman and Littlefield [3]. As mentioned before, *spline.m* applies not-a-knot algorithm for calculating the polynomial coefficients.

4.6 Integration Formulas

In the following sections we develop the integration formulas. This operation is represented by

$$I = \int_{x_0}^{x_n} f(x)\,dx \tag{4.66}$$

which is the integral of the function $y = f(x)$, or *integrand*, with respect to the independent variable x, evaluated between the limits $x = x_0$ to $x = x_n$. If the function $f(x)$ is such that it can be integrated analytically, the numerical methods are not needed for this problem. However, in many cases, the function $f(x)$ is very complicated, or the function is only a set of tabulated values of x and y, such as experimental data. Under these circumstances, the integral in Eq. (4.66) must be developed numerically. This operation is known as *numerical quadrature*.

It is known from differential calculus that the integral of a function $f(x)$ is equivalent to the area between the function and the x axis enclosed within the limits of integration, as shown in Fig. 4.1*a*. Any portion of the area that is below the x axis is counted as negative area (Fig. 4.1*b*). Therefore, one way of evaluating the integral

$$\int_{x_0}^{x_n} y\,dx$$

is to plot the function graphically and then simply measure the area enclosed by the function. However, this is a very impractical and inaccurate way of evaluating integrals.

A more accurate and systematic way of evaluating integrals is to perform the integration numerically. In the next two sections, we derive Newton-Cotes integration formulas for equally spaced intervals and Gauss quadrature for unequally spaced points.

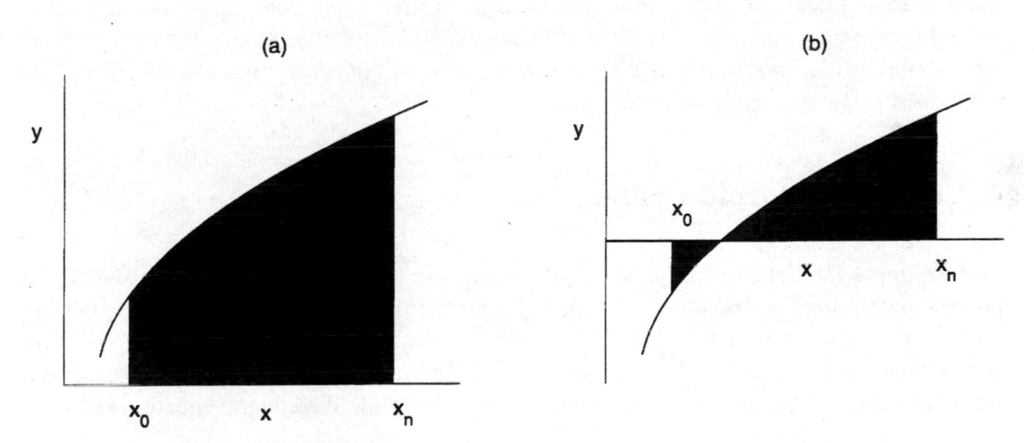

Figure 4.1 Graphical representation of the integral. (*a*) Positive area only (*b*) Positive and negative areas.

4.7 Newton-Cotes Formulas of Integration

This method is accomplished by first replacing the function $y = f(x)$ with a polynomial approximation, such as the Gregory-Newton forward interpolation formula [Eq. (3.122)]. In practice, the interval $[x_0, x_n]$ is being divided into several segments, each of width h, and the Gregory-Newton forward interpolation formula becomes (note that $x_{i+1} = x_i + h$):

$$y = y_0 + \frac{(x - x_0)}{h}\Delta y_0 + \frac{(x - x_0)(x - x_1)}{2!\,h^2}\Delta^2 y_0 + \frac{(x - x_0)(x - x_1)(x - x_2)}{3!\,h^3}\Delta^3 y_0 + \cdots \tag{4.67}$$

Because this interpolation formula fits the function exactly at a finite number of points (n + 1), we divide the total interval of integration $[x_0, x_n]$ into n segments, each of width h. In the next step, by using Eq. (4.67), Eq. (4.66) can be integrated. The upper limits of integration can be chosen to include an increasing set of segments of integration, each of width h. In each case, we retain a number of finite differences in the finite series of Eq. (4.67) equal to the number of segments of integration. This operation yields the well-known *Newton-Cotes formulas of integration*. The first three of the Newton-Cotes formulas are also known by the names *trapezoidal rule*, *Simpson's 1/3 rule* and *Simpson's 3/8 rule*, respectively. These are developed in the next three sections.

4.7.1 The Trapezoidal Rule

In developing the first Newton-Cotes formula, we use one segment of width h and fit the polynomial through two points (x_0, y_0) and (x_1, y_1) (see Fig. 4.2). This is tantamount to fitting a straight line between these points. We retain the first two terms of the Gregory-Newton polynomial (up to, and including, the first forward finite difference) and group together the rest of the terms of the polynomial into *remainder* term. Thus, the integral equation becomes

$$I_1 = \int_{x_0}^{x_1}\left[y_0 + \frac{(x - x_0)}{h}\Delta y_0\right]dx + \int_{x_0}^{x_1} R_n(x)\,dx \tag{4.68}$$

The first integral on the right-hand side is integrated with respect to x and the first forward difference is replaced with its definition of $\Delta y_0 = y_1 - y_0$, to obtain

$$I_1 = \frac{h}{2}(y_0 + y_1) + \int_{x_0}^{x_1} R_n(x)\,dx \tag{4.69}$$

The remainder term is evaluated as follows:

$$\int_{x_0}^{x_1} R_n(x)\,dx$$

$$= \int_{x_0}^{x_1} \left[\frac{(x - x_0)(x - x_1)}{2!\,h^2} \Delta^2 y_0 + \frac{(x - x_0)(x - x_1)(x - x_2)}{3!\,h^3} \Delta^3 y_0 + \dots \right] dx$$

$$= \left[-\frac{1}{12} h \Delta^2 y_0 + \frac{1}{24} h \Delta^3 y_0 - \dots \right] \tag{4.70}$$

The forward difference operators, Δ^2, Δ^3, . . . , are replaced by their equivalent in terms of differential operators [Eqs. (3.57) and (3.58)], and the remainder term becomes

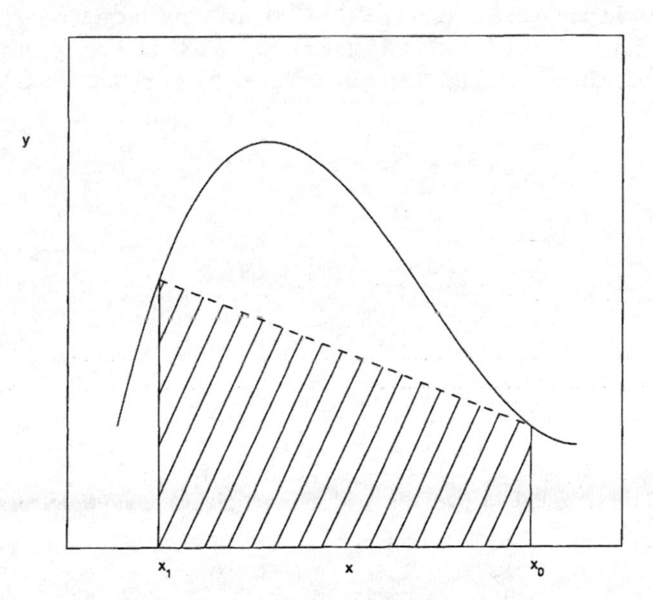

Figure 4.2 Enlargement of segment showing the application of the trapezoidal rule.

$$\int_{x_0}^{x_1} R_n(x)\,dx = -\frac{1}{12}h^3 D^2 y_0 - \frac{1}{24}h^4 D^3 y_0 + \dots \tag{4.71}$$

The remainder series can be replaced by one term evaluated at ξ_1; therefore,

$$\int_{x_0}^{x_1} R_n(x)\,dx = -\frac{1}{12}h^3 D^2 f(\xi_1) \tag{4.72}$$

This is a term of order h^3 and is abbreviated by $O(h^3)$. Therefore, Eq. (4.69) can be written as

$$I_1 = \frac{h}{2}(y_0 + y_1) + O(h^3) \tag{4.73}$$

This equation is known as the *trapezoidal rule*, because the term $(h/2)(y_0 + y_1)$ is essentially the formula for calculating the area of a trapezoid. In this case, the segment of integration is a trapezoid standing on its side. It was mentioned earlier that fitting a polynomial through only two points is equivalent to fitting a straight line through these points. This causes the shape of the integration segment to be a trapezoid, shown as the shaded area in Fig. 4.2. The area between $y = f(x)$ and the straight line represents the truncation error of the trapezoidal rule. If the function $f(x)$ is actually linear, then the trapezoidal rule calculates the integral exactly, because $D^2 f(\xi_1) = 0$, which causes the remainder term to vanish.

The trapezoidal rule in the form of Eq. (4.73) gives the integral of only one integration segment of width h. To obtain the total integral, Eq. (4.68) must be applied over each of the n segment (with the appropriate limits of integration) to obtain the following series of equations:

$$I_1 = \frac{h}{2}(y_0 + y_1) + O(h^3) \tag{4.73}$$

$$\cdot$$

$$I_2 = \frac{h}{2}(y_1 + y_2) + O(h^3) \tag{4.74}$$

$$\vdots$$

$$I_n = \frac{h}{2}(y_{n-1} + y_n) + O(h^3) \tag{4.75}$$

Addition of all these equations over the total interval gives the *multiple-segment trapezoidal rule*:

$$I = \frac{h}{2}\left(y_0 + 2\sum_{i=1}^{n-1} y_i + y_n \right) + nO(h^3) \tag{4.76}$$

For simplicity, the error term has been shown as $nO(h^3)$. This is only an approximation because the remainder term includes the second-order derivative of y evaluated at unknown values of ξ_i, each ξ_i being specific for that interval of integration. The absolute value of the error term cannot be calculated, but its relative magnitude can be measured by the order of the term. Because n is inversely proportional to h:

$$n = \frac{x_n - x_0}{h} \tag{4.77}$$

the error term for the multiple-segment trapezoidal rule becomes

$$nO(h^3) = \frac{x_n - x_0}{h} O(h^3) \simeq O(h^2) \tag{4.78}$$

That is, the repeated application of the trapezoidal rule over multiple segments has lowered the error term by approximately one order of magnitude. A more rigorous analysis of the truncation error is given in the next chapter.

4.7.2 Simpson's 1/3 Rule

In the derivation of the second Newton-Cotes formula of integration we use two segments of width h (see Fig. 4.3) and fit the polynomial through three points, (x_0, y_0), (x_1, y_1), and (x_2, y_2). This is equivalent to fitting a parabola through these points. We retain the first three terms of the Gregory-Newton polynomial (up to, and including, the second forward finite difference) and group together the rest of the terms of the polynomial into the remainder term. The integral equation becomes

$$I_1 = \int_{x_0}^{x_2} \left[y_0 + \frac{(x - x_0)}{h} \Delta y_0 + \frac{(x - x_0)(x - x_1)}{2!\,h^2} \Delta^2 y_0 \right] dx + \int_{x_0}^{x_2} R_n(x)\,dx \tag{4.79}$$

Integration of Eq. (4.79) and substitution of the relevant finite difference relations simplify this equation to

$$I_1 = \frac{h}{3}(y_0 + 4y_1 + y_2) - \frac{1}{90} h^5 D^4 f(\xi_1) \tag{4.80}$$

The error term is of order h^5 and may be abbreviated by $O(h^5)$. We would have expected to obtain an error term of $O(h^4)$ because three terms were retained in the Gregory-Newton polynomial. However, the term containing h^4 in the remainder has a zero coefficient, thus giving this fortuitous result. The final form of the second Newton-Cotes formula, which is better known as *Simpson's 1/3 rule*, is

$$I_1 = \frac{h}{3}(y_0 + 4y_1 + y_2) + O(h^5) \tag{4.81}$$

This equation calculates the integral over two segments of integration. Repeated application of Simpson's 1/3 rule over subsequent pairs of segments, and summation of all formulas over the total interval, gives the *multiple-segment Simpson's 1/3 rule*:

$$I = \frac{h}{3}\left(y_0 + 4\sum_{i=1}^{n/2} y_{2i-1} + 2\sum_{i=1}^{n/2-1} y_{2i} + y_n \right) + O(h^4) \tag{4.82}$$

Simpson's 1/3 rule fits *pairs* of segments, therefore the total interval must be subdivided into an *even* number of segments. The first summation term in Eq. (4.82) sums up the odd-subscripted terms, and the second summation adds up the even-subscripted terms.

The order of error of the multiple-segment Simpson's 1/3 rule was reduced by one order of magnitude to $O(h^4)$ for the same reason as in Sec. 4.7.1. Simpson's 1/3 rule is more accurate than the trapezoidal rule but requires additional arithmetic operations.

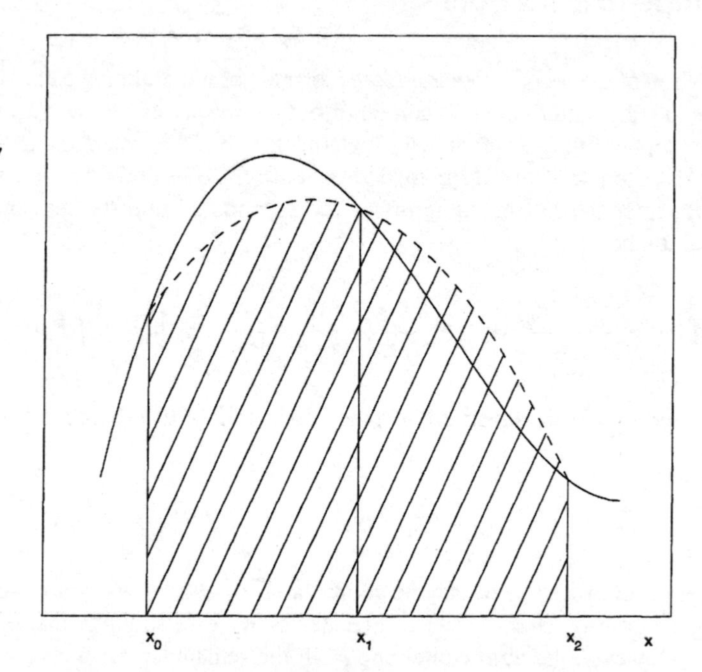

Figure 4.3 Application of Simpson's 1/3 rule over two segments of integration.

4.7.3 Simpson's 3/8 Rule

In the derivation of the third Newton-Cotes formula of integration we use three segments of width h (see Fig. 4.4) and fit the polynomial through four points, (x_0, y_0), (x_1, y_1), (x_2, y_2), and (x_3, y_3). This, in fact, is equivalent to fitting a cubic equation through the four points. We retain the first four terms of the Gregory-Newton polynomial (up to, and including, the third forward finite difference) and group together the rest of the terms of the polynomial into the remainder term. The integral equation becomes

$$I_1 = \int_{x_0}^{x_2}\left[y_0 + \frac{(x-x_0)}{h}\Delta y_0 + \frac{(x-x_0)(x-x_1)}{2!\,h^2}\Delta^2 y_0 + \frac{(x-x_0)(x-x_1)(x-x_2)}{3!\,h^3}\Delta^3 y_0\right]dx$$

$$+ \int_{x_0}^{x_2} R_n(x)\,dx \tag{4.83}$$

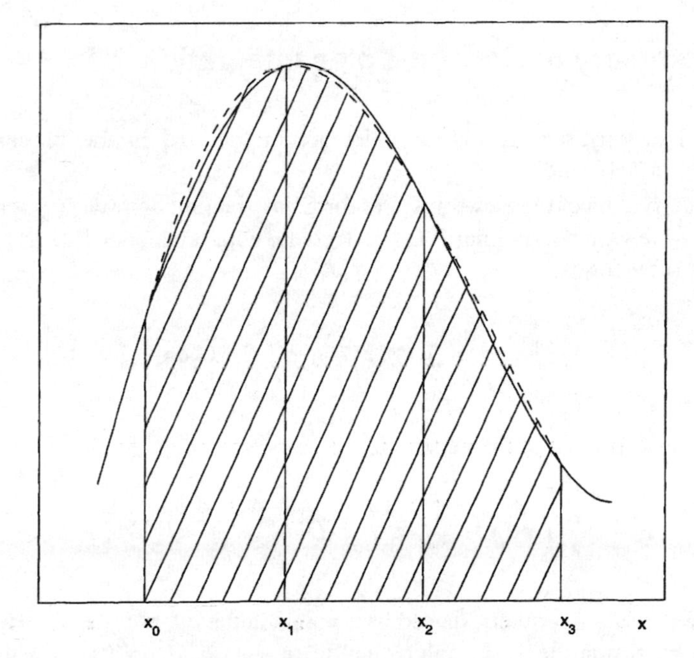

Figure 4.4 Application of Simpson's 3/8 rule over three segments of integration.

Integration of Eq. (4.83) and substitution of the relevant finite difference relations simplify the equation to

$$I_1 = \frac{3h}{8}(y_0 + 3y_1 + 3y_2 + y_3) - \frac{3}{80}h^5 D^4 f(\xi_1) \tag{4.84}$$

The error term is of order h^5 and may be abbreviated by $O(h^5)$. The final form of this equation, which is better known as *Simpson's 3/8 rule*, is given by

$$I_1 = \frac{3h}{8}(y_0 + 3y_1 + 3y_2 + y_3) + O(h^5) \tag{4.85}$$

The *multiple-segment Simpson's 3/8 rule* is obtained by repeated application of Eq. (4.83) over triplets of segments and summation over the total interval of integration:

$$I = \frac{3h}{8}\left(y_0 + 3\sum_{i=1}^{n/3}(y_{3i-2} + y_{3i-1}) + 2\sum_{i=1}^{n/3-1} y_{3i} + y_n \right) + O(h^4) \tag{4.86}$$

Comparison of the error terms of Simpson's 1/3 rule and Simpson's 3/8 rule shows that they are both of the same order, with the latter being only slightly more accurate. For this reason, Simpson's 1/3 rule is usually preferred, because it achieves the same order of accuracy with three points rather than the four points required by the 3/8 rule.

4.7.4 Summary of Newton-Cotes Integration

The three Newton-Cotes formulas of integration derived in the previous sections are summarized in Table 4.4.

In the derivation of the Newton-Cotes formulas, the function $y = f(x)$ is approximated by the Gregory-Newton polynomial $P_n(x)$ of degree n with remainder $R_n(x)$. The evaluation of the integral is performed:

$$\int_a^b y\,dx = \int_a^b P_n(x)\,dx + \int_a^b R_n(x)\,dx \tag{4.87}$$

This results in a formula of the general form:

$$\int_a^b y\,dx = \sum_{i=0}^{n} w_i y_i + O[h^{n+2}, D^{n+1}f(\xi)] \tag{4.88}$$

where the x_i are $(n + 1)$ equally spaced base points in the interval $[a, b]$. The weights w_i are determined by fitting the $P_n(x)$ polynomial to $(n + 1)$ base points. The integral is exact, that is,

$$\int_a^b y\,dx = \sum_{i=0}^{n} w_i y_i \tag{4.89}$$

for any function $y = f(x)$ that is of polynomial form up to degree n, because the derivative $D^{n+1}f(\xi)$ is zero for polynomials of degree $\le n$; thus, the error term $O[h^{n+2}, D^{n+1}f(\xi)]$ vanishes.

There are three functions in MATLAB, *trapz.m*, *quad.m*, and *quad8.m*, that numerically evaluate the integral of a vector or a function using different Newton-Cotes formulas:

- The function *trapz*(x, y) calculates the integral of y (vector of function values) with respect to x (vector of variables) using the trapezoidal rule.
- The function *quad*('*file_name*', a, b) evaluates the integral of the function represented in the m-file *file_name.m*, over the interval $[a, b]$ by Simpson's 1/3 rule.
- The function *quad8*('*file_name*', a, b) evaluates the integral of the function introduced
- in the m-file *file_name.m* from a to b using 8-interval (9-point) Newton-Cotes formula.

Table 4.4 Summary of the Newton-Cotes numerical integration formulas

Trapezoidal rule	$\int_{x_0}^{x_1} y\,dx = \frac{h}{2}(y_0 + y_1) - \frac{1}{12}h^3 D^2 f(\xi)$
Simpson's 1/3 rule	$\int_{x_0}^{x_2} y\,dx = \frac{h}{3}(y_0 + 4y_1 + y_2) - \frac{1}{90}h^5 D^4 f(\xi)$
Simpson's 3/8 rule	$\int_{x_0}^{x_3} y\,dx = \frac{3h}{8}(y_0 + 3y_1 + 3y_2 + y_3) - \frac{3}{80}h^5 D^4 f(\xi)$
General quadrature formula	$\int_{x_0}^{x_n} y\,dx = \sum_{i=0}^{n} w_i y_i + O[h^{n+2}, D^{n+1}f(\xi)]$

Example 4.3: Integration formulas–Trapezoidal and Simpson's 1/3 Rules. Write a general MATLAB function for integrating experimental data using Simpson's 1/3 rule. Compare the results of this function and the existing MATLAB function *trapz* (trapezoidal rule) for solution of the following problem:

Two very important quantities in the study of fermentation processes are the carbon dioxide evolution rate and the oxygen uptake rate. These are calculated from experimental analysis of the inlet and exit gases of the fermentor, and the flow rates, temperature, and pressure of these gases. The ratio of carbon dioxide evolution rate to oxygen uptake rate yields the respiratory quotient, which is a good barometer of the metabolic activity of the microorganism. In addition, the above rates can be integrated to obtain the total amounts of carbon dioxide produced and oxygen consumed during the fermentation. These total amounts form the basis of the material balancing technique described in Sec. 2.1. Table E4.3*a* shows a set of rates calculated from the fermentation of *Penicillium chrysogenum*, which produces penicillin antibiotics.

Using Simpson's 1/3 rule, calculate the total amounts of carbon dioxide produced and oxygen consumed during this 10-h period of fermentation. Repeat this using the trapezoidal rule and compare the results obtained from the two methods.

Table E4.3*a* Fermentation data

Time of fermentation (h)	Carbon dioxide evolution rate (g/h)	Oxygen uptake rate (g/h)
140	15.72	15.49
141	15.53	16.16
142	15.19	15.35
143	16.56	15.13
144	16.21	14.20
145	17.39	14.23
146	17.36	14.29
147	17.42	12.74
148	17.60	14.74
149	17.75	13.68
150	18.95	14.51

Method of Solution: In this problem, the carbon dioxide evolution rate data and the oxygen uptake rate data are integrated separately. There are 11 data points (10 intervals) for each rate; therefore, we can use either the trapezoidal rule or Simpson's 1/3 rule for this integration. We first use Simpson's 1/3 rule and then repeat using the trapezoidal rule, as the problem specifies.

Program Description: The MATLAB function *Simpson.m* first tests the input arguments, which are the vector of independent variable (x) and the vector of function values (y). These two vectors should be of the same length. Elements of vector x have to be equally spaced values. Also, the number of elements of these vectors (n) should be odd (even number of intervals). If the vectors contain an even number of elements (odd number of intervals), the function calculates the value of the integral up to the point (n - 1) and adds the value of the integral, approximated by the trapezoidal rule, for the last interval. The user should pay special attention to this case because the truncation errors for Simpson's 1/3 rule and trapezoidal rule are not of the same order. After checking the above conditions, the function calculates the value of the integral based on Eq. (4.82). If necessary, the function adds the value of the integral for the last segment according to Eq. (4.75).

The main program *Example4_3.m* asks the user to input the data from the keyboard, calls the functions *trapz* and *Simpson* for integration, and displays the results.

Program

Example4_3.m

```
% Example4_3.m
% Solution to Example 4.3. It calculates carbon dioxide evolved and
% oxygen uptaken in a fermentation process using TRAPZ (trapezoidal
% rule) and SIMPSON (Simpson's 1/3 rule) functions.

clear
clc

% Input data
t = input(' Vector of time = ');
r_CO2 = input(' Carbon dioxide evolution rate (g/h) = ');
r_O2 = input(' Oxygen uptake rate (g/h) = ');

% Integration
m1CO2 = trapz(t,r_CO2);
m2CO2 = Simpson(t,r_CO2);

m1O2 = trapz(t,r_O2);
m2O2 = Simpson(t,r_O2);

% Output
fprintf('\n Total carbon dioxide evolution   = %9.4f (evaluated by
the trapezoidal rule)',m1CO2)
fprintf('\n Total carbon dioxide evolution   = %9.4f (evaluated by
the Simpson 1/3 rule)',m2CO2)
```

```
fprintf('\n Total oxygen uptake                    = %9.4f (evaluated by
the trapezoidal rule)',m1O2)
fprintf('\n Total oxygen uptake                    = %9.4f (evaluated by
the Simpson 1/3 rule)\n',m2O2)
```

Simpson.m

```
function Q = Simpson(x , y)
%SIMPSON Numerical evaluation of integral by Simpson's 1/3 rule.
%
%     SIMPSON(X,Y) numerically evaluates the integral of the
%     vector of function values Y with respect to X by
%     Simpson's 1/3 rule. X is the vector of equally spaced
%     independent variable. Length of Y has to be odd (even
%     number of intervals). If length of Y is even, the function
%     calculates the integral for [LENGTH(Y)-1] points by
%     Simpson's 1/3 rule and adds to it the value of the
%     integral for the last interval by trapezoidal rule.
%
%     See also TRAPZ , QUAD , QUAD8, GAUSSLEGENDRE

% (c) N. Mostoufi & A. Constantinides
% January 1, 1999

points = length(x);
if length(y) ~= points
   error('x and y are not of the same length')
   break
end

dx = diff(x);
maxi = max([min(abs(x))/1000 , 1e-10]);
if max(dx)-min(dx) > maxi
   error('X is not equally spaced.')
   break
end
h = dx(1);

if mod(points,2) == 0
   warning('Odd number of intervals; Trapezoidal rule will be used
for the last interval.')
   n = points - 1;
else
   n = points;
end

% Integration
y1 = y(2 : 2 : n - 1);
y2 = y(3 : 2 : n - 2);
Q = (y(1) + 4 * sum(y1) + 2 * sum(y2) + y(n)) * h /3;

if n ~= points
   Q = Q + (y(points) + y(n)) * h / 2;
end
```

Input and Results

```
>>Example4_3

 Vector of time = [140:150]
 Carbon dioxide evolution rate (g/h) = [15.72, 15.53, 15.19, 16.56,
16.21, 17.39, 17.36, 17.42, 17.60, 17.75, 18.95]
 Oxygen uptake rate (g/h) = [15.49, 16.16, 15.35, 15.13, 14.20,
14.23, 14.29, 12.74, 14.74, 13.68, 14.51]

 Total carbon dioxide evolution   =  168.3450 (evaluated by the
trapezoidal rule)
 Total carbon dioxide evolution   =  168.6633 (evaluated by the
Simpson 1/3 rule)
 Total oxygen uptake              =  145.5200 (evaluated by the
trapezoidal rule)
 Total oxygen uptake              =  144.9733 (evaluated by the
Simpson 1/3 rule)
```

Discussion of Results: The integration of the experimental data, using both Simpson's 1/3 rule and the trapezoidal rule, yield the total amounts of carbon dioxide and oxygen shown in Table E4.3*b*.

Table E4.3*b*

	Simpson's 1/3	Trapezoidal
Total CO_2 (g)	168.6633	168.3450
Total O_2 (g)	144.9733	145.5200

4.8 Gauss Quadrature

In the development of the Newton-Cotes formulas, we have assumed that the interval of integration could be divided into segments of equal width. This is usually possible when integrating continuous functions. However, if experimental data are to be integrated, such data may be used with a variable-width segment. It has been suggested by Chapra and Canale [4] that a combination of the trapezoidal rule with Simpson's rules may be feasible for integrating certain sets of unevenly spaced data points.

Gauss quadrature is a powerful method of integration that employs unequally spaced base points. This method uses the Lagrange polynomial to approximate the function and then applies orthogonal polynomials to locate the loci of the base points. If no restrictions are placed on the location of the base points, they may be chosen to be the locations of the roots of certain orthogonal polynomials in order to achieve higher accuracy than the Newton-Cotes formulas for the same number of base points. This concept is used in the Gauss quadrature method, which is discussed in this section.

4.8.1 Two-Point Gauss-Legendre Quadrature

In order to illustrate the approach, we first develop the integration formula for the two-point problem. In Newton-Cotes method, the location of the base points is determined, and integration is done based on the values of the function at these base points. This is shown in Fig. 4.2, for the trapezoidal rule that approximates the integral by taking the area under the straight line connecting the function values at the ends of the integration interval.

Now, consider the case that the restriction of fixed points is withdrawn, and we are able to estimate the integral from the area under a straight line that joins any two points on the curve. By choosing these points in proper positions, a straight line that balances the positive and negative errors can be drawn, as illustrated in Fig. 4.5. As a result, we obtain an improved estimate of the integral.

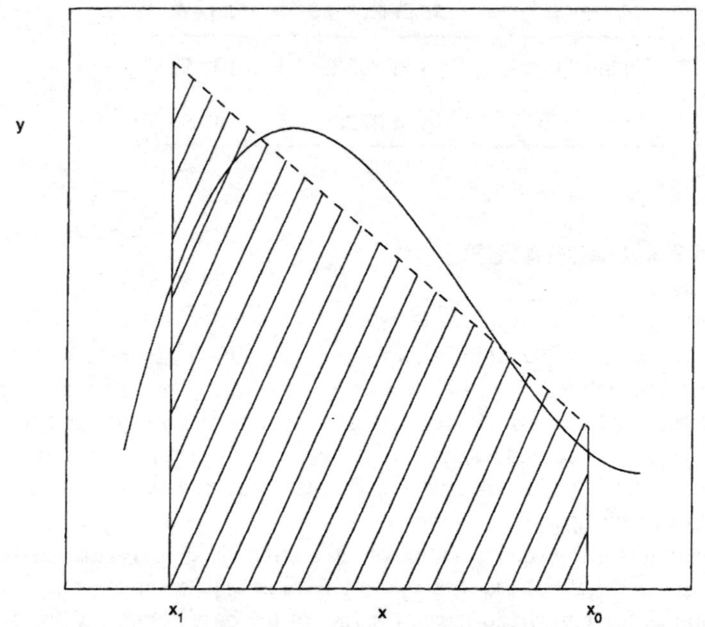

Figure 4.5 Application of two-point Gauss quadrature to improve integral estimation.

In order to derive the two-point Gauss quadrature, the function $y = f(x)$ is replaced by a linear polynomial and a remainder:

$$y = \left[\frac{x - x_1}{x_0 - x_1} y_0 + \frac{x - x_0}{x_1 - x_0} y_1 \right] + R(x) \tag{4.90}$$

The integral $\int_a^b y\,dx$ is evaluated by

$$\int_a^b y\,dx = \int_a^b \left[\frac{x - x_1}{x_0 - x_1} y_0 + \frac{x - x_0}{x_1 - x_0} y_1 \right] dx + \int_a^b R(x)\,dx \tag{4.91}$$

Without loss of generality, the interval $[a, b]$ is changed to $[-1, 1]$. The general transformation equation for converting between x in interval $[a, b]$ and z in interval $[c, d]$ is the following:

$$z = \frac{2x - (a + b)}{b - a} \tag{4.93}$$

For converting to the interval [-1, 1], this equation becomes
Using Eq. (4.93), the transformed integral is given by

$$\int_a^b y(x)\,dx = \frac{b - a}{2} \int_{-1}^{1} Y(z)\,dz \tag{4.94}$$

and

$$\int_{-1}^{1} Y\,dz = w_0 Y_0 + w_1 Y_1 + \int_{-1}^{1} R(z)\,dz \tag{4.95}$$

where use of Y (instead of y) indicates that the function value at the variable z (rather than x) should be used. The weights w_0 and w_1 are calculated from

$$w_0 = \int_{-1}^{1} \frac{z - z_1}{z_0 - z_1} dz = \frac{-2z_1}{z_0 - z_1} \tag{4.96}$$

and

$$w_1 = \int_{-1}^{1} \frac{z - z_0}{z_1 - z_0} dz = \frac{-2z_0}{z_1 - z_0} \tag{4.97}$$

Up to this point, the development of this method is equivalent to that of the trapezoidal rule. The Gauss quadrature method goes a step beyond this in order to make the error term in Eq. (4.95) vanish. To do so, the integral of the error term is expanded in terms of 2nd-degree Legendre polynomial (see Sec. 3.10):

$$\int_{-1}^{1} R(z)\,dz = \frac{3}{2}z^2 - \frac{1}{2} \tag{4.98}$$

The values of z_0 and z_1 are chosen as the root of the 2nd-degree Legendre polynomial, that is,

$$z_{0,1} = \pm\frac{1}{\sqrt{3}}$$

This choice of roots causes the error term to vanish. Therefore, Eq. (4.95) becomes

$$\int_{-1}^{1} Y\,dz = w_0 Y_0 + w_1 Y_1 = Y_0 + Y_1 \tag{4.99}$$

Calculation of the integral through Eqs. (4.94) and (4.98) imply that instead of evaluating the function at $z_0 = -1$ and $z_1 = 1$ (using function values at base points), which is the case in the trapezoidal rule, function values at $z_0 = -1/\sqrt{3}$ and $z_1 = 1/\sqrt{3}$ should be used in the Gauss quadrature method. This results in improving the precision of calculation, as illustrated in Fig. 4.5. This is roughly equivalent to the application of five-point trapezoidal rule.

The Gauss quadrature formula developed in this section is known as the *Gauss-Legendre quadrature* because of the use of the Legendre polynomials. Other orthogonal polynomials, such as Chebychev, Laguerre, or Hermite, may be used in a similar manner to develop a variety of Gauss quadrature formulas.

4.8.2 Higher-Point Gauss-Legendre Formulas

The function $y = f(x)$ is replaced by the Lagrange polynomial (see Sec. 3.9.1) and its remainder,

$$y = \sum_{i=0}^{n} L_i y_i + R_n(x) \tag{4.100}$$

where

$$L_i = \prod_{\substack{j=0 \\ j\neq i}}^{n} \frac{x - x_j}{x_i - x_j} \tag{4.101}$$

and

$$R_n(x) = \prod_{i=0}^{n} (x - x_i) \frac{f^{(n+1)}(\xi)}{(n+1)!} \qquad a < \xi < b \tag{3.137}$$

The integral $\int_a^b y\,dx$ is evaluated by

$$\int_a^b y\,dx = \int_a^b \left[\sum_{i=0}^{n} L_i y_i \right] dx + \int_a^b R_n(x)\,dx \tag{4.102}$$

Converting the interval from [a, b] to [-1, 1] through Eqs. (4.92) and (4.93), the transformed integral is given by

$$\int_{-1}^{1} Y\,dz = \sum_{i=0}^{n} w_i Y_i + \int_{-1}^{1} R_n(z)\,dz \tag{4.103}$$

where the weights w_i are calculated from

$$w_i = \int_{-1}^{1} L_i(z)\,dz = \int_{-1}^{1} \prod_{\substack{j=0 \\ j \neq i}}^{n} \frac{z - z_j}{z_i - z_j}\,dz \tag{4.104}$$

and the error term is given by

$$\int_{-1}^{1} R_n(z)\,dz = \int_{-1}^{1} \prod_{i=0}^{n} (z - z_i)\, q_n(z)\,dz \tag{4.105}$$

The $q_n(z)$ and $\prod_{i=0}^{n} (z - z_i)$ are polynomials of degree n and $(n + 1)$, respectively.

Up to this point, the development of this method is different from that of the Newton-Cotes formulas in only one respect: the use of Lagrange interpolation formula for unequally spaced points instead of the Gregory-Newton formula. The Gauss quadrature method goes a step beyond this in order to make the error term [Eq. (4.105)] vanish. To do so, the two polynomials in the error term are expanded in terms of Legendre orthogonal polynomials (see Sec. 3.10). The values of z_i are chosen as the roots of the $(n + 1)$st-degree Legendre polynomial. This choice of roots combined with the orthogonality property [Eq. (3.149)] of the Legendre polynomials causes the error term to vanish. Therefore, Eq. (4.103) becomes

$$\int_{-1}^{1} Y\,dz = \sum_{i=0}^{n} w_i Y_i \tag{4.106}$$

Since the vanishing error term was of degree $(n + 1)$, Eq. (4.106) yields the integral of the function Y *exactly* when Y is a polynomial of degree $(2n + 1)$ or less. In effect, the judicious choice of the $(n + 1)$ base points at the $(n + 1)$ roots of the Legendre polynomial has increased the accuracy of the integration from n to $(2n + 1)$. As usual, however, the increase in accuracy has been obtained at the cost of having to perform a larger number of arithmetic calculations. The error of Gauss-Legendre formulas is given by [5]

$$\int_a^b R_n(x)\,dx = \frac{2^{2n+3}\,[(n+1)!]^4}{(2n+3)\,[(2n+2)!]^3} f^{(2n+2)}(\xi) \qquad a < \xi < b \tag{4.107}$$

The roots z_i of the Legendre polynomials can be evaluated after calculating the coefficients of the polynomial from the formula given in Table 3.7. The values of the weights w_i corresponding to these roots have been calculated for the integration interval [-1, 1]. Table 4.5 lists the roots and weights of the Gauss-Legendre quadrature for selected values of n.

Example 4.4: Integration Formulas - Gauss-Legendre Quadrature. Write a general MATLAB function for integrating a function using a general Gauss-Legendre quadrature. Apply this function for the solution of the following problem:

A cold liquid film, initially at temperature T_0, is falling down (in z-direction) a vertical solid wall (xz-plane). The solid wall is maintained at a temperature (T_S) higher than that of the falling film. It is desired to know the temperature profile of the fluid as a function of y and z, near the wall. The partial differential equation that describes the temperature of the liquid for this problem is

$$\rho C_p v_z \frac{\partial T}{\partial z} = k \frac{\partial^2 T}{\partial y^2}$$

where ρ is the density of the liquid, C_p is heat capacity of the liquid, v_z is the velocity of the liquid, k is the thermal conductivity of the liquid, and T is the temperature of the liquid.

The velocity profile of the falling liquid is given by Bird et al. [2]:

$$v_z = \frac{\delta^2 \rho g}{2\mu}\left[2\frac{y}{\delta} - \left(\frac{y}{\delta}\right)^2\right]$$

where δ is thickness of the film, g is gravity acceleration, and μ is the viscosity of the liquid. Therefore, near the wall, where $y \ll \delta$, the velocity simplifies to:

$$v_z = \frac{\rho g \delta}{\mu} y$$

Putting this velocity profile into the energy balance equation, we get

$$y \frac{\partial T}{\partial z} = \beta \frac{\partial^2 T}{\partial y^2}$$

Table 4.5 Roots of Legendre polynomials $P_{n+1}(z)$ and the weight factors for the Gauss-Legendre quadrature

Number of points	Roots (z_i)	Weight factors (w_i)
Two-point formula	±0.57735026918926	1.00000000000000
($n + 1 = 2$)		
Three-point formula	0	0.888888888888888
($n + 1 = 3$)	±0.774596669241483	0.555555555555555
Four-point formula	±0.339981043584856	0.652145154862546
($n + 1 = 4$)	±0.861136311594053	0.347854845137454
Five-point formula	0	0.568888888888889
($n + 1 = 5$)	±0.538469310105683	0.478628670499366
	±0.906179845938664	0.236926885056189
Six-point formula	±0.238619186083197	0.467913934572691
($n + 1 = 6$)	±0.661209386466265	0.360761573048139
	±0.932469514203152	0.171324492379170
Ten-point formula	±0.148874338981631	0.295524224714753
($n + 1 = 10$)	±0.433395394129247	0.269266719309996
	±0.679409568299024	0.219086362515982
	±0.865063366688985	0.149451349150581
	±0.973906528517172	0.066671344308688
Fifteen-point formula	0	0.202578241925561
($n + 1 = 15$)	±0.201194093997435	0.198431485327111
	±0.394151347077543	0.186161000115562
	±0.570972172608539	0.166269205816994
	±0.724417731360170	0.139570677926154
	±0.848206583410427	0.107159220467172
	±0.937273392400706	0.070366047488108
	±0.987992518020485	0.030753241996117

in which $\beta = \mu k / \rho^2 C_p g \delta$. For short contact time, we may write the boundary conditions as

$$\begin{array}{lll} \text{At } z = 0, & T = T_0 & \text{for } y > 0 \\ \text{At } y = 0, & T = T_s & \text{for } z > 0 \\ \text{At } y = \infty, & T = T_0 & \text{for } z \text{ finite} \end{array}$$

The analytical solution to this problem is [2]

$$\Theta = \frac{T - T_0}{T_s - T_0} = 1 - \frac{1}{\Gamma(\frac{4}{3})} \int_0^{\eta} e^{-\eta^3} d\eta$$

where $\eta = y / \sqrt[3]{9\beta z}$ is a dimensionless variable, and the Gamma function $\Gamma(x)$ is defined as

$$\Gamma(x) = \int_0^{\infty} t^{x-1} e^{-t} dt \qquad x > 0$$

Using Gauss-Legendre quadrature calculate the above temperature profile and plot it against η.

Method of Solution: In order to evaluate the temperature profile (Θ), we first have to integrate the function $e^{-\eta^3}$ for several values of $\eta \geq 0$. The temperature profile itself, then, can be calculated from the equation described above.

Program Description: The function *GaussLegendre.m* numerically evaluates the integral of a function by n-point Gauss-Legendre quadrature. The program checks the inputs to the function to be sure that they have valid values. If no value is introduced for the integration step, the function sets it to the integration interval. Also, the default value for the number of points of the Gauss-Legendre formula is two.

The next step in the function is the calculation of the coefficients of the *n*th degree Legendre polynomial. Once these coefficients are calculated, the program evaluates the roots of the Legendre polynomial (z_1 to z_n) using the MATLAB function *roots*. Then, the function calculates the coefficients of the Lagrange polynomial terms (L_1 to L_n) and evaluates the weight factors, w_i, as defined in Eq. (4.104). Finally, using the values of z_i and w_i, the integral is numerically evaluated by Eq. (4.106).

In order to solve the problem described in this example, the main program *Example4_4.m* is written to calculate the temperature profile for specific range of the dimensionless number η. The function to be integrated is introduced in the MATLAB function *Ex4_4_func.m.*

Program

Example4_4.m

```
% Example4_4.m
% Solution to Example 4.4. It calculates and plots the temperature
```

```
% profile of a liquid falling along a wall of different temperature.
% The program uses GAUSSLEGENDRE function to evaluate the integral.

clear
clc
clf

% Input data
eta = input(' Vector of independent variable (eta) = ');
h = eta(2) - eta(1);    % Step size
fname = input(' Name of m-file containing the function subject to
integration = ');

% Calculation of the temperature profile
for k = 1 : length(eta)
   theta(k) = GaussLegendre(fname,0,eta(k),h);
end
theta = 1 - theta / gamma(4/3);

% Plotting the results
plot(eta,theta)
xlabel('\eta')
ylabel('\theta')
```

Ex4_4_func.m

```
function y = Ex4_4_func(x)
% Function Ex4_4_func.m
% Function to be integrated in Example 4.4.

y = exp(-x^3);
```

GaussLegendre.m

```
function Q = GaussLegendre(fnctn,a,b,h,n,varargin)
%GAUSSLEGENDRE Gauss-Legendre quadrature
%
%   GAUSSLEGENDRE('F',A,B,H,N) numerically evaluates the
%   integral of the function described by M-file F.M from A to B,
%   using interval spacing H, by a N-point Gauss-Legendre
%   quadrature.
%
%   GAUSSLEGENDRE('F',A,B,[],[],P1,P2,...) calculates the
%   integral using interval spacing H=B-A and N=2 and also allows
%   parameters P1, P2, ... to pass directly to function F.M
%
%   See also QUAD, QUAD8, TRAPZ, SIMPSON

% (c) N. Mostoufi & A. Constantinides
% January 1, 1999
```

```
% Checking input arguments
if nargin < 4 | isempty(h) | h == 0 | abs(b - a) < abs(h)
   h = b - a;
end
if nargin < 5 | isempty(n) | n < 2
   n = 2;
end
if sign(h) ~= sign(b-a)
   h = - h;
end
n = fix(n);

% Coefficients of the Legendre polynomial
for k = 0 : n/2
   cl(2*k+1) = (-1)^k * gamma(2*n - 2*k + 1) / ...
      (2^n * gamma(k + 1) * gamma(n - k + 1) * gamma(n - 2*k + 1));
   if k < n/2
      cl(2*k+2) = 0;
   end
end
z = roots(cl);   % Roots of the Legendre polynomial (Zi)
% Weight factors
for p = 1 : n
   B = [1 0];
   k = 0;
   denom = 1;
   A(1) = B(1);
   % Constructing vector of coefficients of the
   % Lagrange polynomial (coefficients of Li)
   for q = 1 : n
      if q ~= p
         k = k + 1;
         for r = 2 : k+1
            A(r) = B(r) - B(r-1) * z(q);    % Vector of coefficients
         end
         denom = denom * (z(p) - z(q));     % Denominator of Li
      end
      B = [A 0];
   end
   % Vector of coefficients of integral of Lagrange polynomial
   for k = 1 : n
      % Ai are coefficients of the integral polynomial
      Ai(k) = A(k) / (n - k + 1);
   end
   Ai(n + 1) = 0;
   % weight factor Wi
   w(p) = (polyval(Ai , 1) - polyval(Ai , -1)) / denom;
   clear A
end
```

```
Q = 0;
% Integration
for x = a : h : b - h
   for p = 1 : n
      xp = x + (z(p) + 1) * h / 2;
      Q = Q + w(p) * feval(fnctn , xp , varargin{:}) * h /2;
   end
end

% Integral of the remainder interval (if (b-a)/h is not an integer)
xr = x + h;
hr = b - xr;
if hr > 0
   for p = 1 : n
      xp = xr + (z(p) + 1) * hr / 2;
      Q = Q + w(p) * feval(fnctn , xp , varargin{:}) * hr /2;
   end
end
```

Input and Results

```
>>Example4_4

Initial value of Vector of independent variable (eta) = [0: 0.2: 2]
Name of m-file containing the function subject to integration =
'Ex4_4_func'
```

Discussion of Results: The temperature profile of the liquid near the wall is calculated by the program *Example4_4.m* for $0 \leq \eta \leq 2$ and is plotted in Fig. E4.4. We can verify the solution at the boundaries of y and z from Fig. E4.4:

- The results represented in Fig. E4.4 show that at $\eta = 0$, the temperature of the liquid is identical to that of the plate (that is, $\Theta = 1$, therefore, $T = T_s$). The variable η attains a value of zero at only two situations:
  a. In the liquid next to the wall (at $y = 0$ and at all values of z).
  b. After an infinite distance from the origin of flow (at $z = \infty$ and at all values of y).
  Situation *a* is consistent with the boundary conditions given in the statement of the problem whereas situation *b* is an expected result, since passing a long-enough distance along the wall, all the liquid will be at the same temperature as the wall.
- Fig. E4.4 also shows that at high-enough dimensionless number η the temperature of the liquid is equal to the initial temperature of the liquid, that is,

$$\lim_{\eta \to \infty} \Theta = 0$$

- The variable η becomes infinity under the following circumstances:
  a. In the fluid far away from the wall (at $y = \infty$ and at all values of z).
  b. At the origin of the flow (at $z = 0$ and at all values of y).

Both these situations are specified as boundary conditions of the problem.

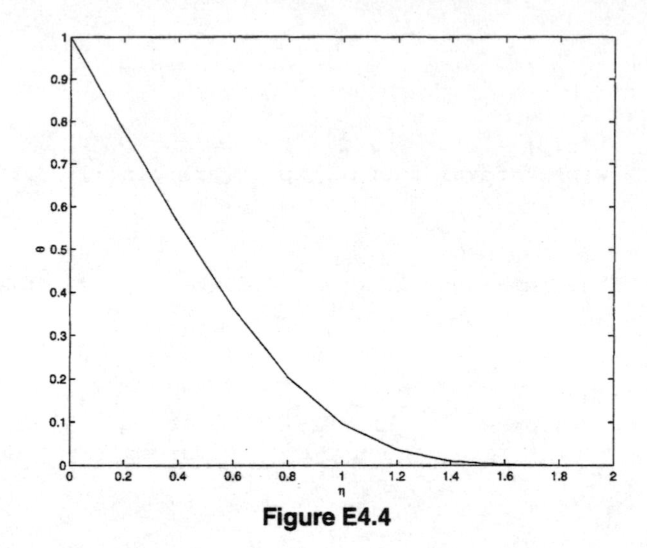

Figure E4.4

4.9 SPLINE INTEGRATION

Another method of integrating unequally spaced data points is to interpolate the data using a suitable interpolation method, such as cubic splines, and then evaluate the integral from the relevant polynomial. Therefore, the integral of Eq. (4.66) may be calculated by integrating Eq. (3.143) over the interval $[x_{i-1}, x_i]$ and summing up these terms for all the intervals:

$$\int_{x_0}^{x_n} y\,dx = \sum_{i=1}^{n}\left[\frac{1}{2}(x_i - x_{i-1})(y_i + y_{i-1}) + \frac{1}{36}(x_{i-1} - x_i)^3(y''_{i-1} + y''_i)\right] \tag{4.108}$$

Prior to calculating the integral from Eq. (4.108), the values of the second derivative at the base points should be calculated from Eq. (3.147). Note that if a natural spline interpolation is employed, the second derivatives for the first and the last intervals are equal to zero. Eq. (4.108) is basically an improved trapezoidal formula in which the value of the integral by trapezoidal rule [the first term in the bracket of Eq. (4.108)] is corrected for the curvature of the function [the second term in the bracket of Eq. (4.108)].

The reader can easily modify the MATLAB function *NaturalSPLINE.m* (see Example 3.2) in order to calculate the integral of a function from a series of tabulated data. It is enough to replace the formula of the interpolation section with the integration formula, Eq. (4.108). Also, the MATLAB function *spline.m* is able to give the piecewise polynomial coefficients from which the integral of the function can be evaluated. A good example of applying such a method can be found in Hanselman and Littlefield [3]. Remember that *spline.m* applies the not-a-knot algorithm for calculating the polynomial coefficients.

4.10 MULTIPLE INTEGRALS

In this section, we discuss the evaluation of double integrals. Evaluation of integrals with more than two dimensions can be obtained in a similar manner. Let us start with a simple case of double integral with constant limits, that is, integration over a rectangle in the xy plane:

$$\begin{aligned} I &= \int_a^b \int_c^d f(x,y)\, dy dx \\ &= \int_a^b \left(\int_c^d f(x,y)\, dy \right) dx \end{aligned} \tag{4.109}$$

The inner integral may be calculated by one of the methods described in Secs. 4.7-4.9. We use the trapezoidal rule [Eq. (4.76)] for simplicity:

$$\int_c^d f(x,y)\, dy = \frac{k}{2}\left[f(x,c) + 2\sum_{j=1}^{m-1} f(x,y_j) + f(x,d) \right] \tag{4.110}$$

where m is the number of divisions and k is the integration step in the y-direction, and x is considered to be constant. Replacing Eq. (4.110) into Eq. (4.109) results in

$$I = \frac{k}{2}\int_a^b f(x,c)\, dx + k\sum_{j=1}^{m-1} \int_a^b f(x,y_j)\, dx + \frac{k}{2}\int_a^b f(x,d)\, dx \tag{4.111}$$

Now we apply the trapezoidal rule to each of the integrals of Eq. (4.111):

$$\int_a^b f(x,y_j)\, dx = \frac{h}{2}\left[f(a,y_j) + 2\sum_{i=1}^{n-1} f(x_i,y_j) + f(b,y_j) \right] \tag{4.112}$$

Here n is the number of divisions and h is the integration step in the x-direction, and y_j is considered to be constant.

Finally, we combine Eqs. (4.111) and (4.112) to calculate the estimated value of the integral (4.109):

$$I = \frac{hk}{4}\left[f(a,c) + 2\sum_{i=1}^{n-1} f(x_i,c) + f(b,c)\right]$$

$$+ \frac{hk}{2}\left[\sum_{j=1}^{m-1} f(a,y_j) + 2\sum_{j=1}^{m-1}\sum_{i=1}^{n-1} f(x_i,y_j) + \sum_{j=1}^{m-1} f(b,y_j)\right]$$

$$+ \frac{hk}{4}\left[f(a,d) + 2\sum_{i=1}^{n-1} f(x_i,d) + f(b,d)\right] \tag{4.113}$$

The method described above may be slightly modified to be applicable to the double integrals with variable inner limits of the form

$$I = \int_a^b \int_{c(x)}^{d(x)} f(x,y)\,dy\,dx \tag{4.114}$$

Because the length of the integration interval for the inner integral (that is, $[c, d]$) changes with the value of x, we may either keep the number of divisions constant in the y-direction and let the integration step change with x $[k = k(x)]$ or keep the integration step in the y-direction constant and use different number of divisions at each x value $[m = m(x)]$. However, in order to maintain the same order of error throughout the calculation, the second condition (that is, constant step size) should be employed. Therefore, Eq. (4.110) can be written at each position x_i in the following form to count for the variable limits:

$$\int_{c(x_i)}^{d(x_i)} f(x_i,y)\,dy = \frac{k}{2}\left[f(x_i,c(x_i)) + 2\sum_{j=1}^{m_i-1} f(x_i,y_j) + f(x_i,d(x_i))\right] \tag{4.115}$$

where m_i indicates that the number of divisions in the y-direction is a function of x. In practice, at each x value, we may have to change the step size k slightly to obtain an integer value for the number of divisions. Although this does not change the order of magnitude of the step size, we have to acknowledge this change at each step of outer integration; therefore, the approximate value of the integral (4.114) is calculated from

$$I = \frac{hk_0}{4}\left[f(a,c(a)) + 2\sum_{i=1}^{n-1} f(x_i,c(x_i)) + f(b,c(b))\right]$$
$$+ \frac{h}{2}\sum_{i=1}^{n-1}\left\{k_i \sum_{j=1}^{m_i-1}\left[f(a,y_j) + 2f(x_i,y_j) + f(b,y_j)\right]\right\}$$
$$+ \frac{hk_n}{4}\left[f(a,d(a)) + 2\sum_{i=1}^{n-1} f(x_i,d(x_i)) + f(b,d(b))\right] \tag{4.116}$$

If writing a computer program for evaluation of double integrals, it is not necessary to apply Eqs. (4.113) and (4.115) in such a program. As a matter of fact, any ordinary integration function may be applied to evaluate the inner integral at each value of the outer variable; then the same function is applied for the second time to calculate the outer integral. This algorithm can be similarly applied to the multiple integrals of any dimension. The MATLAB function *dblquad* evaluates double integral of a function with fixed inner integral limits.

PROBLEMS

4.1 Derive the equation that expresses the third-order derivative of y in terms of backward finite differences, with

(a) Error of order h
(b) Error of order h^2.

4.2 Repeat Prob. 4.1, using forward finite differences.

4.3 Derive the equations for the first, second, and third derivatives of y in terms of backward finite differences with error of order h^3.

4.4 Repeat Prob. 4.3, using forward finite differences.

4.5 Derive the equation which expresses the third-order derivative of y in terms of central finite differences, with

(a) Error of order h^2
(b) Error of order h^4.

4.6 Derive the equations for the first, second, and third derivatives of y in terms of central finite differences with error of order h^6.

4.7 Velocity profiles of solids in a bed of sand particles fluidized with air at the superficial velocity of 1 m/s are given in Tables P4.7*a* and *b*. Calculate the axial gradient of velocities (that is, $\partial V_z / \partial z$ and $\partial V_r / \partial z$). Plot the z-averaged gradients versus radial position and compare their order of magnitude.

Table P4.7*a* Radial velocity profile (mm/s)

		Radial position (mm)							
		4.7663	14.2988	23.8313	33.3638	42.8962	52.4288	61.9612	71.4938
Axial position, mm	25	-13.09	-37.66	-52.41	-54.44	-58.21	-41.35	-23.97	-7.21
	75	-15.81	-15.99	-27.81	-25.37	-22.3	-11.1	-2.26	1.63
	125	1.77	1.17	3.45	5.5	1.63	-1.79	-0.26	1.09
	175	1.43	-0.57	4.86	2.44	0.2	-0.65	0.35	2.21
	225	-5.07	-7.26	-18.43	-18.17	-17.3	-10	-2.65	0.29
	275	13.11	16.51	19.32	21	20.29	15.64	0.98	-9.81
	325	11.7	34.5	58.3	71.44	73.49	64.88	50.91	19.14
	375	8.18	25.29	31.18	37.07	30.05	2.61	-17.06	-15.88
	425	3.35	-0.39	-18	-42.22	-57.42	-82.36	-69.34	-17.35
	475	-27.05	-22.25	-49.45	-79.45	-110.08	-116.62	-128.25	-76.49

4.8 In studying the mixing characteristics of chemical reactors, a sharp pulse of a nonreacting tracer is injected into the reactor at time $t = 0$. The concentration of material in the effluent from the reactor is measured as a function of time $c(t)$. The residence time distribution (RTD) function for the reactor is defined as

$$E(t) = \frac{c(t)}{\int_0^\infty c(t)\,dt}$$

and the cumulative distribution function is defined as

$$F(t) = \int_0^t E(t)\,dt$$

The mean residence time of the reactor is calculated from

$$t_m = \frac{V}{q} = \int_0^\infty tE(t)\,dt$$

Table P4.7*b* Axial velocity profile (mm/s)

Axial position, mm	Radial position (mm) 4.7663	14.2988	23.8313	33.3638	42.8962	52.4288	61.9612	71.4938
25	93.33	74.12	69.35	43.68	18.8	-6.9	-21.56	-22.65
75	244.73	217.07	177.09	103.79	16.87	-39.74	-74.91	-59.48
125	304.34	260.58	201.15	118.82	22.76	-52.23	-82.86	-51.9
175	308.81	281.67	209.18	133.9	53.88	-51.92	-98.47	-41.94
225	379.66	328.52	279.3	165.61	53.25	-65.97	-133.92	-46.69
275	416.08	366.96	314.09	203.08	44.97	-76.93	-160.04	-91.33
325	184.46	157.25	111.99	63.23	1.03	-63.66	-71.23	-31.4
375	55.74	-12.28	-18.74	-47.26	-42.1	-9.95	125.57	271.16
425	-67.81	-118.77	-108.46	-89.68	9.24	61.78	175.43	309.21
475	-136.25	-32.33	-65.5	-111.72	38.74	115.6	84.88	191.37

where V is the volume of the reactor and q is the flow rate. The variance of the RTD function is defined by

$$\sigma^2 = \int_0^{\infty} (t - t_m)^2 E(t)\, dt$$

The exit concentration data shown in Table P4.8 were obtained from a tracer experiment studying the mixing characteristics of a continuous flow reactor. Calculate the RTD function, cumulative distribution function, mean residence time, and the variance of the RTD function of this reactor.

Table P4.8

Time (s)	$c(t)$ (mg/L)	Time (s)	$c(t)$ (mg/L)
0	0	5	5
1	2	6	2
2	4	7	1
3	7	8	0
4	6		

4.9 The following catalytic reaction is carried out in an isothermal circulating fluidized bed reactor:

$$A_{(g)} \rightarrow B_{(g)}$$

For a surface-reaction limited mechanism, in which both A and B are absorbed on the surface of the catalyst, the rate law is

$$-r_A = \frac{k_1 C_A}{1 + k_2 C_A + k_3 C_B}$$

where r_A is the rate of the reaction in kmol/m^3.s, C_A and C_B are concentrations of A and B, respectively, in kmol/m^3, and k_1, k_2, and k_3 are constants.

Assume that the solids move in plug flow at the same velocity of the gas (U). Evaluate the height of the reactor at which the conversion of A is 60%. Additional data are as follows:

$C_{A0} = 0.2$ kmol/m^3 $\quad C_{B0} = 0$ $\quad U = 7.5$ m/s

$k_1 = 8$ s^{-1} $\quad k_2 = 3$ m^3/kmol $\quad k_3 = 0.01$ m^3/kmol

4.10 A gaseous feedstock containing 40% A, 40% B, and 20% inert will be processed in a reactor, where the following chemical reaction takes place:

$$A + 2B \rightarrow C$$

The reaction rate is

$$-r_A = kC_A C_B^2$$

where $k = 0.01$ s^{-1}(gmol/L)$^{-2}$ at 500°C

C_A = concentration of A, gmol/L

C_B = concentration of B, gmol/L

Choose a basis of 100 gmol of feed and assume that all gases behave as ideal gases. Calculate the following:

(a) The time needed to produce a product containing 11.8% B in a batch reactor operating at 500°C and at constant pressure of 10 atm.

(b) The time needed to produce a product containing 11.8% B in a batch reactor operating at 500°C and constant volume. The temperature of the reactor is 500°C and the initial pressure is 10 atm.

4.11 Derive the numerical approximation of double integrals using Simpson's 1/3 rule in both dimensions.

REFERENCES

1. Larachi, F., Chaouki, J., Kennedy, G., and Duduković, M. P., "Radioactive Particle Tracking in Multiphase Reactors: Principles and Applications," in Chaouki, J., Larachi, F., and Duduković, M. P., (eds.), *Non-Invasive Monitoring of Multiphase Flows*, Elsevier, Amsterdam, 1997.

2. Bird, R. B., Stewart, W. E., and Lightfoot, E. N., *Transport Phenomena*, Wiley, New York, 1960.

3. Hanselman, D., and Littlefield, B., *Mastering MATLAB 5. A Comprehensive Tutorial and Reference*, Prentice Hall, Upper Saddle River, NJ, 1998.

4. Chapra, S. C., and Canale, R. P., *Numerical Methods for Engineers*, 3rd ed., McGraw-Hill, New York, 1998.

5. Carnahan, B., Luther, H. A., and Wilkes, J. O., *Applied Numerical Methods*, Wiley, New York, 1969.

CHAPTER 5

Numerical Solution of Ordinary Differential Equations

5.1 Introduction

Ordinary differential equations arise from the study of the dynamics of physical and chemical systems that have one independent variable. The latter may be either the space variable x or the time variable t depending on the geometry of the system and its boundary conditions.

For example, when a chemical reaction of the type

$$A + B \underset{k_2}{\overset{k_1}{\rightleftharpoons}} C + D \overset{k_3}{\rightarrow} E \tag{5.1}$$

takes place in a reactor, the material balance can be applied:

$$\text{Input} + \text{Generation} = \text{Output} + \text{Accumulation} \tag{5.2}$$

For a batch reactor, the input and output terms are zero; therefore, the material balance simplifies to

$$\text{Accumulation} = \text{Generation} \tag{5.3}$$

Assuming that reaction (5.1) takes place in the liquid phase with negligible change in volume, Eq. (5.3) written for each component of the reaction will have the form

$$\begin{aligned}
\frac{dC_A}{dt} &= -k_1 C_A C_B + k_2 C_C C_D \\
\frac{dC_B}{dt} &= -k_1 C_A C_B + k_2 C_C C_D \\
\frac{dC_C}{dt} &= k_1 C_A C_B - k_2 C_C C_D - k_3 C_C^n C_D^m \\
\frac{dC_D}{dt} &= k_1 C_A C_B - k_2 C_C C_D - k_3 C_C^n C_D^m \\
\frac{dC_E}{dt} &= k_3 C_C^n C_D^m
\end{aligned} \tag{5.4}$$

where C_A, C_B, C_C, C_D, and C_E represent the concentrations of the five chemical components of this reaction. This is a set of simultaneous *first-order nonlinear ordinary differential equations,* which describe the dynamic behavior of the chemical reaction. With the methods to be developed in this chapter, these equations, with a set of initial conditions, can be integrated to obtain the time profiles of all the concentrations.

Consider the growth of a microorganism, say a yeast, in a continuous fermentor of the type shown in Fig. 5.1. The volume of the liquid in the fermentor is V. The flow rate of nutrients into the fermentor is F_{in}, and the flow rate of products out of the fermentor is F_{out}. The material balance for the cells X is

$$\text{Input} + \text{Generation} = \text{Output} + \text{Accumulation}$$

$$F_{in} X_{in} + r_X V = F_{out} X_{out} + \frac{d(VX)}{dt} \tag{5.5}$$

The material balance for the substrate S is given by

$$F_{in} S_{in} + r_S V = F_{out} S_{out} + \frac{d(VS)}{dt} \tag{5.6}$$

The overall volumetric balance is

$$F_{in} = F_{out} + \frac{dV}{dt} \tag{5.7}$$

If we make the assumption that the fermentor is perfectly mixed, that is, the concentrations at every point in the fermentor are the same, then

$$X = X_{out}$$
$$S = S_{out} \tag{5.8}$$

and the equations simplify to

$$\frac{d(VX)}{dt} = (F_{in}X_{in} - F_{out}X) + r_X V \tag{5.9}$$

$$\frac{d(VS)}{dt} = (F_{in}S_{in} - F_{out}S) + r_S V \tag{5.10}$$

$$\frac{dV}{dt} = F_{in} - F_{out} \tag{5.11}$$

Further assumptions are made that the flow rates in and out of the fermentor are identical, and that the rates of cell formation and substrate utilization are given by

$$r_X = \frac{\mu_{max} SX}{K + S} \tag{5.12}$$

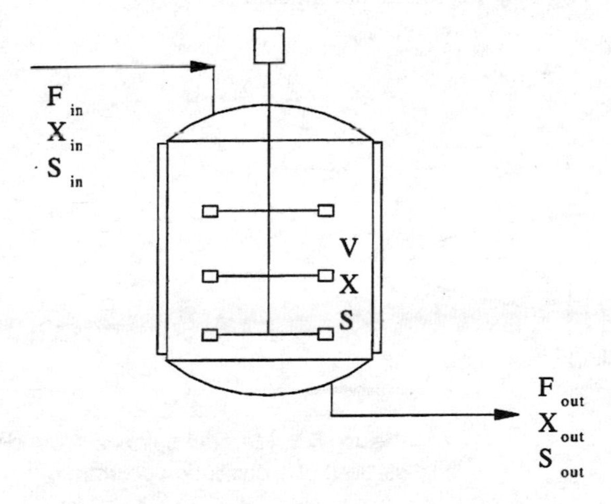

Figure 5.1 Continuous fermentor.

and
$$r_S = -\frac{1}{Y_S}\frac{\mu_{max}SX}{K + S} \tag{5.13}$$

The set of equations becomes

$$\frac{dX}{dt} = \left(\frac{F_{out}}{V}\right)(X_{in} - X) + \frac{\mu_{max}SX}{K + S} \tag{5.14}$$

$$\frac{dS}{dt} = \left(\frac{F_{out}}{V}\right)(S_{in} - S) - \frac{1}{Y_S}\frac{\mu_{max}SX}{K + S} \tag{5.15}$$

This is a set of simultaneous ordinary differential equations, which describe the dynamics of a continuous culture fermentation.

The dynamic behavior of a distillation column may be examined by making material balances around each stage of the column. Fig. 5.2 shows a typical stage n with a liquid flow into the stage L_{n+1} and out of the stage L_n and a vapor flow into the stage V_{n-1} and out of the stage V_n. The liquid holdup on the stage is designated as H_n. There is no generation of material in this process, so the material balance [Eq. (5.2)] becomes

Accumulation = Input - Output

$$\frac{dH_n}{dt} = V_{n-1} + L_{n+1} - V_n - L_n \tag{5.16}$$

The liquids and vapors in this operation are multicomponent mixtures of k components. The mole fractions of each component in the liquid and vapor phases are designated by x_i and y_i, respectively. Therefore, the material balance for the ith component is

$$\frac{d(H_n x_{i,n})}{dt} = V_{n-1}y_{i,n-1} + L_{n+1}x_{i,n+1} - V_n y_{i,n} - L_n x_{i,n} \tag{5.17}$$

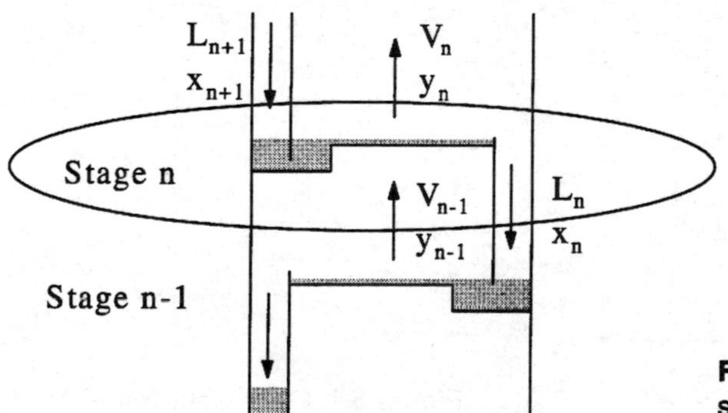

Figure 5.2 Material balance around stage *n* of a distillation column.

The concentrations of liquid and vapor are related by the equilibrium relationship

$$y_{i,n} = f(x_{i,n}) \tag{5.18}$$

If the assumptions of constant molar overflow and negligible delay in vapor flow are made, then $V_{n-1} = V_n$. The delay in liquid flow is

$$\tau \frac{dL_n}{dt} = L_{n-1} - L_n \tag{5.19}$$

where τ is the hydraulic time constant.

The above equations applied to each stage in a multistage separation process result in a large set of simultaneous ordinary differential equations.

In all the above examples, the systems were chosen so that the models resulted in sets of *simultaneous first-order ordinary differential equations*. These are the most commonly encountered types of problems in the analysis of multicomponent and/or multistage operations. *Closed-form* solutions for such sets of equations are not usually obtainable. However, *numerical methods* have been thoroughly developed for the solution of sets of simultaneous differential equations. In this chapter, we discuss the most useful techniques for the solution of such problems. We first show that higher-order differential equations can be reduced to first order by a series of substitutions.

5.2 Classifications of Ordinary Differential Equations

Ordinary differential equations are classified according to their *order*, their *linearity*, and their *boundary conditions*.

The order of a differential equation is the order of the highest derivative present in that equation. Examples of first-, second-, and third-order differential equations are given below:

First order:

$$\frac{dy}{dx} + y = kx \tag{5.20}$$

Second order:

$$\frac{d^2y}{dx^2} + y\frac{dy}{dx} = kx \tag{5.21}$$

Third order:

$$\frac{d^3y}{dx^3} + a\frac{d^2y}{dx^2} + b\left(\frac{dy}{dx}\right)^2 = kx \tag{5.22}$$

Ordinary differential equations may be categorized as *linear* and *nonlinear* equations. A differential equation is nonlinear if it contains products of the dependent variable or its derivatives or of both. For example, Eqs. (5.21) and (5.22) are nonlinear because they contain the terms $y(dy/dx)$ and $(dy/dx)^2$, respectively, whereas Eq. (5.20) is linear. The general form of a linear differential equation of order n may be written as

$$b_0(x)\frac{d^n y}{dx^n} + b_1(x)\frac{d^{n-1}y}{dx^{n-1}} + \ldots + b_{n-1}(x)\frac{dy}{dx} + b_n(x)y = R(x) \tag{5.23}$$

If $R(x) = 0$, the equation is called *homogeneous*. If $R(x) \neq 0$, the equation is *nonhomogeneous*. The coefficients $\{b_i \mid i = 1, \ldots, n\}$ are called *variable coefficients* when they are functions of x and *constant coefficients* when they are scalars. A differential equation is *autonomous* if the independent variable does not appear explicitly in that equation. For example, if Eq. (5.23) is homogeneous with constant coefficients, it is also autonomous.

To obtain a unique solution of an nth-order differential equation or of a set of n simultaneous first-order differential equations, it is necessary to specify n values of the dependent variables (or their derivatives) at specific values of the independent variable.

Ordinary differential equations may be classified as *initial-value* problems or *boundary-value* problems. In initial-value problems, the values of the dependent variables and/or their derivatives are *all* known at the initial value of the independent variable.[1] In boundary-value problems, the dependent variables and/or their derivatives are known at more than one point of the independent variable. If some of the dependent variables (or their derivatives) are specified at the initial value of the independent variable, and the remaining variables (or their derivatives) are specified at the final value of the independent variable, then this is a *two-point boundary-value* problem.

The methods of solution of initial-value problems are developed in Sec. 5.5, and the methods for boundary-value problems are discussed in Sec. 5.6.

[1] A problem whose dependent variables and/or their derivatives are all known at the final value of the independent variable (rather than the initial value) is identical to the initial-value problem, because only the direction of integration must be reversed. Therefore, the term initial-value problem refers to both cases.

5.3 TRANSFORMATION TO CANONICAL FORM

Numerical integration of ordinary differential equations is most conveniently performed when the system consists of a set of n simultaneous first-order ordinary differential equations of the form:

$$\begin{aligned} \frac{dy_1}{dx} &= f_1(y_1, y_2, \ldots, y_n, x) \\ \frac{dy_2}{dx} &= f_2(y_1, y_2, \ldots, y_n, x) \\ &\vdots \\ \frac{dy_n}{dx} &= f_n(y_1, y_2, \ldots, y_n, x) \end{aligned} \tag{5.24}$$

This is called the *canonical* form of the equations. When the initial conditions are given at a common point x_0:

$$\begin{aligned} y_1(x_0) &= y_{1,0} \\ y_2(x_0) &= y_{2,0} \\ &\vdots \\ y_n(x_0) &= y_{n,0} \end{aligned} \tag{5.25}$$

then the system equations (5.24) have solutions of the form

$$\begin{aligned} y_1 &= F_1(x) \\ y_2 &= F_2(x) \\ &\vdots \\ y_n &= F_n(x) \end{aligned} \tag{5.26}$$

The above problem can be condensed into matrix notation, where the system equations are represented by

$$\frac{d\mathbf{y}}{dx} = \mathbf{f}(x, \mathbf{y}) \tag{5.27}$$

the vector of initial conditions is

$$\mathbf{y}(x_0) = \mathbf{y}_0 \tag{5.28}$$

and the vector of solutions is

$$\mathbf{y} = \mathbf{F}(x) \tag{5.29}$$

Differential equations of higher order, or systems containing equations of mixed order, can be transformed to the canonical form by a series of substitutions. For example, consider the nth-order differential equation

$$\frac{d^n z}{dx^n} = G\left(z, \frac{dz}{dx}, \frac{d^2 z}{dx^2}, \ldots, \frac{d^{n-1} z}{dx^{n-1}}, x \right) \tag{5.30}$$

The following transformations

$$\begin{aligned}
z &= y_1 \\
\frac{dz}{dx} &= \frac{dy_1}{dx} = y_2 \\
\frac{d^2 z}{dx^2} &= \frac{dy_2}{dx} = y_3 \\
&\cdot \\
&\cdot \\
&\cdot \\
\frac{d^{n-1} z}{dx^{n-1}} &= \frac{dy_{n-1}}{dx} = y_n \\
\frac{d^n z}{dx^n} &= \frac{dy_n}{dx}
\end{aligned} \tag{5.31}$$

when substituted into the nth-order equation (5.30), give the equivalent set of n first-order equations of canonical form

$$\begin{aligned}
\frac{dy_1}{dx} &= y_2 \\
\frac{dy_2}{dx} &= y_3 \\
&\cdot \\
&\cdot \\
&\cdot \\
\frac{dy_n}{dx} &= G(y_1, y_2, y_3, \ldots, y_n, x)
\end{aligned} \tag{5.32}$$

If the right-hand side of the differential equations is not a function of the independent variable, that is,

$$\frac{dy}{dx} = f(y) \tag{5.33}$$

then the set is *autonomous*. A *nonautonomous* set may be transformed to an autonomous set by an appropriate substitution. See Example 5.1 (*b*) and (*d*). If the functions $f(y)$ are linear in terms of y, then the equations can be written in matrix form:

$$y' = Ay \tag{5.34}$$

as in Example 5.1 (*a*) and (*b*). Solutions for linear sets of ordinary differential equations are developed in Sec. 5.4. The methods for solution of nonlinear sets are discussed in Secs. 5.5 and 5.6.

A more restricted form of differential equation is

$$\frac{dy}{dx} = f(x) \tag{5.35}$$

where $f(x)$ are functions of the independent variable only. Solution methods for these equations were developed in Chap. 4.

The next example demonstrates the technique for converting higher-order linear and nonlinear differential equations to canonical form.

Example 5.1: Transformation of Ordinary Differential Equations into Their Canonical Form: Apply the transformations defined by Eqs. (5.31) and (5.32) to the following ordinary differential equations:

(*a*)
$$\frac{d^4z}{dt^4} + 5\frac{d^3z}{dt^3} - 2\frac{d^2z}{dt^2} - 6\frac{dz}{dt} + 3z = 0$$

(*b*)
$$\frac{d^4z}{dt^4} + 5\frac{d^3z}{dt^3} - 2\frac{d^2z}{dt^2} - 6\frac{dz}{dt} + 3z = e^{-t}$$

(*c*)
$$\frac{d^3z}{dx^3} + z^2\frac{d^2z}{dx^2} - \left(\frac{dz}{dx}\right)^3 - 2z = 0$$

(*d*)
$$\frac{d^3z}{dt^3} + t^3\frac{d^2z}{dt^2} - t^2\frac{dz}{dt} + 5z = 0$$

Solution: (*a*) Apply the transformation according to Eqs. (5.31):

$$z = y_1$$

$$\frac{dz}{dt} = \frac{dy_1}{dt} = y_2$$

$$\frac{d^2z}{dt^2} = \frac{dy_2}{dt} = y_3$$

$$\frac{d^3z}{dt^3} = \frac{dy_3}{dt} = y_4$$

$$\frac{d^4z}{dt^4} = \frac{dy_4}{dt}$$

Make these substitutions into Eq. (*a*) to obtain the following four equations:

$$\frac{dy_1}{dt} = y_2$$

$$\frac{dy_2}{dt} = y_3$$

$$\frac{dy_3}{dt} = y_4$$

$$\frac{dy_4}{dt} = -3y_1 + 6y_2 + 2y_3 - 5y_4$$

This is a set of linear ordinary differential equations which can be represented in matrix form

$$\boldsymbol{y}' = \boldsymbol{A}\boldsymbol{y} \tag{5.34}$$

where matrix $\boldsymbol{A}$ is given by

$$\boldsymbol{A} = \begin{bmatrix} 0 & 1 & 0 & 0 \\ 0 & 0 & 1 & 0 \\ 0 & 0 & 0 & 1 \\ -3 & 6 & 2 & -5 \end{bmatrix}$$

The method of obtaining the solution of sets of linear ordinary differential equations is discussed in Sec. 5.4.

(*b*) The presence of the term e^{-t} on the right-hand side of this equation makes it a nonhomogeneous equation. The left-hand side is identical to that of Eq. (*a*), so that the transformations of Eq. (*a*) are applicable. An additional transformation is needed to replace the e^{-t} term. This transformation is

$$y_5 = e^{-t}$$

$$\frac{dy_5}{dt} = -e^{-t} = -y_5$$

Make the substitutions into Eq. (*b*) to obtain the following set of five linear ordinary differential equations:

$$\frac{dy_1}{dt} = y_2$$
$$\frac{dy_2}{dt} = y_3$$
$$\frac{dy_3}{dt} = y_4$$
$$\frac{dy_4}{dt} = -3y_1 + 6y_2 + 2y_3 - 5y_4 + y_5$$
$$\frac{dy_5}{dt} = -y_5$$

which also condenses into the matrix form of Eq. (5.34), with the matrix $\boldsymbol{A}$ given by

$$\boldsymbol{A} = \begin{bmatrix} 0 & 1 & 0 & 0 & 0 \\ 0 & 0 & 1 & 0 & 0 \\ 0 & 0 & 0 & 1 & 0 \\ -3 & 6 & 2 & -5 & 1 \\ 0 & 0 & 0 & 0 & -1 \end{bmatrix}$$

(*c*) Apply the following transformations:

$$z = y_1$$
$$\frac{dz}{dx} = \frac{dy_1}{dx} = y_2$$
$$\frac{d^2z}{dx^2} = \frac{dy_2}{dx} = y_3$$
$$\frac{d^3z}{dx^3} = \frac{dy_3}{dt}$$

Make the substitutions into Eq. (c) to obtain the set

$$\frac{dy_1}{dx} = y_2$$
$$\frac{dy_2}{dx} = y_3$$
$$\frac{dy_3}{dx} = 2y_1 + y_2^3 - y_1^2 y_3$$

This is a set of *nonlinear* differential equations which cannot be expressed in matrix form. The methods of solution of nonlinear differential equations are developed in Secs. 5.5 and 5.6.

(d) Apply the following transformations:

$$z = y_1$$
$$\frac{dz}{dt} = \frac{dy_1}{dt} = y_2$$
$$\frac{d^2z}{dt^2} = \frac{dy_2}{dt} = y_3$$
$$\frac{d^3z}{dt^3} = \frac{dy_3}{dt}$$
$$y_4 = t$$
$$\frac{dy_4}{dt} = 1$$

Make the substitutions into Eq. (d) to obtain the set

$$\frac{dy_1}{dt} = y_2$$
$$\frac{dy_2}{dt} = y_3$$
$$\frac{dy_3}{dt} = -5y_1 + y_4^2 y_2 - y_4^3 y_3$$
$$\frac{dy_4}{dt} = 1$$

This is a set of autonomous nonlinear differential equations. Note that the above set of substitutions converted the nonautonomous Eq. (d) to a set of autonomous equations.

5.4 LINEAR ORDINARY DIFFERENTIAL EQUATIONS

The analysis of many *physicochemical* systems yields mathematical models that are sets of *linear* ordinary differential equations with constant coefficients and can be reduced to the form

$$\boldsymbol{y}' = \boldsymbol{A}\boldsymbol{y} \tag{5.34}$$

with given initial conditions

$$\boldsymbol{y}(0) = \boldsymbol{y}_0 \tag{5.36}$$

Such examples abound in chemical engineering. The *unsteady-state* material and energy balances of multiunit processes, without chemical reaction, often yield linear differential equations.

Sets of linear ordinary differential equations with constant coefficients have closed-form solutions that can be readily obtained from the eigenvalues and eigenvectors of the matrix $\boldsymbol{A}$.

In order to develop this solution, let us first consider a single linear differential equation of the type

$$\frac{dy}{dt} = ay \tag{5.37}$$

with the given initial condition

$$y(0) = y_0 \tag{5.38}$$

Eq. (5.37) is essentially the scalar form of the matrix set of Eq. (5.34). The solution of the scalar equation can be obtained by separating the variables and integrating both sides of the equation

$$\begin{aligned} \int_{y_0}^{y} \frac{dy}{y} &= \int_0^t a\,dt \\ \ln\frac{y}{y_0} &= at \\ y &= e^{at}y_0 \end{aligned} \tag{5.39}$$

In an analogous fashion, the matrix set can be integrated to obtain the solution

$$\boldsymbol{y} = e^{\boldsymbol{A}t}\boldsymbol{y}_0 \tag{5.40}$$

In this case, $\boldsymbol{y}$ and $\boldsymbol{y}_0$ are *vectors* of the dependent variables and the initial conditions, respectively. The term e^{At} is the matrix exponential function, which can be obtained from Eq. (2.83):

$$e^{At} = I + At + \frac{A^2t^2}{2!} + \frac{A^3t^3}{3!} + \dots \tag{5.41}$$

It can be demonstrated that Eq. (5.40) is a solution of Eq. (5.34) by differentiating it:

$$\begin{aligned}
\frac{d\boldsymbol{y}}{dt} &= \frac{d}{dt}(e^{At})\boldsymbol{y}_0 \\
&= \frac{d}{dt}\left(I + At + \frac{A^2t^2}{2!} + \frac{A^3t^3}{3!} + \dots\right)\boldsymbol{y}_0 \\
&= \left(A + A^2t + \frac{A^3t^2}{2!} + \dots\right)\boldsymbol{y}_0 \\
&= A\left(I + At + \frac{A^2t^2}{2!} + \dots\right)\boldsymbol{y}_0 \\
&= A(e^{At})\boldsymbol{y}_0 \\
&= A\boldsymbol{y}
\end{aligned}$$

The solution of the set of linear ordinary differential equations is very cumbersome to evaluate in the form of Eq. (5.40), because it requires the evaluation of the infinite series of the exponential term e^{At}. However, this solution can be modified by further algebraic manipulation to express it in terms of the eigenvalues and eigenvectors of the matrix A.

In Chap. 2, we showed that a nonsingular matrix A of order n has n eigenvectors and n nonzero eigenvalues, whose definitions are given by

$$\begin{aligned}
A\boldsymbol{x}_1 &= \lambda_1\boldsymbol{x}_1 \\
A\boldsymbol{x}_2 &= \lambda_2\boldsymbol{x}_2 \\
&\vdots \\
A\boldsymbol{x}_n &= \lambda_n\boldsymbol{x}_n
\end{aligned} \tag{5.42}$$

All the above eigenvectors and eigenvalues can be represented in a more compact form as follows:

$$AX = X\Lambda \tag{5.43}$$

where the columns of matrix X are the individual eigenvectors:

$$X = [x_1, x_2, x_3, \ldots, x_n] \tag{5.44}$$

and Λ is a diagonal matrix with the eigenvalues of A on its diagonal:

$$\Lambda = \begin{bmatrix} \lambda_1 & 0 & 0 & \ldots & 0 \\ 0 & \lambda_2 & 0 & \ldots & 0 \\ 0 & 0 & \lambda_3 & \ldots & 0 \\ \ldots & \ldots & \ldots & \ldots & \ldots \\ 0 & 0 & 0 & \ldots & \lambda_n \end{bmatrix} \tag{5.45}$$

If we postmultiply each side of Eq. (5.43) by X^{-1}, we obtain

$$AXX^{-1} = A = X\Lambda X^{-1} \tag{5.46}$$

Squaring Eq. (5.46):

$$\begin{aligned} A^2 &= [X\Lambda X^{-1}][X\Lambda X^{-1}] \\ &= X\Lambda^2 X^{-1} \end{aligned} \tag{5.47}$$

Similarly, raising Eq. (5.46) to any power n we obtain

$$A^n = X\Lambda^n X^{-1} \tag{5.48}$$

Starting with Eq. (5.41) and replacing the matrices $A, A^2, \ldots, A^n$ with their equivalent from Eqs. (5.46)-(5.48), we obtain

$$e^{At} = I + X\Lambda X^{-1}t + X\Lambda^2 X^{-1}\frac{t^2}{2!} + \ldots \tag{5.49}$$

The identity matrix I can be premultiplied by X and postmultiplied by X^{-1} without changing it. Therefore, Eq. (5.49) rearranges to

$$e^{At} = X\left(I + \Lambda t + \frac{\Lambda^2 t^2}{2!} + \ldots\right)X^{-1} \tag{5.50}$$

which simplifies to

$$e^{At} = Xe^{\Lambda t}X^{-1} \tag{5.51}$$

where the exponential matrix $e^{\Lambda t}$ is defined as

$$e^{\Lambda t} = \begin{bmatrix} e^{\lambda_1 t} & 0 & 0 & \dots & 0 \\ 0 & e^{\lambda_2 t} & 0 & \dots & 0 \\ 0 & 0 & e^{\lambda_3 t} & \dots & 0 \\ \dots & \dots & \dots & \dots & \dots \\ 0 & 0 & 0 & \dots & e^{\lambda_n t} \end{bmatrix} \tag{5.52}$$

The solution of the linear differential equations can now be expressed in terms of eigenvalues and eigenvectors by combining Eqs. (5.40) and (5.51):

$$y = [Xe^{\Lambda t}X^{-1}]y_0 \tag{5.53}$$

The eigenvalues and eigenvectors of matrix A can be calculated using the techniques developed in Chap. 2 or simply by applying the built-in function *eig* in MATLAB. This is demonstrated in Example 5.2.

Example 5.2: Solution of a Chemical Reaction System. Develop a general MATLAB function to solve the set of linear differential equations. Apply this function to determine the concentration profiles of all components of the following chemical reaction system:

$$A \underset{k_2}{\overset{k_1}{\rightleftharpoons}} B \underset{k_4}{\overset{k_3}{\rightleftharpoons}} C$$

Assume that all steps are first-order reactions and write the set of linear ordinary differential equations that describe the kinetics of these reactions. Solve the problem numerically for the following values of the kinetic rate constants:

$$k_1 = 1\ \text{min}^{-1} \qquad k_2 = 0\ \text{min}^{-1} \qquad k_3 = 2\ \text{min}^{-1} \qquad k_4 = 3\ \text{min}^{-1}$$

The value of $k_2 = 0$ reveals that the first reaction is irreversible in this special case. The initial concentrations of the three components are

$$C_{A_0} = 1 \qquad C_{B_0} = 0 \qquad C_{C_0} = 0$$

Plot the graph of concentrations versus time.

Method of Solution: Assuming that all steps are first-order reactions, the set of differential equations that give the rate of formation of each compound is:

$$\frac{dC_A}{dt} = -k_1 C_A + k_2 C_B$$

$$\frac{dC_B}{dt} = k_1 C_A - k_2 C_B - k_3 C_B + k_4 C_C$$

$$\frac{dC_C}{dt} = k_3 C_B - k_4 C_C$$

In matrix form, this set reduces to

$$\dot{c} = Kc$$

where

$$\dot{c} = \begin{bmatrix} \frac{dC_A}{dt} \\ \frac{dC_B}{dt} \\ \frac{dC_C}{dt} \end{bmatrix} \qquad c = \begin{bmatrix} C_A \\ C_B \\ C_C \end{bmatrix}$$

and

$$K = \begin{bmatrix} -k_1 & k_2 & 0 \\ k_1 & -k_2 - k_3 & k_4 \\ 0 & k_3 & -k_4 \end{bmatrix}$$

The solution of a set of linear ordinary differential equations can be obtained either by applying Eq. (5.40):

$$c = e^{Kt} c_0$$

or by Eq. (5.53):

$$c = [X e^{\Lambda t} X^{-1}] c_0$$

where the matrix X consists of the eigenvectors of K and c_0 is the vector of initial concentrations:

$$c_0 = \begin{bmatrix} C_{A_0} \\ C_{B_0} \\ C_{C_0} \end{bmatrix}$$

Program Description: The MATLAB function *LinearODE.m* solves a set of linear ordinary differential equations. The first part of the function checks the number of inputs and their sizes, or values. The next section of the function performs the solution of the set of ordinary differential equations, which can be done by either the matrix exponential method [Eq. (5.40)] or the eigenvector method [Eq. (5.53)]. The method of solution may be introduced to the function through the fifth input argument. The default method of solution is the matrix exponential method.

The main program *Example5_2.m* solves the particular problem posed in this example by applying *LinearODE.m*. This program gets the required input data, including the rate constants and initial concentrations of the components, from the keyboard. Then, it builds the matrix of coefficients and the vector of times at which the concentrations are to be calculated. In the last section, the program asks the user to select the method of solution and calls the function *LinearODE* to solve the set of equations for obtaining the concentrations and plots the results. The reader may try another method of solution and repeat solving the set of linear differential equations in this part.

Program

Example5_2.m

```
% Example5_2.m
% Solution to Example 5.2. This program calculates and plots
% concentrations of the components of the system A<->B<->C vs
% time. It calls the function LinearODE to solve the set of
% linear ordinary differential equations.

clear
clc
clf

% Input data
k1 = input(' A->B   , k1 = ');
k2 = input(' B->A   , k2 = ');
k3 = input(' B->C   , k3 = ');
k4 = input(' C->B   , k4 = ');
disp(' ')
c0(1) = input(' Initial concentration of A = ');
c0(2) = input(' Initial concentration of B = ');
c0(3) = input(' Initial concentration of C = ');
disp(' ')
tmax = input(' Maximum time = ');
dt = input(' Time interval = ');
disp(' ')

% Matrix of coefficients
```

```
K = [-k1, k2, 0; k1, -k2-k3, k4; 0, k3, -k4];
t = [0:dt:tmax];      % Vector of time
if t(end) ~= tmax
   t(end+1) = tmax;
end

disp(' ')
disp('  1 ) Matrix exponential method')
disp('  2 ) Eigenvector method')
disp('  0 ) Exit')
method = input('\n Choose the method of solution : ');

% Solution
method = 1;
while method
   c = LinearODE(K,c0,t,[],method);% Solving the set of equations
   plot(t,c(1,:),t,c(2,:),'.-',t,c(3,:),'--')% Plotting the results
   xlabel('Time')
   ylabel('Concentration')
   legend('C_A','C_B','C_C')
   method = input('\n Choose the method of solution : ');
end
```

LinearODE.m

```
function y = LinearODE(A,y0,t,t0,method)
% LINEARODE Solves a set of linear ordinary differential equations.
%
%   Y=LINEARODE(A,Y0,T) solves a set of linear ordinary
%   differential equations whose matrix of coefficients
%   is A and its initial conditions are Y0. The function
%   returns the values of the solution Y at times T.
%
%   Y=LINEARODE(A,Y0,T,T0,METHOD) takes T0 as the time in
%   which the initial conditions Y0 are given. Default value
%   for T0 is zero. METHOD is the method of solution.
%      Use METHOD = 1 for matrix exponential method
%      Use METHOD = 2 for eigenvector method
%   Default value for METHOD is 1.
%
%  See also ODE23, ODE45, ODE113, ODE15S, ODE23S, EULER, MEULER,
%           RK, ADAMS, ADAMSMOULTON

% (c) N. Mostoufi & A. Constantinides
% January 1, 1999

% Checking inputs
if nargin<3 | isempty(t)
```

```
   error('Vector of independent variable is empty.')
end

if nargin<4 | isempty(t0)
   t0 = 0;
end
t = t - t0;
nt = length(t);

if nargin<5 | isempty(method) | method < 1 | method > 2
   method = 1;
end

nA = length(A);
if nA ~= length(y0)
   error('Matrix of coefficients and vector of initial values are
not of the same order.');
end

y0 = (y0(:).')';      % Make sure it's a column vector

switch method
case 1 % Matrix exponential method
   for k = 1:nt
      if t(k) > 0
         y(:,k) = expm(A*t(k))*y0;
      else
         y(:,k) = y0;
      end
   end
case 2                % Eigenvector method
   [X,D] = eig(A);   % Eigenvectors and eigenvalues
   IX = inv(X);
   e_lambda_t = zeros(nA,nA,nt);
   % Building the matrix exp(LAMBDA.t)
   for k = 1:nA
      e_lambda_t(k,k,:) = exp(D(k,k) * t);
   end
   % Solving the set of equations
   for k = 1:nt
      if t(k) > 0
         y(:,k) = X * e_lambda_t(:,:,k) * IX * y0;
      else
         y(:,k) = y0;
      end
   end
end
```

Input and Results

```
>>Example5_2

 A->B   , k1 = 1
 B->A   , k2 = 0
 B->C   , k3 = 2
 C->B   , k4 = 3

 Initial concentration of A = 1
 Initial concentration of B = 0
 Initial concentration of C = 0

 Maximum time = 5
 Time interval = 0.1

  1 ) Matrix exponential method
  2 ) Eigenvector method
  0 ) Exit

 Choose the method of solution : 2

 Choose the method of solution : 0
```

Discussion of Results: The results of solution of this problem are shown in Fig. E5.2. It is seen from this figure, as expected for this special case, that after long enough time, all the component A is consumed and the components B and C satisfy the equilibrium condition $C_B/C_C = k_4/k_3$. These results also confirm the conservation of mass principle:

$$C_{A_0} + C_{B_0} + C_{C_0} = C_A + C_B + C_C$$

Because both methods of solution are exact, results obtained by these methods would be identical. However, when dealing with a large number of equations and/or a long time vector, the matrix exponential method is appreciably faster in the MATLAB environment than the eigenvector method. This is because the exponential of a matrix is performed by the built-in MATLAB function *expm*, whereas the eigenvector method involves several element-by-element operations when building the matrix $e^{\Lambda t}$. The reader is encouraged to verify the difference between the methods by repeating the solution and choosing a smaller time interval, say 0.001, and applying this to both solution methods.

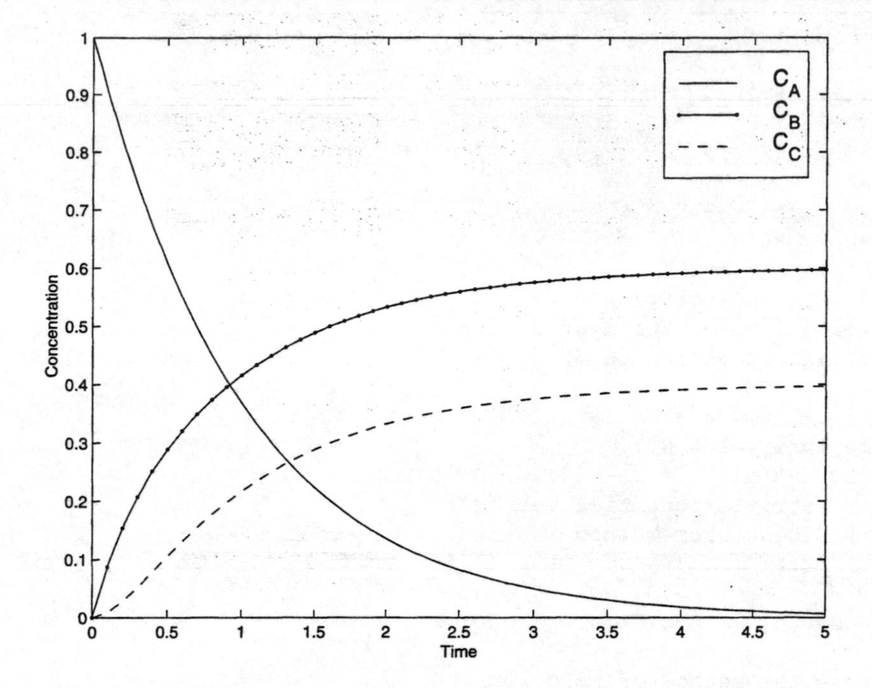

Figure E5.2 Concentration profiles.

5.5 NONLINEAR ORDINARY DIFFERENTIAL EQUATIONS–INITIAL-VALUE PROBLEMS

In this section, we develop numerical solutions for a set of ordinary differential equations in their canonical form:

$$\frac{d\mathbf{y}}{dx} = \mathbf{f}(x, \mathbf{y}) \tag{5.27}$$

with the vector of initial conditions given by

$$\mathbf{y}(x_0) = \mathbf{y}_0 \tag{5.28}$$

In order to be able to illustrate these methods graphically, we treat $\mathbf{y}$ as a single variable rather than as a vector of variables. The formulas developed for the solution of a single differential equation are readily expandable to those for a set of differential equations, which must be solved *simultaneously*. This concept is demonstrated in Sec. 5.5.4.

We begin the development of these methods by first rearranging Eq. (5.27) and integrating both sides between the limits of $x_i \le x \le x_{i+1}$ and $y_i \le y \le y_{i+1}$:

$$\int_{y_i}^{y_{i+1}} dy = \int_{x_i}^{x_{i+1}} f(x, y)\, dx \tag{5.54}$$

The left side integrates readily to obtain

$$y_{i+1} - y_i = \int_{x_i}^{x_{i+1}} f(x, y)\, dx \tag{5.55}$$

One method for integrating Eq. (5.55) is to take the left-hand side of this equation and use finite differences for its approximation. This technique works directly with the tangential trajectories of the dependent variable y rather than with the areas under the function $f(x, y)$. This is the technique applied in Secs. 5.5.1 and 5.5.2.

In Chap. 4, we developed the integration formulas by first replacing the function $f(x)$ with an interpolating polynomial and then evaluating the integral $f(x)dx$ between the appropriate limits. A similar technique could be followed here to integrate the right-hand side of Eq. (5.55). This approach is followed in Sec. 5.5.3.

There are several functions in MATLAB for the solution of a set of ordinary differential equations. These solvers, along with their method of solution, are listed in Table 5.1. The solver that one would want to try first on a problem is *ode45*. The statement $[x, y]$ = *ode45*('*y_prime*', $[x_0, x_f]$, y_0) solves the set of ordinary differential equations described in the MATLAB function *y_prime.m*, from x_0 to x_f, with the initial values given in the vector y_0, and returns the values of independent and dependent variables in the vectors x and y, respectively. The vector of dependent variable, x, is not equally spaced, because the function controls the step size. If the solution is required at specified points of x, the interval $[x_0, x_f]$ should be replaced by a vector containing the values of the independent variable at these

Table 5.1 Ordinary differential equation solvers in MATLAB

Solver	Method of solution
ode23	Runge-Kutta lower-order (2nd order-3 stages)
ode45	Runge-Kutta higher-order (4th order-5 stages)
ode113	Adams-Bashforth-Moulton of varying order (1-13)
ode23s	Modified Rosenbrock of order 2
ode15s	Implicit, multistep of varying order (1-5)

points. For example, [x, y] = *ode45*('*y_prime*', [x_0: h: x_f], y_0) returns the solution of the set of ordinary differential equations from x_0 to x_f at intervals of the width h. The vector x in this case would be monotonic (with the exception of, perhaps, its last interval). The basic syntax for applying the other MATLAB ordinary differential equation solvers is the same as that described above for *ode45*.

The function *y_prime.m* should return the value of derivative(s) as a column vector. The first input to this function has to be the independent variable, x, even if it is not explicitly used in the definition of the derivative. The second input argument to *y_prime* is the vector of dependent variable, y. It is possible to pass additional parameters to the derivative function. It should be noted, however, that in this case, the third input to *y_prime.m* has to be an empty variable, flag, and the additional parameters are introduced in the fourth argument.

5.5.1 The Euler and Modified Euler Methods

One of the earliest techniques developed for the solution of ordinary differential equations is the *Euler method*. This is simply obtained by recognizing that the left side of Eq. (5.55) is the first forward finite difference of y at position i:

$$y_{i+1} - y_i = \Delta y_i \tag{5.56}$$

which, when rearranged, gives a "forward marching" formula for evaluating y:

$$y_{i+1} = y_i + \Delta y_i \tag{5.57}$$

The forward difference term Δy_i is obtained from Eq. (3.53) applied to y at position i:

$$\Delta y_i = hDy_i + \frac{h^2 D^2 y_i}{2} + \frac{h^3 D^3 y_i}{6} + \dots \tag{5.58}$$

In the Euler method, the above series is truncated after the first term to obtain

$$\Delta y_i = hDy_i + O(h^2) \tag{5.59}$$

The combination of Eqs. (5.57) and (5.59) gives the *explicit Euler formula* for integrating differential equations

$$y_{i+1} = y_i + hDy_i + O(h^2) \tag{5.60}$$

The derivative Dy_i is replaced by its equivalent y'_i or $f(x_i, y_i)$ to give the more commonly used form of the explicit Euler method[2]

$$y_{i+1} = y_i + hf(x_i, y_i) + O(h^2) \tag{5.61}$$

This equation simply states that the next value of y is obtained from the previous value by moving a step of width h in the tangential direction of y. This is demonstrated graphically in Fig 5.3*a*. This Euler formula is rather inaccurate because it has a truncation error of only $O(h^2)$. If h is large the trajectory of y can quickly deviate from its true value, as shown in Fig. 5.3*b*.

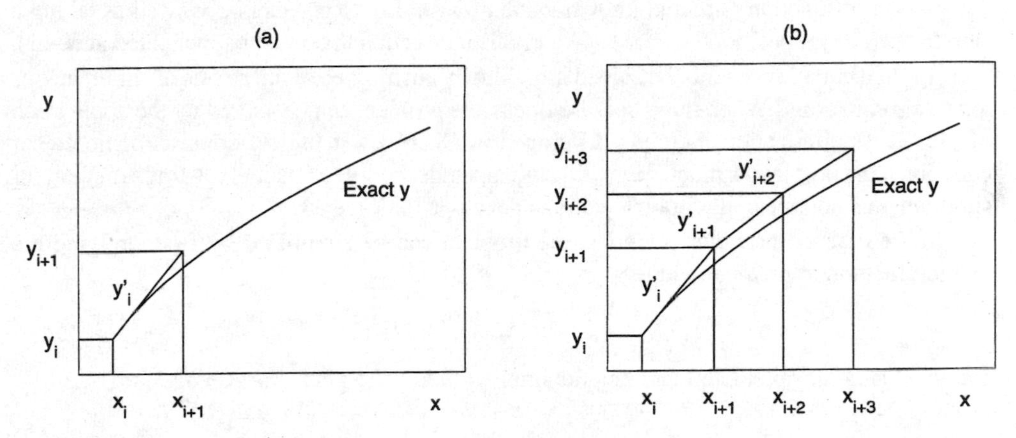

Figure 5.3 The explicit Euler method of integration. (*a*) Single step. (*b*) Several steps.

The accuracy of the Euler method can be improved by utilizing a combination of forward and backward differences. Note that the first forward difference of y at i is equal to the first backward difference of y at $(i + 1)$:

$$\Delta y_i = y_{i+1} - y_i = \nabla y_{i+1} \tag{5.62}$$

Therefore, the forward marching formula in terms of backward differences is

$$y_{i+1} = y_i + \nabla y_{i+1} \tag{5.63}$$

---

[2] From here on the term y'_i and $f(x_i, y_i)$ will be used interchangeably. The reader should remember that these are equal to each other through the differential equation (5.27).

The backward difference term ∇y_{i+1} is obtained from Eq. (3.32) applied to y at position $(i+1)$:

$$\nabla y_{i+1} = hDy_{i+1} - \frac{h^2 D^2 y_{i+1}}{2} + \frac{h^3 D^3 y_{i+1}}{6} - \ldots \tag{5.64}$$

Combining Eqs. (5.63) and (5.64):

$$y_{i+1} = y_i + hf(x_{i+1}, y_{i+1}) + O(h^2) \tag{5.65}$$

This is called the *implicit Euler formula* (or backward Euler), because it involves the calculation of function f at an unknown value of y_{i+1}. Eq. (5.65) can be viewed as taking a step forward from position i to $(i+1)$ in a gradient direction that must be evaluated at $(i+1)$.

Implicit equations cannot be solved individually but must be set up as sets of simultaneous algebraic equations. When these sets are linear, the problem can be solved by the application of the Gauss elimination methods developed in Chap. 2. If the set consists of nonlinear equations, the problem is much more difficult and must be solved using Newton's method for simultaneous nonlinear algebraic equations developed in Chap. 1.

In the case of the Euler methods, the problem can be simplified by first applying the explicit method to *predict* a value y_{i+1}:

$$(y_{i+1})_{Pr} = y_i + hf(x_i, y_i) + O(h^2) \tag{5.66}$$

and then using this predicted value in the implicit method to get a *corrected* value:

$$(y_{i+1})_{Cor} = y_i + hf(x_{i+1}, (y_{i+1})_{Pr}) + O(h^2) \tag{5.67}$$

This combination of steps is known as the *Euler predictor-corrector* (or *modified Euler*) method, whose application is demonstrated graphically in Fig. 5.4. Correction by Eq. (5.67) may be applied more than once until the corrected value converges, that is, the difference between the two consecutive corrected values becomes less than the convergence criterion. However, not much more accuracy is achieved after the second application of the corrector.

The explicit, as well as the implicit, forms of the Euler methods have error of order (h^2). However, when used in combination, as predictor-corrector, their accuracy is enhanced, yielding an error of order (h^3). This conclusion can be reached by adding Eqs. (5.57) and (5.63):

$$y_{i+1} = y_i + \frac{1}{2}(\Delta y_i + \nabla y_{i+1}) \tag{5.68}$$

and utilizing (5.58) and (5.64) to obtain

$$y_{i+1} = y_i + \frac{h}{2}[f(x_i, y_i) + f(x_{i+1}, y_{i+1})] + O(h^3) \tag{5.69}$$

The terms of order (h^2) cancel out because they have opposite sign, thus giving a formula of higher accuracy. Eq. (5.69) is essentially the same as the trapezoidal rule [Eq. (4.73)], the only difference being in the way the function is evaluated at (x_{i+1}, y_{i+1}).

It has been shown [1] that the Euler implicit formula is more stable than the explicit one. The stability of these methods will be discussed in Sec. 5.7.

It can be seen by writing Eq. (5.69) in the form

$$y_{i+1} = y_i + \frac{1}{2}hf(x_i, y_i) + \frac{1}{2}hf(x_{i+1}, y_{i+1}) + O(h^3) \tag{5.70}$$

that this Euler method uses the weighted trajectories of the function y evaluated at two positions that are located one full step of width h apart and weighted equally. In this form, Eq. (5.70) is also known as the Crank-Nicolson method.

Eq. (5.70) can be written in a more general form as

$$y_{i+1} = y_i + w_1k_1 + w_2k_2 \tag{5.71}$$

where, in this case:

$$k_1 = hf(x_i, y_i) \tag{5.72}$$

$$k_2 = hf(x_i + c_2h, y_i + a_{21}k_1) \tag{5.73}$$

The choice of the weighting factors w_1 and w_2 and the positions i and $(i + 1)$ at which to evaluate the trajectories is dictated by the accuracy required of the integration formula, that is, by the number of terms retained in the infinite series expansion.

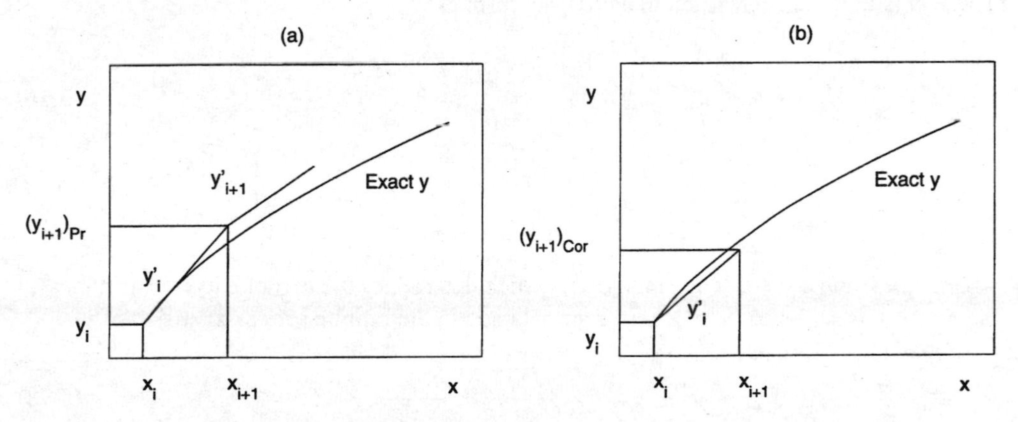

Figure 5.4 The Euler predictor-corrector method. (*a*) Value of y_{i+1} is predicted and y'_{i+1} is calculated. (*b*) Value of y_{i+1} is corrected.

This concept forms the basis for a whole series of integration formulas, with increasingly higher accuracies, for ordinary differential equations. These are discussed in the following section.

5.5.2 The Runge-Kutta Methods

The most widely used methods of integration for ordinary differential equations are the series of methods called Runge-Kutta second, third, and fourth order, plus a number of other techniques that are variations on the Runge-Kutta theme. These methods are based on the concept of weighted trajectories formulated at the end of Sec. 5.5.1. In a more general fashion, the forward marching integration formula for the differential equation (5.27) is given by the recurrence equation

$$y_{i+1} = y_i + w_1 k_1 + w_2 k_2 + w_3 k_3 + \dots + w_m k_m \tag{5.74}$$

where each of the trajectories k_i are evaluated by

$$\begin{aligned} k_1 &= hf(x_i, y_i) \\ k_2 &= hf(x_i + c_2 h,\ y_i + a_{21} k_1) \\ k_3 &= hf(x_i + c_3 h,\ y_i + a_{31} k_1 + a_{32} k_2) \\ &\ \ \vdots \\ k_m &= hf(x_i + c_m h,\ y_i + a_{m1} k_1 + a_{m2} k_2 + \dots + a_{m,m-1} k_{m-1}) \end{aligned} \tag{5.75}$$

These equations can be written in a compact form as

$$y_{i+1} = y_i + \sum_{i=1}^{m} w_i k_i \tag{5.76}$$

$$k_j = hf\left(x_i + c_j h,\ y_i + \sum_{l=1}^{j-1} a_{jl} k_l \right) \tag{5.77}$$

where $c_1 = 0$ and $a_{1j} = 0$. The value of m, which determines the complexity and accuracy of the method, is set when $(m + 1)$ terms are retained in the infinite series expansion of y_{i+1}

$$y_{i+1} = y_i + hy_i' + \frac{h^2 y_i''}{2!} + \frac{h^3 y_i'''}{3!} + \dots \tag{5.78}$$

or

$$y_{i+1} = y_i + hDy_i + \frac{h^2D^2y_i}{2!} + \frac{h^3D^3y_i}{3!} + \ldots \tag{5.79}$$

The procedure for deriving the Runge-Kutta methods can be divided into five steps which are demonstrated below in the derivation of the *second-order Runge-Kutta* formulas.

Step 1: Choose the value of m, which fixes the accuracy of the formula to be obtained. For second-order Runge-Kutta, $m = 2$. Truncate the series (5.79) after the $(m + 1)$ term:

$$y_{i+1} = y_i + hDy_i + \frac{h^2D^2y_i}{2!} + O(h^3) \tag{5.80}$$

Step 2: Replace each derivative of y in (5.80) by its equivalent in f, remembering that f is a function of both x and $y(x)$:

$$Dy_i = f_i \tag{5.81}$$

$$D^2y_i = \frac{df}{dx} = \left(\frac{\partial f}{\partial x}\frac{dx}{dx} + \frac{\partial f}{\partial y}\frac{dy}{dx}\right)_i$$

$$= (f_x + ff_y)_i \tag{5.82}$$

Combine Eqs. (5.80) to (5.82) and regroup the terms:

$$y_{i+1} = y_i + hf_i + \frac{h^2}{2}f_{x_i} + \frac{h^2}{2}f_if_{y_i} + O(h^3) \tag{5.83}$$

Step 3: Write Eq. (5.76) with m terms in the summation:

$$y_{i+1} = y_i + w_1k_1 + w_2k_2 \tag{5.84}$$

where

$$k_1 = hf(x_i, y_i) \tag{5.85}$$

$$k_2 = hf(x_i + c_2h, y_i + a_{21}k_1) \tag{5.86}$$

Step 4: Expand the f function in Taylor series:

$$f(x_i + c_2h, y_i + a_{21}k_1) = f_i + c_2hf_{x_i} + a_{21}hf_{y_i}f_i + O(h^2) \tag{5.87}$$

Combine Eqs. (5.84) to (5.87) and regroup the terms:

$$y_{i+1} = y_i + (w_1 + w_2)hf_i + (w_2c_2)h^2f_{x_i} + (w_2a_{21})h^2f_if_{y_i} + O(h^3) \tag{5.88}$$

Step 5: In order for Eqs. (5.83) and (5.88) to be identical, the coefficients of the corresponding terms must be equal to one another. This results in a set of simultaneous nonlinear algebraic equations in the unknown constants w_j, c_j, and a_{jl}. For this second-order Runge-Kutta method, there are three equations and four unknowns:

$$\begin{aligned} w_1 + w_2 &= 1 \\ w_2c_2 &= \frac{1}{2} \\ w_2a_{21} &= \frac{1}{2} \end{aligned} \tag{5.89}$$

It turns out that there are always more unknowns than equations. The degree of freedom allows us to choose some of the parameters. For second-order Runge-Kutta, there is one degree of freedom. For third- and fourth-order Runge-Kutta, there are two degrees of freedom. For fifth-order Runge-Kutta, there are at least five degrees of freedom. This freedom of choice of parameters gives rise to a very large number of different forms of the Runge-Kutta formulas. It is usually desirable to first choose the values of the c_j constants, thus fixing the positions along the independent variable, where the functions

$$f\left(x_i + c_jh,\ y_i + \sum_{l=1}^{j-1} a_{jl}k_l\right)$$

are to be evaluated. An important consideration in choosing the free parameters is to minimize the *roundoff error* of the calculation. Discussion of the effect of the roundoff error will be given in Sec. 5.7.

For the second-order Runge-Kutta method, which we are currently deriving, let us choose $c_2 = 1$. The rest of the parameters are evaluated from Eqs. (5.89):

$$w_1 = w_2 = \frac{1}{2} \qquad a_{21} = 1 \tag{5.90}$$

With this set of parameters, the second-order Runge-Kutta formula is

$$\left.\begin{aligned} y_{i+1} &= y_i + \frac{1}{2}(k_1 + k_2) \\ k_1 &= hf(x_i, y_i) \\ k_2 &= hf(x_i + h, y_i + k_1) \end{aligned}\right\} O(h^3) \tag{5.91}$$

This method is essentially identical to the Crank-Nicolson method [see Eq. (5.70)].

A different version of the second-order Runge-Kutta is obtained by choosing to evaluate the function at the midpoints (that is, $c_2 = 1/2$). This yields the formula

$$\left.\begin{aligned} y_{i+1} &= y_i + k_2 \\ k_1 &= hf(x_i, y_i) \\ k_2 &= hf\left(x_i + \frac{1}{2}h, y_i + \frac{1}{2}k_1\right) \end{aligned}\right\} O(h^3) \tag{5.92}$$

Higher-order Runge-Kutta formulas are derived in an analogous manner. Several of these are listed in Table 5.2. The fourth-order Runge-Kutta, which has an error of $O(h^5)$, is probably the most widely used numerical integration method for ordinary differential equations.

5.5.3 The Adams and Adams-Moulton Methods

The Runge-Kutta family of integration techniques, developed above, are called *single-step* methods. The value of y_{i+1} is obtained from y_i and the trajectories of y within the single step from (x_i, y_i) to (x_{i+1}, y_{i+1}). This procedure marches forward, taking single step of width h, over the entire interval of integration. These methods are very suitable for solving initial-value problems because they are *self-starting* from a given initial point of integration.

Other categories of integration techniques, called *multiple-step* methods, have been developed. These compute the value of y_{i+1} utilizing several previously unknown, or calculated, values of y (y_i, y_{i-1}, y_{i-2}, etc.) as the base points. For this reason, the multiple-step methods are *nonself-starting*. For the solution of initial-value problems, where only y_0 is known, the multiple-step methods must be "primed" by first utilizing a self-starting procedure to obtain the requisite number of base points. There are several multiple-step methods, two of these, the Adams and Adams-Moulton methods, will be covered in this section.

Once again, let us start by evaluating y_{i+1} by integrating the derivative function over the interval $[x_i, x_{i+1}]$

$$y_{i+1} - y_i = \int_{x_i}^{x_{i+1}} f(x, y)\, dx \tag{5.55}$$

In order to evaluate the right-hand side of Eq. (5.55), $f(x, y)$ may be approximated by an nth-degree polynomial. In the Adams method, a quadratic polynomial is passed through the three past points, that is, (x_{i-2}, y_{i-2}), (x_{i-1}, y_{i-1}), and (x_i, y_i), and is used to extrapolate the value

Table 5.2 Summary of the Runge-Kutta integration formulas

Second order

$$y_{i+1} = y_i + \frac{1}{2}(k_1 + k_2) + O(h^3)$$

$$k_1 = hf(x_i, y_i)$$

$$k_2 = hf(x_i + h, y_i + k_1)$$

Third order

$$y_{i+1} = y_i + \frac{1}{6}(k_1 + 4k_2 + k_3) + O(h^4)$$

$$k_1 = hf(x_i, y_i)$$

$$k_2 = hf\left(x_i + \frac{h}{2}, y_i + \frac{k_1}{2}\right)$$

$$k_3 = hf(x_i + h, y_i + 2k_2 - k_1)$$

Fourth order

$$y_{i+1} = y_i + \frac{1}{6}(k_1 + 2k_2 + 2k_3 + k_4) + O(h^5)$$

$$k_1 = hf(x_i, y_i)$$

$$k_2 = hf\left(x_i + \frac{h}{2}, y_i + \frac{k_1}{2}\right)$$

$$k_3 = hf\left(x_i + \frac{h}{2}, y_i + \frac{k_2}{2}\right)$$

$$k_4 = hf(x_i + h, y_i + k_3)$$

Table 5.2 Summary of the Runge-Kutta integration formulas (cont'd)

Fifth order

$$y_{i+1} = y_i + \frac{1}{90}(7k_1 + 32k_3 + 12k_4 + 32k_5 + 7k_6) + O(h^6)$$

$$k_1 = hf(x_i, y_i)$$

$$k_2 = hf\left(x_i + \frac{h}{2}, y_i + \frac{k_1}{2}\right)$$

$$k_3 = hf\left(x_i + \frac{h}{4}, y_i + \frac{3k_1}{16} + \frac{k_2}{16}\right)$$

$$k_4 = hf\left(x_i + \frac{h}{2}, y_i + \frac{k_3}{2}\right)$$

$$k_5 = hf\left(x_i + \frac{3h}{4}, y_i - \frac{3k_2}{16} + \frac{6k_3}{16} + \frac{9k_4}{16}\right)$$

$$k_6 = hf\left(x_i + h, y_i + \frac{k_1}{7} + \frac{4k_2}{7} + \frac{6k_3}{7} - \frac{12k_4}{7} + \frac{8k_5}{7}\right)$$

Runge-Kutta-Fehlberg

$$y_{i+1} = y_i + \left(\frac{25}{216}k_1 + \frac{1408}{2565}k_3 + \frac{2197}{4104}k_4 - \frac{1}{5}k_5\right) + O(h^5)$$

$$k_1 = hf(x_i, y_i)$$

$$k_2 = hf\left(x_i + \frac{h}{4}, y_i + \frac{k_1}{4}\right)$$

$$k_3 = hf\left(x_i + \frac{3}{8}h, y_i + \frac{3}{32}k_1 + \frac{9}{32}k_2\right)$$

$$k_4 = hf\left(x_i + \frac{12}{13}h, y_i + \frac{1932}{2197}k_1 - \frac{7200}{2197}k_2 + \frac{7296}{2197}k_3\right)$$

$$k_5 = hf\left(x_i + h, y_i + \frac{439}{216}k_1 - 8k_2 + \frac{3860}{513}k_3 - \frac{845}{4104}k_4\right)$$

$$k_6 = hf\left(x_i + \frac{h}{2}, y_i - \frac{8}{27}k_1 + 2k_2 - \frac{3544}{2565}k_3 + \frac{1859}{4104}k_4 - \frac{11}{40}k_5\right)$$

$$T_E \simeq \frac{1}{360}k_1 - \frac{128}{4275}k_3 - \frac{2197}{75240}k_4 + \frac{1}{50}k_5 + \frac{2}{55}k_6$$

of $f(x_{i+1}, y_{i+1})$. If we choose a uniform step size, a second-degree backward Gregory-Newton interpolating polynomial may be applied to this problem and Eq. (5.55) becomes

$$y_{i+1} = y_i + \int_{x_i}^{x_{i+1}} \left[f_i - \frac{(x - x_i)}{h} \nabla f_i + \frac{(x - x_i)(x - x_{i-1})}{2!\, h^2} \nabla^2 f_i \right] dx + \int_{x_i}^{x_{i+1}} R_n(x)\, dx \tag{5.93}$$

where $f_i = f(x_i, y_i)$, and it may be considered a function of x only. Noting that $(x_{i+1} - x_i) = h$, Eq. (5.93) reduces to

$$y_{i+1} = y_i + h\left(f_i + \frac{1}{2}\nabla f_i + \frac{5}{12}\nabla^2 f_i \right) + O(h^4) \tag{5.94}$$

This equation would be easier to use by expanding the backward differences in terms of the function values given in Table 3.5. Replacing the backward differences followed by further rearrangements results in the following formula known as the Adams method for solution of the ordinary differential equations:

$$y_{i+1} = y_i + \frac{h}{12}[23f(x_i,y_i) - 16f(x_{i-1},y_{i-1}) + 5f(x_{i-2},y_{i-2})] + O(h^4) \tag{5.95}$$

Eq. (5.95) shows that prior to evaluating y_{i+1}, the values of the function at three points before that have to be known. Because in an initial-value problem only the value of the function at the start of the solution interval is known, another two succeeding values should be calculated by a single-step method, such as Runge-Kutta. Solution of the ordinary differential equation from the fourth point may then be continued with Eq. (5.95).

In order to derive the Adams-Moulton technique, we repeat the same procedure by applying a third-degree interpolating polynomial (using four past points) instead of a second-degree polynomial to approximate $f(x, y)$ in Eq. (5.55). This procedure results in prediction of y_{i+1}

$$(y_{i+1})_{Pr} = y_i + \frac{h}{24}[55f(x_i,y_i) - 59f(x_{i-1},y_{i-1}) + 37f(x_{i-2},y_{i-2}) - 9f(x_{i-3},y_{i-3} \\ + O(h^5) \tag{5.96}$$

In the Adams-Moulton method, we do not stop here and correct y_{i+1} before moving to the next step. The value of y_{i+1} calculated from Eq. (5.96) is a good approximation of the dependent variable at position $(i + 1)$; therefore, almost the correct value of $f(x_{i+1}, y_{i+1})$ may be evaluated from $f(x_{i+1}, (y_{i+1})_{Pr})$ at this stage. We now interpolate the function $f(x, y)$, using a cubic Gregory-Newton backward interpolating polynomial over the range from x_{i-2} to x_{i+1} and calculate the corrected value of y_{i+1} by the integral of Eq. (5.55):

$$(y_{i+1})_{Cor} = y_i + \frac{h}{24}[9f(x_{i+1}, (y_{i+1})_{Pr}) + 19f(x_i, y_i) - 5f(x_{i-1}, y_{i-1}) + f(x_{i-2}, y \\ + O(h^5) \tag{5.97}$$

Eqs. (5.96) and (5.97) should be used as predictor and corrector, respectively. Correction by Eq. (5.97) may be applied more than once until the corrected value converges; that is, the difference between the two consecutive corrected value becomes less than the convergence criterion. However, two applications of the corrector is probably optimum in terms of computer time and the accuracy gained. Once again, solution of the ordinary differential equation by this technique may start from the fifth point; therefore, some other technique should be applied at the beginning of the solution to evaluate y_1 to y_3.

5.5.4 Simultaneous Differential Equations

It was mentioned at the beginning of Sec. 5.5 that the methods of solution of a single differential equation are readily adaptable for solving sets of simultaneous differential equations. To illustrate this, we use the set of n simultaneous ordinary differential equations:

$$\begin{aligned}
\frac{dy_1}{dx} &= f_1(x,\ y_1,\ y_2,\ \ldots,\ y_n)\\
\frac{dy_2}{dx} &= f_2(x,\ y_1,\ y_2,\ \ldots,\ y_n)\\
&\vdots\\
\frac{dy_n}{dx} &= f_n(x,\ y_1,\ y_2,\ \ldots,\ y_n)
\end{aligned} \tag{5.98}$$

and expand, for example, the fourth-order Runge-Kutta formulas to

$$\begin{aligned}
y_{i+1,j} &= y_{ij} + \frac{1}{6}(k_{1j} + 2k_{2j} + 2k_{3j} + k_{4j}) && j = 1, 2, \ldots, n\\
k_{1j} &= hf_j(x_i,\ y_{i1},\ y_{i2},\ \ldots,\ y_{in}) && j = 1, 2, \ldots, n\\
k_{2j} &= hf_j\left(x_i + \frac{h}{2},\ y_{i1} + \frac{k_{11}}{2},\ y_{i2} + \frac{k_{12}}{2},\ \ldots,\ y_{in} + \frac{k_{1n}}{2}\right) && j = 1, 2, \ldots, n\\
k_{3j} &= hf_j\left(x_i + \frac{h}{2},\ y_{i1} + \frac{k_{21}}{2},\ y_{i2} + \frac{k_{22}}{2},\ \ldots,\ y_{in} + \frac{k_{2n}}{2}\right) && j = 1, 2, \ldots, n\\
k_{4j} &= hf_j(x_i + h,\ y_{i1} + k_{31},\ y_{i2} + k_{32},\ \ldots,\ y_{in} + k_{3n}) && j = 1, 2, \ldots, n
\end{aligned} \tag{5.99}$$

This method is easily programmable using nested loops. In MATLAB, the values of k and y_i can be put in vectors and easily perform Eq. (5.99) in matrix form.

Example 5.3: Solution of Nonisothermal Plug-Flow Reactor. Write general MATLAB functions for integrating simultaneous nonlinear differential equations using the Euler, Euler predictor-corrector (modified Euler), Runge-Kutta, Adams, and Adams-Moulton methods. Apply these functions for the solution of differential equations that simulate a nonisotherm plug flow reactor, as described below.[3]

Vapor-phase cracking of acetone, described by the following endothermic reaction:

$$CH_3COCH_3 \rightarrow CH_2CO + CH_4$$

takes place in a jacketed tubular reactor. Pure acetone enters the reactor at a temperature of $T_0 = 1035$ K and pressure of $P_0 = 162$ kPa, and the temperature of external gas in the heat exchanger is constant at $T_a = 1150$ K. Other data are as follows:

Volumetric flow rate: $v_0 = 0.002 \text{ m}^3\text{/s}$

Volume of the reactor: $V_R = 1 \text{ m}^3$

Overall heat transfer coefficient: $U = 110 \text{ W/m}^2\text{.K}$

Heat transfer area: $a = 150 \text{ m}^2\text{/m}^3$ reactor

Reaction constant: $k = 3.58 \exp\left[34222\left(\dfrac{1}{1035} - \dfrac{1}{T}\right)\right] \text{ s}^{-1}$

Heat of reaction:

$$\Delta H_R = 80770 + 6.8(T - 298) - 5.75\times10^{-3}(T^2 - 298^2) - 1.27\times10^{-6}(T^3 - 298^3) \text{ J/mol}$$

Heat capacity of acetone: $C_{p_A} = 26.63 + 0.1830T - 45.86\times10^{-6}T^2$ J/mol.K

Heat capacity of ketene: $C_{p_B} = 20.04 + 0.0945T - 30.95\times10^{-6}T^2$ J/mol.K

Heat capacity of methane: $C_{p_C} = 13.39 + 0.0770T - 18.71\times10^{-6}T^2$ J/mol.K

Determine the temperature profile of the gas along the length of the reactor. Assume constant pressure throughout the reactor.

Method of Solution: In order to calculate the temperature profile in the reactor, we have to solve the material balance and energy balance equations simultaneously:

Mole balance:
$$\frac{dX}{dV} = \frac{-r_A}{F_{A_0}}$$

Energy balance:
$$\frac{dT}{dV} = \frac{Ua(T_a - T) + r_A \Delta H_R}{F_{A_0}(C_{p_A} + X\Delta C_p)}$$

---

[3] This problem was adopted from Fogler [2] by permission of the author.

where X is the conversion of acetone, V is the volume of the reactor, $F_{A_0} = C_{A_0} v_0$ is the molar flow rate of acetone at the inlet, T is the temperature of the reactor, $\Delta C_p = C_{p_B} + C_{p_C} - C_{p_A}$, and C_{A_0} is the concentration of acetone vapor at the inlet. The reaction rate is given as

$$-r_A = kC_{A_0} \frac{1 - X}{1 + X} \cdot \frac{T_0}{T}$$

In order to introduce the pair of differential equations as a MATLAB function the following definitions are assumed:

Program Description: Five general MATLAB functions are written for the solution of a set of simultaneous nonlinear ordinary differential equations. They are *Euler.m, MEuler.m, RK.m, Adams.m,* and *AdamsMoulton.m.* All these functions consist of two main sections. The first part is initialization, in which specific input arguments are checked, and some vectors to be used in the second part are initiated. The next section of the function is solution of the set of nonlinear ordinary differential equations according to the specified method, which is done simultaneously in vector form. Brief descriptions of the method of solution of these five functions are given below:

Euler.m – The Euler method: This function solves the set of differential equations based on Eq. (5.61).

MEuler.m – The Euler predictor-corrector (modified Euler) method: This function solves the set of differential equations based on Eqs. (5.66) and (5.67).

RK.m – The Runge-Kutta methods: This function is capable of solving the set of differential equations by a second-, third-, fourth-, or fifth-order Runge-Kutta method. The formulas that appeared in Table 5.2 are used for calculating a Runge-Kutta solution of the differential equations.

Adams.m – The Adams method: This function solves the set of differential equations using Eq. (5.95). The required starting points are evaluated by the third-order Runge-Kutta (using the function *RK.m*) which has the same order of truncation error as the Adams method.

AdamsMoulton.m – The Adams-Moulton method: This function solves the set of differential equations using Eqs. (5.96) and (5.97). The required starting points are evaluated by the fourth-order Runge-Kutta method (using the function *RK.m*), which has the same order of truncation error as the Adams-Moulton method.

The first input argument to all the above method functions is the name of the MATLAB function containing the set of differential equations. Note that the first input argument to this function has to be the independent variable, even if it is not used explicitly in the equations. It is important that this function returns the values of the derivatives (f_i) as a column vector. The other inputs to the method functions are initial and final values of the independent variable, interval width, and the initial value of the dependent variable. In *RK.m,* the order of the method may also be specified. It is possible to pass, through the above functions, additional arguments to the M-file describing the set of differential equations.

Program

Example5_3.m

```
% Example5_3.m
% Solution to Example 5.3. This program calculates and plots
% the temperature and conversion profile of a plug-flow reactor
% in which the endothermic cracking of acetone takes place.
% It can call Euler, MEuler, RK, Adams, or AdamsMoulton solvers
% for solution of the pair of energy and material balances.
% It is also capable of comparing different solvers.

change = 1;
while change
   clear
   clc
   % Input data
   T0 = input(' Inlet temperature (K)                          = ');
   P0 = input(' Inlet pressure (Pa)                            = ');
   v0 = input(' Inlet volumetric flow rate (m3/s)              = ');
   X0 = input(' Inlet conversion of acetone                    = ');
   VR = input(' Volume of the reactor (m3)                     = ');
   Ta = input(' External gas temperature (K)                   = ');
   U  = input(' Overall heat transfer coefficient (W/m2.K) = ');
   a  = input(' Heat transfer area (m2/m3)                     = ');

   CA0 = P0 * (1-X0) / (8.314 * T0);% Input concentration (mol/m3)
   FA0 = v0 * CA0;                    % Input molar flow rate (mol/s)
   fprintf('\n')
   fname=input(' M-file containing the set of differential equations
: ');
   h = input(' Step size = ');

   met = 1;
   while met
     clc
     fprintf('\n')
     disp('  1 ) Euler')
     disp('  2 ) Modified Euler')
     disp('  3 ) Runge-Kutta')
     disp('  4 ) Adams')
     disp('  5 ) Adams-Moulton')
     disp('  6 ) Comparison of methods')
     disp('  0 ) End')
```

```
    met = input('\n Choose the method of solution : ');
    if met == 6
      method=input('\n Input the methods to be compared, as a
vector : ');
    else
      method = met;
    end
    lgnd = 'legend(';
    lmethod = length(method);
    for k = 1:lmethod
       switch method(k)
       case 1  % Euler
          [V,y] = Euler(fname,0,VR,h,[X0,T0],T0,CA0,FA0,U,a,Ta);
          if k > 1
             lgnd = [lgnd ','];
          end
          lgnd = [lgnd '''Euler'''];
       case 2  % Modified Euler
          [V,y] = MEuler(fname,0,VR,h,[X0,T0],T0,CA0,FA0,U,a,Ta);
          if k > 1
             lgnd = [lgnd ','];
          end
          lgnd = [lgnd '''Modified Euler'''];
       case 3  % Runge-Kutta
          n = input('\n Order of the Runge-Kutta method (2-5) = ');
           if n < 2 | n > 5
             n = 2;
           end
         [V,y] = RK(fname,0,VR,h,[X0,T0],n,T0,CA0,FA0,U,a,Ta);
          if k > 1
             lgnd = [lgnd ','];
          end
          lgnd = [lgnd '''RK' int2str(n) ''''];
       case 4   % Adams
          [V,y] = Adams(fname,0,VR,h,[X0,T0],T0,CA0,FA0,U,a,Ta);
          if k > 1
             lgnd = [lgnd ','];
          end
          lgnd = [lgnd '''Adams'''];
       case 5   % Adams-Moulton
        [V,y] = AdamsMoulton(fname,0,VR,h,[X0,T0],T0,CA0, ...
                FA0,U, a,Ta);
          if k > 1
             lgnd = [lgnd ','];
          end
          lgnd = [lgnd '''Adams-Moulton'''];
       end
       x(k,:) = y(1,:);              % Conversion
```

```
        t(k,:) = y(2,:);               % Temperature
      end
      if met
         clf
         % Plotting the results
         subplot(2,1,1), plot(V/VR,x(1:lmethod,:))
         ylabel('Conversion, X(%)')
         title('(a) Acetone Conversion Profile')
         subplot(2,1,2), plot(V/VR,t(1:lmethod,:))
         xlabel('V/V_R')
         ylabel('Temperature, T(K)')
         title('(b) Temperature Profile')
         lgnd = [lgnd ')'];
         eval(lgnd)
      end
   end
   change=input('\n\n Do you want to repeat the solution with
different input data (0/1)? ');
end
```

Ex5_3_func.m

```
function fnc = Ex5_3_func(V,y,T0,CA0,FA0,U,a,Ta)
% Function Ex5_3_func.M
% This function contains the pair of ordinary differential
% equations introduced in Example 5.3. The name of this function
% is an input to the main program Example5_3.m and will be called
% by the selected ODE solver.

X = y(1);                                        % Conversion
T = y(2);                                        % Temperature
k = 3.58*exp(34222*(1/1035-1/T));                % Rate constant
dHR = 80770+6.8*(T-298)-5.75e-3*(T^2-298^2)-1.27e-6*(T^3-298^3);
                                                 % Heat of reaction
CpA = 26.63 + .183*T - 45.86e-6*T^2;             % Heat capacity of A
CpB = 20.04 + .0945*T - 30.95e-6*T^2;            % Heat capacity of B
CpC = 13.39 + .077*T - 18.71e-6*T^2;             % Heat capacity of C
dCp = CpB + CpC - CpA;
rA = -k * CA0 * (1-X)/(1+X) * T0/T;              % Reaction rate
% Mole balance and energy balance
fnc = [-rA/FA0; (U*a*(Ta-T)+rA*dHR)/(FA0*(CpA+X*dCp))];
```

Euler.m

```
function [x,y] = Euler(ODEfile,xi,xf,h,yi,varargin)
% EULER Solves a set of ordinary differential equations by
%   the Euler method.
%
%   [X,Y]=EULER('F',XI,XF,H,YI) solves a set of ordinary
%   differential equations by the Euler method, from XI to XF.
```

```
%   The equations are given in the M-file F.M.   H is the length
%   of interval and YI is the vector of initial values of the
%   dependent variable at XI.
%
%   [X,Y]=EULER('F',XI,XF,H,YI,P1,P2,...) allows for additional
%   arguments which are passed to the function F(X,P1,P2,...).
%
%  See also ODE23, ODE45, ODE113, ODE15S, ODE23S, MEULER, RK,
%           ADAMS, ADAMSMOULTON

% (c) N. Mostoufi & A. Constantinides
% January 1, 1999

% Initialization
if isempty (h) | h == 0
   h = linspace(xi,xf);
end

yi = (yi(:).')';          % Make sure it's a column vector

x = [xi:h:xf];            % Vector of x values
if x(end) ~= xf
   x(end+1) = xf;
end
d = diff(x);              % Vector of x-increments

y(:,1) = yi;              % Initial condition
% Solution
for i = 1:length(x)-1
   y(:,i+1) = y(:,i) + d(i) *
feval(ODEfile,x(i),y(:,i),varargin{:});
end
```

MEuler.m

```
function [x,y] = MEuler(ODEfile,xi,xf,h,yi,varargin)
% MEULER Solves a set of ordinary differential equations by
%   the modified Euler (predictor-corrector) method.
%
%   [X,Y]=MEULER('F',XI,XF,H,YI) solves a set of ordinary
%   differential equations by the modified Euler (the Euler
%   predictor-corrector)  method, from XI to XF.
%   The equations are given in the M-file F.M.  H is the length of
%   interval and YI is the vector of initial values of the dependent
%   variable at XI.
%
%   [X,Y]=MEULER('F',XI,XF,H,YI,P1,P2,...) allows for additional
%   arguments which are passed to the function F(X,P1,P2,...).
%
```

```
%  See also ODE23, ODE45, ODE113, ODE15S, ODE23S, EULER, RK,
%           ADAMS, ADAMSMOULTON

% (c) N. Mostoufi & A. Constantinides
% January 1, 1999

% Initialization
if isempty (h) | h == 0
   h = linspace(xi,xf);
end

yi = (yi(:).')';          % Make sure it's a column vector

x = [xi:h:xf];            % Vector of x values
if x(end) ~= xf
   x(end+1) = xf;
end
d = diff(x);              % Vector of x-increments

y(:,1) = yi;              % Initial condition
% Solution
for i = 1:length(x)-1
  % Predictor
  y(:,i+1)=y(:,i) + d(i) * feval(ODEfile,x(i),y(:,i), varargin{:});
  % Corrector
  y(:,i+1)=y(:,i)+d(i) * feval(ODEfile,x(i+1),y(:,i+1),varargin{:});
end
```

RK.m

```
function [x,y] = RK(ODEfile,xi,xf,h,yi,n,varargin)
% RK Solves a set of ordinary differential equations by the
%   Runge-Kutta method.
%
%   [X,Y]=RK('F',XI,XF,H,YI,N) solves a set of ordinary differential
%   equations by the Nth-order Runge-Kutta method, from XI to XF.
%   The equations are given in the M-file F.M.  H is the length of
%   interval. YI is the vector of initial values of the dependent
%   variable at XI. N should be an integer from 2 to 5. If there
%   are only five input arguments or the sixth input argument is an
%   empty matrix, the 2nd-order Runge-Kutta method will be
%   performed.
%
%   [X,Y]=RK('F',XI,XF,H,YI,N,P1,P2,...) allows for additional
%   arguments which are passed to the function F(X,P1,P2,...).
%
%  See also ODE23, ODE45, ODE113, ODE15S, ODE23S, EULER, MEULER,
%           ADAMS, ADAMSMOULTON
```

```
% (c) N. Mostoufi & A. Constantinides
% January 1, 1999

% Initialization
if isempty (h) | h == 0
   h = linspace(xi,xf);
end

if nargin == 5 | isempty(n) | n < 2 | n > 5
   n = 2;
end
n = fix(n);

yi = (yi(:).')';        % Make sure it's a column vector

x = [xi:h:xf];          % Vector of x values
if x(end) ~= xf
   x(end+1) = xf;
end
d = diff(x);            % Vector of x-increments

y(:,1) = yi;            % Initial condition
% Solution

switch n
case 2                  % 2nd-order Runge-Kutta
   for i = 1:length(x)-1
      k1 = d(i) * feval(ODEfile,x(i),y(:,i),varargin{:});
      k2 = d(i) * feval(ODEfile,x(i+1),y(:,i)+k1,varargin{:});
      y(:,i+1) = y(:,i) + (k1+k2)/2;
   end
case 3                  % 3rd-order Runge-Kutta
   for i = 1:length(x)-1
      k1 = d(i) * feval(ODEfile,x(i),y(:,i),varargin{:});
      k2 = d(i) * feval(ODEfile,x(i)+d(i)/2,y(:,i)+k1/2,...
         varargin{:});
      k3 = d(i) * feval(ODEfile,x(i+1),y(:,i)+2*k2-k1,varargin{:});
      y(:,i+1) = y(:,i) + (k1+4*k2+k3)/6;
   end
case 4                  % 4th-order Runge-Kutta
   for i = 1:length(x)-1
      k1 = d(i) * feval(ODEfile,x(i),y(:,i),varargin{:});
      k2 = d(i) * feval(ODEfile,x(i)+d(i)/2,y(:,i)+k1/2, ...
         varargin{:});
      k3 = d(i) * feval(ODEfile,x(i)+d(i)/2,y(:,i)+k2/2, ...
         varargin{:});
      k4 = d(i) * feval(ODEfile,x(i+1),y(:,i)+k3,varargin{:});
      y(:,i+1) = y(:,i) + (k1+2*k2+2*k3+k4)/6;
```

```
   end
case 5                  % 5th-order Runge-Kutta
   for i = 1:length(x)-1
      k1 = d(i) * feval(ODEfile,x(i),y(:,i),varargin{:});
      k2 = d(i) * feval(ODEfile,x(i)+d(i)/2,y(:,i) +k1/2, ...
         varargin{:});
      k3 = d(i) * feval(ODEfile,x(i)+d(i)/4,y(:,i)+3*k1/16+k2/16,...
         varargin{:});
      k4 = d(i) * feval(ODEfile,x(i)+d(i)/2,y(:,i)+k3/2, ...
         varargin{:});
      k5 = d(i) * feval(ODEfile,x(i)+3*d(i)/4,y(:,i)-3*k2/16+ ...
         6*k3/16+9*k4/16, varargin{:});
      k6 = d(i) * feval(ODEfile,x(i+1),y(:,i)+k1/7+4*k2/7+ ...
         6*k3/7-12*k4/7+8*k5/7, varargin{:});
      y(:,i+1) = y(:,i) + (7*k1+32*k3+12*k4+32*k5+7*k6)/90;
   end
end
```

Adams.m

```
function [x,y] = Adams(ODEfile,xi,xf,h,yi,varargin)
% ADAMS Solves a set of ordinary differential equations by the
%   Adams method.
%
%   [X,Y]=ADAMS('F',XI,XF,H,YI) solves a set of ordinary
%   differential equations by the Adams method, from XI to XF.
%   The equations are given in the M-file F.M.  H is the length
%   of the interval and YI is the vector of initial values of
%   the dependent variable at XI.
%
%   [X,Y]=ADAMS('F',XI,XF,H,YI,P1,P2,...) allows for additional
%   arguments which are passed to the function F(X,P1,P2,...).
%
%  See also ODE23, ODE45, ODE113, ODE15S, ODE23S, EULER, MEULER,
%           RK, ADAMSMOULTON

% (c) N. Mostoufi & A. Constantinides
% January 1, 1999

% Initialization
if isempty (h) | h == 0
   h = linspace(xi,xf);
end

yi = (yi(:).'); % Make sure it's a row vector

x = [xi:h:xf]'; % Vector of x values
if x(end) ~= xf
   x(end+1) = xf;
```

```
end
d = diff(x); % Vector of x-increments

% Starting values
[a,b]=RK(ODEfile,x(1),x(3),h,yi,3,varargin{:});
y(:,1:3) = b;
for i = 1:3
   f(:,i) = feval(ODEfile,x(i),y(:,i),varargin{:});
end

% Solution
for i = 3:length(x)-1
   y(:,i+1) = y(:,i) + d(i)/12 * (23*f(:,i) - 16*f(:,i-1) +...
        5*f(:,i-2));
   f(:,i+1) = feval(ODEfile,x(i+1),y(:,i+1),varargin{:});
end
```

AdamsMoulton.m

```
function [x,y] = AdamsMoulton(ODEfile,xi,xf,h,yi,varargin)
% ADAMSMOULTON Solves a set of ordinary differential equations by
%   the Adams-Moulton method.
%
%   [X,Y]=ADAMSMOULTON('F',XI,XF,H,YI) solves a set of ordinary
%   differential equations by the Adams-Moulton method, from XI to
%   XF.  The equations are given in the M-file F.M.   H is the
%   length of interval and YI is the vector of initial values of
%   the dependent variable at XI.
%
%   [X,Y]=ADAMSMOULTON('F',XI,XF,H,YI,P1,P2,...) allows for
%   additional arguments which are passed to the function
%   F(X,P1,P2,...).
%
%  See also ODE23, ODE45, ODE113, ODE15S, ODE23S, EULER, MEULER,
%           RK, ADAMS

% (c) N. Mostoufi & A. Constantinides
% January 1, 1999

% Initialization
if isempty (h) | h == 0
   h = linspace(xi,xf);
end

yi = (yi(:).')';      % Make sure it's a column vector

x = [xi:h:xf]';       % Vector of x values
if x(end) ~= xf
   x(end+1) = xf;
```

```
end
d = diff(x);            % Vector of x-increments

% Starting values
[a,b] = RK(ODEfile,x(1),x(4),h,yi,4,varargin{:});
y(:,1:4) = b;
for i = 1:4
   f(:,i) = feval(ODEfile,x(i),y(:,i),varargin{:});
end

% Solution
for i = 4:length(x)-1
   % Predictor
   y(:,i+1) = y(:,i) + d(i)/24 * (55*f(:,i) - 59*f(:,i-1) ...
      + 37*f(:,i-2) - 9*f(:,i-3));
   f(:,i+1) = feval(ODEfile,x(i+1),y(:,i+1),varargin{:});
   % Corrector
   y(:,i+1) = y(:,i) + d(i)/24 * (9*f(:,i+1) + 19*f(:,i) ...
      - 5*f(:,i-1) + f(:,i-2));
   f(:,i+1) = feval(ODEfile,x(i+1),y(:,i+1),varargin{:});
end

% Solution
for i = 4:length(x)-1
   % Predictor
   y(:,i+1) = y(:,i) + d(i)/24 * (55*f(:,i) - 59*f(:,i-1) ...
      + 37*f(:,i-2) - 9*f(:,i-3));
   f(:,i+1) = feval(ODEfile,x(i+1),y(:,i+1),varargin{:});
   % Corrector
   y(:,i+1) = y(:,i) + d(i)/24 * (9*f(:,i+1) + 19*f(:,i) ...
      - 5*f(:,i-1) + f(:,i-2));
   f(:,i+1) = feval(ODEfile,x(i+1),y(:,i+1),varargin{:});
end
```

Input and Results

```
>>Example5_3

 Inlet temperature (K)                              = 1035
 Inlet pressure (Pa)                                = 162e3
 Inlet volumetric flow rate (m3/s)                  = 0.002
 Inlet conversion of acetone                        = 0
 Volume of the reactor (m3)                         = 0.001
 External gas temperature (K)                       = 1200
 Overall heat transfer coefficient (W/m2.K)         = 110
 Heat transfer area (m2/m3)                         = 150

 M-file containing the set of differential equations : 'Ex5_3_func'
```

```
Step size = 0.00003

  1 ) Euler
  2 ) Modified Euler
  3 ) Runge-Kutta
  4 ) Adams
  5 ) Adams-Moulton
  6 ) Comparison of methods
  0 ) End

Choose the method of solution : 6

Input the methods to be compared, as a vector : [1, 3, 4]

Order of the Runge-Kutta method (2-5) = 2

  1 ) Euler
  2 ) Modified Euler
  3 ) Runge-Kutta
  4 ) Adams
  5 ) Adams-Moulton
  6 ) Comparison of methods
  0 ) End

Choose the method of solution : 0
Do you want to repeat the solution with different input data (0/1)? 0
```

Discussion of Results: The mole and energy balance equations are solved by three different methods of different order of error: Euler [$O(h^2)$], second-order Runge-Kutta [$O(h^3)$], and Adams [$O(h^4)$]. Graphical results are given in Figs. E5.3*a* and *b*.[4] At the beginning the temperature of the reactor decreases because the reaction is endothermic. However, it starts to increase steadily at about 10% of the length of the reactor, due to the heat transfer from the hot gas circulation around the reactor.

It can be seen from Figs. E5.3*a* and *b* that there are visible differences between the three methods in the temperature profile where the temperature reaches minimum. This region is where the change in the derivative of temperature (energy balance formula) is greater than the other parts of the curve, and as a result, different techniques for approximation of this derivative give different values for it. The reader is encouraged to repeat this example with different methods of solution and step sizes.

---

[4] When running *Example5_3.m,* solution results will be shown on the screen by solid lines of different color. However, results for the three different methods used here are illustrated by different line type in Figs. E5.3*a* and *b* in order to make them identifiable.

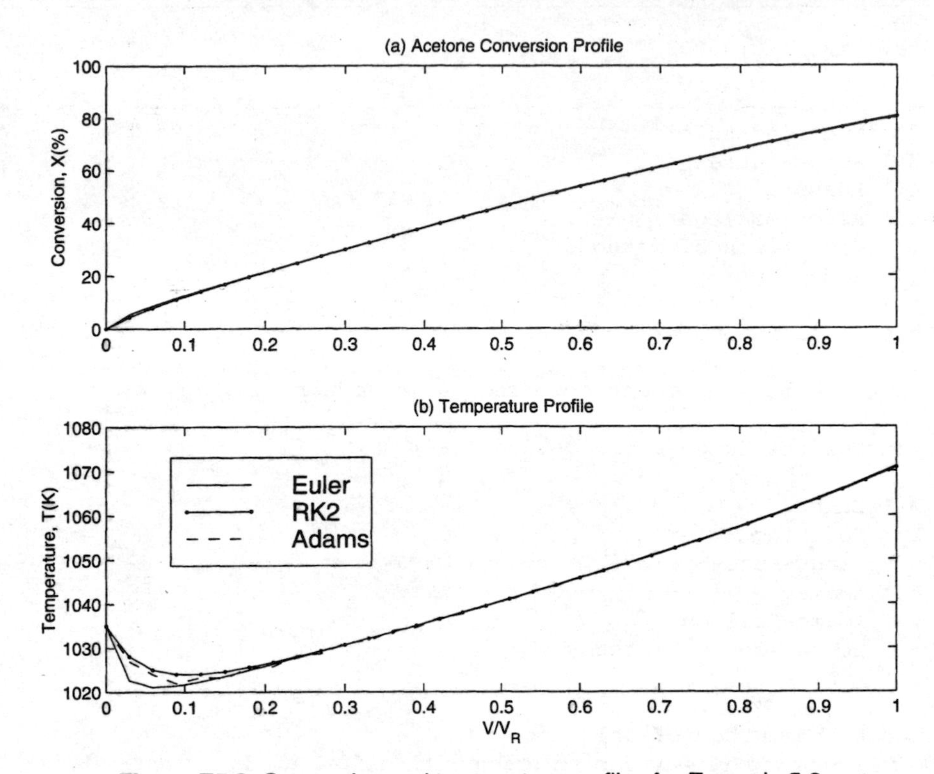

Figure E5.3 Conversion and temperature profiles for Example 5.3.

5.6 Nonlinear Ordinary Differential Equations–Boundary-Value Problems

Ordinary differential equations with boundary conditions specified at two or more points of the independent variable are classified as boundary-value problems. There are many chemical engineering applications that result in ordinary differential equations of the boundary-value type. To mention only a few examples:

1. Diffusion with chemical reaction in the study of chemical catalysis or enzyme catalysis
2. Heat and mass transfer in boundary-layer problems
3. Application of rigorous optimization methods, such as Pontryagin's maximum principle or the calculus of variations
4. Discretization of nonlinear elliptic partial differential equations [3].

The diversity of problems of the boundary-value type have generated a variety of methods for their solution. The system equations in these problems could be linear or nonlinear, and the boundary conditions could be linear or nonlinear, separated or mixed, two-point or multipoint. Comprehensive discussions of the solutions of boundary-value problems are given by Kubíček and Hlaváček [3] and by Aziz [4]. In this section, we have chosen to discuss algorithms that are applicable to the solution of nonlinear (as well as linear) boundary-value problems. These are the *shooting* method, the *finite difference* method, and the *collocation* methods. The last two methods will be discussed again in Chap. 6 in connection with the solution of partial differential equations of the boundary-value type.

The canonical form of a two-point boundary-value problem with linear boundary conditions is

$$\frac{dy_j}{dx} = f_j(x, y_1, y_2, \ldots, y_n) \qquad x_0 \le x \le x_f \; ; \; j = 1, 2, \ldots, n \tag{5.100}$$

where the boundary conditions are split between the initial point x_0 and the final point x_f. The first r equations have initial conditions specified and the last $(n - r)$ equations have final conditions given:

$$y_j(x_0) = y_{j,0} \qquad j = 1, 2, \ldots, r \tag{5.101}$$

$$y_j(x_f) = y_{j,f} \qquad j = r+1, \ldots, n \tag{5.102}$$

A second-order two-point boundary-value problem may be expressed in the form:

$$\frac{d^2y}{dx^2} = f\left(x, y, \frac{dy}{dx}\right) \qquad x_0 \le x \le x_f \tag{5.103}$$

subject to the boundary conditions

$$a_0 y(x_0) + b_0 y'(x_0) = \gamma_0 \tag{5.104}$$

$$a_f y(x_f) + b_f y'(x_f) = \gamma_f \tag{5.105}$$

where the subscript 0 designates conditions at the left boundary (initial) and the subscript f identifies conditions at the right boundary (final).

This problem can be transformed to the canonical form (5.100) by the appropriate substitutions described in Sec. 5.3.

5.6.1 The Shooting Method

The shooting method converts the boundary-value problem to an initial-value one to take advantage of the powerful algorithms available for the integration of initial-value problems (see Sec. 5.5). In this method, the unspecified initial conditions of the system differential equations are guessed and the equations are integrated forward as a set of simultaneous initial-value differential equations. At the end, the calculated final values are compared with the boundary conditions and the guessed initial conditions are corrected if necessary. This procedure is repeated until the specified terminal values are achieved within a small convergence criterion. This general algorithm forms the basis for the family of shooting methods. These may vary in their choice of initial or final conditions and in the integration of the equations in one direction or two directions. In this section, we develop Newton's technique, which is the most widely known of the shooting methods and can be applied successfully to boundary-value problem of any complexity as long as the resulting initial-value problem is stable and a set of good guesses for unspecified conditions can be made [3].

We develop the Newton method for a set of two differential equations

$$\frac{dy_1}{dx} = f_1(x, y_1, y_2)$$
$$\frac{dy_2}{dx} = f_2(x, y_1, y_2) \tag{5.106}$$

with split boundary conditions

$$y_1(x_0) = y_{1,0} \tag{5.107}$$

$$y_2(x_f) = y_{2,f} \tag{5.108}$$

We guess the initial condition for y_2

$$y_2(x_0) = \gamma \tag{5.109}$$

If the system equations are integrated forward, the two trajectories may look like those in Fig. 5.5. Since the value of $y_2(x_0)$ was only a guess, the trajectory of y_2 misses its target at x_f; that is, it does not satisfy the boundary condition of (5.108). For the given guess of γ, the calculated value of y_2 at x_f is designated as $y_2(x_f, \gamma)$. The desirable objective is to find the value of γ which forces $y_2(x_f, \gamma)$ to satisfy the specified boundary condition, that is,

$$y_2(x_f, \gamma) = y_{2,f} \tag{5.110}$$

Rearrange Eq. (5.110) to

$$\phi(\gamma) = y_2(x_f, \gamma) - y_{2,f} = 0 \tag{5.111}$$

The function $\phi(\gamma)$ can be expressed in a Taylor series around γ:

$$\phi(\gamma + \Delta\gamma) = \phi(\gamma) + \frac{\partial\phi}{\partial\gamma}\Delta\gamma + O[(\Delta\gamma)^2] \tag{5.112}$$

In order for the system to converge, that is, for the trajectory of y_2 to hit the specified boundary value of x_f:

$$\lim_{\Delta\gamma \to 0} \phi(\gamma + \Delta\gamma) = 0 \tag{5.113}$$

Therefore, Eq. (5.112) becomes

$$0 = \phi(\gamma) + \frac{\partial\phi}{\partial\gamma}\Delta\gamma + O[(\Delta\gamma)^2] \tag{5.114}$$

Truncation and rearrangement gives

$$\Delta\gamma = \frac{-\phi(\gamma)}{\left[\dfrac{\partial\phi}{\partial\gamma}\right]} \tag{5.115}$$

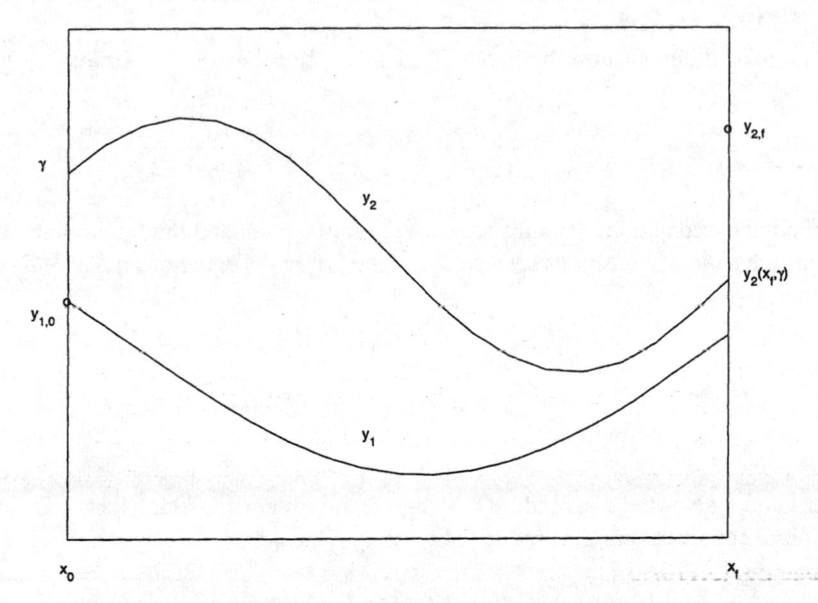

Figure 5.5 Forward integration using a guessed initial condition γ. The $\circ$ designates the known boundary points.

The reader should be able to recognize this equation as a form of the Newton-Raphson equation of Chap. 1. Using the definition of $\phi(\gamma)$ [Eq. (5.111)], taking its partial derivative, and combining with Eq. (5.115), we obtain

$$\Delta\gamma = \frac{-[y_2(x_f, \gamma) - y_{2,f}]}{\left[\dfrac{\partial y_2(x_f, \gamma)}{\partial \gamma}\right]} = \frac{\delta y}{\left[\dfrac{\partial y_2(x_f, \gamma)}{\partial \gamma}\right]} \tag{5.116}$$

where δy is the difference between the specified final boundary value $y_{2,f}$ and the calculated final value $y_2(x_f, \gamma)$ obtained from using the guessed γ:

$$\delta y = -[y_2(x_f, \gamma) - y_{2,f}] \tag{5.117}$$

The value of $\Delta\gamma$ is the correction to be applied to the guessed γ to obtain a new guess:

$$(\gamma)_{new} = (\gamma)_{old} + \Delta\gamma \tag{5.118}$$

In order to avoid divergence it may sometimes be necessary to take a fractional correction step by using relaxation, that is,

$$(\gamma)_{new} = (\gamma)_{old} + \rho\Delta\gamma \qquad 0 < \rho \leq 1 \tag{5.119}$$

Solution of the set of differential equations continues with the new value of γ [calculated by Eq. (5.119)] until $|\Delta\gamma| \leq \epsilon$.

The algorithm can now be generalized to apply to a set of n simultaneous system equations:

$$\frac{dy_j}{dx} = f_j(x, y_1, y_2, \ldots, y_n) \qquad x_0 \leq x \leq x_f \;;\; j = 1, 2, \ldots, n \tag{5.100}$$

whose boundary conditions are split between the initial point and the final point. The first r equations have *initial* conditions specified, and the last $(n - r)$ equations have *final* conditions given:

$$y_j(x_0) = y_{j,0} \qquad j = 1, 2, \ldots, r \tag{5.101}$$

$$y_j(x_f) = y_{j,f} \qquad j = r+1, \ldots, n \tag{5.102}$$

In order to apply Newton's procedure to integrate the system equations forward, the missing $(n - r)$ initial conditions are guessed as follows:

$$y_j(x_0) = \gamma_j \qquad j = r+1, \ldots, n \tag{5.120}$$

The system equations (5.100) with the given initial conditions (5.101) and the guessed initial conditions (5.120) are integrated simultaneously in the forward direction. At the right-hand boundary (x_f), the *Jacobian matrix* [equivalent to the derivative term in Eq. (5.116)] is evaluated:

$$J(x_f, \gamma) = \begin{bmatrix} \frac{\partial y_{r+1}}{\partial \gamma_{r+1}}\Big|_{x_f} & \frac{\partial y_{r+1}}{\partial \gamma_{r+2}}\Big|_{x_f} & \cdots & \frac{\partial y_{r+1}}{\partial \gamma_n}\Big|_{x_f} \\ \cdots & \cdots & \cdots & \cdots \\ \frac{\partial y_n}{\partial \gamma_{r+1}}\Big|_{x_f} & \frac{\partial y_n}{\partial \gamma_{r+2}}\Big|_{x_f} & \cdots & \frac{\partial y_n}{\partial \gamma_n}\Big|_{x_f} \end{bmatrix} \tag{5.121}$$

The correction of the guessed initial values is implemented by the equation

$$\Delta\gamma = [J(x_f, \gamma)]^{-1}\,\delta y \tag{5.122}$$

where the vector δy is the difference between the specified final boundary values and the calculated final values using the guessed initial conditions

$$\delta y = -\begin{bmatrix} y_{r+1}(x_f, \gamma) - y_{r+1,f} \\ \cdots\cdots\cdots \\ y_n(x_f, \gamma) - y_{n,f} \end{bmatrix} \tag{5.123}$$

The new estimate of the guessed initial conditions is then evaluated from Eq. (5.118) in the vector form.

$$(\gamma)_{new} = (\gamma)_{old} + \rho\Delta\gamma \qquad 0 < \rho \le 1 \tag{5.119}$$

The shooting method algorithm using the Newton technique is outlined in the following five steps:

1. The missing initial conditions of the system equations are guessed by Eq. (5.120).
2. The system equations (5.100) are integrated forward simultaneously.
3. Evaluate the Jacobian matrix from Eq. (5.121) either numerically or analytically.
4. The correction $\Delta\gamma$ to be applied to γ is calculated from Eq. (5.122). The new value of γ is obtained from Eq. (5.119).
5. Steps 2 and 3 are repeated, each time with a corrected value of γ, until $min(|\Delta\gamma_j|) \le \epsilon$, where ϵ is the convergence criterion.

Note that the number of differential equations with final boundary conditions is not in any case more than half of the total number of equations. In the case when final conditions are specified for more than half the total number of differential equations, we may simply reverse the integrating direction and as a result obtain fewer number of final conditions. The application of the Newton technique in the shooting method by the above algorithm is demonstrated in Example 5.4.

Example 5.4: Flow of a Non-Newtonian Fluid. Write a general MATLAB function for solution of a boundary value problem by the shooting method using the Newton's technique. Apply this function to find the velocity profile of a non-Newtonian fluid that is flowing through a circular tube as shown in Fig. E5.4*a*. Also calculate the volumetric flow rate of the fluid. The viscosity of this fluid can be described by the Carreau model [5]:

$$\frac{\mu}{\mu_0} = \left[1 + (t_1\dot{\gamma})^2\right]^{(n-1)/2}$$

where μ is the viscosity of the fluid, μ_0 is the zero shear rate viscosity, $\dot{\gamma}$ is the shear rate, t_1 is the characteristic time, and n is a dimensionless constant.

The momentum balance for this flow, assuming the tube is very long so that end effect is negligible, results in

$$\frac{d}{dr}(r\tau_{rz}) = -\frac{\Delta P}{L}r \tag{1}$$

where $\Delta P/L$ is the pressure drop gradient along the pipe and the shear stress is expressed as

$$\tau_{rz} = -\mu\dot{\gamma} = -\mu\frac{dv_z(r)}{dr}$$

Therefore, Eq. (1) is a second-order ordinary differential equation, which should be solved with the following boundary conditions:

No slip at the wall: $r = R, \quad v_z = 0$

Symmetry: $r = 0, \quad \dfrac{dv_z}{dr} = 0$

The required data for the solution of this problem are:

$\mu_0 = 102.0$ Pa.s $\quad t_1 = 4.36$ s $\quad n = 0.375 \quad R = 0.1$ m $\quad -\Delta P/L = 20$ kPa/m

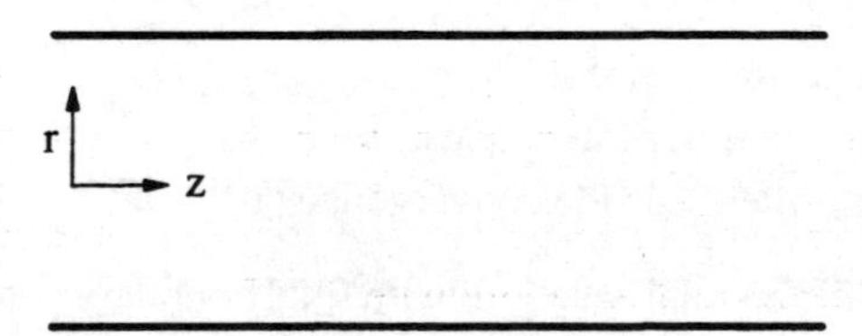

Figure E5.4*a*

Method of Solution: First we define the following two variables:

Dimensionless distance: $\eta = r/R$
Dimensionless velocity: $\phi = v_z/v^*$

where $v^* = (-\Delta P)R^2/L\mu_0$. Eq. (1) can be expanded and rearranged in its dimensionless form into the following second-order differential equation:

$$\frac{d^2\phi}{d\eta^2} = -\frac{\dfrac{1}{\eta}\dfrac{d\phi}{d\eta} + \left[1 + \lambda^2\left(\dfrac{d\phi}{d\eta}\right)^2\right]^{(1-n)/2}}{1 - \dfrac{(1-n)\lambda^2\left(\dfrac{d\phi}{d\eta}\right)}{\left[1 + \lambda^2\left(\dfrac{d\phi}{d\eta}\right)^2\right]}} \tag{2}$$

where $\lambda = t_1 v^*/R$.

In order to obtain the canonical form of Eq. (2), we apply the following transformation:

$$y_1 = \frac{d\phi}{d\eta}$$

$$y_2 = \phi$$

The canonical form of Eq. (2) is the given as

$$\frac{dy_1}{d\eta} = -\frac{\dfrac{1}{\eta}y_1 + \left[1 + \lambda^2 y_1^2\right]^{(1-n)/2}}{1 - \dfrac{(1-n)\lambda^2 y_1}{\left[1 + \lambda^2 y_1^2\right]}} \tag{3}$$

$$\frac{dy_2}{d\eta} = y_1 \tag{4}$$

The set of nonlinear ordinary differential equations (3) and (4) should be solved with the following boundary conditions:

$$y_1(0) = y_{1,0} = 0 \tag{5}$$

$$y_2(1) = y_{2,f} = 0 \tag{6}$$

The initial value $y_1(0)$ is known, but the initial value $y_2(0)$ must be guessed. We designate this guess, in accordance with Eq. (5.120), as follows:

$$y_2(0) = \gamma = [(-\Delta P)R^2/4L\mu_0]/v^* = 1/4 \tag{7}$$

The right-hand side of Eq. (7) corresponds to the velocity of the fluid at the center of the pipe if it was a Newtonian fluid with the viscosity μ_0.

The complete set of equations for the solution of this two-point boundary-value problem consists of:

1. The four system equations with their known boundary values [Eqs. (3)-(6)]
2. The guessed initial condition for y_2 [Eq. (7)]
3. Eq. (5.121) for construction of the Jacobian matrix
4. Eq. (5.123) for calculation of the δy vector
5. Eqs. (5.122) and (5.119) for correcting the guessed initial conditions.

Once the velocity profile is determined, the flow rate of the fluid can be calculated from the following integral formula:

$$Q = \int_0^R 2\pi r v_z dr$$

Program Description: The MATLAB function *shooting.m* is developed to solve a set of first-order ordinary differential equations in a boundary-value problem using the shooting method. The structure of this function is very similar to that of the function *Newton.m* developed in Example 1.4.

The function *shooting.m* begins with checking the input arguments. The inputs to the function are the name of the file containing the set of differential equations, lower and upper limits of integration interval, the integration step size, the vector of initial conditions, the vector of final conditions, the vector of guesses of initial conditions for those equations who have final conditions, the order of Runge-Kutta method, the relaxation factor, and the convergence criterion. From the above list, introducing the integration step size, the order of Runge-Kutta method, the relaxation factor, and the convergence criterion are optional, and the function assumes default values for each of the above variables, if necessary. The number of guessed initial conditions has to be equal to the number of final conditions; also, the number of equations should be equal to the total number of boundary conditions (initial and final). If these conditions are not met, the function gives a proper error message on the screen and stops execution.

The next section in the function is Newton's technique. This procedure begins with solving the set of differential equations by the Runge-Kutta method, using the known and guessed initial conditions, in forward direction. It then sets the differentiation increment for the approximate initial conditions and consequently evaluates the elements of the Jacobian matrix, column-wise, by differentiating using forward finite differences method. At the end of this section, the approximate initial conditions are corrected according to Eqs. (5.119) and (5.122). This procedure is repeated until the convergence is reached for all the final conditions.

It is important to note that the function *shooting.m* requires to receive the values of the set of ordinary differential equations at each point in a column vector with the values of the equations whose initial conditions are known to be at the top, followed by those whose final conditions are fixed. It is also important to pass the initial and final conditions to the function in the order corresponding to the order of equations appearing in the file that introduces the ordinary differential equations.

The main program *Example5_4.m* asks the reader to input the parameters required for solution of the problem. The program then calls the function *shooting* to solve the set of equations and finally, it shows the value of the flow rate on the screen and plots the calculated velocity profile. The default values of the relaxation factor and the convergence criterion are used in this example.

The function *Ex5_4_func.m* evaluates the values of the set of Eqs. (3) and (4) at a given point. The first function evaluated is that of Eq. (3), the initial condition of which is known.

Program

Example5_4.m

```
% Example5_4.m
% Solution to Example 5.4. This program calculates and plots
% the velocity profile of a non-Newtonian fluid flowing in a
% circular pipe. It uses the function SHOOTING to solve the
% one-dimentional equation of motion which is rearranged as
% a set of boundary-value ordinary differential equations.

clear
clc
clf

% Input data
R = input(' Inside diameter of the pipe (m) = ');
dP = input(' Pressure drop gradient (Pa/m) = ');
mu0 = input(' Zero shear rate viscosity of the fluid (Pa.s) = ');
t1 = input(' Characteristic time of the fluid (s) = ');
n = input(' The exponent n from the power-law = ');
fname = input('\n M-file containing the set of differential
equations : ');
order = input(' Order of Runge-Kutta method = ');
h = input(' Step size = ');

vmax0 = dP*R^2/(4*mu0);    % Initial guess of velocity
vstar = 4*vmax0;
lambda = t1*vstar/R;

% Solution of the set of differential equations
[eta,y] = shooting(fname,h/100,1,h,0,0,vmax0/vmax, order, ...
    [],[],n,lambda);
r = eta*R;                 % Radial position
vz = y(2,:)*vstar;         % Velocity profile
```

```
Q = 2*pi*trapz(r,r.*vz);   % Flow rate
fprintf('\n Volumetric flow rate = %4.2f lit/s \n',Q*1000)

% Plotting the results
plot(1000*r,vz)
ylabel('v_z (m/s)')
xlabel('r (mm)')
```

Ex5_4_func.m

```
function f = Ex5_4_func(eta,y,n,lambda)
% Function Ex5_4_func.M
% This function introduces the set of ordinary differential
% equations used in Example 5.4.

f(1) = -(y(1)/eta+(1+lambda^2*y(1)^2)^((1-n)/2))/ ...
   (1-(1-n)*lambda^2*y(1)/(1+lambda^2*y(1)^2));
f(2) = y(1);

f = f';   % Make it a column vector
```

shooting.m

```
function [x,y] = shooting(ODEfile,x0,xf,h,y0,yf,gamma0,order, ...
   rho,tol,varargin)
%SHOOTING Solves a boundary value set of ordinary differential
%   equations by shooting method using Newton's technique.
%
%   [X,Y]=SHOOTING('F',X0,XF,H,Y0,YF,GAMMA) integrates the set of
%   ordinary differential equations from X0 to XF, using the
%   4th-order Runge-Kutta method.  The equations are described in
%   the M-file F.M.  H is the step size. Y0, YF, and GAMMA are the
%   vectors of initial conditions, final conditions, and starting
%   guesses, respectively. The function returns the independent
%   variable in the vector X and the set of dependent variables in
%   the matrix Y.
%
%   [X,Y]=SHOOTING('F',X0,XF,H,Y0,YF,GAMMA,ORDER,RHO,TOL,P1,P2,...)
%   applies the ORDERth-order Runge-Kutta method for forward
%   integration, and uses relaxation factor RHO and tolerance TOL
%   for convergence test.  Additional parameters P1, P2, ... are
%   passed directly to the function F.  Pass an empty matrix for
%   ORDER, RHO, or TOL to use the default value.
%
%   See also COLLOCATION, RK

% (c) N. Mostoufi & A. Constantinides
% January 1, 1999

% Initialization
if isempty(h) | h == 0
```

```
    h = (xf - xi)/99;
end
if nargin < 8 | isempty(order)
    order = 4;
end
if nargin < 9 | isempty(rho)
    rho = 1;
end
if nargin < 10 | isempty(tol)
    tol = 1e-6;
end

y0 = (y0(:).')';                  % Make sure it's a column vector
yf = (yf(:).')';                  % Make sure it's a column vector
gamma0 = (gamma0(:).')';          % Make sure it's a column vector

% Checking the number of guesses
if length(yf) ~= length(gamma0)
    error(' The number of guessed conditions is not equal to the
number of final conditions.')
end

r = length(y0);                   % Number of initial conditions
n = r + length(yf);               % Number of boundary conditions
% Checking the number of equations
ftest = feval(ODEfile,x0,[y0 ; gamma0],varargin{:});
if length(ftest) ~= n
    error(' The number of equations is not equal to the number of
boundary conditions.')
end

gamma1 = gamma0 * 1.1;
gammanew = gamma0;
iter = 0;
maxiter = 100;

% Newton's technique
while max(abs(gamma1 - gammanew)) > tol & iter < maxiter
    iter = iter + 1;
    gamma1 = gammanew;
    [x,y] = RK(ODEfile,x0,xf,h,[y0 ; gamma1],order,varargin{:});
    fnk = y(r+1:n,end);

    % Set d(gamma) for derivation
    for k = 1:length(gamma1)
        if gamma1(k) ~= 0
            dgamma(k) = gamma1(k) / 100;
        else
            dgamma(k) = 0.01;
```

```
        end
    end

    % Calculation of the Jacobian matrix
    a = gamma1;
    for k = 1:n-r
        a(k) = gamma1(k) + dgamma(k);
        [xa,ya] = RK(ODEfile,x0,xf,h,[y0 ; a],order,varargin{:});
        fnka = ya(r+1:n,end);
        jacob(:,k) = (fnka - fnk) / dgamma(k);
        a(k) = gamma1(k) - dgamma(k);
    end

    % Next approximation of the roots
    if det(jacob) == 0
        gammanew = gamma1 + max([abs(dgamma), 1.1*tol]);
    else
        gammanew = gamma1 - rho * inv(jacob) * (fnk - yf);
    end
end

if iter >= maxiter
    disp('Warning : Maximum iterations reached.')
end
```

Input and Results

```
>>Example5_4

 Inside diameter of the pipe (m) = 0.1
 Pressure drop gradient (Pa/m) = 20e3
 Zero shear rate viscosity of the fluid (Pa.s) = 102
 Characteristic time of the fluid (s) = 4.36
 The exponent n from the power-law = 0.375

 M-file containing the set of differential equations : 'Ex5_4_func'
 Order of Runge-Kutta method = 4
 Step size = 0.01

 Volumetric flow rate = 2.91 lit/s
```

Discussion of Results: The volumetric flow rate of the fluid in this condition is calculated to be 2.91 L/s, and the velocity profile is shown in Fig. E5.4*b*. It should be noted that because there is the term $1/\eta$ in Eq. (2), the lower limit of numerical integration cannot be zero; instead, a very small value close to zero should be used in such a situation. In the main program *Example5_4*, the lower limit of integration is set to $h/100$, which is negligible with respect to the dimension of the pipe.

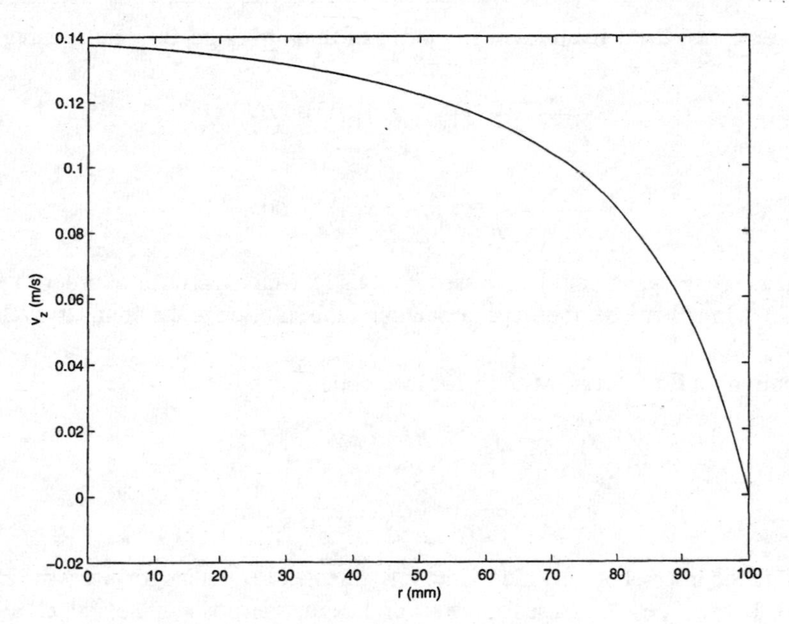

Figure E5.4*b* Velocity profile for non-Newtonian fluid.

5.6.2 The Finite Difference Method

The *finite-difference* method replaces the derivatives in the differential equations with finite difference approximations at each point in the interval of integration, thus converting the differential equations to a large set of simultaneous nonlinear algebraic equations. To demonstrate this method, we use, as before, the set of two differential equations:

$$\frac{dy_1}{dx} = f_1(x, y_1, y_2)$$
$$\frac{dy_2}{dx} = f_2(x, y_1, y_2) \tag{5.106}$$

with split boundary conditions:

$$y_1(x_0) = y_{1,0} \tag{5.107}$$

$$y_2(x_f) = y_{2,f} \tag{5.108}$$

Next, we express the derivatives of y in terms of forward finite differences using Eq. (4.33):

$$\frac{dy_{1,i}}{dx} = \frac{1}{h}(y_{1,i+1} - y_{1,i}) + O(h) \tag{5.124a}$$

$$\frac{dy_{2,i}}{dx} = \frac{1}{h}(y_{2,i+1} - y_{2,i}) + O(h) \tag{5.124b}$$

For higher accuracy, we could have used Eq. (4.41), which has error of order (h^2), instead of Eq. (4.33). In either case, the steps of obtaining the solution to the boundary-value problem are identical.

Combining Eqs. (5.124) with (5.106) we obtain

$$y_{1,i+1} - y_{1,i} = hf_1(x, y_{1,i}, y_{2,i}) \tag{5.125a}$$

$$y_{2,i+1} - y_{2,i} = hf_2(x, y_{1,i}, y_{2,i}) \tag{5.125b}$$

We divide the interval of integration into n segments of equal length and write Eqs. (5.125) for $i = 0, 1, 2, \ldots, n - 1$. These form a set of $2n$ simultaneous nonlinear algebraic equations in $(2n + 2)$ variables. The two boundary conditions provide values for two of these variables:

$$y_1(x_0) = y_{1,0} \tag{5.107}$$

$$y_2(x_f) = y_{2,f} = y_{2,n} \tag{5.108}$$

Therefore, the system of $2n$ equations in $2n$ unknown can be solved using Newton's method for simultaneous nonlinear algebraic equations, described in Chap. 1. It should be emphasized, however, that the problem of solving a large set of nonlinear algebraic equations is not a trivial task. It requires, first, a good initial guess of all the values of y_{ij}, and it involves the evaluation of the $(2n \times 2n)$ Jacobian matrix. Kubíček and Hlaváček [3] state that computational experience with the finite difference technique has shown that, for a practical engineering problem, this method is more difficult to apply than the shooting method. They recommend that the finite difference method be used only for problems that are too unstable to integrate by the shooting methods. On the other hand, if the differential equations are *linear*, the resulting set of simultaneous algebraic equations will also be linear. In such a case, the solution can be obtained by straightforward application of matrix inversion or the Gauss elimination procedure.

5.6.3 Collocation Methods

These methods are based on the concept of interpolation of unequally spaced points; that is, choosing a function, usually a polynomial, that approximates the solution of a differential equation in the range of integration, $x_0 \le x \le x_f$, and determining the coefficients of that function from a set of base points.

Let us again consider the set of two differential equations:

$$\frac{dy_1}{dx} = f_1(x, y_1, y_2)$$
$$\frac{dy_2}{dx} = f_2(x, y_1, y_2) \tag{5.106}$$

with split boundary conditions:

$$y_1(x_0) = y_{1,0} \tag{5.107}$$

$$y_2(x_f) = y_{2,f} \tag{5.108}$$

Suppose that the solutions $y_1(x)$ and $y_2(x)$ of Eq. (5.106) can be approximated by the following polynomials, which we call *trial functions*:

$$y_1(x) \cong P_{1,n}(x) = c_{1,0} + c_{1,1}x + c_{1,2}x^2 + \ldots + c_{1,n}x^n \tag{5.126a}$$

$$y_2(x) \cong P_{2,n}(x) = c_{2,0} + c_{2,1}x + c_{2,2}x^2 + \ldots + c_{2,n}x^n \tag{5.126b}$$

We take the derivatives of both sides of Eq. (5.126) and substitute in Eqs. (5.106):

$$P'_{m,n}(x) \cong f_m(x, P_{1,n}(x), P_{2,n}(x)) \qquad m = 1, 2 \tag{5.127}$$

We then form the residuals:

$$R_m(x) = P'_{m,n}(x) - f_m(x, P_{1,n}(x), P_{2,n}(x)) \qquad m = 1, 2 \tag{5.128}$$

The objective is to determine the coefficients $\{c_{m,i} \mid i = 0, 1, \ldots, n;\ m = 1, 2\}$ of the polynomials $P_{m,n}(x)$ to make the residuals as small as possible over the range of integration of the differential equation. This is accomplished by making the following integral vanish:

$$\int_{x_0}^{x_f} W_k R_m(x)\, dx = 0 \tag{5.129}$$

where W_k are weighting functions to be chosen. This technique is called the *method of weighted residuals.*

The collocation method chooses the weighting functions to be the *Dirac delta* (*unit impulse*) function:

$$W_k = \delta(x - x_k) \qquad x_0 \le x_k \le x_f \tag{5.130}$$

which has the property that

$$\int_{x_0}^{x_f} a(x)\delta(x - x_k)\, dx = a(x_k) \tag{5.131}$$

Therefore, the integral (5.129) becomes

$$\int_{x_0}^{x_f} W_k R_m(x)\, dx = R_m(x_k) = 0 \tag{5.132}$$

Combining Eqs. (5.128) and (5.132), we have

$$P'_{m,n}(x_k) - f_m(x_k, P_{1,n}(x_k), P_{2,n}(x_k)) = 0 \qquad m = 1, 2 \tag{5.133}$$

This implies that at a given number of *collocation points*, $\{x_k \mid k = 0, 1, \ldots, n\}$, the coefficients of the polynomials (5.126) are chosen so that Eq. (5.133) is satisfied; that is, the polynomials are *exact* solutions of the differential equations at those collocation points (note that $x_n = x_f$). The larger the number of collocation points, the closer the trial function would resemble the true solution $y_m(x)$ of the differential equations.

Eq. (5.133) contains the $(2n + 2)$ yet-to-be-determined coefficients $\{c_{m,i} \mid i = 0, 1, \ldots, n;\ m = 1, 2\}$ of the polynomials. These can be calculated by choosing $(2n + 2)$ collocation points. Because it is necessary to satisfy the boundary conditions of the problem, two collocation points are already fixed in this case of boundary-value problem. At $x = x_0$:

$$y_1(x_0) = y_{1,0} = c_{1,0} + c_{1,1}x_0 + \ldots + c_{1,n}x_0^n = \sum_{i=0}^{n} c_{1,i}x_0^i \tag{5.134}$$

and at $x = x_f$:

$$y_2(x_f) = y_{2,f} = c_{2,0} + c_{2,1}x_f + \ldots + c_{2,n}x_f^n = \sum_{i=0}^{n} c_{2,i}x_f^i \tag{5.135}$$

Therefore, we have the freedom to choose the remaining $(2n)$ internal collocation points and then write Eq. (5.133) for each of these points:

$$\begin{aligned} P'_{1,n}(x_1) - f_1(x_1, P_{1,n}(x_1), P_{2,n}(x_1)) &= 0 \\ &\vdots \\ P'_{1,n}(x_n) - f_1(x_n, P_{1,n}(x_n), P_{2,n}(x_n)) &= 0 \end{aligned} \tag{5.136a}$$

$$\begin{aligned} P'_{2,n}(x_0) - f_2(x_0, P_{1,n}(x_0), P_{2,n}(x_0)) &= 0 \\ &\vdots \\ P'_{2,n}(x_{n-1}) - f_2(x_{n-1}, P_{1,n}(x_{n-1}), P_{2,n}(x_{n-1})) &= 0 \end{aligned} \tag{5.136b}$$

Note that we have also written Eq. (5.133) for $x = x_f = x_n$ in Eq. (5.136*a*) and for $x = x_0$ in Eq. (5.136*b*) because the values $y_{1,f}$ and $y_{2,0}$ are yet unknown. Eqs. (5.134)-(5.136) constitute a complete set of $(2n + 2)$ simultaneous nonlinear equations in $(2n + 2)$ unknowns. The solution of this problem requires the application of Newton's method (see Chap. 1) for simultaneous nonlinear equations.

If the collocation points are chosen at equidistant intervals within the interval of integration, then the collocation method is equivalent to polynomial interpolation of equally spaced points and to the finite difference method. This is not at all surprising, as the development of interpolating polynomials and finite differences were all based on expanding the function in Taylor series (see Chap. 3). It is not necessary, however, to choose the collocation points at equidistant intervals. In fact, it is more advantageous to locate the collocation points at the roots of appropriate orthogonal polynomials, as the following discussion shows.

The *orthogonal collocation method*, which is an extension of the method just described, provides a mechanism for automatically picking the collocation points by making use of orthogonal polynomials.[5] This method chooses the trial functions $y_1(x)$ and $y_2(x)$ to be the linear combination

$$y_m(x) = \sum_{i=0}^{n+1} a_{m,i}P_i(x) \qquad m = 1, 2 \tag{5.137}$$

[5] For a more complete discussion of orthogonal collocation methods see Finlayson [1].

of a series of orthogonal polynomials $P_i(x)$:

$$P_0(x) = c_{0,0}$$
$$P_1(x) = c_{1,0} + c_{1,1}x$$
$$P_2(x) = c_{2,0} + c_{2,1}x + c_{2,2}x^2 \tag{5.138}$$
$$\vdots$$
$$P_i(x) = c_{i,0} + c_{i,1}x + c_{i,2}x^2 + \ldots + c_{i,n}x^i$$

This set of polynomials can be written in a condensed form:

$$P_i(x) = \sum_{k=0}^{i} c_{ik}x^k \qquad i = 0, 1, \ldots, n+1 \tag{5.139}$$

The coefficients c_{ik} are chosen so that the polynomials obey the orthogonality condition defined in Sec. 3.10:

$$\int_a^b w(x)P_i(x)P_j(x)\,dx = 0 \qquad i \neq j \tag{5.140}$$

When $P_i(x)$ is chosen to be the *Legendre* set of orthogonal polynomials [see Table 3.7], the weight $w(x)$ is unity. The standard interval of integration for Legendre polynomials is $[-1, 1]$. The transformation equation (4.92) is used to transform the Legendre polynomials to the interval $[x_0, x_f]$, which applies to our problem at hand:

$$x = \frac{(x_f - x_0)}{2}z + \frac{(x_f + x_0)}{2} \tag{5.141}$$

Eq. (5.141) relates the variables x and z so that every value of x in the interval $[x_0, x_f]$ corresponds to a value of z in the interval $[-1, 1]$ and vice versa. Therefore, using x or z as independent variables is equivalent. Hereafter, we use z as the independent variable of the Legendre polynomials to stress that the domain under study is the interval $[-1, 1]$. The derivatives with respect to x and z are related to each other by the following relation:

$$\frac{dy_m}{dx} = \frac{2}{(x_f - x_0)}\frac{dy_m}{dz} \tag{5.142}$$

The two-point boundary-value problem given by Eqs. (5.106)-(5.108) has $(2n + 2)$ collocation points, $\{z_j \mid j = 0, 1, \ldots, n+1\}$, including the two known boundary values ($z_0 = -1$ and $z_{n+1} = 1$). The location of the n internal collocation points (z_1 to z_n) are determined from the roots of the polynomial $P_n(z) = 0$. The coefficients $a_{m,i}$ in Eq. (5.137) must be determined so that the boundary conditions are satisfied. Eq. (5.137) can be written for the $(n+2)$ points (z_0 to z_{n+1}) as

$$y_1(z_j) = \sum_{i=0}^{n+1} d_{1,i} z_j^i \tag{5.143a}$$

$$y_2(z_j) = \sum_{i=0}^{n+1} d_{2,i} z_j^i \tag{5.143b}$$

where the terms of the polynomials have been regrouped. Eqs. (5.143) may be presented in matrix notation as

$$\boldsymbol{y}_1 = \boldsymbol{Q}\boldsymbol{d}_1 \tag{5.144a}$$

$$\boldsymbol{y}_2 = \boldsymbol{Q}\boldsymbol{d}_2 \tag{5.144b}$$

where $\boldsymbol{d}_1$ and $\boldsymbol{d}_2$ are the matrices of coefficients and

$$Q_{j+1,i+1} = z_j^i \qquad \begin{cases} i = 0, 1, \ldots, n+1 \\ j = 0, 1, \ldots, n+1 \end{cases} \tag{5.145}$$

Solving Eqs. (5.144) for $\boldsymbol{d}_1$ and $\boldsymbol{d}_2$, we find

$$\boldsymbol{d}_1 = \boldsymbol{Q}^{-1}\boldsymbol{y}_1 \tag{5.146a}$$

$$\boldsymbol{d}_2 = \boldsymbol{Q}^{-1}\boldsymbol{y}_2 \tag{5.146b}$$

The derivatives of ys are taken as

$$\frac{dy_1(z_j)}{dz} = \sum_{i=0}^{n+1} d_{1,i} i z_j^{i-1} \tag{5.147a}$$

$$\frac{dy_2(z_j)}{dz} = \sum_{i=0}^{n+1} d_{2,i} i z_j^{i-1} \tag{5.147b}$$

which in matrix form become

$$\frac{d\boldsymbol{y}_1}{dz} = \boldsymbol{C}\boldsymbol{d}_1 = \boldsymbol{C}\boldsymbol{Q}^{-1}\boldsymbol{y}_1 = \boldsymbol{A}\boldsymbol{y}_1 \tag{5.148a}$$

$$\frac{d\boldsymbol{y}_2}{dz} = \boldsymbol{C}\boldsymbol{d}_2 = \boldsymbol{C}\boldsymbol{Q}^{-1}\boldsymbol{y}_2 = \boldsymbol{A}\boldsymbol{y}_2 \tag{5.148b}$$

where

$$C_{j+1,i+1} = iz_j^{i-1} \qquad \begin{cases} i = 0,1,\ldots,n+1 \\ j = 0,1,\ldots,n+1 \end{cases} \tag{5.149}$$

The two-point boundary-value problem of Eq. (5.106) can now be expressed in terms of the orthogonal collocation method as

$$\begin{aligned} \boldsymbol{A}\boldsymbol{y}_1 &= \boldsymbol{f}_1(z, \boldsymbol{y}_1, \boldsymbol{y}_2) \\ \boldsymbol{A}\boldsymbol{y}_2 &= \boldsymbol{f}_2(z, \boldsymbol{y}_1, \boldsymbol{y}_2) \end{aligned} \tag{5.150}$$

or

$$\sum_{j=0}^{n+1} A_{ij} y_{1,j} = f_1(z_j, y_{1,j}, y_{2,j}) \tag{5.151a}$$

$$\sum_{j=0}^{n+1} A_{ij} y_{2,j} = f_2(z_j, y_{1,j}, y_{2,j}) \tag{5.151b}$$

with the boundary conditions

$$y_1(z_0) = y_{1,0} \qquad \text{and} \qquad y_{2,n+1} = y_2(z_f) = y_{2,f} \tag{5.152}$$

Eqs. (5.151) and (5.152) constitute a set of $(2n+4)$ simultaneous nonlinear equations whose solution can be obtained using Newton's method for nonlinear equations. It is possible to combine Eqs. (5.151) and present them in matrix form:

$$\boldsymbol{A}_2 \boldsymbol{Y} = \boldsymbol{F} \tag{5.153}$$

where

$$\boldsymbol{A}_2 = \begin{bmatrix} \boldsymbol{A} & \boldsymbol{0} \\ \boldsymbol{0} & \boldsymbol{A} \end{bmatrix} \tag{5.154}$$

$$\begin{aligned} \boldsymbol{Y} &= \begin{bmatrix} \boldsymbol{y}_1 \\ \boldsymbol{y}_2 \end{bmatrix} \\ &= \left[y_{1,0}, \ldots, y_{1,n+1}, y_{2,0}, \ldots, y_{2,n+1} \right]' \end{aligned} \tag{5.155}$$

$$\boldsymbol{F} = \begin{bmatrix} f_1 \\ f_2 \end{bmatrix} = \begin{bmatrix} f_1(z_0, y_{1,0}, y_{2,0}) \\ \cdot \\ \cdot \\ \cdot \\ f_1(z_{n+1}, y_{1,n+1}, y_{2,n+1}) \\ f_2(z_0, y_{1,0}, y_{2,0}) \\ \cdot \\ \cdot \\ \cdot \\ f_2(z_{n+1}, y_{1,n+1}, y_{2,n+1}) \end{bmatrix} \tag{5.156}$$

The bold zeros in Eq. (5.154) are zero matrices of size $(n+2) \times (n+2)$, the same size as that of matrix $\boldsymbol{A}$.

It should be noted that Eq. (5.153) is solved for the unknown collocation points which means that we should exclude the equations corresponding to the boundary conditions. In the problem described above, the first and the last equations in the set of equations (5.153) will not be used because the corresponding dependent values are determined by a boundary condition rather than by the collocation method.

The above formulation of solution for a two-equation boundary-value problem can be extended to the solution of m simultaneous first-order ordinary differential equations. For this purpose, we define the following matrices:

$$\boldsymbol{A}_m = \begin{bmatrix} \boldsymbol{A} & \boldsymbol{0} & \dots & \boldsymbol{0} \\ \boldsymbol{0} & \boldsymbol{A} & \dots & \boldsymbol{0} \\ \dots & \dots & \dots & \dots \\ \boldsymbol{0} & \boldsymbol{0} & \dots & \boldsymbol{A} \end{bmatrix} \tag{5.157}$$

$$\boldsymbol{Y} = [\boldsymbol{y}_1, \boldsymbol{y}_2, \dots, \boldsymbol{y}_m]' \tag{5.158}$$

$$\boldsymbol{F} = [\boldsymbol{f}_1, \boldsymbol{f}_2, \dots, \boldsymbol{f}_m]' \tag{5.159}$$

Note that the matrix $\boldsymbol{A}$ in Eq. (5.157) is defined by Eq. (5.148) and appears m times on the diagonal of the matrix $\boldsymbol{A}_m$. The values of the dependent variables $\{y_{ij} \mid i = 1, 2, \dots, m;$

$j = 0, 2, \ldots, n+1\}$ are then evaluated from the simultaneous solution of the following set of nonlinear equations plus boundary conditions:

$$\boldsymbol{A}_m \boldsymbol{Y} - \boldsymbol{F} = \boldsymbol{0} \tag{5.160}$$

The equations corresponding to the boundary conditions have to be excluded from Eq. (5.160) at the time of solution.

If the problem to be solved is a second-order two-point boundary-value problem in the form

$$y'' = f(x, y, y') \tag{5.161}$$

with the boundary conditions

$$y(x_0) = y_0 \qquad \text{and} \qquad y(x_f) = y_f \tag{5.162}$$

we may follow the similar approach as described above and approximate the function $y(x)$ at $(n+2)$ points, after transforming the independent variable from x to z, as

$$y(z_j) = \sum_{i=0}^{n+1} d_i z_j^i \tag{5.163}$$

The derivatives of y are then taken as

$$\frac{dy(z_j)}{dz} = \sum_{i=0}^{n+1} d_i i z_j^{i-1} \tag{5.164}$$

$$\frac{d^2y(z_j)}{dz^2} = \sum_{i=0}^{n+1} d_i i(i-1) z_j^{i-2} \tag{5.165}$$

These equations can be written in matrix form:

$$\frac{d\boldsymbol{y}}{dz} = \boldsymbol{C}\boldsymbol{Q}^{-1}\boldsymbol{y} = \boldsymbol{A}\boldsymbol{y} \tag{5.166}$$

$$\frac{d^2\boldsymbol{y}}{dz^2} = \boldsymbol{D}\boldsymbol{Q}^{-1}\boldsymbol{y} = \boldsymbol{B}\boldsymbol{y} \tag{5.167}$$

where

$$\boldsymbol{D}_{j+1,i+1} = i(i-1)z_j^{i-2} \qquad \begin{cases} i = 0, 1, \ldots, n+1 \\ j = 0, 1, \ldots, n+1 \end{cases} \tag{5.168}$$

The two-point boundary-value problem of Eq. (5.161) can now be expressed in terms of the orthogonal collocation method as

$$\boldsymbol{B}\boldsymbol{y} = \boldsymbol{f}(z, \boldsymbol{y}, \boldsymbol{A}\boldsymbol{y}) \tag{5.169}$$

Eq. (5.169) represents a set of $(n + 2)$ simultaneous nonlinear equations, two of which correspond to the boundary conditions (the first and the last equation) and should be neglected when solving the set. The solution of the remaining n nonlinear equations can be obtained using Newton's method for nonlinear equations.

The orthogonal collocation method is more accurate than either the finite difference method or the collocation method. The choice of collocation points at the roots of the orthogonal polynomials reduces the error considerably. In fact, instead of the user choosing the collocation points, the method locates them automatically so that the best accuracy is achieved.

Example 5.5: Solution of the Optimal Temperature Profile for Penicillin Fermentation. Apply the orthogonal collocation method to solve the two-point boundary-value problem arising from the application of the *maximum principle of Pontryagin* to a batch penicillin fermentation. Obtain the solution of this problem, and show the profiles of the state variables, the adjoint variables, and the optimal temperature. The equations that describe the state of the system in a batch penicillin fermentation, developed by Constantinides et al.[6], are:

Cell mass production:
$$\frac{dy_1}{dt} = b_1 y_1 - \frac{b_1}{b_2} y_1^2 \qquad y_1(0) = 0.03 \tag{1}$$

Penicillin synthesis:
$$\frac{dy_2}{dt} = b_3 y_1 \qquad y_2(0) = 0.0 \tag{2}$$

where y_1 = dimensionless concentration of cell mass
y_2 = dimensionless concentration of penicillin
t = dimensionless time, $0 \le t \le 1$.

The parameters b_i are functions of temperature, θ:

$$b_1 = w_1\left[\frac{1.0 - w_2(\theta - w_3)^2}{1.0 - w_2(25 - w_3)^2}\right] \qquad b_2 = w_4\left[\frac{1.0 - w_2(\theta - w_3)^2}{1.0 - w_2(25 - w_3)^2}\right]$$

$$b_3 = w_5\left[\frac{1.0 - w_2(\theta - w_6)^2}{1.0 - w_2(25 - w_6)^2}\right] \qquad b_i \ge 0 \tag{3}$$

where $w_1 = 13.1$ (value of b_1 at 25°C obtained from fitting the model to experimental data)
$w_2 = 0.005$
$w_3 = 30°C$
$w_4 = 0.94$ (value of b_2 at 25°C)
$w_5 = 1.71$ (value of b_3 at 25°C)
$w_6 = 20°C$
θ = temperature, °C.

These parameter-temperature functions are inverted parabolas that reach their peak at 30°C for b_1 and b_2, at 20°C for b_3. The values of the parameters decrease by a factor of 2 over a 10°C change in temperature on either side of the peak. The inequality, $b_i \geq 0$, restricts the values of the parameters to the positive regime. These functions have shapes typical of those encountered in microbial or enzyme-catalyzed reactions.

The maximum principle has been applied to the above model to determine the optimal temperature profile (see Ref. [7]), which maximizes the concentration of penicillin at the final time of the fermentation, $t_f = 1$. The maximum principle algorithm when applied to the state equations, (1) and (2), yields the following additional equations:

The *adjoint equations*:

$$\frac{dy_3}{dt} = -b_1y_3 + 2\frac{b_1}{b_2}y_1y_3 - b_3y_4 \qquad y_3(1) = 0 \tag{4}$$

$$\frac{dy_4}{dt} = 0 \qquad y_4(1) = 1.0 \tag{5}$$

The *Hamiltonian*:

$$H = y_3\left(b_1y_1 - \frac{b_1}{b_2}y_1^2\right) + y_4(b_3y_1)$$

The *necessary condition* for maximum:

$$\frac{\partial H}{\partial \theta} = 0 \tag{6}$$

Eqs. (1)-(6) form a two-point boundary-value problem. Apply the orthogonal collocation method to obtain the solution of this problem, and show the profiles of the state variables, the adjoint variables, and the optimal temperature.

Method of Solution: The fundamental numerical problem of optimal control theory is the solution of the two-point boundary-value problem, which invariably arises from the application of the maximum principle to determine optimal control profiles. The state and adjoint equations, coupled together through the necessary condition for optimality, constitute a set of simultaneous differential equations that are often unstable. This difficulty is further complicated, in certain problems, when the necessary condition is not solvable explicitly for the control variable θ. Several numerical methods have been developed for the solution of this class of problems.

We first consider the second adjoint equation, Eq. (5), which is independent of the other variables and, therefore, may be integrated directly:

$$y_4 = 1 \qquad 0 \le t \le 1 \tag{7}$$

This reduces the number of differential equations to be solved by one. The remaining three differential equations, Eqs. (1), (2), and (4), are solved by Eq. (5.160), where $m = 3$.

Finally, we express the necessary condition [Eq. (6)] in terms of the system variables:

$$\frac{\partial H}{\partial \theta} = y_3\left[y_1\left(\frac{\partial b_1}{\partial \theta}\right) - y_1^2\frac{\partial(b_1/b_2)}{\partial \theta}\right] + y_1 y_4\left(\frac{\partial b_3}{\partial \theta}\right) = 0 \tag{8}$$

The temperature θ can be calculated from Eq. (8) once the system variables have been determined.

Program Description: The MATLAB function *collocation.m* is developed to solve a set of first-order ordinary differential equations in a boundary-value problem by the orthogonal collocation method. It starts with checking the input arguments and assigning the default values, if necessary. The number of guessed initial conditions has to be equal to the number of final conditions, and also the number of equations should be equal to the total number of boundary conditions (initial and final). If these conditions are not met, the function gives a proper error message on the screen and stops execution.

In the next section, the function builds the coefficients of the Lagrange polynomial and finds its roots, z_j. The vector of x_j is then calculated from Eq. (5.141). The function applies Newton's method for solution of the set of nonlinear equations (5.160). Therefore, the starting values for this technique are generated by the second-order Runge-Kutta method, using the guessed initial conditions. The function continues with building the matrices $\boldsymbol{Q}, \boldsymbol{C}, \boldsymbol{A}, \boldsymbol{A}_m$, and vectors $\boldsymbol{Y}$ and $\boldsymbol{F}$.

Just before entering the Newton's technique iteration loop, the function keeps track of the equations to be solved; that is, all the equations excluding those corresponding to the boundary conditions. The last part of the function is the solution of the set of equations (5.160) by Newton's method. This procedure begins with evaluating the differential equations function values followed by calculating the Jacobian matrix, by differentiating using forward finite differences method and, finally, correcting the dependent variables. This procedure is repeated until the convergence is reached at all the collocation points.

It is important to note that the *collocation.m* function must receive the values of the set of ordinary differential equations at each point in a column vector, with the initial value equations at the top, followed by the final value equations. It is also important to pass the initial and final conditions to the function in the order corresponding to the order of equations appearing in the file that introduces the ordinary differential equations.

The main program *Example5_5.m* asks the reader to input the parameters required for solution of the problem. The program then calls the function *collocation* to solve the set of equations. Knowing the system variables, the program calls the function *fzero* to find the temperature at each point. At the end, the program plots the calculated cell concentration, penicillin concentration, first adjoint variable, and the temperature against time.

The function *Ex5_5_func.m* evaluates the values of the set of Eqs. (1), (2), and (4) at a given point. It is important to note that the first input argument to *Ex5_5_func* is the independent variable, though it does not appear in the differential equations in this case. This function also calls the MATLAB function *fzero* to calculate the temperature from Eq. (8), which is introduced in the function *Ex5_5_theta.m.*

Program

Example5_5.m

```
% Example5_5.m
% Solution to Example 5.5. This program calculates and plots
% the concentration of cell mass, concentration of penicillin,
% optimal temperature profile, and adjoint variable of a batch
% penicillin fermentor.  It uses the function COLLOCATION to
% solve the set of system and adjoint equations.

clear
clc
clf

% Input data
w = input(' Enter w''s as a vector : ');
y0 = input(' Vector of known initial conditions = ');
yf = input(' Vector of final conditions = ');
guess = input(' Vector of guessed initial conditions = ');
fname = input('\n M-file containing the set of differential
equations : ');
fth=input(' M-file containing the necessary condition function : ');
n = input(' Number of internal collocation points = ');
rho = input(' Relaxation factor = ');

% Solution of the set of differential equations
[t,y] = collocation(fname,0,1,y0,yf,guess,n,rho,[],w,fth);
% Temperature changes
for k = 1:n+2
   theta(k) = fzero(fth,30,1e-6,0,y(:,k),w);
end

% Plotting the results
```

```
subplot(2,2,1), plot(t,y(1,:))
xlabel('Time')
ylabel('Cell')
title('(a)')

subplot(2,2,2), plot(t,y(2,:))
xlabel('Time')
ylabel('Penicillin')
title('(b)')

subplot(2,2,3), plot(t,y(3,:))
xlabel('Time')
ylabel('First Adjoint')
title('(c)')

subplot(2,2,4), plot(t,theta)
xlabel('Time')
ylabel('Temperature (deg C)')
title('(d)')
```

Ex5_5_func.m

```
function f = Ex5_5_func(t,y,w,fth)
% Function Ex5_5_func.M
% This function introduces the set of ordinary differential
% equations used in Example 5.5.

% Temperature
theta = fzero(fth,30,1e-6,0,y,w);

% Calculating the b's
b1 = w(1) * (1-w(2)*(theta-w(3))^2) / (1-w(2)*(25-w(3))^2);
if b1<0, b1=0; end
b2 = w(4) * (1-w(2)*(theta-w(3))^2) / (1-w(2)*(25-w(3))^2);
if b2<0, b2=1e-6; end
b3 = w(5) * (1-w(2)*(theta-w(6))^2) / (1-w(2)*(25-w(6))^2);
if b3<0, b3=0; end

% Evaluating the function values
f(1) = b1*y(1) - b1/b2*y(1)^2;
f(2) = b3*y(1);
f(3) = -b1*y(3) + 2*b1/b2*y(1)*y(3) - b3;

f = f';   % Make it a column vector
```

Ex5_5_theta.m

```
function ftheta = Ex5_5_theta(theta,y,w)
% Function Ex5_5_theta.M
% This function calculates the value of the necessary condition
% as a function of the temperature (theta). It is used in solving
% Example 5.5.
```

```
% Calculating the b's
b1 = w(1) * (1-w(2)*(theta-w(3))^2) / (1-w(2)*(25-w(3))^2);
db1 = w(1)*(-w(2))*2*(theta-w(3)) / (1-w(2)*(25-w(3))^2);

b2 = w(4) * (1-w(2)*(theta-w(3))^2) / (1-w(2)*(25-w(3))^2);
db2 = w(4)*(-w(2))*2*(theta-w(3)) / (1-w(2)*(25-w(3))^2);

b3 = w(5) * (1-w(2)*(theta-w(6))^2) / (1-w(2)*(25-w(6))^2);
db3 = w(5)*(-w(2))*2*(theta-w(6)) / (1-w(2)*(25-w(6))^2);

% The function
ftheta = y(3)*(y(1)*db1-y(1)^2*(db1*b2-db2*b1)/b2^2)+y(1)*db3;
```

collocation.m

```
function [x,y] = collocation(ODEfile,x0,xf,y0,yf,guess,n,rho, ...
    tol,varargin)
%COLLOCATION Solves a boundary value set of ordinary differential
% equations by the orthogonal collocation method.
%
%   [X,Y]=COLLOCATION('F',X0,XF,Y0,YF,GAMMA,N) integrates the set of
%   ordinary differential equations from X0 to XF by the Nth-degree
%   orthogonal collocation method.  The equations are contained in
%   the M-file F.M.  Y0, YF, and GAMMA are the vectors of initial
%   conditions, final conditions, and starting guesses respectively.
%   The function returns the independent variable in the vector X
%   and the set of dependent variables in the matrix Y.
%
%   [X,Y]=COLLOCATION('F',X0,XF,Y0,YF,GAMMA,N,RHO,TOL,P1,P2,...)
%   uses relaxation factor RHO and tolerance TOL for convergence
%   test.  Additional parameters P1, P2, ... are passed directly to
%   the function F.  Pass an empty matrix for RHO or TOL to use the
%   default value.
%
%   See also SHOOTING

% (c) N. Mostoufi & A. Constantinides
% January 1, 1999

% Initialization
if nargin < 7 | isempty(n)
   n = 1;
end
if nargin < 8 | isempty(rho)
   rho = 1;
end
if nargin < 9 | isempty(tol)
   tol = 1e-6;
end
```

```
y0 = (y0(:).')';          % Make sure it's a column vector
yf = (yf(:).')';          % Make sure it's a column vector
guess = (guess(:).')'; % Make sure it's a column vector

% Checking the number of guesses
if length(yf) ~= length(guess)
   error(' The number of guessed conditions is not equal to the
number of final conditions.')
end

r = length(y0);           % Number of initial conditions
m = r + length(yf);       % Number of boundary conditions
% Checking the number of equations
ftest = feval(ODEfile,x0,[y0 ; guess],varargin{:});
if length(ftest) ~= m
   error(' The number of equations is not equal to the number of
boundary conditions.')
end

fprintf('\n Integrating. Please wait.\n\n')

% Coefficients of the Legendre polynomial
for k = 0 : n/2
   cl(2*k+1) = (-1)^k * gamma(2*n-2*k+1) / ...
      (2^n * gamma(k+1) * gamma(n-k+1) * gamma(n-2*k+1));
   if k < n/2
      cl(2*k+2) = 0;
   end
end
zl = roots(cl);                    % Roots of the Legendre polynomial
z = [-1; sort(zl); 1];             % Collocation points (z)
x = (xf-x0)*z/2+(x0+xf)/2;         % Collocation points (x)

% Bulding the vector of starting values of the dependent variables
[p,q] = RK(ODEfile,x0,xf,(xf-x0)/20,[y0 ; guess],2,varargin{:});
for k = 1:m
   y(k,:) = spline(p,q(k,:),x');
end
y(r+1:m,end) = yf(1:m-r);

% Building the matrix A
Q(:,1) = ones(n+2,1);
C(:,1) = zeros(n+2,1);
for i = 1:n+1
   Q(:,i+1) = x.^i;
   C(:,i+1) = i*x.^(i-1);
end
A = C*inv(Q);
for k = 1:m
   k1 = (k-1)*(n+2)+1;
```

```
   k2 = k1 + n+1;
   Am(k1:k2,k1:k2) = A;          % Building the matrix Am
   Y(k1:k2) = y(k,:);            % Building the vector Y
end
Y = Y';                          % Make it a column vector

Y1 = Y * 1.1;
iter = 0;
maxiter = 100;
F = zeros(m*(n+2),1);
Fa = zeros(m*(n+2),1);
dY = zeros(m*(n+2),1);

position = [];   % Collocation points excluding boundary conditions
for k = 1:m
   if k <= r
      position = [position, (k-1)*(n+2)+[2:n+2] ];
   else
      position = [position, (k-1)*(n+2)+[1:n+1] ];
   end
end

% Newton's method
while max(abs(Y1 - Y)) > tol & iter < maxiter
   iter = iter + 1;
   fprintf(' Iteration %3d\n',iter)
   Y1 = Y;
   % Building the vector F
   for k = 1:n+2
      F(k : n+2 : (m-1)*(n+2)+k) = feval(ODEfile,x(k),...
          Y(k : n+2 : (m-1)*(n+2)+k),varargin{:});
   end
   fnk = Am * Y - F;

   % Set dY for derivation
   for k = 1:m*(n+1)
      if Y(position(k)) ~= 0
         dY(position(k)) = Y(position(k)) / 100;
      else
         dY(position(k)) = 0.01;
      end
   end

   % Calculation of the Jacobian matrix
   for k = 1:m
      for kk = 1:n+1
         a = Y;
         nc = (k-1)*(n+1)+kk;
         a(position(nc)) = Y(position(nc)) + dY(position(nc));
         for kkk = 1:n+2
```

```
            Fa(kkk : n+2 : (m-1)*(n+2)+kkk) = ...
       feval(ODEfile,x(kkk),a(kkk:n+2:(m-1)*(n+2)+kkk),varargin{:});
          end
          fnka = Am * a - Fa;
          jacob(:,nc) = (fnka(position) - fnk(position)) ...
             / dY(position(nc));
       end
    end

    % Next approximation of the roots
    if det(jacob) == 0
       Y(position) = Y(position) + max([abs(dY(position)); 1.1*tol]);
    else
       Y(position) = Y(position) - rho * inv(jacob) * fnk(position);
    end
end

% Rearranging the y's
for k = 1:m
    k1 = (k-1)*(n+2)+1;
    k2 = k1 + n+1;
    y(k,:) = Y(k1:k2)';
end
x = x';

if iter >= maxiter
    disp('Warning : Maximum iterations reached.')
end
```

Input and Results

```
>>Example5_5

 Enter w's as a vector : [13.1, 0.005, 30, 0.94, 1.71, 20]
 Vector of known initial conditions = [0.03, 0]
 Vector of final conditions = 0
 Vector of guessed initial conditions = 3

 M-file containing the set of differential equations : 'Ex5_5_func'
 M-file containing the necessary condition function : 'Ex5_5_theta'
 Number of internal collocation points = 10
 Relaxation factor = 0.9

 Integrating. Please wait.

 Iteration   1
 Iteration   2
 Iteration   3
 Iteration   4
```

```
Iteration    5
Iteration    6
Iteration    7
Iteration    8
Iteration    9
Iteration   10
Iteration   11
```

Discussion of Results: The choice of the value of the missing initial condition for y_3 is an important factor in the convergence of the collocation method, because it generates the starting values to the technique. The value of $y_3(0) = 3$ was chosen as the guessed initial condition after some trial and error. The collocation method converged to the correct solution in 11 iterations.

Figs. E5.5*a* to E5.5*d* show the profiles of the system variables and the optimal control variable (temperature). For this particular formulation of the penicillin fermentation, the maximum principle indicates that the optimal temperature profile varies from 30 to 20 °C in the pattern shown in Fig. E5.5*d*.

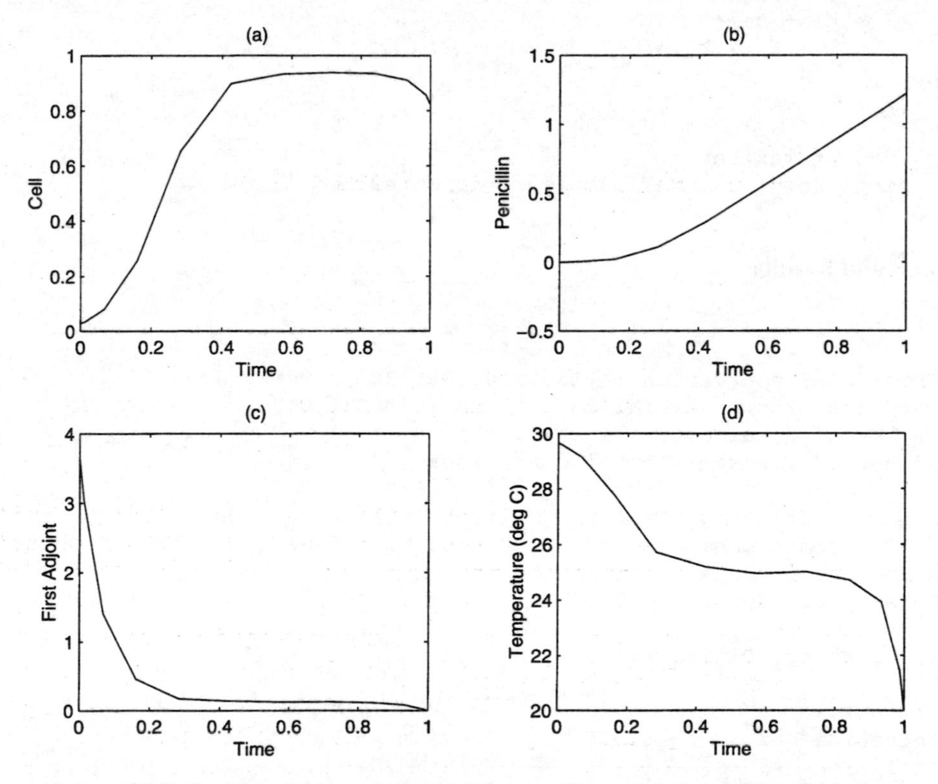

Figure E5.5 Profiles of the system variables and the optimal control variable for penicillin fermentation.

5.7 Error Propagation, Stability, and Convergence

Topics of paramount importance in the numerical integration of differential equations are the *error propagation*, *stability*, and *convergence* of these solutions. Two types of stability considerations enter in the solution of ordinary differential equations: *inherent stability* (or instability) and *numerical stability* (or instability). Inherent stability is determined by the mathematical formulation of the problem and is dependent on the eigenvalues of the Jacobian matrix of the differential equations. On the other hand, numerical stability is a function of the error propagation in the numerical integration method. The behavior of error propagation depends on the values of the characteristic roots of the difference equations that yield the numerical solution. In this section, we concern ourselves with numerical stability considerations as they apply to the numerical integration of ordinary differential equations.

There are three types of errors present in the application of numerical integration methods. These are the *truncation error*, the *roundoff error*, and the *propagation error*. The truncation error is a function of the number of terms that are retained in the approximation of the solution from the infinite series expansion. The truncation error may be reduced by retaining a larger number of terms in the series or by reducing the step size of integration h. The plethora of available numerical methods of integration of ordinary differential equations provides a choice of increasingly higher accuracy (lower truncation error), at an escalating cost in the number of arithmetic operations to be performed, and with the concomitant accumulation of roundoff errors.

Computers carry numbers using a finite number of significant figures. A roundoff error is introduced in the calculation when the computer rounds up or down (or just chops) the number to n significant figures. Roundoff errors may be reduced significantly by the use of double precision. However, even a very small roundoff error may affect the accuracy of the solution, especially in numerical integration methods that march forward (or backward) for hundreds or thousands of steps, each step being performed using rounded numbers.

The truncation and roundoff errors in numerical integration accumulate and propagate, creating the propagation error, which, in some cases, may grow in exponential or oscillatory pattern, thus causing the calculated solution to deviate drastically from the correct solution.

Fig. 5.6 illustrates the propagation of error in the Euler integration method. Starting with a known initial condition y_0, the method calculates the value y_1, which contains the truncation error for this step and a small roundoff error introduced by the computer. The error has been magnified in order to illustrate it more clearly. The next step starts with y_1 as the initial point and calculates y_2. But because y_1 already contains truncation and roundoff errors, the value obtained for y_2 contains these errors propagated, in addition to the new truncation and roundoff errors from the second step. The same process occurs in subsequent steps.

Error propagation in numerical integration methods is a complex operation that depends on several factors. Roundoff error, which contributes to propagation error, is entirely determined by the accuracy of the computer being used. The truncation error is fixed by the

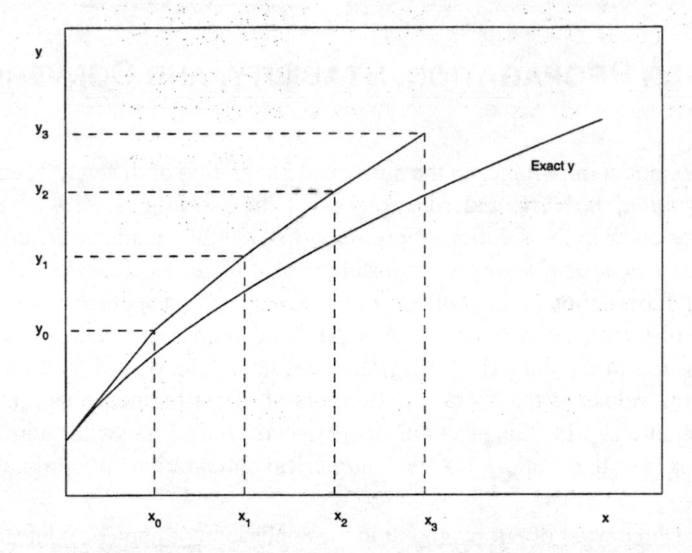

Figure 5.6 Error propagation of the Euler method.

choice of method being applied, by the step size of integration, and by the values of the derivatives of the functions being integrated. For these reasons, it is necessary to examine the error propagation and stability of each method individually and in connection with the differential equations to be integrated. Some techniques work well with one class of differential equations but fail with others.

In the sections that follow, we examine systematically the error propagation and stability of several numerical integration methods and suggest ways of reducing these errors by the appropriate choice of step size and integration algorithm.

5.7.1 Stability and Error Propagation of Euler Methods

Let us consider the initial-value differential equation in the linear form:

$$\frac{dy}{dx} = \lambda y \tag{5.170}$$

where the initial condition is given as

$$y(x_0) = y_0 \tag{5.171}$$

We assume that λ is real and y_0 is finite. The analytical solution of this differential equation is

$$y(x) = y_0 e^{\lambda x} \tag{5.172}$$

This solution is *inherently stable* for $\lambda < 0$. Under these conditions:

$$\lim_{x \to \infty} y(x) = 0 \tag{5.173}$$

Next, we examine the stability of the numerical solution of this problem obtained from using the explicit Euler method. Momentarily we ignore the truncation and roundoff errors. Applying Eq. (5.60), we obtain the recurrence equation

$$y_{n+1} = y_n + h\lambda y_n \tag{5.174}$$

which rearranges to the following *first-order homogeneous difference equation*

$$y_{n+1} - (1 + h\lambda)y_n = 0 \tag{5.175}$$

Using the methods described in Sec. 3.6, we obtain the characteristic equation

$$E - (1 + h\lambda) = 0 \tag{5.176}$$

whose root is

$$\mu_1 = (1 + h\lambda) \tag{5.177}$$

From this, we obtain the solution of the difference equation (5.175) as

$$y_n = C(1 + h\lambda)^n \tag{5.178}$$

The constant C is calculated from the initial condition, at $x = x_0$:

$$n = 0 \qquad y_n = y_0 = C \tag{5.179}$$

Therefore, the final form of the solution is

$$y_n = y_0(1 + h\lambda)^n \tag{5.180}$$

The differential equation is an initial-value problem; therefore, n can increase without bound. Because the solution y_n is a function of $(1 + h\lambda)^n$, its behavior is determined by the value of $(1 + h\lambda)$. A numerical solution is said to be *absolutely stable* if

$$\lim_{n \to \infty} y_n = 0 \tag{5.181}$$

The solution of the differential equation (5.170) using the explicit Euler method is absolutely stable if

$$|1 + h\lambda| \le 1 \tag{5.182}$$

Because $(1 + h\lambda)$ is the root of the characteristic equation (5.176), an alternative definition of absolute stability is

$$|\mu_i| \le 1 \qquad i = 1, 2, \ldots, k \tag{5.183}$$

where more than one root exists in the multistep numerical methods.

Returning to the problem at hand, the inequality (5.182) is rearranged to

$$-2 \le h\lambda \le 0 \tag{5.184}$$

This inequality sets the limits of the integration step size for a stable solution as follows: Because h is positive, then $\lambda < 0$ and

$$h \le \frac{2}{|\lambda|} \tag{5.185}$$

Inequality (5.185) is a finite *general stability boundary*, and for this reason, the explicit Euler method is called *conditionally stable*. Any method with an infinite general stability boundary can be called *unconditionally stable*.

At the outset of our discussion, we assumed that λ was real in order to simplify the derivation. This assumption is not necessary: λ can be a complex number. In the earlier discussion of the stability of difference equations (Sec. 3.6), we mentioned that a solution is stable, converging with damped oscillations, when complex roots are present, and the moduli of the roots are less than or equal to unity:

$$|r| \le 1 \tag{5.186}$$

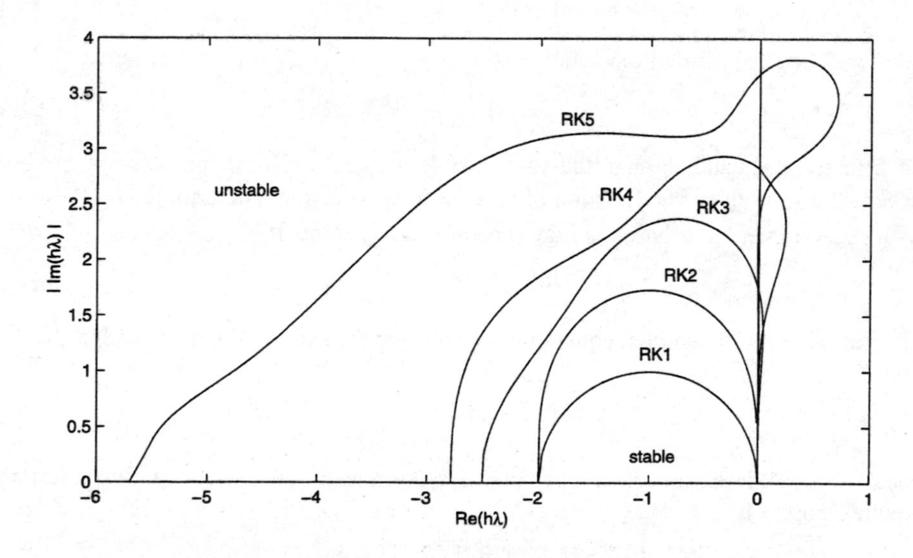

Figure 5.7 Stability region in the complex plane for Runge-Kutta methods of order 1 (explicit Euler), 2, 3, 4, and 5.

The two inequalities (5.184) and (5.186) describe the circle with a radius of unity on the complex plane shown in Fig. 5.7. Since the explicit Euler method can be categorized as a first-order Runge-Kutta method, the corresponding curve in this figure is marked by RK1. The set of values of $h\lambda$ inside the circle yields stable numerical solutions of Eq. (5.170) using the Euler integration method.

We now return to the consideration of the truncation and roundoff errors of the Euler method and develop a difference equation, which describes the propagation of the error in the numerical solution. We work with the nonlinear form of the initial-value problem

$$\frac{dy}{dx} = f(x, y) \tag{5.27}$$

where the initial condition is given by

$$y(x_0) = y_0 \tag{5.28}$$

We define the accumulated error of the numerical solution at step $(n + 1)$ as

$$\epsilon_{n+1} = y_{n+1} - y(x_{n+1}) \tag{5.187}$$

where $y(x_{n+1})$ is the *exact* value of y, and y_{n+1} is the *calculated* value of y at x_{n+1}. We then write the exact solution $y(x_{n+1})$ as a Taylor series expansion, showing as many terms as needed for the Euler method:

$$y(x_{n+1}) = y(x_n) + hf(x_n, y(x_n)) + T_{E,n+1} \tag{5.188}$$

where $T_{E,n+1}$ is the local truncation error for step $(n + 1)$. We also write the calculated value y_{n+1} obtained from the implicit Euler formula

$$y_{n+1} = y_n + hf(x_n, y_n) + R_{E,n+1} \tag{5.189}$$

where $R_{E,n+1}$ is the roundoff error introduced by the computer in step $(n + 1)$.

Combining Eqs. (5.187)-(5.189) we have

$$\epsilon_{n+1} = y_n - y(x_n) + h[f(x_n, y_n) - f(x_n, y(x_n))] - T_{E,n+1} + R_{E,n+1} \tag{5.190}$$

which simplifies to

$$\epsilon_{n+1} = \epsilon_n + h[f(x_n, y_n) - f(x_n, y(x_n))] - T_{E,n+1} + R_{E,n+1} \tag{5.191}$$

The mean-value theorem

$$f(x_n, y_n) - f(x_n, y(x_n)) = \frac{\partial f}{\partial y}\Big|_{\alpha,x_n}[y_n - y(x_n)] \qquad y_n < \alpha < y(x_n) \tag{5.192}$$

can be used to further modify the error equation (5.191) to

$$\epsilon_{n+1} - \left[1 + h\frac{\partial f}{\partial y}\Big|_{\alpha,x_n}\right]\epsilon_n = -T_{E,n+1} + R_{E,n+1} \tag{5.193}$$

This is a *first-order nonhomogeneous difference equation with varying coefficients,* which can be solved only by iteration. However, by making the following simplifying assumptions:

$$T_{E,n+1} = T_E = \text{constant}$$
$$R_{E,n+1} = R_E = \text{constant} \tag{5.194}$$
$$\frac{\partial f}{\partial y}\Big|_{\alpha,x_n} = \lambda = \text{constant}^6$$

Eq. (5.193) simplifies to

$$\epsilon_{n+1} - (1 + h\lambda)\epsilon_n = -T_E + R_E \tag{5.195}$$

whose solution is given by the sum of the homogeneous and particular solutions [8]:

$$\epsilon_n = C_1(1 + h\lambda)^n + \frac{-T_E + R_E}{1 - (1 + h\lambda)} \tag{5.196}$$

Comparison of Eqs. (5.175) and (5.195) reveals that the characteristic equations for the solution y_n and the error ϵ_n are identical. The truncation and roundoff error terms in Eq. (5.195) introduce the particular solution. The constant C_1 is calculated by assuming that the initial condition of the differential equation has no error; that is, $\epsilon_0 = 0$. The final form of the equation that describes the behavior of the propagation error is

$$\epsilon_n = \frac{-T_E + R_E}{h\lambda}[(1 + h\lambda)^n - 1] \tag{5.197}$$

A great deal of insight can be gained by thoroughly examining Eq. (5.197). As expected, the value of $(1 + h\lambda)$ is the determining factor in the behavior of the propagation error. Consider first the case of a fixed finite step size h, with the number of integration steps increasing to a very large n. The limit on the error as $n \to \infty$ is

$$\lim_{n\to\infty} |\epsilon_n| = \frac{-T_E + R_E}{h\lambda} \qquad \text{for } |1 + h\lambda| < 1 \tag{5.198}$$

$$\lim_{n\to\infty} |\epsilon_n| = \infty \qquad \text{for } |1 + h\lambda| > 1 \tag{5.199}$$

In the first situation [Eq. (5.198)], $\lambda < 0$, $0 < h < 2/|\lambda|$, the error is bounded, and the numerical solution is stable. The numerical solution differs from the exact solution by only the finite quantity $(-T_E + R_E)/h\lambda$, which is a function of the truncation error, the roundoff error, the step size, and the eigenvalue of the differential equation.

---

[6] Under this assumption, Eq. (5.27) becomes identical to Eq. (5.170).

In the second situation [Eq. (5.199)], $\lambda > 0$, $h > 0$, the error is unbounded and the numerical solution is unstable. For $\lambda > 0$, however, the exact solution is *inherently unstable.* For this reason we introduce the concept of *relative error* defined as

$$\text{relative error} = \frac{\epsilon_n}{y_n} \tag{5.200}$$

Utilizing Eqs. (5.180) and (5.197), we obtain the relative error as

$$\frac{\epsilon_n}{y_n} = \frac{-T_E + R_E}{y_0 h\lambda}\left[1 - \frac{1}{(1 + h\lambda)^n}\right] \tag{5.201}$$

The relative error is bounded for $\lambda > 0$ and unbounded for $\lambda < 0$. So we conclude that for inherently stable differential equations, the absolute propagation error is the pertinent criterion for numerical stability, whereas for inherently unstable differential equations, the relative propagation error must be investigated.

Let us now consider a fixed interval of integration, $0 \le x \le \alpha$, so that

$$h = \frac{\alpha}{n} \tag{5.202}$$

and we increase the number of integration steps to a very large n. This, of course, causes $h \to 0$. A numerical method is said to be *convergent* if

$$\lim_{h \to 0} |\epsilon_n| = 0 \tag{5.203}$$

In the absence of roundoff error, the Euler method, and most other integration methods, are convergent because

$$\lim_{h \to 0} T_E = 0 \tag{5.204}$$

and

$$\lim_{h \to 0} |\epsilon_n| = 0 \tag{5.203}$$

However, roundoff error is *never* absent in numerical calculations. As $h \to 0$ the roundoff error is the crucial factor in the propagation of error:

$$\lim_{h \to 0} |\epsilon_n| = R_E \lim_{h \to 0} \frac{(1 + h\lambda)^n - 1}{h\lambda} \tag{5.205}$$

Application of L'Hôpital's rule shows that the roundoff error propagates unbounded as the number of integration steps becomes very large:

$$\lim_{h \to 0} \epsilon_n = R_E[\infty] \tag{5.206}$$

This is the "catch 22" of numerical methods: A smaller step size of integration reduces the truncation error but requires a large number of steps, thereby increasing the roundoff error.

A similar analysis of the *implicit Euler method* (backward Euler) results in the following two equations, for the solution

$$y_{n+1} = \frac{y_0}{(1 - h\lambda)^n} \tag{5.207}$$

and the propagation error

$$\epsilon_{n+1} = \frac{-T_E + R_E}{h\lambda}(1 - h\lambda)\left[\frac{1}{(1 - h\lambda)^n} - 1\right] \tag{5.208}$$

For $\lambda < 0$ and $0 < h < \infty$, the solution is stable:

$$\lim_{n \to \infty} y_n = 0 \tag{5.209}$$

and the error is bounded:

$$\lim_{n \to \infty} \epsilon_n = -\frac{-T_E + R_E}{h\lambda}(1 - \lambda h) \tag{5.210}$$

No limitation is placed on the step size; therefore, the implicit Euler method is *unconditionally stable* for $\lambda < 0$. On the other hand, when $\lambda > 0$, the following inequality must be true for a stable solution:

$$|1 - h\lambda| \leq 1 \tag{5.211}$$

This imposes the limit on the step size:

$$-2 \leq h\lambda \leq 0 \tag{5.212}$$

It can be concluded that the implicit Euler method has a wider range of stability than the explicit Euler method (see Table 5.3).

5.7.2 Stability and Error Propagation of Runge-Kutta Methods

Using methods parallel to those of the previous section, the recurrence equations and the corresponding roots for the Runge-Kutta methods can be derived [9]. For the differential equation (5.170), these are:

Second-order Runge-Kutta:

$$y_{n+1} = \left(1 + h\lambda + \frac{1}{2}h^2\lambda^2\right)y_n \tag{5.213}$$

$$\mu_1 = 1 + h\lambda + \frac{1}{2}h^2\lambda^2 \tag{5.214}$$

Third-order Runge-Kutta:

$$y_{n+1} = \left(1 + h\lambda + \frac{1}{2}h^2\lambda^2 + \frac{1}{6}h^3\lambda^3\right) y_n \tag{5.215}$$

$$\mu_1 = 1 + h\lambda + \frac{1}{2}h^2\lambda^2 + \frac{1}{6}h^3\lambda^3 \tag{5.216}$$

Fourth-order Runge-Kutta:

$$y_{n+1} = \left(1 + h\lambda + \frac{1}{2}h^2\lambda^2 + \frac{1}{6}h^3\lambda^3 + \frac{1}{24}h^4\lambda^4\right) y_n \tag{5.217}$$

$$\mu_1 = 1 + h\lambda + \frac{1}{2}h^2\lambda^2 + \frac{1}{6}h^3\lambda^3 + \frac{1}{24}h^4\lambda^4 \tag{5.218}$$

Table 5.3 Real stability boundaries

Method	Boundary
Explicit Euler	$-2 \le h\lambda \le 0$
Implicit Euler	$0 < h < \infty$ *for* $\lambda < 0$ $-2 < h\lambda < 0$ *for* $\lambda > 0$
Modified Euler (predictor-corrector)	$-1.077 \le h\lambda \le 0$
Second-order Runge-Kutta	$-2 \le h\lambda \le 0$
Third-order Runge-Kutta	$-2.5 \le h\lambda \le 0$
Fourth-order Runge-Kutta	$-2.785 \le h\lambda \le 0$
Fifth-order Runge-Kutta	$-5.7 \le h\lambda \le 0$
Adams	$-0.546 \le h\lambda \le 0$
Adams-Moulton	$-1.285 \le h\lambda \le 0$

Fifth-order Runge-Kutta:

$$y_{n+1} = \left(1 + h\lambda + \frac{1}{2}h^2\lambda^2 + \frac{1}{6}h^3\lambda^3 + \frac{1}{24}h^4\lambda^4 + \frac{1}{120}h^5\lambda^5 + \frac{0.5625}{720}h^6\lambda^6\right)y_n \tag{5.219}$$

$$\mu_1 = 1 + h\lambda + \frac{1}{2}h^2\lambda^2 + \frac{1}{6}h^3\lambda^3 + \frac{1}{24}h^4\lambda^4 + \frac{1}{120}h^5\lambda^5 + \frac{0.5625}{720}h^6\lambda^6 \tag{5.220}$$

The last term in the right-hand side of Eqs. (5.219) and (5.220) is specific to the fifth-order Runge-Kutta, which appears in Table 5.2 and varies for different fifth-order formulas. The condition for absolute stability:

$$|\mu_i| \le 1 \qquad i = 1, 2, \ldots, k \tag{5.183}$$

applies to all the above methods. The absolute real stability boundaries for these methods are listed in Table 5.3, and the regions of stability in the complex plane are shown on Fig. 5.7. In general, as the order increases, so do the stability limits.

5.7.3 Stability and Error Propagation of Multistep Methods

Using methods parallel to those of the previous section, the recurrence equations and the corresponding roots for the modified Euler, Adams, and Adams-Moulton methods can be derived [9]. For the differential equation (5.170), these are:

Modified Euler (combination of predictor and corrector):

$$y_{n+1} = (1 + h\lambda + h^2\lambda^2)y_n \tag{5.221}$$

$$\mu_1 = 1 + h\lambda + h^2\lambda^2 \tag{5.222}$$

Adams:

$$y_{n+2} = \left(1 + \frac{23}{12}h\lambda\right)y_n - \frac{4h\lambda}{3}y_{n-1} + \frac{5h\lambda}{12}y_{n-2} \tag{5.223}$$

$$\mu^3 - \left(1 + \frac{23}{12}h\lambda\right)\mu^2 + \left(\frac{4}{3}h\lambda\right)\mu - \frac{5}{12}h\lambda = 0 \tag{5.224}$$

Adams-Moulton (combination of predictor and corrector):

$$y_{n+1} = \left(1 + \frac{7h\lambda}{6} + \frac{55h^2\lambda^2}{64}\right)y_n - \left(\frac{5h\lambda}{24} + \frac{59h^2\lambda^2}{64}\right)y_{n-1}$$

$$+ \left(\frac{h\lambda}{24} + \frac{37h^2\lambda^2}{64}\right)y_{n-2} - \frac{9h^2\lambda^2}{64}y_{n-3} \tag{5.225}$$

$$\mu^4 - \left(1 + \frac{7h\lambda}{6} + \frac{55h^2\lambda^2}{64}\right)\mu^3 + \left(\frac{5h\lambda}{24} + \frac{59h^2\lambda^2}{64}\right)\mu^2$$

$$- \left(\frac{h\lambda}{24} + \frac{37h^2\lambda^2}{64}\right)\mu + \frac{9h^2\lambda^2}{64} = 0 \tag{5.226}$$

The condition for absolute stability,

$$|\mu_i| \le 1 \qquad i = 1, 2, \ldots, k \tag{5.183}$$

applies to all the above methods. The absolute real stability boundaries for these methods are also listed in Table 5.3, and the regions of stability in the complex plane are shown on Fig. 5.8.

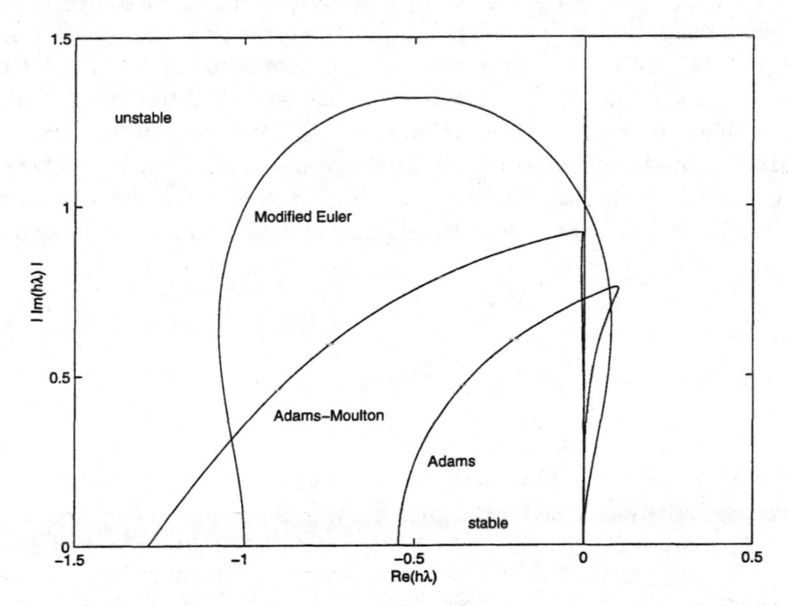

Figure 5.8 Stability region in the complex plane for the modified Euler (Euler predictor-corrector), Adams, and Adams-Moulton methods.

5.8 STEP SIZE CONTROL

The discussion of stability analysis in the previous sections made the simplifying assumption that the value of λ remains constant throughout the integration. This is true for linear equations such as Eq. (5.170); however, for the nonlinear equation (5.27), the value of λ may vary considerably over the interval of integration. The step size of integration must be chosen using the maximum possible value of λ, thus resulting in the minimum step size. This, of course, will guarantee stability at the expense of computation time. For problems in which computation time becomes excessive, it is possible to develop strategies for automatically adjusting the step size at each step of the integration.

A simple test for checking the step size is to do the calculations at each interval twice: Once with the full step size, and then repeat the calculations over the same interval with a smaller step size, usually half that of the first one. If at the end of the interval, the difference between the predicted value of y by both approaches is less than the specified convergence criterion, the step size may be increased. Otherwise, a larger than acceptable difference between the two calculated y values suggests that the step size is large, and it should be shortened in order to achieve an acceptable truncation error.

Another method of controlling the step size is to obtain an estimation of the truncation error at each interval. A good example of such an approach is the Runge-Kutta-Fehlberg method (see Table 5.2), which provides the estimation of the local truncation error. This error estimate can be easily introduced into the computer program, and let the program automatically change the step size at each point until the desired accuracy is achieved.

As mentioned before, the optimum number of application of corrector is two. Therefore, in the case of using a predictor-corrector method, if the convergence is achieved before the second corrected value, the step size may be increased. On the other hand, if the convergence is not achieved after the second application of the corrector, the step size should be reduced.

5.9 STIFF DIFFERENTIAL EQUATIONS

In Sec. 5.7, we showed that the stability of the numerical solution of differential equations depends on the value of $h\lambda$, and that λ together with the stability boundary of the method determine the step size of integration. In the case of the linear differential equation

$$\frac{dy}{dx} = \lambda y \tag{5.170}$$

λ is the eigenvalue of that equation, and it remains a constant throughout the integration. The nonlinear differential equation

$$\frac{dy}{dx} = f(x, y) \tag{5.27}$$

can be linearized at each step using the mean-value theorem (5.192), so that λ can be obtained from the partial derivative of the function with respect to y:

$$\lambda = \left.\frac{\partial f}{\partial y}\right|_{\alpha, x_n} \tag{5.227}$$

The value of λ is no longer a constant but varies in magnitude at each step of the integration.

This analysis can be extended to a set of simultaneous nonlinear differential equations:

$$\begin{aligned}
\frac{dy_1}{dx} &= f_1(x, y_1, y_2, \ldots\ y_n) \\
\frac{dy_2}{dx} &= f_2(x, y_1, y_2, \ldots, y_n) \\
&\vdots \\
\frac{dy_n}{dx} &= f_n(x, y_1, y_2, \ldots, y_n)
\end{aligned} \tag{5.98}$$

Linearization of the set produces the Jacobian matrix

$$\boldsymbol{J} = \begin{bmatrix} \dfrac{\partial f_1}{\partial y_1} & \cdots & \dfrac{\partial f_1}{\partial y_n} \\ & \cdots\cdots\cdots & \\ \dfrac{\partial f_n}{\partial y_1} & \cdots & \dfrac{\partial f_n}{\partial y_n} \end{bmatrix} \tag{5.228}$$

The eigenvalues $\{\lambda_i \mid i = 1, 2, \ldots, n\}$ of the Jacobian matrix are the determining factors in the stability analysis of the numerical solution. The step size of integration is determined by the stability boundary of the method and the maximum eigenvalue.

When the eigenvalues of the Jacobian matrix of the differential equations are all of the same order of magnitude, no unusual problems arise in the integration of the set. However, when the maximum eigenvalue is several orders of magnitude larger than the minimum eigenvalue, the equations are said to be *stiff.* The *stiffness ratio* (SR) of such a set is defined as

$$SR = \frac{\max\limits_{1 \le i \le n} |\text{Real}(\lambda_i)|}{\min\limits_{1 \le i \le n} |\text{Real}(\lambda_i)|} \tag{5.229}$$

The step size of integration is determined by the largest eigenvalue, and the final time of integration is usually fixed by the smallest eigenvalue; therefore, integration of differential equations using explicit methods may be time intensive. Finlayson [1] recommends using implicit methods for integrating stiff differential equations in order to reduce computation time.

The MATLAB functions *ode23s* and *ode15s* are solvers suitable for solution of stiff ordinary differential equations (see Table 5.1).

PROBLEMS

5.1 Derive the second-order Runge-Kutta method of Eq. (5.92) using central differences.

5.2 The solution of the following second-order linear ordinary differential equation should be determined using numerical techniques:

$$\frac{d^2x}{dt^2} - 3\frac{dx}{dt} - 10x = 0$$

The initial conditions for this equation are, at $t = 0$:

$$x|_0 = 3 \quad \text{and} \quad \frac{dx}{dt}\Big|_0 = 15$$

(a) Transform the above differential equation into a set of first-order linear differential equations with appropriate initial conditions.

(b) Find the solution using eigenvalues and eigenvectors, and evaluate the variables in the range $0 \le t \le 1.0$.

(c) Use the fourth-order Runge-Kutta method to verify the results of part (b).

5.3 A radioactive material (A) decomposes according to the series reaction:

$$A \xrightarrow{k_1} B \xrightarrow{k_2} C$$

where k_1 and k_2 are the rate constants and B and C are the intermediate and final products, respectively. The rate equations are

$$\frac{dC_A}{dt} = -k_1 C_A$$

$$\frac{dC_B}{dt} = k_1 C_A - k_2 C_B$$

$$\frac{dC_C}{dt} = k_2 C_B$$

where C_A, C_B, and C_C are the concentrations of materials *A*, *B*, and *C*, respectively. The values of the rate constants are

$$k_1 = 3 \text{ s}^{-1} \qquad k_2 = 1 \text{ s}^{-1}$$

Initial conditions are

$$C_A(0) = 1 \text{ mol/m}^3 \qquad C_B(0) = 0 \qquad C_C(0) = 0$$

(a) Use the eigenvalue-eigenvector method to determine the concentrations C_A, C_B, and C_C as a function of time *t*.
(b) At time $t = 1$ s and $t = 10$ s, what are the concentrations of *A*, *B*, and *C*?
(c) Sketch the concentration profiles for *A*, *B*, and *C*.

5.4 (a) Integrate the following differential equations:

$$\frac{dC_A}{dt} = -4C_A + C_B \qquad C_A(0) = 100.0$$

$$\frac{dC_B}{dt} = 4C_A - 4C_B \qquad C_B(0) = 0.0$$

for the time period $0 \leq t \leq 5$, using (1) the Euler predictor-corrector method, (2) the fourth-order Runge-Kutta method.
(b) Which method would give a solution closer to the analytical solution?
(c) Why do these methods give different results?

5.5 In the study of fermentation kinetics, the logistic law

$$\frac{dy_1}{dt} = k_1 y_1 \left(1 - \frac{y_1}{k_2} \right)$$

has been used frequently to describe the dynamics of cell growth. This equation is a modification of the logarithmic law

$$\frac{dy_1}{dt} = k_1 y_1$$

The term $(1 - y_1/k_2)$ in the logistic law accounts for cessation of growth due to a limiting nutrient.

The logistic law has been used successfully in modeling the growth of *penicillium chryscogenum*, a penicillin-producing organism [6]. In addition, the rate of production of penicillin has been mathematically quantified by the equation

$$\frac{dy_2}{dt} = k_3 y_1 - k_4 y_2$$

Penicillin (y_2) is produced at a rate proportional to the concentration of the cell (y_1) and is degraded by hydrolysis, which is proportional to the concentration of the penicillin itself.

(a) Discuss other possible interpretations of the logistic law.

(b) Show that k_2 is equivalent to the maximum cell concentration that can be reached under given conditions.

(c) Apply the fourth-order Runge-Kutta integration method to find the numerical solution of the cell and penicillin equations. Use the following constants and initial conditions:

$$k_1 = 0.03120 \qquad k_2 = 47.70 \qquad k_3 = 3.374 \qquad k_4 = 0.01268$$

at $t = 0$, $y_1(0) = 5.0$, and $y_2(0) = 0.0$; the range of t is $0 \le t \le 212$ h.

5.6 The conversion of glucose to gluconic acid is a simple oxidation of the aldehyde group of the sugar to a carboxyl group. This transformation can be achieved by a microorganism in a fermentation process. The enzyme glucose oxidase, present in the microorganism, converts glucose to gluconolactone. In turn, the gluconolactone hydrolyzes to form the gluconic acid. The overall mechanism of the fermentation process that performs this transformation can be described as follows:

Cell growth:

$$\text{Glucose} + \text{Cells} \rightarrow \text{Cells}$$

Glucose oxidation:

$$\text{Glucose} + O_2 \xrightarrow{\text{Glucose oxidase}} \text{Gluconolactone} + H_2O_2$$

Gluconolactone hydrolysis:

$$\text{Gluconolactone} + H_2O \longrightarrow \text{Gluconic acid}$$

Peroxide decomposition:

$$H_2O_2 \xrightarrow{\text{Catalyst}} H_2O + \frac{1}{2}O_2$$

A mathematical model of the fermentation of the bacterium *Pseudomonas ovalis*, which produces gluconic acid, has been developed by Rai and Constantinides [10]. This model, which describes the dynamics of the logarithmic growth phases, can be summarized as follows:

Rate of cell growth:

$$\frac{dy_1}{dt} = b_1 y_1 \left(1 - \frac{y_1}{b_2}\right)$$

.1 Rate of gluconolactone formation:

$$\frac{dy_2}{dt} = \frac{b_3 y_1 y_4}{b_4 + y_4} - 0.9082 b_5 y_2$$

Rate of gluconic acid formation:

$$\frac{dy_3}{dt} = b_5 y_2$$

Rate of glucose consumption:

$$\frac{dy_4}{dt} = -1.011\left(\frac{b_3 y_1 y_4}{b_4 + y_4}\right)$$

where y_1 = concentration of cell
y_2 = concentration of glucononctone
y_3 = concentration of gluconic acid
y_4 = concentration of glucose
b_1-b_5 = parameters of the system which are functions of temperature and pH.

At the operating conditions of 30°C and pH 6.6, the values of the five parameters were determined from experimental data to be

$$b_1 = 0.949 \qquad b_2 = 3.439 \qquad b_3 = 18.72 \qquad b_4 = 37.51 \qquad b_5 = 1.169$$

At these conditions, develop the time profiles of all variables, y_1 to y_4, for the period $0 \le t \le 9$ h. The initial conditions at the start of this period are

$$y_1(0) = 0.5 \text{ U.O.D./mL} \qquad y_3(0) = 0.0 \text{ mg/mL}$$
$$y_2(0) = 0.0 \text{ mg/mL} \qquad y_4(0) = 50.0 \text{ mg/mL}$$

5.7 The best-known mathematical representation of population dynamics between interacting species is the Lokta-Volterra model [11]. For the case of two competing species, these equations take the general form

$$\frac{dN_1}{N_1 dt} = f_1(N_1, N_2) \qquad \frac{dN_2}{N_2 dt} = f_2(N_1, N_2)$$

where N_1 is the population density of species 1 and N_2 is the population density of species 2. The functions f_1 and f_2 describe the specific growth rates of the two populations. Under certain assumptions, these functions can be expressed in terms of N_1, N_2, and a set of constants whose values depend on natural birth and death rates and on the interactions between the two species. Numerous examples of such interactions can be cited from ecological and microbiological studies. The predator-prey problem, which has been studied extensively, presents a very interesting ecological example of population dynamics. On the other hand, the interaction between bacteria and phages in a fermentor is a well-known nemesis to industrial microbiologists.

Let us now consider in detail the classical predator-prey problem, that is, the interaction between two wild-life species, the prey, which is a herbivore, and the predator, a carnivore. These two animals coinhabit a region where the prey have an abundant supply of natural vegetation for food, and the predators depend on the prey for their entire supply of food. This is a simplification of the real ecological system where more than two species coexist, and where predators usually feed on a variety of prey. The Lotka-Volterra equations have also been formulated for such complex systems; however, for the sake of this problem, our ecological system will contain only two interacting species. An excellent example of such an ecological system is Isle Royale National Park, a 210-square mile archipelago in Lake Superior. The park comprises a single large island and many small islands which extend off the main island. According to a very interesting article in *National Geographic* [12], moose arrived on Isle Royale around 1900, probably swimming in from Canada. By 1930, their unchecked numbers approached 3000, ravaging vegetation. In 1949, across an ice bridge from Ontario, came a predator–the wolf. Since 1958, the longest study of its kind [13]-[15] still seeks to define the complete cycle in the ebb and flow of predator and prey populations, with wolves fluctuating from 11 to 50 and moose from 500 to 2400 [see Table P5.7*a*].

In order to formulate the predator-prey problem, we make the following assumptions:

(a) In the absence of the predator, the prey has a natural birth rate b and a natural death rate d. Because an abundant supply of natural vegetation for food is available, and assuming that no catastrophic diseases plague the prey, the birth rate is higher than the death rate; therefore, the net specific growth rate α is positive; that is,

$$\frac{dN_1}{N_1 dt} = b - d = \alpha$$

(b) In the presence of the predator the prey is consumed at a rate proportional to the number of predators present βN_2:

$$\frac{dN_1}{N_1 dt} = \alpha - \beta N_2$$

(c) In the absence of the prey, the predator has a negative specific growth rate ($-\gamma$), as the inevitable consequence of such a situation is the starvation of the predator:

$$\frac{dN_2}{N_2 dt} = -\gamma$$

(d) In the presence of the prey, the predator has an ample supply of food, which enables it to survive and produce at a rate proportional to the abundance of the prey, δN_1. Under these circumstances, the specific growth rate of the predator is

$$\frac{dN_2}{N_2 dt} = -\gamma + \delta N_1$$

The equations in parts (b) and (d) constitute the Lokta-Volterra model for the one-predator-one-prey problem. Rearranging these two equations to put them in the canonical form,

$$\frac{dN_1}{dt} = \alpha N_1 - \beta N_1 N_2 \tag{1}$$

$$\frac{dN_2}{dt} = -\gamma N_2 + \delta N_1 N_2 \tag{2}$$

This is a set of simultaneous first-order nonlinear ordinary differential equations. The solution of these equations first requires the determination of the constants α, β, γ, and δ, and the specification

Table P5.7*a* Population of moose and wolves on Isle Royale

Year	Moose	Wolves	Year	Moose	Wolves
1959	522	20	1979	738	43
1960	573	22	1980	705	50
1961	597	22	1981	544	30
1962	603	23	1982	972	14
1963	639	20	1983	900	23
1964	726	26	1984	1041	24
1965	762	28	1985	1062	22
1966	900	26	1986	1025	20
1967	1008	22	1987	1380	16
1968	1176	22	1988	1653	12
1969	1191	17	1989	1397	11
1970	1320	18	1990	1216	15
1971	1323	20	1991	1313	12
1972	1194	23	1992	1596	12
1973	1137	24	1993	1880	13
1974	1026	31	1994	1770	17
1975	915	41	1995	2422	16
1976	708	44	1996	1200	22
1977	573	34	1997	500	24
1978	905	40	1998	700	14

of boundary conditions. The latter could be either initial or final conditions. In population dynamics, it is more customary to specify initial population densities, because actual numerical values of the population densities may be known at some point in time, which can be called the initial starting time. However, it is conceivable that one may want to specify final values of the population densities to be accomplished as targets in a well-managed ecological system. In this problem, we will specify the initial population densities of the prey and predator to be

$$N_1(t_0) = N_1^0 \qquad \text{and} \qquad N_2(t_0) = N_2^0 \tag{3}$$

Equations (1)-(3) constitute the complete mathematical formulation of the predator-prey problem based on assumptions (a) to (d). Different assumptions would yield another set of differential equations [see Problem (5.8)]. In addition, the choice of constants and initial conditions influence the solution of the differential equations and generate a diverse set of qualitative behavior patterns for the two populations. Depending on the form of the differential equations and the values of the constants chosen, the solution patterns may vary from stable, damped oscillations, where the species reach their respective stable symbiotic population densities, to highly unstable situations, in which one of the species is driven to extinction while the other explodes to extreme population density.

The literature on the solution of the Lotka-Volterra problems is voluminous. Several references on this topic are given at the end of this chapter. A closed-form analytical solution of this system of nonlinear ordinary differential equations is not possible. The equations must be integrated numerically using any of the numerical integration methods covered in this chapter. However, before numerical integration is attempted, the stability of these equations must be examined thoroughly. In a recent treatise on this subject, Vandermeer [16] examined the stability of the solutions of these equations around equilibrium points. These points are located by setting the derivatives in Eqs. (1) and (2) to zero:

$$\frac{dN_1}{dt} = \alpha N_1 - \beta N_1 N_2 = 0$$

$$\frac{dN_2}{dt} = -\gamma N_2 + \delta N_1 N_2 = 0$$

and rearranging these equations to obtain the values of N_1 and N_2 at the equilibrium point in terms of the constants

$$N_1^* = \frac{\gamma}{\delta} \qquad N_2^* = \frac{\alpha}{\beta}$$

where * denotes the equilibrium values of the population densities. Vandermeer stated that: "Sometimes only one point (N_1^*, N_2^*) will satisfy the equilibrium equations. At other times, multiple points will satisfy the equilibrium equations. . . . The neighborhood stability analysis is undertaken in the neighborhood of a single equilibrium point." The stability is determined by examining the eigenvalues of the Jacobian matrix evaluated at equilibrium:

$$J = \begin{bmatrix} \left(\frac{\partial f_1}{\partial N_1}\right)^* & \left(\frac{\partial f_1}{\partial N_2}\right)^* \\ \left(\frac{\partial f_2}{\partial N_1}\right)^* & \left(\frac{\partial f_2}{\partial N_2}\right)^* \end{bmatrix}$$

where f_1 and f_2 are the right-hand sides of Eqs. (1) and (2), respectively.

The eigenvalues of the Jacobian matrix can be obtained by the solution of the following equation (as described in Chap. 2):

$$|J - \lambda I| = 0$$

For the problem of two differential equations, there are two eigenvalues that can possibly have both real and imaginary parts. These eigenvalues take the general form

$$\lambda_1 = a_1 + b_1 i$$

$$\lambda_2 = a_2 + b_2 i$$

where $i = \sqrt{-1}$. The values of the real parts (a_1, a_2) and imaginary parts (b_1, b_2) determine the nature of the stability (or instability) in the neighborhood of the equilibrium points. These possibilities are summarized in Table P5.7*b*.

Table P5.7*b*

a_1, a_2	b_1, b_2	Stability analysis
Negative	Zero	Stable, nonoscillatory
Positive	Zero	Unstable, nonoscillatory
One positive, one negative	Zero	Metastable, saddle point
Negative	Nonzero	Stable, oscillatory
Positive	Nonzero	Unstable, oscillatory
Zero	Nonzero	Neutrally stable, oscillatory

Many combinations of values of constants and initial conditions exist that would generate solutions to Eqs. (1) and (2). In order to obtain a realistic solution to these equations, we utilize the data of Allen [13] and Peterson [14] on the moose-wolf populations of Isle Royale National Park given in Table P5.7*a*. From these data, which cover the period 1959–1998, we estimate the average values of the moose and wolf populations (over the entire 40-year period) and use these as equilibrium values:

$$N_1^* = \frac{\gamma}{\delta} = 1045 \qquad N_2^* = \frac{\alpha}{\beta} = 23$$

In addition, we estimate the period of oscillation to be 25 years. This was based on the moose data; the wolf data show a shorter period. For this reason, we predict that the predator equation may not be a good representation of the data. Lotka has shown that the period of oscillation around the equilibrium point is approximated by

$$\tau = \frac{2\pi}{\sqrt{\alpha\gamma}}$$

These three equations have four unknowns. By assuming the value of α to be 0.3 (this is an estimate of the net specific growth rate of the prey in the absence of the predator), the complete set of constants is

$$\alpha = 0.3 \qquad \beta = 0.0130$$
$$\gamma = 0.2106 \qquad \delta = 0.0002015$$

This initial conditions are taken from Allen [13] for 1959, the earliest date for which complete data are available. These are

$$N_1(1959) = 522 \quad \text{and} \quad N_2(1959) = 20$$

Integrate the predator-prey equations for the period 1959–1999 using the above constants and initial conditions and compare the simulation with the actual data. Draw the phase plot of N_1 versus N_2, and discuss the stability of these equations with the aid of the phase plot.

5.8 It can be shown that whenever the Lotka-Volterra problem has the form of Eqs. (1) and (2) in Prob. 5.7, the real parts of the eigenvalues of the Jacobian matrix are zero. This implies that the solution always has neutrally stable oscillatory behavior. This is explained by the fact that assumptions (a) to (d) of Prob. 5.7 did not include the crowding effect each population may have on its own fertility or mortality. For example, Eq. (1) can be rewritten with the additional term ϵN_1^2:

$$\frac{dN_1}{dt} = \alpha N_1 - \beta N_1 N_2 - \epsilon N_1^2$$

The new term introduces a negative density-dependency of the specific growth rate of the prey on its own population. This term can be viewed as either a contribution to the death rate or a reduction of the birth rate caused by overcrowding of the species.

In this problem, modify the Lotka-Volterra equations by introducing the effect of overcrowding, account for at least one additional source of food for the predator (a second prey), or attempt to quantify other interferences you believe are important in the life cycle of these two species. Choose the constants and initial conditions of your equations carefully in order to obtain an ecologically feasible situation. Integrate the resulting equations and obtain the time profiles of the populations of all the species involved. In addition, draw phase plots of N_1 versus N_2, N_1 versus N_3, and so on, and discuss the stability considerations with the aid of the phase plots.

5.9 The steady-state simulation of continuous contact countercurrent processes involving simultaneous heat and mass transfer may be described as a nonlinear boundary-value problem [3]. For instance, for a continuous adiabatic gas absorption contactor unit, the model can be written in the following form:

$$\frac{dY_A}{dt} = N\left[\frac{x_A J_A}{P}\exp\left(-\frac{A_A}{T_L}\right) - \frac{Y_A}{1 + Y_A + Y_B}\right]$$

$$\frac{dY_B}{dt} = GN\left[\frac{(1 - x_A)J_B}{P}\exp\left(-\frac{A_B}{T_L}\right) - \frac{Y_B}{1 + Y_A + Y_B}\right]$$

$$\frac{dT_G}{dt} = HN(T_L - T_G)$$

$$\frac{dT_L}{dt} = \frac{\phi}{RC_L}\frac{dY_A}{dt} + \frac{r_{B_0} + \mu}{RC_L}\frac{dY_B}{dt} + \frac{C_G}{C_L}\frac{dT_G}{dt}$$

$$\frac{dR}{dt} = \frac{dY_A}{dt} + \frac{dY_B}{dt}$$

$$\frac{dx_A}{dt} = \frac{1}{R}\frac{dY_A}{dt} - \frac{x_A}{R}\frac{dR}{dt}$$

Thermodynamic and physical property data for the system ammonia-air-water are

$$\begin{aligned} J_A &= 1.36\times10^{11}\ \text{N/m}^2 & \phi &= 1.08\times10^5\ \text{J/kmol} \\ A_A &= 4.212\times10^3\ \text{K} & r_{B_0} &= 1.36\times10^5\ \text{J/kmol} \\ J_B &= 6.23\times10^{10}\ \text{N/m}^2 & \mu &= 0.0\ \text{J/mol} \\ A_B &= 5.003\times10^3\ \text{K} & C_L &= 232\ \text{J/kmol} \\ G &= 1.41 & C_G &= 93\ \text{J/kmol} \\ H &= 1.11 & P &= 10^5\ \text{N/m}^2 \\ N &= 10 \end{aligned}$$

The inlet conditions are

$$\begin{aligned} Y_A(0) &= 0.05 & Y_B(0) &= 0.0 & T_G(0) &= 298 \\ T_L(1) &= 293 & R(1) &= 1.0 & x_A(1) &= 0.0 \end{aligned}$$

Calculate the profiles of all dependent variables using the shooting method.

5.10 A plug-flow reactor is to be designed to produce the product D from A according to the following reaction:

$$A \rightarrow D \qquad r_D = 60C_A \text{ mole } D/\text{L.s}$$

In the operating condition of this reactor, the following undesired reaction also takes place:

$$A \rightarrow U \qquad r_U = \frac{0.003C_A}{1 + 10^5 C_A} \text{ mole } U/\text{L.s}$$

The undesired product U is a pollutant and it costs 10 \$/mol U to dispose it, whereas the desired product D has a value of 35 \$/mol D. What size of reactor should be chosen in order to obtain an effluent stream at its maximum value?

Pure reactant A with volumetric flow rate of 15 L/s and molar flow rate of 0.1 mol/s enters the reactor. Value of A is 5 \$/mol A.

REFERENCES

1. Finlayson, B. A., *Nonlinear Analysis in Chemical Engineering*, McGraw-Hill, New York, 1980.

2. Fogler, H. S., *Elements of Chemical Reaction Engineering*, 3rd ed., Prentice Hall, Upper Saddle River, NJ, 1999.

3. Kubíček, M., and Hlavàček, V., *Numerical Solution of Nonlinear Boundary Value Problems with Applications*, Prentice Hall, New York, 1975.

4. Aziz, A. K. (ed.), *Numerical Solutions of Boundary Value Problems for Ordinary Differential Equations*, Academic, New York, 1975.

5. Carreau, P. J., De Kee, D. C. R., and Chhabra, R. P., *Rheology of Polymeric Systems: Principles and Applications*, Hanser, Munich, Germany, 1998.

6. Constantinides, A., Spencer, J. L., and Gaden, E. L., Jr., "Optimization of Batch Fermentation Processes, I. Development of Mathematical Models for Batch Penicillin Fermentations," *Biotech. Bioeng.*, vol. 12, 1970, p. 803.

7. Constantinides, A., Spencer, J. L., and Gaden, E. L., Jr., "Optimization of Batch Fermentation Processes, II. Optimum Temperature Profiles for Batch Penicillin Fermentations," *Biotech. Bioeng.*, vol. 12, 1970, p. 1081.

8. Lapidus, L., *Digital Computation for Chemical Engineering*, McGraw-Hill, New York, 1962.

9. Lapidus, L., and Sienfeld, J. H., *Numerical Solution of Ordinary Differential Equations*, Academic, New York, 1971.

10. Rai, V. R., and Constantinides, A., "Mathematical Modeling and Optimization of the Gluconic Acid Fermentation," *AIChE Symp. Ser.*, vol. 69, no. 132, 1973, p. 114.

11. Lotka, A. J., *Elements of Mathematical Biology*, Dover, Inc., New York, 1956.

12. Elliot, J. L., "Isle Royale: A North Woods Park Primeval," *National Geographic*, vol. 167, April 1985, p. 534.

13. Allen, D. L., *Wolves of Minong*, Houghton Mifflin, Boston, 1973.

14. Peterson, R. O., *The Wolves of Isle Royale, A Broken Balance*, Willow Creek Press, Minocqua, WI, 1995.

15. Peterson, R. O., *Ecological Studies of Wolves on Isle Royale,* Annual Reports, Michigan Technological University, Houghton, MI, 1984–1998.

16. Vandermeer, J., *Elementary Mathematical Ecology*, Wiley, New York, 1981.

CHAPTER 6

Numerical Solution of Partial Differential Equations

6.1 Introduction

The laws of conservation of mass, momentum, and energy form the basis of the field of transport phenomena. These laws applied to the flow of fluids result in the *equations of change,* which describe the change of velocity, temperature, and concentration with respect to time and position in the system. The dynamics of such systems, which have more than one independent variable, are modeled by *partial differential equations*. For example, the mass balance:

$$\begin{pmatrix} \text{Rate of mass} \\ \text{accumulation} \end{pmatrix} = \begin{pmatrix} \text{Rate of} \\ \text{mass in} \end{pmatrix} - \begin{pmatrix} \text{Rate of} \\ \text{mass out} \end{pmatrix} \tag{6.1}$$

applied to a stationary volume element $\Delta x\Delta y\Delta z$, through which pure fluid is flowing (Fig. 6.1) results in the *equation of continuity*[1]:

$$\frac{\partial \rho}{\partial t} = -\left(\frac{\partial}{\partial x}\rho v_x + \frac{\partial}{\partial y}\rho v_y + \frac{\partial}{\partial z}\rho v_z\right) \tag{6.2}$$

where ρ is the density of the fluid, and v_x, v_y, and v_z are the velocity components in the three rectangular coordinates.

The application of a momentum balance:

$$\begin{pmatrix}\text{Rate of}\\ \text{momentum}\\ \text{accumulation}\end{pmatrix} = \begin{pmatrix}\text{Rate of}\\ \text{momentum}\\ \text{in}\end{pmatrix} - \begin{pmatrix}\text{Rate of}\\ \text{momentum}\\ \text{out}\end{pmatrix} + \begin{pmatrix}\text{Sum of forces}\\ \text{acting on}\\ \text{system}\end{pmatrix} \tag{6.3}$$

on the volume element $\Delta x\Delta y\Delta z$, for isothermal flow of fluid, yields the *equation of motion* in the three directions:

$$\frac{\partial}{\partial t}\rho v_j = -\left(\frac{\partial}{\partial x}\rho v_x v_j + \frac{\partial}{\partial y}\rho v_y v_j + \frac{\partial}{\partial z}\rho v_z v_j\right)$$

$$-\left(\frac{\partial}{\partial x}\tau_{xj} + \frac{\partial}{\partial y}\tau_{yj} + \frac{\partial}{\partial z}\tau_{zj}\right) - \frac{\partial p}{\partial j} + \rho g_j \qquad j = x, y, \text{or } z \tag{6.4}$$

where τ_{ij} are the components of the shear-stress tensor, p is pressure, and g_j are the components of the gravitational acceleration.

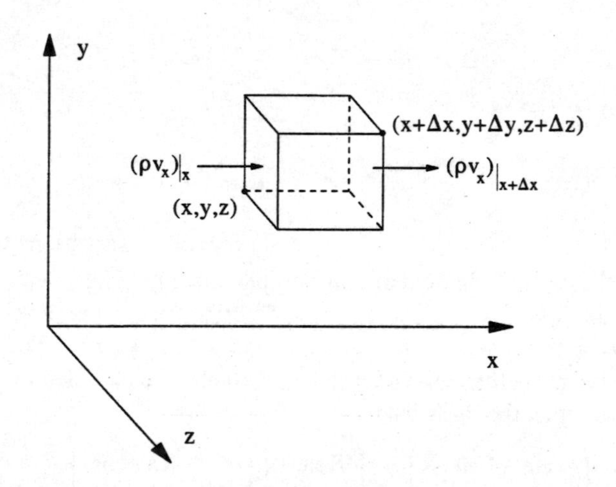

Figure 6.1 Volume element $\Delta x\Delta y\Delta z$ for fluid flow.

[1] For detailed derivation of these equations see Ref. [1].

The application of the following energy balance:

$$\begin{pmatrix}\text{Rate of}\\ \text{accumulation}\\ \text{of energy}\end{pmatrix} = \begin{pmatrix}\text{Rate of}\\ \text{energy in}\\ \text{by convection}\end{pmatrix} - \begin{pmatrix}\text{Rate of}\\ \text{energy out}\\ \text{by convection}\end{pmatrix}$$

$$+ \begin{pmatrix}\text{Net rate of}\\ \text{heat addition}\\ \text{by conduction}\end{pmatrix} - \begin{pmatrix}\text{Net rate of work}\\ \text{done by system}\\ \text{on surroundings}\end{pmatrix} \tag{6.5}$$

on the volume element $\Delta x \Delta y \Delta z$, for nonisothermal flow of fluid, results in the *equation of energy:*

$$\rho C_v \left(\frac{\partial T}{\partial t} + v_x \frac{\partial T}{\partial x} + v_y \frac{\partial T}{\partial y} + v_z \frac{\partial T}{\partial z} \right) = -\left(\frac{\partial q_x}{\partial x} + \frac{\partial q_y}{\partial y} + \frac{\partial q_z}{\partial z} \right)$$

$$-T\left(\frac{\partial p}{\partial T}\right)_\rho \left(\frac{\partial v_x}{\partial x} + \frac{\partial v_y}{\partial y} + \frac{\partial v_z}{\partial z} \right) - \left(\tau_{xx}\frac{\partial v_x}{\partial x} + \tau_{yy}\frac{\partial v_y}{\partial y} + \tau_{zz}\frac{\partial v_z}{\partial z} \right)$$

$$-\left[\tau_{xy}\left(\frac{\partial v_x}{\partial y} + \frac{\partial v_y}{\partial x} \right) + \tau_{xz}\left(\frac{\partial v_x}{\partial z} + \frac{\partial v_z}{\partial x} \right) + \tau_{yz}\left(\frac{\partial v_y}{\partial z} + \frac{\partial v_z}{\partial y} \right) \right] \tag{6.6}$$

where T is the temperature, C_v is the heat capacity at constant volume, and q_i are the components of the energy flux given by *Fourier's law of heat conduction*:

$$q_i = -k\frac{\partial T}{\partial i} \qquad i = x, y, \text{or } z \tag{6.7}$$

where k is the thermal conductivity.

For heat conduction in solids, where the velocity terms are zero, Eq. (6.6) simplifies considerably. When combined with Eq. (6.7), it gives the well-known three-dimensional unsteady-state heat conduction equation

$$\rho C_p \frac{\partial T}{\partial t} = k\left(\frac{\partial^2 T}{\partial x^2} + \frac{\partial^2 T}{\partial y^2} + \frac{\partial^2 T}{\partial z^2} \right) \tag{6.8}$$

where C_p, the heat capacity at constant pressure, replaces C_v, and k has been assumed to be constant within the solid.

The equation of continuity for component A in a binary mixture (components A and B) of constant fluid density ρ and constant diffusion coefficient D_{AB} is

$$\frac{\partial c_A}{\partial t} + \left(v_x \frac{\partial c_A}{\partial x} + v_y \frac{\partial c_A}{\partial y} + v_z \frac{\partial c_A}{\partial z} \right) = D_{AB}\left(\frac{\partial^2 c_A}{\partial x^2} + \frac{\partial^2 c_A}{\partial y^2} + \frac{\partial^2 c_A}{\partial z^2} \right) + R_A \tag{6.9}$$

where c_A = molar concentration of A, and R_A = molar rate of production of component A. This equation reduces to *Fick's second law of diffusion* when $R_A = 0$ and $v_x = v_y = v_z = 0$:

$$\frac{\partial c_A}{\partial t} = D_{AB}\left(\frac{\partial^2 c_A}{\partial x^2} + \frac{\partial^2 c_A}{\partial y^2} + \frac{\partial^2 c_A}{\partial z^2} \right) \tag{6.10}$$

Eq. (6.10) is the three-dimensional unsteady-state diffusion equation, which has the same form as the respective heat conduction equation (6.8).

The most commonly encountered partial differential equations in chemical engineering are of first and second order. Our discussion in this chapter focuses on these two categories. In the next two sections, we attempt to classify these equations and their boundary conditions, and in the remainder of the chapter we develop the numerical methods, using finite difference and finite element analysis, for the numerical solution of first- and second-order partial differential equations.

6.2 Classification of Partial Differential Equations

Partial differential equations are classified according to their *order*, *linearity*, and *boundary conditions*.

The order of a partial differential equation is determined by the highest-order partial derivative present in that equation. Examples of first-, second-, and third-order partial differential equations are:

First order:
$$\frac{\partial u}{\partial x} - \alpha \frac{\partial u}{\partial y} = 0 \tag{6.11}$$

Second order:
$$\frac{\partial^2 u}{\partial x^2} + u \frac{\partial u}{\partial y} = 0 \tag{6.12}$$

Third order:
$$\left(\frac{\partial^3 u}{\partial x^3} \right)^2 + \frac{\partial^2 u}{\partial x \partial y} + \frac{\partial u}{\partial y} = 0 \tag{6.13}$$

Partial differential equations are categorized into *linear*, *quasilinear*, and *nonlinear* equations. Consider, for example, the following second-order equation:

$$a(.)\frac{\partial^2 u}{\partial y^2} + 2b(.)\frac{\partial^2 u}{\partial x \partial y} + c(.)\frac{\partial^2 u}{\partial x^2} + d(.) = 0 \tag{6.14}$$

If the coefficients are constants or functions of the independent variables only $[(.) \equiv (x, y)]$, then Eq. (6.14) is linear. If the coefficients are functions of the dependent variable and/or any of its derivatives of lower order than that of the differential equation $[(.) \equiv (x, y, u, \partial u/\partial x, \partial u/\partial y)]$, then the equation is quasilinear. Finally, if the coefficients are functions of derivatives of the same order as that of the equation $[(.) \equiv (x, y, u, \partial^2 u/\partial x^2, \partial u^2/\partial y^2, \partial^2 u/\partial x\partial y)]$, then the equation is nonlinear. In accordance with these definitions, Eq. (6.11) is linear, (6.12) is quasilinear, and (6.13) is nonlinear.

Linear second-order partial differential equations in two independent variables are further classified into three canonical forms: *elliptic*, *parabolic*, and *hyperbolic*. The general form of this class of equations is

$$a\frac{\partial^2 u}{\partial x^2} + 2b\frac{\partial^2 u}{\partial x \partial y} + c\frac{\partial^2 u}{\partial y^2} + d\frac{\partial u}{\partial x} + e\frac{\partial u}{\partial y} + fu + g = 0 \tag{6.15}$$

where the coefficients are either constants or functions of the independent variables only. The three canonical forms are determined by the following criterion:

$$b^2 - ac < 0, \quad \text{elliptic} \tag{6.16a}$$

$$b^2 - ac = 0, \quad \text{parabolic} \tag{6.16b}$$

$$b^2 - ac > 0, \quad \text{hyperbolic.} \tag{6.16c}$$

If $g = 0$, then Eq. (6.15) is a *homogeneous* differential equation.

The classic examples of second-order partial differential equations that conform to the three canonical forms are

Laplace's equation (elliptic):

$$\frac{\partial^2 u}{\partial x^2} + \frac{\partial^2 u}{\partial y^2} = 0 \tag{6.17}$$

Heat conduction or diffusion equation (parabolic):

$$\alpha\frac{\partial^2 u}{\partial x^2} = \frac{\partial u}{\partial t} \tag{6.18}$$

Wave equation (hyperbolic):

$$a^2\frac{\partial^2 u}{\partial x^2} = \frac{\partial^2 u}{\partial t^2} \tag{6.19}$$

A similar classification for second-order partial differential equations with *three* independent variables is given by Tychonov and Samarski [2]. This classification includes elliptic, parabolic, hyperbolic, and *ultrahyperbolic*. The majority of partial differential equations in engineering and physics are of second-order with two, three, or four independent variables. Most of these equations have canonical forms; however, the names elliptic, parabolic, and hyperbolic have been also applied to equations that are not of second-order but which possess similar properties [3].

The methods of solution of partial differential equations depend on their canonical form, as will be demonstrated in the rest of this chapter. Because the coefficients of these equations can be functions of the independent variables, it is possible that an equation may shift from one canonical form to another over the range of integration of (x, y).

6.3 Initial and Boundary Conditions

The initial and boundary conditions associated with the partial differential equations must be specified in order to obtain unique numerical solutions to these equations. In general, boundary conditions for partial differential equations are divided into three categories. These are demonstrated below, using the one-dimensional unsteady-state heat conduction equation

$$\alpha \frac{\partial^2 T}{\partial x^2} = \frac{\partial T}{\partial t} \tag{6.20}$$

This is identical to Eq. (6.18). It is derived from Eq. (6.8) by assuming that the temperature gradients in the y and z dimensions are zero. Eq. (6.20) essentially describes the change in temperature within a solid slab (e.g., the wall of a furnace), where heat transfer takes place in the x-direction (see Fig. 6.2).

Following are the three categories of conditions:

Dirichlet conditions (first kind): The values of the dependent variable are given at fixed values of the independent variable. Examples of Dirichlet conditions for the heat conduction equation are

$$T = f(x) \qquad \text{at } t = 0 \text{ and } 0 \le x \le 1$$

or

$$T = T_0 \qquad \text{at } t = 0 \text{ and } 0 \le x \le 1$$

These are alternative initial conditions that specify that the initial temperature inside the slab (wall) is a function of position $f(x)$ or a constant T_0 (Fig. 6.2*a*).

Boundary conditions of the first kind are expressed as

$$T = f(t) \qquad \text{at } x = 0 \text{ and } t > 0$$

and

$$T = T_1 \qquad \text{at } x = 1 \text{ and } t > 0$$

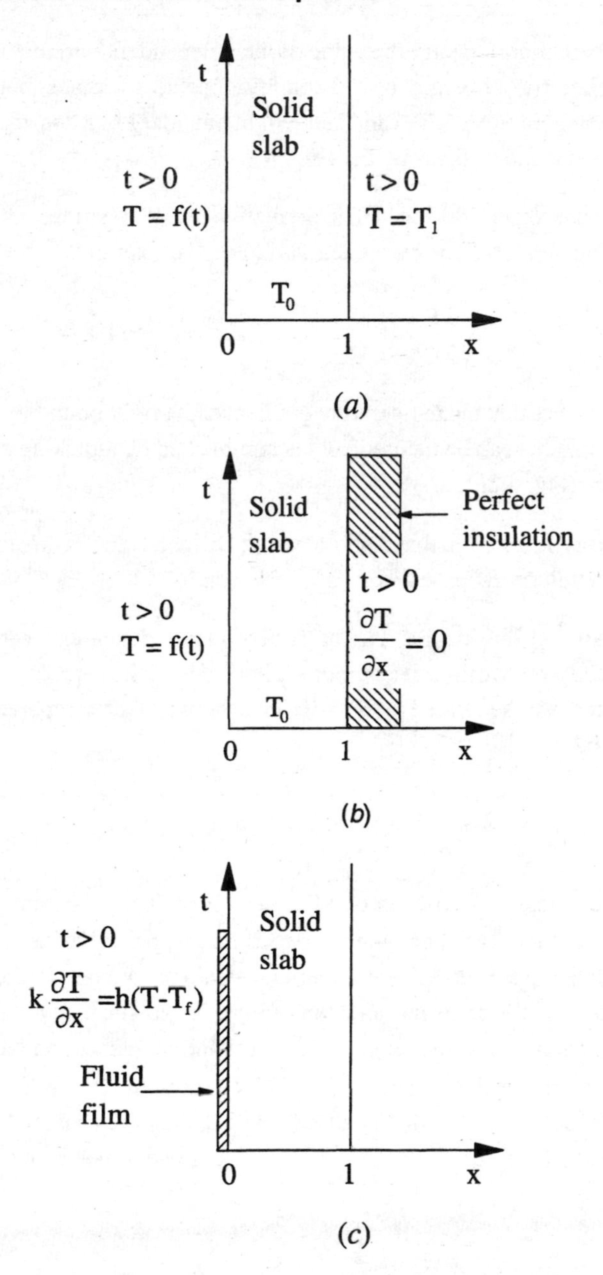

Figure 6.2 Examples of initial and boundary conditions for the heat conduction problem. (*a*) Dirichlet conditions. (*b*) Cauchy conditions (Dirichlet and Neumann). (*c*) Robbins condition.

These boundary conditions specify the value of the independent variable at the left boundary as a function of time $f(t)$ (this may be the condition inside a furnace that is maintained at a preprogrammed temperature profile) and at the right boundary as a constant T_1 (e.g., the room temperature at the outside of the furnace) (Fig. 6.2*a*).

Neumann conditions (second kind): The derivative of the dependent variable is given as a constant or as a function of the independent variable. For example:

$$\frac{\partial T}{\partial x} = 0 \qquad \text{at } x = 1 \text{ and } t \geq 0$$

This condition specifies that the temperature gradient at the right boundary is zero. In the heat conduction problem, this can be theoretically accomplished by attaching perfect insulation at the right boundary (Fig. 6.2*b*).

Cauchy conditions: A problem that combines both Dirichlet and Neumann conditions is said to have Cauchy conditions (Fig. 6.2*b*).

Robbins conditions (third kind): The derivative of the dependent variable is given as a function of the dependent variable itself. For the heat conduction problem, the heat flux at the solid-fluid interface may be related to the difference between the temperature at the interface and that in the fluid, that is,

$$k\frac{\partial T}{\partial x} = h(T - T_f) \qquad \text{at } x = 0 \text{ and } t \geq 0$$

where h is the heat transfer coefficient of the fluid (Fig. 6.2*c*).

On the basis of their initial and boundary conditions, partial differential equations may be further classified into *initial-value* or *boundary-value* problems. In the first case, at least one of the independent variables has an *open region*. In the unsteady-state heat conduction problem, the time variable has the range $0 \leq t \leq \infty$, where no condition has been specified at $t = \infty$; therefore, this is an initial-value problem. When the region is *closed* for all independent variables and conditions are specified at all boundaries, then the problem is of the boundary-value type. An example of this is the three-dimensional steady-state heat conduction problem described by the equation

$$\frac{\partial^2 T}{\partial x^2} + \frac{\partial^2 T}{\partial y^2} + \frac{\partial^2 T}{\partial z^2} = 0 \tag{6.21}$$

with the boundary conditions given at all boundaries:

$$\left.\begin{matrix} T(0,y,z) \\ T(1,y,z) \\ T(x,0,z) \\ T(x,1,z) \\ T(x,y,0) \\ T(x,y,1) \end{matrix}\right\} = \text{specified.} \tag{6.22}$$

6.4 SOLUTION OF PARTIAL DIFFERENTIAL EQUATIONS USING FINITE DIFFERENCES

In Chaps. 3 and 4, we developed the methods of finite differences and demonstrated that ordinary derivatives can be approximated, with any degree of desired accuracy, by replacing the differential operators with finite difference operators.

In this section, we apply similar procedures in expressing partial derivatives in terms of finite differences. Since partial differential equations involve more than one independent variable, we first establish two-dimensional and three-dimensional grids, in two and three independent variables, respectively, as shown in Fig. 6.3.

The notation (i, j) is used to designate the pivot point for the two-dimensional space and (i, j, k) for the three-dimensional space, where i, j, and k are the counters in the x, y, and z directions, respectively. For unsteady-state problems, in which time is one of the independent variables, the counter n is used to designate the time dimension. In order to keep the notation as simple as possible, we add subscripts only when needed.

The distances between grid points are designated as Δx, Δy, and Δz. When time is one of the independent variables, the time step is shown by Δt.

We now express first, second, and mixed partial derivatives in terms of finite differences. We show the development of these approximations using central differences, and in addition we summarize in tabular form the formulas obtained from using forward and backward differences.

The partial derivative of u with respect to x implies that y and z are held constant; therefore:

$$\left.\frac{\partial u}{\partial x}\right|_{i,j,k} \equiv \left.\frac{du}{dx}\right|_{i,j,k} \tag{6.23}$$

Using Eq. (4.50), which is the approximation of the first-order derivative in terms of central differences, and converting it to the three-dimensional space, we obtain

$$\left.\frac{\partial u}{\partial x}\right|_{i,j,k} = \frac{1}{2\Delta x}(u_{i+1,j,k} - u_{i-1,j,k}) + O(\Delta x^2) \tag{6.24}$$

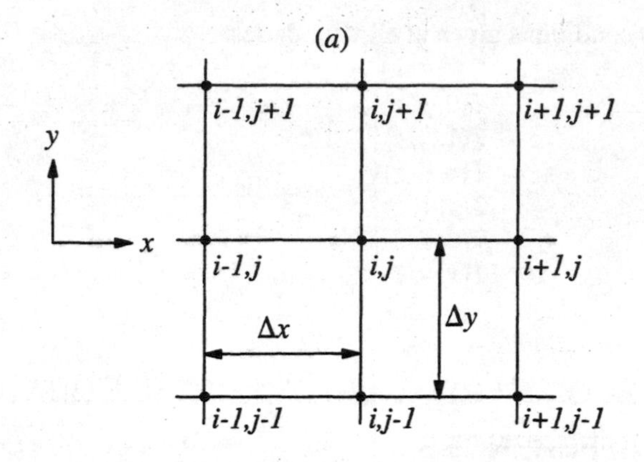

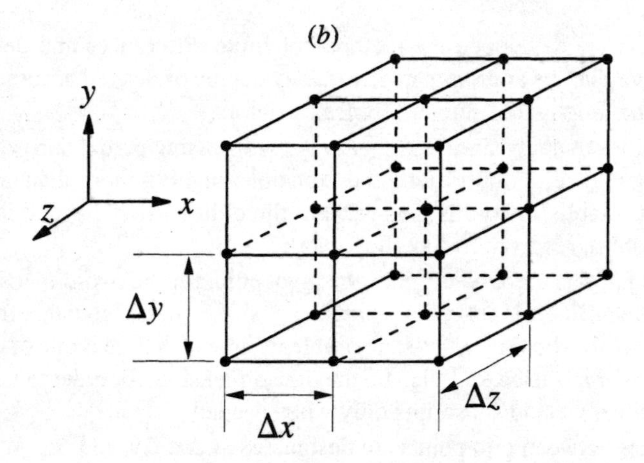

Figure 6.3 Finite difference grids. (*a*) Two-dimensional grid. (*b*) Three-dimensional grid.

Similarly, the first-order partial derivatives in the *y*- and *z*-directions are given by

$$\frac{\partial u}{\partial y}\bigg|_{i,j,k} = \frac{1}{2\Delta y}(u_{i,j+1,k} - u_{i,j-1,k}) + O(\Delta y^2) \tag{6.25}$$

$$\frac{\partial u}{\partial z}\bigg|_{i,j,k} = \frac{1}{2\Delta z}(u_{i,j,k+1} - u_{i,j,k-1}) + O(\Delta z^2) \tag{6.26}$$

In an analogous manner, the second-order partial derivatives are expressed in terms of central differences by using Eq. (4.54):

$$\frac{\partial^2 u}{\partial x^2}\bigg|_{i,j,k} = \frac{1}{\Delta x^2}(u_{i+1,j,k} - 2u_{i,j,k} + u_{i-1,j,k}) + O(\Delta x^2) \tag{6.27}$$

$$\frac{\partial^2 u}{\partial y^2}\Big|_{i,j,k} = \frac{1}{\Delta y^2}(u_{i,j+1,k} - 2u_{i,j,k} + u_{i,j-1,k}) + O(\Delta y^2) \tag{6.28}$$

$$\frac{\partial^2 u}{\partial z^2}\Big|_{i,j,k} = \frac{1}{\Delta z^2}(u_{i,j,k+1} - 2u_{i,j,k} + u_{i,j,k-1}) + O(\Delta z^2) \tag{6.29}$$

Finally, the mixed partial derivative is developed as follows:

$$\frac{\partial^2 u}{\partial y\,\partial x}\Big|_{i,j,k} = \frac{\partial}{\partial y}\left[\frac{\partial u}{\partial x}\Big|_{i,j,k}\right] \tag{6.30}$$

This is equivalent to applying $\partial u/\partial x$ at points $(i, j+1, k)$ and $(i, j-1, k)$, so

$$\begin{aligned}\frac{\partial^2 u}{\partial y\,\partial x}\Big|_{i,j,k} &\\ &= \frac{1}{2\Delta y}\left[\frac{1}{2\Delta x}(u_{i+1,j+1,k} - u_{i-1,j+1,k}) - \frac{1}{2\Delta x}(u_{i+1,j-1,k} - u_{i-1,j-1,k})\right] + O(\Delta x^2 + \Delta y^2)\\ &= \frac{1}{4\Delta x\Delta y}(u_{i+1,j+1,k} - u_{i-1,j+1,k} - u_{i+1,j-1,k} + u_{i-1,j-1,k}) + O(\Delta x^2 + \Delta y^2)\end{aligned} \tag{6.31}$$

The above central difference approximations of partial derivatives are summarized in Table 6.1. The corresponding approximations obtained from using forward and backward differences are shown in Tables 6.2 and 6.3, respectively. Equivalent sets of formulas, which are more accurate than the above, may be developed by using finite difference approximations that have higher accuracies [such as Eqs. (4.59) and (4.64) for central differences, Eqs. (4.41) and (4.46) for forward differences, and Eqs. (4.24) and (4.29) for backward differences]. However, the more accurate formulas are not commonly used, because they involve a larger number of terms and require more extensive computation times.

The use of finite difference approximations is demonstrated in the following sections of this chapter in setting up the numerical solutions of elliptic, parabolic, and hyperbolic partial differential equations.

6.4.1 Elliptic Partial Differential Equations

Elliptic differential equations are often encountered in steady-state heat conduction and diffusion operations. For example, in three-dimensional steady-state heat conduction in solids, Eq. (6.8) becomes

Table 6.1 Finite difference approximations of partial derivatives using central differences

Derivative	Central difference	Error
$\frac{\partial u}{\partial x}\Big\vert_{i,j,k}$	$\frac{1}{2\Delta x}(u_{i+1,j,k} - u_{i-1,j,k})$	$O(\Delta x^2)$
$\frac{\partial u}{\partial y}\Big\vert_{i,j,k}$	$\frac{1}{2\Delta y}(u_{i,j+1,k} - u_{i,j-1,k})$	$O(\Delta y^2)$
$\frac{\partial u}{\partial z}\Big\vert_{i,j,k}$	$\frac{1}{2\Delta z}(u_{i,j,k+1} - u_{i,j,k-1})$	$O(\Delta z^2)$
$\frac{\partial^2 u}{\partial x^2}\Big\vert_{i,j,k}$	$\frac{1}{\Delta x^2}(u_{i+1,j,k} - 2u_{i,j,k} + u_{i-1,j,k})$	$O(\Delta x^2)$
$\frac{\partial^2 u}{\partial y^2}\Big\vert_{i,j,k}$	$\frac{1}{\Delta y^2}(u_{i,j+1,k} - 2u_{i,j,k} + u_{i,j-1,k})$	$O(\Delta y^2)$
$\frac{\partial^2 u}{\partial z^2}\Big\vert_{i,j,k}$	$\frac{1}{\Delta z^2}(u_{i,j,k+1} - 2u_{i,j,k} + u_{i,j,k-1})$	$O(\Delta z^2)$
$\frac{\partial^2 u}{\partial y \partial x}\Big\vert_{i,j,k}$	$\frac{1}{4\Delta x \Delta y}(u_{i+1,j+1,k} - u_{i-1,j+1,k} - u_{i+1,j-1,k} + u_{i-1,j-1,k})$	$O(\Delta x^2 + \Delta y^2)$

$$\frac{\partial^2 T}{\partial x^2} + \frac{\partial^2 T}{\partial y^2} + \frac{\partial^2 T}{\partial z^2} = 0 \tag{6.21}$$

Similarly, Fick's second law of diffusion [Eq. (6.10)] simplifies to

$$\frac{\partial^2 c_A}{\partial x^2} + \frac{\partial^2 c_A}{\partial y^2} + \frac{\partial^2 c_A}{\partial z^2} = 0 \tag{6.32}$$

when steady state is assumed.

We begin our discussion of numerical solutions of elliptic differential equations by first examining the two-dimensional problem in its general form (Laplace's equation):

$$\frac{\partial^2 u}{\partial x^2} + \frac{\partial^2 u}{\partial y^2} = 0 \tag{6.17}$$

Table 6.2 Finite difference approximations of partial derivatives using forward differences

Derivative	Forward difference	Error
$\frac{\partial u}{\partial x}\Big\vert_{i,j,k}$	$\frac{1}{\Delta x}(u_{i+1,j,k} - u_{i,j,k})$	$O(\Delta x)$
$\frac{\partial u}{\partial y}\Big\vert_{i,j,k}$	$\frac{1}{\Delta y}(u_{i,j+1,k} - u_{i,j,k})$	$O(\Delta y)$
$\frac{\partial u}{\partial z}\Big\vert_{i,j,k}$	$\frac{1}{\Delta z}(u_{i,j,k+1} - u_{i,j,k})$	$O(\Delta z)$
$\frac{\partial^2 u}{\partial x^2}\Big\vert_{i,j,k}$	$\frac{1}{\Delta x^2}(u_{i+2,j,k} - 2u_{i+1,j,k} + u_{i,j,k})$	$O(\Delta x)$
$\frac{\partial^2 u}{\partial y^2}\Big\vert_{i,j,k}$	$\frac{1}{\Delta y^2}(u_{i,j+2,k} - 2u_{i,j+1,k} + u_{i,j,k})$	$O(\Delta y)$
$\frac{\partial^2 u}{\partial z^2}\Big\vert_{i,j,k}$	$\frac{1}{\Delta z^2}(u_{i,j,k+2} - 2u_{i,j,k+1} + u_{i,j,k})$	$O(\Delta z)$
$\frac{\partial^2 u}{\partial y \partial x}\Big\vert_{i,j,k}$	$\frac{1}{\Delta x \Delta y}(u_{i+1,j+1,k} - u_{i,j+1,k} - u_{i+1,j,k} + u_{i,j,k})$	$O(\Delta x + \Delta y)$

We replace each second-order partial derivative by its approximation in central differences, Eqs. (6.27) and (6.28), to obtain

$$\frac{1}{\Delta x^2}(u_{i+1,j,k} - 2u_{i,j,k} + u_{i-1,j,k}) + \frac{1}{\Delta y^2}(u_{i,j+1,k} - 2u_{i,j,k} + u_{i,j-1,k}) = 0 \tag{6.33}$$

which rearranges to

$$-2\left(\frac{1}{\Delta x^2} + \frac{1}{\Delta y^2}\right)u_{i,j} + \left(\frac{1}{\Delta x^2}\right)u_{i+1,j} + \left(\frac{1}{\Delta x^2}\right)u_{i-1,j} + \left(\frac{1}{\Delta y^2}\right)u_{i,j+1} + \left(\frac{1}{\Delta y^2}\right)u_{i,j-1} = 0 \tag{6.34}$$

This is a linear algebraic equation involving the value of the dependent variable at five adjacent grid points.

A rectangular-shaped object divided into p segments in the x-direction and q segments in the y-direction has $(p + 1) \times (q + 1)$ total grid points and $(p - 1) \times (q - 1)$

Table 6.3 Finite difference approximations of partial derivatives using backward differences

Derivative	Backward difference	Error
$\frac{\partial u}{\partial x}\Big\vert_{i,j,k}$	$\frac{1}{\Delta x}(u_{i,j,k} - u_{i-1,j,k})$	$O(\Delta x)$
$\frac{\partial u}{\partial y}\Big\vert_{i,j,k}$	$\frac{1}{\Delta y}(u_{i,j,k} - u_{i,j-1,k})$	$O(\Delta y)$
$\frac{\partial u}{\partial z}\Big\vert_{i,j,k}$	$\frac{1}{\Delta z}(u_{i,j,k} - u_{i,j,k-1})$	$O(\Delta z)$
$\frac{\partial^2 u}{\partial x^2}\Big\vert_{i,j,k}$	$\frac{1}{\Delta x^2}(u_{i,j,k} - 2u_{i-1,j,k} + u_{i-2,j,k})$	$O(\Delta x)$
$\frac{\partial^2 u}{\partial y^2}\Big\vert_{i,j,k}$	$\frac{1}{\Delta y^2}(u_{i,j,k} - 2u_{i,j-1,k} + u_{i,j-2,k})$	$O(\Delta y)$
$\frac{\partial^2 u}{\partial z^2}\Big\vert_{i,j,k}$	$\frac{1}{\Delta z^2}(u_{i,j,k} - 2u_{i,j,k-1} + u_{i,j,k-2})$	$O(\Delta z)$
$\frac{\partial^2 u}{\partial y \partial x}\Big\vert_{i,j,k}$	$\frac{1}{\Delta x \Delta y}(u_{i,j,k} - u_{i,j-1,k} - u_{i-1,j,k} + u_{i-1,j-1,k})$	$O(\Delta x + \Delta y)$

internal grid points. Eq. (6.34), written for each of the internal points, constitutes a set of $(p-1) \times (q-1)$ simultaneous linear algebraic equations in $(p+1) \times (q+1) - 4$ unknowns (the four corner points do not appear in these equations). The boundary conditions provide the additional information for the solution of the problem. If the boundary conditions are of Dirichlet type the values of the dependent variable are known at all the external grid points. On the other hand, if the boundary conditions at any of the external surfaces are of the Neumann or Robbins type, which specify partial derivatives at the boundaries, these conditions must also be replaced by finite difference approximations.

We demonstrate this by specifying a Neumann condition at the left boundary, that is,

$$\frac{\partial u}{\partial x} = \beta \qquad \text{at } x = 0 \text{ and all } y \tag{6.35}$$

where β is a constant. Replacing the partial derivative in Eq. (6.35) with a central difference approximation, we obtain

$$\frac{1}{2\Delta x}(u_{i+1,j} - u_{i-1,j}) = \beta \tag{6.36}$$

This is valid only at $x = 0$ where $i = 0$; therefore, Eq. (6.36) becomes

$$u_{-1,j} = u_{1,j} - 2\beta\Delta x \tag{6.37}$$

The points (-1, j) are located outside the object; therefore, $u_{-1,j}$ have fictitious values. Their calculation, however, is necessary for the evaluation of the Neumann boundary condition. Eq. (6.37), written for all y ($j = 0, 1, \ldots, q$), provides ($q + 1$) additional equations but at the same time introduces ($q + 1$) additional variables. To counter this, Eq. (6.34) is also written for ($q + 1$) points along this boundary (at $x = 0$), thus providing the necessary number of independent equations for the solution of the problem.

Replacing the partial derivative in Eq. (6.35) with a forward difference does not require the use of fictitious points. However, it is important to use the forward difference formula with the same accuracy as the other equations. In this case, Eq. (4.41) should be used for evaluation of the partial derivative at $x = 0$ ($i = 0$):

$$\frac{1}{2\Delta x}(-u_{2,j} + 4u_{1,j} - 3u_{0,j}) = \beta \tag{6.38}$$

or

$$-3u_{0,j} + 4u_{1,j} - u_{2,j} = 2\beta\Delta x \tag{6.39}$$

Eq. (6.39) provides ($q + 1$) additional equations without introducing additional variables.

In the case of Robbins condition at the left boundary in the form

$$\frac{\partial u}{\partial x} = \beta + \gamma u \qquad \text{at } x = 0 \text{ and all } y \tag{6.40}$$

where β and γ are constants, a similar derivation as above shows that the following equation should be used at the boundary:

$$-(3 + 2\gamma\Delta x)u_{0,j} + 4u_{1,j} - u_{2,j} = 2\beta\Delta x \tag{6.41}$$

Eq. (6.34) and the appropriate boundary conditions constitute a set of linear algebraic equations, so the Gauss methods for the solution of such equations may be used. Eq. (6.34) is actually a predominantly diagonal system; therefore, the Gauss-Seidel method (see Sec. 2.7) is especially suitable for the solution of this problem. Rearranging Eq. (6.34) to solve for $u_{i,j}$:

$$u_{i,j} = \frac{\dfrac{1}{\Delta x^2}(u_{i+1,j} + u_{i-1,j}) + \dfrac{1}{\Delta y^2}(u_{i,j+1} + u_{i,j-1})}{2\left(\dfrac{1}{\Delta x^2} + \dfrac{1}{\Delta y^2}\right)} \tag{6.42}$$

which can be used in the iterative Gauss-Seidel substitution method. An initial estimate of all $u_{i,j}$ is needed, but this can be easily obtained from averaging the Dirichlet boundary conditions.

The Gauss-Seidel method is guaranteed to converge for a predominantly diagonal system of equations. However, its convergence may be quit slow in the solution of elliptic differential equations. The *overrelaxation* method can be used to accelerate the rate of the convergence. This technique applies the following weighting algorithm in evaluating the new values of $u_{i,j}$ at each iteration of the Gauss-Seidel method:

$$(u_{i,j})_{new} = w(u_{i,j})_{from\ Eq.\ (6.42)} + (1-w)(u_{i,j})_{old} \tag{6.43}$$

Special care should be taken when processing the nodes at the boundaries, if $u_{i,j}$ at these nodes is calculated by a different method of finite differences. In such a case, when calculating the new value of $u_{i,j}$, the proper equation should be applied instead of Eq. (6.42). The *relaxation parameter* w can be assigned values from the following ranges:

$0 < w < 1$ for underrelaxation

$1 < w \le 2$ for overrelaxation

When $w = 1$, this method is exactly the same as the unmodified Gauss-Seidel. Methods for estimating the optimal w are given by Lapidus and Pinder [4], who also show that the overrelaxation method is five- to one-hundred times faster (depending on step size and convergence criterion) than the Gauss-Seidel method.

In the case when an equidistant grid can be used, that is, when $\Delta x = \Delta y$, Eq. (6.42) simplifies to

$$u_{i,j} = \frac{u_{i+1,j} + u_{i-1,j} + u_{i,j+1} + u_{i,j-1}}{4} \tag{6.44}$$

which simply shows that the value of the dependent variable at the pivotal point (i, j) in the Laplace equation is the arithmetic average of the values at the grid points to the right and left of and above and below the pivot point. This is demonstrated by the computational molecule of Fig. 6.4, which is sometimes referred to as a "5-point star."

The three-dimensional elliptic partial differential equation

$$\frac{\partial^2 u}{\partial x^2} + \frac{\partial^2 u}{\partial y^2} + \frac{\partial^2 u}{\partial z^2} = 0 \tag{6.45}$$

can be similarly converted to linear algebraic equations using finite difference approximations in three-dimensional space. Applying Eqs. (6.27)-(6.29) to replace the three partial derivatives of Eq. (6.45), we obtain

$$\frac{1}{\Delta x^2}(u_{i+1,j,k} - 2u_{i,j,k} + u_{i-1,j,k}) + \frac{1}{\Delta y^2}(u_{i,j+1,k} - 2u_{i,j,k} + u_{i,j-1,k})$$

$$+ \frac{1}{\Delta z^2}(u_{i,j,k+1} - 2u_{i,j,k} + u_{i,j,k-1}) = 0 \tag{6.46}$$

For the equidistant grid ($\Delta x = \Delta y = \Delta z$), the above equation reduces to

$$u_{i,j,k} = \frac{u_{i+1,j,k} + u_{i-1,j,k} + u_{i,j+1,k} + u_{i,j-1,k} + u_{i,j,k+1} + u_{i,j,k-1}}{6} \tag{6.47}$$

In parallel with the two-dimensional case, the value of the dependent variable at the pivot point (i, j, k) is the arithmetic average of the values at the grid points adjacent to the pivot point. The computational molecule for the three-dimensional elliptic equation is shown in Fig. 6.5.

The *nonhomogeneous* form of the Laplace equation is the *Poisson* equation

$$\frac{\partial^2 u}{\partial x^2} + \frac{\partial^2 u}{\partial y^2} = f(x, y) \tag{6.48}$$

which also belongs to the class of elliptic partial differential equations. A form of the Poisson equation

$$\frac{\partial^2 T}{\partial x^2} + \frac{\partial^2 T}{\partial y^2} = -\frac{Q'(x,y)}{k} \tag{6.49}$$

is used to describe heat conduction in a two-dimensional solid plate with an internal heat source. $Q'(x, y)$ is the heat generated per unit volume per time and k is the thermal conductivity of the material. The finite difference formulation of the Poisson equation is

$$-2\left(\frac{1}{\Delta x^2} + \frac{1}{\Delta y^2}\right)u_{i,j} + \left(\frac{1}{\Delta x^2}\right)u_{i+1,j} + \left(\frac{1}{\Delta x^2}\right)u_{i-1,j} + \left(\frac{1}{\Delta y^2}\right)u_{i,j+1} + \left(\frac{1}{\Delta y^2}\right)u_{i,j-1} = f_{i,j} \tag{6.50}$$

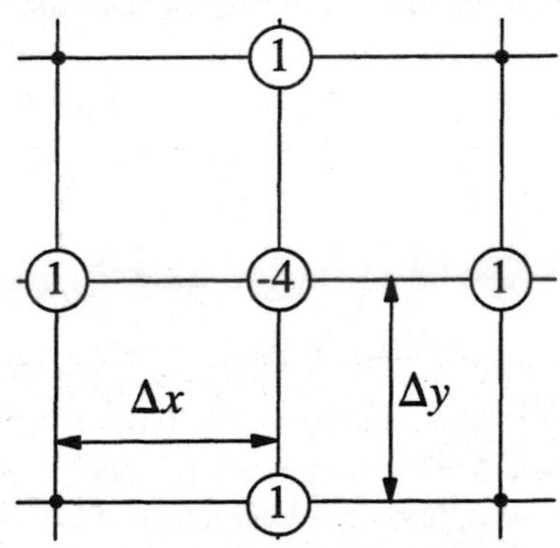

Figure 6.4 Computational molecule for the Laplace equation using equidistant grid. The number in each circle is the coefficient of that point in the difference equation.

or

$$u_{i,j} = \frac{\frac{1}{\Delta x^2}(u_{i+1,j} + u_{i-1,j}) + \frac{1}{\Delta y^2}(u_{i,j+1} + u_{i,j-1})}{2\left(\frac{1}{\Delta x^2} + \frac{1}{\Delta y^2}\right)} - \frac{f_{i,j}}{2\left(\frac{1}{\Delta x^2} + \frac{1}{\Delta y^2}\right)} \tag{6.51}$$

The numerical solution of the Laplace and Poisson elliptic partial differential equations is demonstrated in Example 6.1.

Example 6.1: Solution of the Laplace and Poisson Equations. Write a general MATLAB function to determine the numerical solution of a two-dimensional elliptic partial differential equation of the general form:

$$\frac{\partial^2 u}{\partial x^2} + \frac{\partial^2 u}{\partial y^2} = f$$

for a rectangular object of variable width and height. The object could have Dirichlet, Neumann, or Robbins boundary conditions. The value of f should be assumed to be a constant. Use this function to find the solution of the following problems ($u = T$):

(*a*) A thin square metal plate of dimensions 1 m × 1 m is subjected to four heat sources which maintain the temperature on its four edges as follows:

$$T(0, y) = 250°\text{C}$$

$$T(1, y) = 100°\text{C}$$

$$T(x, 0) = 500°\text{C}$$

$$T(x, 1) = 25°\text{C}$$

The flat sides of the plate are insulated so that no heat is transferred through these sides. Calculate the temperature profiles within the plate.

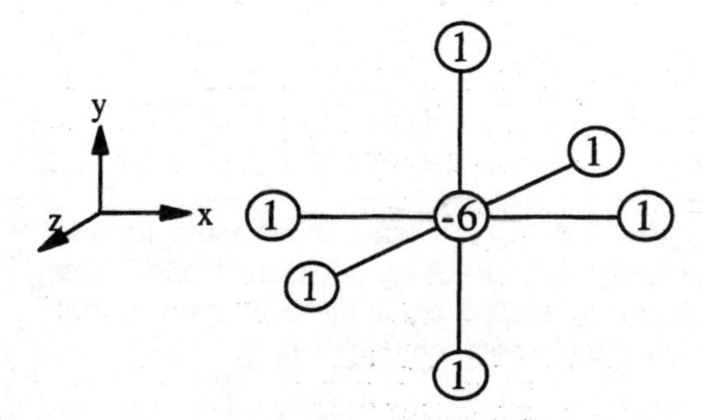

Figure 6.5 Computational molecule for the three-dimensional elliptic differential equation using equidistant grid.

(*b*) Perfect insulation is installed on two edges (right and top) of the plate of part (*a*). The other two edges are maintained at constant temperatures. The set of Dirichlet and Neumann boundary conditions is

$$T(0, y) = 250^\circ\text{C}$$

$$\frac{\partial T}{\partial x}\Big|_{1,y} = 0$$

$$T(x, 0) = 500^\circ\text{C}$$

$$\frac{\partial T}{\partial x}\Big|_{x,1} = 0$$

Calculate the temperature profiles within the plate and compare these with the results of part (*a*).

(*c*) The thin metal plate of part (*a*) is made of an alloy that has a melting point of 800°C and a thermal conductivity of 16 W/m.K. The plate is subject to an electric current that creates a uniform heat source within the plate. The amount of heat generated is $Q' = 100$ kW/m^3. All four edges of the plate are in contact with a fluid at 25°C. The set of Robbins boundary conditions is

$$\frac{\partial T}{\partial x}\Big|_{0,y} = 5[T(0, y) - 25]$$

$$\frac{\partial T}{\partial x}\Big|_{1,y} = 5[25 - T(1, y)]$$

$$\frac{\partial T}{\partial y}\Big|_{x,0} = 5[T(x, 0) - 25]$$

$$\frac{\partial T}{\partial y}\Big|_{x,1} = 5[25 - T(x, 1)]$$

Examine the temperature profiles within the plate to ascertain whether the alloy will begin to melt under these conditions.

Method of Solution: We solve this problem by matrix inversion because matrix operations are much faster than element-by-element operations in MATLAB, especially when a large number of equations are to be solved. In order to solve the set of equations in matrix format, $u_{i,j}$ values have to be rearranged as a column vector. Therefore, we put in order all the dependent variables in a vector and renumber them from 1 to $(p + 1)(q + 1)$, as illustrated in Fig. E6.1*a*. Using this single numbering system, Eq. (6.50) can be written as

$$-2\left(\frac{1}{\Delta x^2} + \frac{1}{\Delta y^2}\right) u_n + \left(\frac{1}{\Delta x^2}\right) u_{n+1} + \left(\frac{1}{\Delta x^2}\right) u_{n-1} + \left(\frac{1}{\Delta y^2}\right) u_{n+p+1} + \left(\frac{1}{\Delta y^2}\right) u_{n-p-1} = f$$

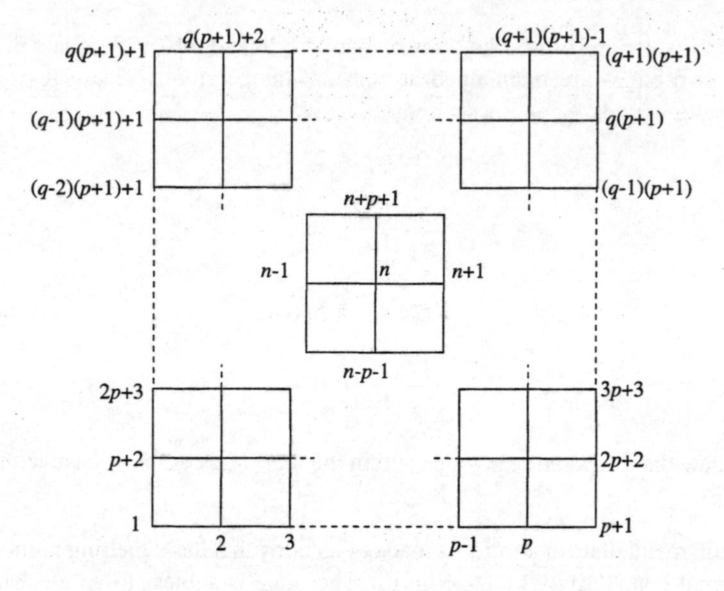

Figure E6.1*a* Scheme of renumbering values of u_{ij} to convert to a column vector.

When the Laplace equation is being solved, f is zero, which makes Eq. (6.50) equivalent to Eq. (6.34). For the Poisson equation, the value of f is assumed constant throughout the plate.

At each boundary, if the condition is of the Dirichlet type, the values of u remain unchanged, provided that the related equation at this point is

$$u_N = \text{constant}$$

where N is a node at the boundary with Dirichlet condition. However, if the condition is of the Neumann or Robbins type, forward or backward difference is used for evaluating the first-order derivative of the order $O(h^2)$ at the boundaries:

Lower x boundary (forward differences):

$$\frac{\partial u}{\partial x}\bigg|_{x=0} = \frac{1}{2\Delta x}(-3u_N + 4u_{N+1} - u_{N+2})$$

where N is a node on the line $x = 0$.

Upper x boundary (backward differences):

$$\frac{\partial u}{\partial x}\bigg|_{x=L} = \frac{1}{2\Delta x}(3u_N - 4u_{N-1} + u_{N-2})$$

where N is a node on the line $x = \text{L}$.

Lower y boundary (forward differences):

$$\frac{\partial u}{\partial y}\bigg|_{y=0} = \frac{1}{2\Delta y}(-3u_N + 4u_{N+p+1} - u_{N+2p+2})$$

where N is a node on the line $y = 0$.

Upper y boundary (backward differences):

$$\frac{\partial u}{\partial y}\bigg|_{y=L} = \frac{1}{2\Delta y}(3u_N - 4u_{N-p-1} + u_{N-2p-2})$$

where N is a node on the line $y = L$.

The value of the dependent variable at the four corner grid points cannot be calculated by this method. The arithmetic average of the values of the dependent variables at the two adjacent points on the boundaries is assigned to the dependent variable at each corner point.

Program Description: The MATLAB function *elliptic.m* is written for the solution of the Laplace equation, or the Poisson equation with constant right hand side value, for a rectangular plate. The first part of the function is initialization and checking of the input arguments. If the last input argument, which introduces the constant of the right-hand side of the elliptic equation to the function, is omitted, it is assumed to be zero and the function solves Laplace's equation. The input argument that carries boundary conditions should consist of four rows, one for each boundary. In each row, the first element is a flag that indicates the type of condition (1 for Dirichlet, 2 for Neumann, and 3 for Robbins), followed by the boundary condition value or parameters.

Next in the function are several sections dealing with building the matrix of coefficients and the vector of constants according to what is discussed in the method of solution. Finally, the function calculates all the u values by matrix inversion method using the built-in MATLAB function *inv*. The outputs of *elliptic.m* are the vectors of x and y and the matrix of $u_{i,j}$ values.

The main program *Example6_1.m* asks the user to input all the necessary parameters for solving the elliptic equation from the keyboard. It then calls the function *elliptic.m* to solve the equation and finally plots the results in a three-dimensional graph.

Program

Example6_1.m

```
% Example6_1.m
% Solution to Example 6.1. This program calculates and plots
% the temperature profiles of a rectangular plate by solving
% Laplace or Poisson equation by finite difference method,
% using the function ELLIPTIC.M.
```

```
clear
clc
clf
disp(' Solution of elliptic partial differential equation.')
bcdialog = [' Lower x boundary condition:'
   ' Upper x boundary condition:'
   ' Lower y boundary condition:'
   ' Upper y boundary condition:'];

redo = 1;
while redo
   disp(' ')
   length = input(' Length of the plate (x-direction) (m) = ');
   width = input(' Width of the plate  (y-direction) (m) = ');
   p = input(' Number of divisions in x-direction = ');
   q = input(' Number of divisions in y-direction = ');
   f = input(' Right hand side of the equation (f) = ');
   disp(' '), disp(' ')
   disp(' Boundary conditions:')
   for k = 1:4
      disp(' ')
      disp(bcdialog(k,:))
      disp(' 1 - Dirichlet')
      disp(' 2 - Neumann')
      disp(' 3 - Robbins')
      bc(k,1) = input(' Enter your choice : ');
      if bc(k,1) < 3
         bc(k,2) = input(' Value = ');
      else
         disp(' u'' = (beta) + (gamma)*u')
         bc(k,2) = input(' Constant    (beta)  = ');
         bc(k,3) = input(' Coefficient (gamma) = ');
      end
   end
   [x,y,T] = elliptic(p,q,length/p,width/q,bc,f);
   surf(y,x,T)
   xlabel('y (m)')
   ylabel('x (m)')
   zlabel('T (deg C)')
   colorbar
   view(135,45)
   disp(' ')
   redo = input(' Repeat calculations (0/1) ? ');
   clc
end
```

elliptic.m

```
function [x,y,U] = elliptic(nx,ny,dx,dy,bc,f)
```

```
%ELLIPTIC solution of a two-dimensional elliptic partial
%   differential equation
%
%   [X,Y,U]=ELLIPTIC(NX,NY,DX,DY,BC) solves the Laplace
%   equation for a rectangular object where
%      X = vector of x values
%      Y = vector of y values
%      U = matrix of dependent variable [U(X,Y)]
%      NX = number of divisions in x-direction
%      NY = number of divisions in y-direction
%      DX = x-increment
%      DY = y-increment
%      BC is a matrix of 4x2 or 4x3 containing the types
%      and values of boundary conditions. The order of
%      appearing boundary conditions are lower x, upper x,
%      lower y, and upper y in rows 1 to 4 of the matrix
%      BC, respectively. The first column of BC determines
%      the type of condition:
%         1 for Dirichlet condition, followed by the set
%         value of U in the second column.
%         2 for Neumann condition, followed by the set value
%         of U' in the second column.
%         3 for Robbins condition, followed by the constant
%         and the coefficient of U in the second and third
%         columns, respectively.
%
%   [X,Y,U]=ELLIPTIC(NX,NY,DX,DY,BC,F) solves the Poisson
%   equation for a rectangular object where F is the constant
%   at the right-hand side of the elliptic partial differential
%   equation.
%
%   See also ADAPTMESH, ASSEMPDE, PDENONLIN, POISOLV

% (c) N. Mostoufi & A. Constantinides
% January 1, 1999

% Initialization
if nargin < 5
   error(' Invalid number of inputs.')
end

[a,b]=size(bc);
if a ~= 4
   error(' Invalid number of boundary conditions.')
end
```

```
if b < 2 | b > 3
   error(' Invalid boundary condition.')
end
if b == 2 & max(bc(:,1)) <= 2
   bc = [bc zeros(4,1)];
end

if nargin < 6 | isempty(f)
   f = 0;
end

nx = fix(nx);
x = [0:nx]*dx;
ny = fix(ny);
y = [0:ny]*dy;
dx2 = 1/dx^2;
dy2 = 1/dy^2;

% Building the matrix of coefficients and the vector of constants
n = (nx+1)*(ny+1);
A = zeros(n);
c = zeros(n,1);
onex = diag(diag(ones(nx-1)));
oney = diag(diag(ones(ny-1)));

% Internal nodes
i = [2:nx];
for j = 2:ny
   ind = (j-1)*(nx+1)+i;
   A(ind,ind) = -2*(dx2+dy2)*onex;
   A(ind,ind+1) = A(ind,ind+1) + dx2*onex;
   A(ind,ind-1) = A(ind,ind-1) + dx2*onex;
   A(ind,ind+nx+1) = A(ind,ind+nx+1) + dy2*onex;
   A(ind,ind-nx-1) = A(ind,ind-nx-1) + dy2*onex;
   c(ind) = f*ones(nx-1,1);
end

% Lower x boundary condition
switch bc(1,1)
case 1
   ind = ([2:ny]-1)*(nx+1)+1;
   A(ind,ind) = A(ind,ind) + oney;
   c(ind) = bc(1,2)*ones(ny-1,1);
case {2, 3}
   ind = ([2:ny]-1)*(nx+1)+1;
```

```
    A(ind,ind) = A(ind,ind) - (3/(2*dx) + bc(1,3))*oney;
    A(ind,ind+1) = A(ind,ind+1) + 2/dx*oney;
    A(ind,ind+2) = A(ind,ind+2) - 1/(2*dx)*oney;
    c(ind) = bc(1,2)*ones(ny-1,1);
end

% Upper x boundary condition
switch bc(2,1)
case 1
    ind = [2:ny]*(nx+1);
    A(ind,ind) = A(ind,ind) + oney;
    c(ind) = bc(2,2)*ones(ny-1,1);
case {2, 3}
    ind = [2:ny]*(nx+1);
    A(ind,ind) = A(ind,ind) + (3/(2*dx) - bc(2,3))*oney;
    A(ind,ind-1) = A(ind,ind-1) - 2/dx*oney;
    A(ind,ind-2) = A(ind,ind-2) + 1/(2*dx)*oney;
    c(ind) = bc(2,2)*ones(ny-1,1);
end

% Lower y boundary condition
switch bc(3,1)
case 1
    ind = [2:nx];
    A(ind,ind) = A(ind,ind) + onex;
    c(ind) = bc(3,2)*ones(nx-1,1);
case {2, 3}
    ind = [2:nx];
    A(ind,ind) = A(ind,ind) - (3/(2*dy) + bc(3,3))*onex;
    A(ind,ind+nx+1) = 2/dy*onex;
    A(ind,ind+2*(nx+1)) = -1/(2*dy)*onex;
    c(ind) = bc(3,2)*ones(nx-1,1);
end

% Upper y boundary condition
switch bc(4,1)
case 1
    ind = ny*(nx+1)+[2:nx];
    A(ind,ind) = A(ind,ind) + onex;
    c(ind) = bc(4,2)*ones(nx-1,1);
case {2, 3}
    ind = ny*(nx+1)+[2:nx];
    A(ind,ind) = A(ind,ind) + (3/(2*dy) - bc(4,3))*onex;
    A(ind,ind-(nx+1)) = A(ind,ind-(nx+1)) - 2/dy*onex;
    A(ind,ind-2*(nx+1)) = A(ind,ind-2*(nx+1)) + 1/(2*dy)*onex;
```

```
   c(ind) = bc(4,2)*ones(nx-1,1);
end

% Corner points
A(1,1) = 1;
A(1,2) = -1/2;
A(1,nx+2) = -1/2;
c(1) = 0;

A(nx+1,nx+1) = 1;
A(nx+1,nx) = -1/2;
A(nx+1,2*(nx+1)) = -1/2;
c(nx+1) = 0;

A(ny*(nx+1)+1,ny*(nx+1)+1) = 1;
A(ny*(nx+1)+1,ny*(nx+1)+2) = -1/2;
A(ny*(nx+1)+1,(ny-1)*(nx+1)+1) = -1/2;
c(ny*(nx+1)+1) = 0;

A(n,n) = 1;
A(n,n-1) = -1/2;
A(n,n-(nx+1)) = -1/2;
c(n) = 0;

u = inv(A)*c; % Solving the set of equations

% Rearranging the final results into matrix format
for k = 1:ny+1
   U(1:nx+1,k) = u((k-1)*(nx+1)+1:k*(nx+1));
end
```

Input and Results[2]

```
>>Example6_1

 Solution of elliptic partial differential equation.

 Length of the plate (x-direction) (m) = 1
 Width of the plate  (y-direction) (m) = 1
 Number of divisions in x-direction = 20
```

[2] Users of the Student Edition of MATLAB will encounter an array size limitation if they use 20 divisions in each direction. They should use 10 divisions instead.

```
Number of divisions in y-direction = 20
Right-hand side of the equation (f) = 0

Boundary conditions:

Lower x boundary condition:
1 - Dirichlet
2 - Neumann
3 - Robbins
Enter your choice : 1
Value = 250

Upper x boundary condition:
1 - Dirichlet
2 - Neumann
3 - Robbins
Enter your choice : 1
Value = 100

Lower y boundary condition:
1 - Dirichlet
2 - Neumann
3 - Robbins
Enter your choice : 1
Value = 500

Upper y boundary condition:
1 - Dirichlet
2 - Neumann
3 - Robbins
Enter your choice : 1
Value = 25

Repeat calculations (0/1) ? 1

Length of the plate (x-direction) (m) = 1
Width of the plate  (y-direction) (m) = 1
Number of divisions in x-direction = 20
Number of divisions in y-direction = 20
Right-hand side of the equation (f) = 0

Boundary conditions:

Lower x boundary condition:
1 - Dirichlet
2 - Neumann
3 - Robbins
Enter your choice : 1
```

```
Value = 250

Upper x boundary condition:
1 - Dirichlet
2 - Neumann
3 - Robbins
Enter your choice : 2
Value = 0

Lower y boundary condition:
1 - Dirichlet
2 - Neumann
3 - Robbins
Enter your choice : 1
Value = 500

Upper y boundary condition:
1 - Dirichlet
2 - Neumann
3 - Robbins
Enter your choice : 2
Value = 0

Repeat calculations (0/1) ? 1

Length of the plate (x-direction) (m) = 1
Width of the plate  (y-direction) (m) = 1
Number of divisions in x-direction = 20
Number of divisions in y-direction = 20
Right-hand side of the equation (f) = -100e3/16

Boundary conditions:

Lower x boundary condition:
1 - Dirichlet
2 - Neumann
3 - Robbins
Enter your choice : 3
u' = (beta) + (gamma)*u
Constant    (beta)  = -5*25
Coefficient (gamma) = 5

Upper x boundary condition:
1 - Dirichlet
2 - Neumann
3 - Robbins
Enter your choice : 3
u' = (beta) + (gamma)*u
```

```
Constant      (beta)    = 5*25
Coefficient (gamma) = -5

Lower y boundary condition:
1 - Dirichlet
2 - Neumann
3 - Robbins
Enter your choice : 3
u' = (beta) + (gamma)*u
Constant      (beta)    = -5*25
Coefficient (gamma) = 5

Upper y boundary condition:
1 - Dirichlet
2 - Neumann
3 - Robbins
Enter your choice : 3
u' = (beta) + (gamma)*u
Constant      (beta)    = 5*25
Coefficient (gamma) = -5

Repeat calculations (0/1) ? 0
```

Discussion of Results: *Part* (*a*) By entering $f = 0$ as input to the program, the Laplace equation with Dirichlet boundary conditions is solved, and the graphical result is shown in Fig. E6.1*b*.

Part (*b*) The result of this part is shown in Fig. E6.1*c*. The effect of insulation on the right and top edges of the plate is evident. The gradient of the temperature near these boundaries approaches zero to satisfy the imposed boundary conditions. Because the insulation stops the flow of heat through these boundaries, the temperature along the insulated edges is higher than that of part (*a*).

Part (*c*) The Poisson equation is solved with a Poisson constant determined from Eq. (6.49):

$$f = -\frac{Q'}{k} = -\frac{100000}{16}$$

The solution is shown in Fig. E6.1*d*. It can be seen from this figure that the temperature within the plate rises sharply to its highest value of 824.9°C at the center point. Under these circumstances, the metal will begin to melt at the center core. Increasing the heat removed from the edges by convection, by either lowering the fluid temperature or increasing the heat transfer coefficient, lowers the internal temperature and can prevent melting of the plate.

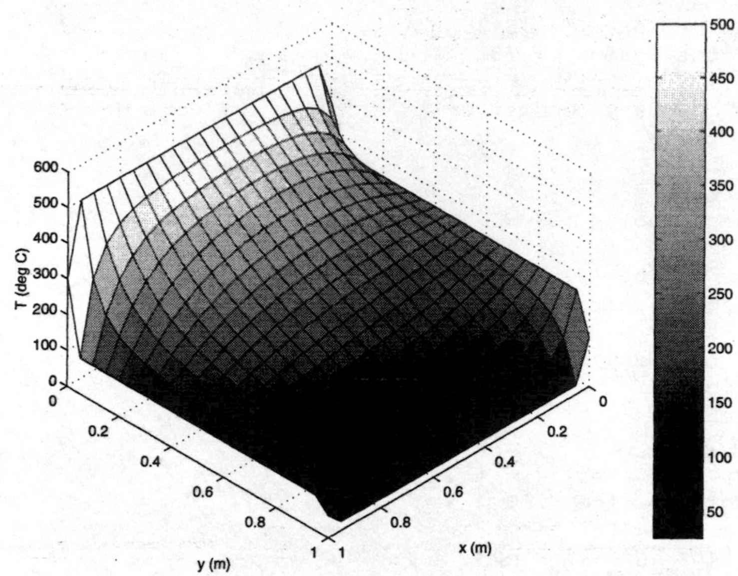

Figure E6.1*b* Solution of the Laplace equation with Dirichlet conditions.

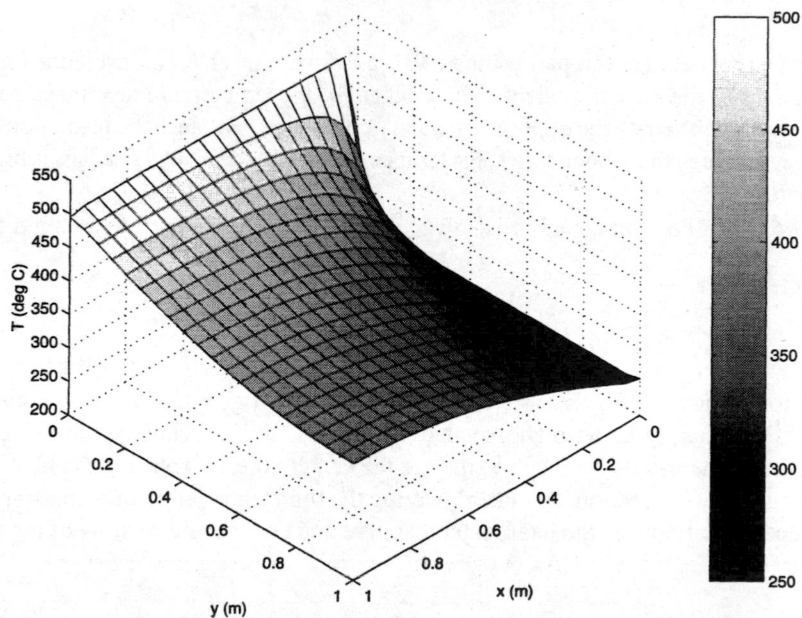

Figure E6.1*c* Solution of the Laplace equation with Neumann conditions (perfect insulation).

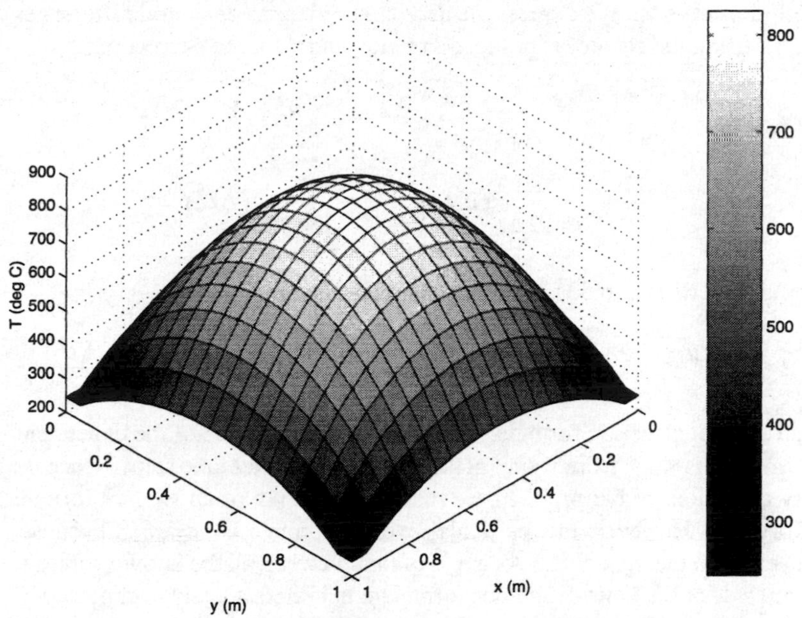

Figure E6.1*d* Solution of the Poisson equation with Robbins conditions.

6.4.2 Parabolic Partial Differential Equations

Classic examples of parabolic differential equations are the unsteady-state heat conduction equation

$$\alpha\left(\frac{\partial^2 T}{\partial x^2} + \frac{\partial^2 T}{\partial y^2} + \frac{\partial^2 T}{\partial z^2}\right) = \frac{\partial T}{\partial t} \tag{6.52}$$

and Fick's second law of diffusion

$$D_{AB}\left(\frac{\partial^2 c_A}{\partial x^2} + \frac{\partial^2 c_A}{\partial y^2} + \frac{\partial^2 c_A}{\partial z^2}\right) = \frac{\partial c_A}{\partial t} \tag{6.10}$$

with Dirichlet, Neumann, or Cauchy boundary conditions.

Let us consider this class of equations in the general one-dimensional form:

$$\frac{\partial u}{\partial t} = \alpha\frac{\partial^2 u}{\partial x^2} \tag{6.18}$$

In this section, we develop several methods of solution of Eq. (6.18) using finite differences.

Explicit methods: We express the derivatives in terms of central differences around the point (i, n), using the counter i for the x-direction and n for the t-direction:

$$\frac{\partial^2 u}{\partial x^2}\bigg|_{i,n} = \frac{1}{\Delta x^2}(u_{i+1,n} - 2u_{i,n} + u_{i-1,n}) + O(\Delta x^2) \tag{6.53}$$

$$\frac{\partial u}{\partial t}\bigg|_{i,n} = \frac{1}{2\Delta t}(u_{i,n+1} - u_{i,n-1}) + O(\Delta t^2) \tag{6.54}$$

Combining Eqs. (6.18), (6.53), and (6.54) and rearranging:

$$u_{i,n+1} = u_{i,n-1} + \frac{2\alpha\Delta t}{\Delta x^2}(u_{i+1,n} - 2u_{i,n} + u_{i-1,n}) + O(\Delta x^2 + \Delta t^2) \tag{6.55}$$

This is an *explicit* algebraic formula, which calculates the value of the dependent variable at the next time step ($u_{j,n+1}$) from values at the current and earlier time steps. Once the initial and boundary conditions of the problem are specified, solution of an explicit formula is usually straightforward. However, this particular explicit formula is ***unstable***, because it contains negative terms on the right side.[3] As a rule of thumb, when all the known values are arranged on the right side of the finite difference formulation, if there are any negative coefficients, the solution is unstable. This is stated more precisely by the positivity rule [5]: "For

$$u_{i,n+1} = Au_{i+1,n} + Bu_{i,n} + Cu_{i-1,n} \tag{6.56}$$

if A, B, C are positive and $A + B + C \le 1$, then the numerical scheme is stable."

In order to eliminate the instability problem, we replace the first-order derivative in Eq. (6.18) with the forward difference:

$$\frac{\partial u}{\partial t}\bigg|_{i,n} = \frac{1}{\Delta t}(u_{i,n+1} - u_{i,n}) + O(\Delta t) \tag{6.57}$$

Combining Eqs. (6.18), (6.53), and (6.57) we obtain the explicit formula:

$$u_{i,n+1} = \left(\frac{\alpha\Delta t}{\Delta x^2}\right)u_{i+1,n} + \left(1 - 2\frac{\alpha\Delta t}{\Delta x^2}\right)u_{i,n} + \left(\frac{\alpha\Delta t}{\Delta x^2}\right)u_{i-1,n} + O(\Delta x^2 + \Delta t) \tag{6.58}$$

For a stable solution, the positivity rule requires that

$$1 - 2\frac{\alpha\Delta t}{\Delta x^2} \ge 0 \tag{6.59}$$

Rearranging Eq. (6.59), we get

$$\frac{\alpha\Delta t}{\Delta x^2} \le \frac{1}{2} \tag{6.60}$$

---

[3] A rigorous discussion of stability analysis is given in Sec. 6.5.

This inequality determines the relationship between the two integration steps, Δx in the x-direction and Δt in the t-direction. As Δx gets smaller, Δt becomes much smaller, thus requiring longer computation times.

If we choose to work with the equality part of Eq. (6.59) or (6.60), that is,

$$\frac{\alpha \Delta t}{\Delta x^2} = \frac{1}{2} \tag{6.61}$$

then Eq. (6.58) simplifies to

$$u_{i,n+1} = \frac{1}{2}(u_{i+1,n} + u_{i-1,n}) + O(\Delta x^2 + \Delta t) \tag{6.62}$$

This explicit formula calculates the value of the dependent variable at position i of the next time step $(n + 1)$ from values to the right and left of i at the present time step n. The computational molecule for this equation is shown in Fig. 6.6.

It should be emphasized that using the forward difference for the first-order derivative introduces the error of order $O(\Delta t)$; therefore, Eq. (6.58) is of order $O(\Delta t)$ in the time direction and $O(\Delta x^2)$ in the x-direction. However, the advantage of gaining stability outweighs the loss of accuracy in this case.

The finite difference solution to the nonhomogeneous parabolic equation

$$\frac{\partial u}{\partial t} = \alpha \frac{\partial^2 u}{\partial x^2} + f(x,t) \tag{6.63}$$

is given by the following explicit formula

$$u_{i,n+1} = \left(\frac{\alpha \Delta t}{\Delta x^2}\right) u_{i+1,n} + \left(1 - 2\frac{\alpha \Delta t}{\Delta x^2}\right) u_{i,n} + \left(\frac{\alpha \Delta t}{\Delta x^2}\right) u_{i-1,n} + (\Delta t) f_{i,n} \tag{6.64}$$

We encounter equations of the type in Eq. (6.63) when there is a source or sink in the physical problem.

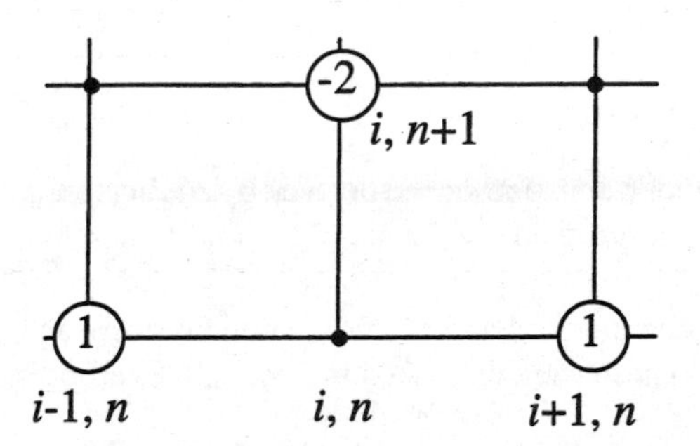

Figure 6.6 Computational molecule for Eq. (6.62).

The same treatment for the two-dimensional parabolic formula:

$$\frac{\partial u}{\partial t} = \alpha\left(\frac{\partial^2 u}{\partial x^2} + \frac{\partial^2 u}{\partial y^2}\right) + f(x, y, t) \tag{6.65}$$

results in

$$u_{i,j,n+1} = \left(\frac{\alpha \Delta t}{\Delta x^2}\right)(u_{i+1,j,n} + u_{i-1,j,n}) + \left(\frac{\alpha \Delta t}{\Delta y^2}\right)(u_{i,j+1,n} + u_{i,j-1,n})$$

$$+ \left(1 - 2\frac{\alpha \Delta t}{\Delta x^2} - 2\frac{\alpha \Delta t}{\Delta y^2}\right) u_{i,j,n} + (\Delta t) f_{i,j,n} \tag{6.66}$$

The stability condition is obtained from the positivity rule:

$$1 - 2\alpha \Delta t\left(\frac{1}{\Delta x^2} + \frac{1}{\Delta y^2}\right) \geq 0 \tag{6.67}$$

which can be rearranged to

$$\frac{1}{\Delta x^2} + \frac{1}{\Delta y^2} \leq \frac{1}{2\alpha \Delta t} \tag{6.68}$$

We also know that

$$(\Delta x^2 - \Delta y^2)^2 \geq 0 \tag{6.69}$$

By adding $4\Delta x^2 \Delta y^2$ to both sides of (6.69), we get

$$(\Delta x^2 + \Delta y^2)^2 \geq 4\Delta x^2 \Delta y^2 \tag{6.70}$$

or

$$\frac{4}{\Delta x^2 + \Delta y^2} \leq \frac{1}{\Delta x^2} + \frac{1}{\Delta y^2} \tag{6.71}$$

Combining inequalities (6.68) and (6.71) followed by further rearrangement simplifies the stability condition to

$$\frac{\alpha \Delta t}{\Delta x^2 + \Delta y^2} \leq \frac{1}{8} \tag{6.72}$$

The formula for the three-dimensional parabolic equation can be derived by adding to Eq. (6.66) the terms that come from $\partial^2 u/\partial z^2$. The right-hand side of the stability condition in this case is 1/18.

Parabolic partial differential equations can have initial and boundary conditions of the Dirichlet, Neumann, Cauchy, or Robbins type. These were discussed in Sec. 6.3. Examples

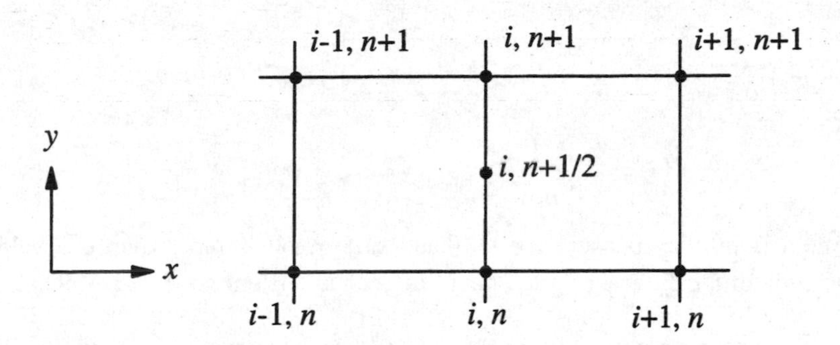

Figure 6.7 Finite difference grid for derivation of implicit formulas.

of these conditions for the heat conduction problem are demonstrated in Fig. 6.2. The boundary conditions must be discretized using the same finite difference grid as used for the differential equation. For Dirichlet conditions, this simply involves setting the values of the dependent variable along the appropriate boundary equal to the given boundary condition. For Neumann and Robbins conditions, the gradient at the boundaries must be replaced by finite difference approximations, resulting in additional algebraic equations that must be incorporated into the overall scheme of solution of the resulting set of algebraic equations.

Implicit methods: Let us now consider some implicit methods for solution of parabolic equations. We utilize the grid of Fig. 6.7, in which the half point in the t-direction $(i, n + ½)$ is shown. Instead of expressing $\partial u/\partial t$ in terms of forward difference around (i, n), as it was done in the explicit form, we express this partial derivative in terms of central difference around the half point:

$$\frac{\partial u}{\partial t}\bigg|_{i,n+½} = \frac{1}{\Delta t}(u_{i,n+1} - u_{i,n}) \tag{6.73}$$

In addition, the second-order partial derivative is expressed at the half point as a weighted average of the central differences at points $(i, n + 1)$ and (i, n):

$$\begin{aligned}\frac{\partial^2 u}{\partial x^2}\bigg|_{i,n+½} &= \theta\frac{\partial^2 u}{\partial x^2}\bigg|_{i,n+1} + (1-\theta)\frac{\partial^2 u}{\partial x^2}\bigg|_{i,n} \\ &= \theta\left[\frac{1}{\Delta x^2}(u_{i+1,n+1} - 2u_{i,n+1} + u_{i-1,n+1})\right] \\ &\quad + (1-\theta)\left[\frac{1}{\Delta x^2}(u_{i+1,n} - 2u_{i,n} + u_{i-1,n})\right]\end{aligned} \tag{6.74}$$

where θ is in the range $0 \le \theta \le 1$. A combination of Eqs. (6.18), (6.73), and (6.74) results in the *variable-weighted implicit* approximation of the parabolic partial differential equation:

$$\alpha\theta\left[\frac{1}{\Delta x^2}(u_{i+1,n+1} - 2u_{i,n+1} + u_{i-1,n+1})\right] - \frac{1}{\Delta t}u_{i,n+1}$$

$$= -\alpha(1-\theta)\left[\frac{1}{\Delta x^2}(u_{i+1,n} - 2u_{i,n} + u_{i-1,n})\right] - \frac{1}{\Delta t}u_{i,n} \tag{6.75}$$

This formula is implicit because the left-hand side involves more than one value at the $(n+1)$ position of the difference grid (that is, more than one unknown at any step in the time domain).

When $\theta = 0$, Eq. (6.75) becomes identical to the classic explicit formula Eq. (6.64). When $\theta = 1$, Eq. (6.75) becomes

$$-\left(\frac{\alpha\Delta t}{\Delta x^2}\right)u_{i-1,n+1} + \left(1 + 2\frac{\alpha\Delta t}{\Delta x^2}\right)u_{i,n+1} - \left(\frac{\alpha\Delta t}{\Delta x^2}\right)u_{i+1,n+1} = u_{i,n} \tag{6.76}$$

This is called the *backward implicit* approximation, which can also be obtained by approximating the first-order partial derivative using the backward difference at $(i, n+1)$ and the second-order partial derivative by the central difference at $(i, n+1)$.

Finally, when $\theta = ½$, Eq. (6.75) yields the well-known *Crank-Nicolson implicit formula*:

$$-\left(\frac{\alpha\Delta t}{\Delta x^2}\right)u_{i-1,n+1} + 2\left(1 + \frac{\alpha\Delta t}{\Delta x^2}\right)u_{i,n+1} - \left(\frac{\alpha\Delta t}{\Delta x^2}\right)u_{i+1,n+1}$$

$$= \left(\frac{\alpha\Delta t}{\Delta x^2}\right)u_{i-1,n} + 2\left(1 - \frac{\alpha\Delta t}{\Delta x^2}\right)u_{i,n} + \left(\frac{\alpha\Delta t}{\Delta x^2}\right)u_{i+1,n} \tag{6.77}$$

For an implicit solution to the nonhomogeneous parabolic equation

$$\frac{\partial u}{\partial t} = \alpha\frac{\partial^2 u}{\partial x^2} + f(x,t) \tag{6.63}$$

by the above method, we also need to calculate the value of f at the midpoint $(i, n+½)$ which we take as the average of the value of f at grid points $(i, n+1)$ and (i, n)

$$f_{i,n+½} = \frac{1}{2}(f_{i,n+1} + f_{i,n}) \tag{6.78}$$

Putting Eqs. (6.73), (6.74) (considering $\theta = ½$), and (6.78) into Eq. (6.63) results in

$$-\left(\frac{\alpha\Delta t}{\Delta x^2}\right)u_{i-1,n+1} + 2\left(1 + \frac{\alpha\Delta t}{\Delta x^2}\right)u_{i,n+1} - \left(\frac{\alpha\Delta t}{\Delta x^2}\right)u_{i+1,n+1} - (\Delta t)f_{i,n+1}$$

$$= \left(\frac{\alpha\Delta t}{\Delta x^2}\right)u_{i-1,n} + 2\left(1 - \frac{\alpha\Delta t}{\Delta x^2}\right)u_{i,n} + \left(\frac{\alpha\Delta t}{\Delta x^2}\right)u_{i+1,n} + (\Delta t)f_{i,n} \tag{6.79}$$

Eq. (6.79) is the Crank-Nicolson implicit formula for the solution of the nonhomogeneous parabolic partial differential equation (6.63).

When written for the entire difference grid, implicit formulas generate sets of simultaneous linear algebraic equations whose matrix of coefficients is usually a tridiagonal matrix. This type of problem may be solved using a Gauss elimination procedure, or more efficiently using the Thomas algorithm [4], which is a variation of Gauss elimination.

Implicit formulas of the type described above have been found to be unconditionally stable. It can be generalized that most explicit finite difference approximations are conditionally stable, whereas most implicit approximations are unconditionally stable. The explicit methods, however, are computationally easier to solve than the implicit techniques.

Method of lines: Another technique for the solution of parabolic partial differential equations is the *method of lines*. This is based on the concept of converting the partial differential equation into a set of ordinary differential equations by discretizing only the spatial derivatives using finite differences and leaving the time derivatives unchanged. This concept applied to Eq. (6.18) results in

$$\frac{du_i}{dt} = \frac{\alpha}{\Delta x^2}(u_{i+1} - 2u_i + u_{i-1})$$

There will be as many of these ordinary differential equations as there are grid points in the x-direction (Fig. 6.8). The complete set of differential equations for $0 \leq i \leq N$ would be

$$\frac{du_0}{dt} = \frac{\alpha}{\Delta x^2}(u_1 - 2u_0 + u_{-1}) \tag{6.80a}$$

$$\vdots$$

$$\frac{du_i}{dt} = \frac{\alpha}{\Delta x^2}(u_{i+1} - 2u_i + u_{i-1}) \tag{6.80b}$$

$$\vdots$$

$$\frac{du_N}{dt} = \frac{\alpha}{\Delta x^2}(u_{N+1} - 2u_N + u_{N-1}) \tag{6.80c}$$

The two equations at the boundaries, (6.80*a*) and (6.80*c*), would have to be modified according to the boundary conditions that are specified in the particular problem. For example, if a Dirichlet condition is given at $x = 0$ and $t > 0$, that is,

$$u_0 = \beta \text{ (constant)} \qquad \text{for } t > 0 \tag{6.81}$$

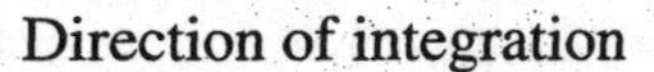

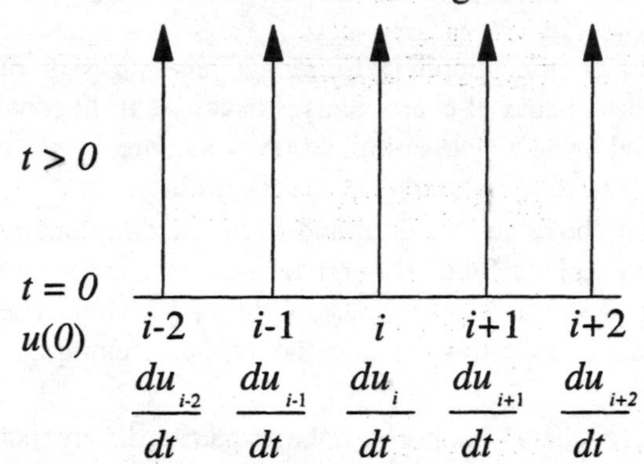

Figure 6.8 Method of lines.

then Eq. (6.80*a*) is modified to

$$\frac{du_0}{dt} = 0 \qquad u_0(0) = \beta \tag{6.82}$$

On the other hand, if a Neumann condition is given at this boundary, that is,

$$\frac{\partial u}{\partial x}\bigg|_{0,t} = 0 \qquad \text{at } x = 0 \text{ and } t > 0 \tag{6.83}$$

the partial derivative is replaced by a central difference approximation:

$$\frac{\partial u}{\partial x}\bigg|_{0,t} = \frac{u_1 - u_{-1}}{2\Delta x} = 0 \tag{6.84}$$

Then Eq. (6.80*a*) becomes

$$\frac{du_0}{dt} = \frac{\alpha}{\Delta x^2}(2u_1 - 2u_0) \tag{6.85}$$

The complete set of simultaneous differential equations must be integrated forward in time (the *n*-direction) starting with the initial conditions of the problem. This method gives stable solutions for parabolic partial differential equations.

Example 6.2: Solution of Parabolic Partial Differential Equation for Diffusion. Write a general MATLAB function to determine the numerical solution of the parabolic partial differential equation

$$\frac{\partial u}{\partial t} = \alpha\frac{\partial^2 u}{\partial x^2} + f(x,t) \tag{6.63}$$

using the Crank-Nicolson implicit formula. The function f may be a constant value or linear with respect to u. Apply this MATLAB function to solve the following problems (In this problem, $u \equiv c_A$ and z is used instead of x to indicate the length):

(*a*) The stagnant liquid B in a container that is 10 cm high (L = 10 cm) is exposed to the nonreactant gas A at time $t = 0$. The concentration of A, dissolved physically in B, reaches c_{A_0} = 0.01 mol/m^3 at the interface instantly and remains constant. The diffusion coefficient of A in B is $D_{AB} = 2\times10^{-9}$ m^2/s. Determine the evolution of concentration of A within the container. Plot the flux of A dissolved in B against time.

(*b*) Repeat part (*a*), but this time consider that A reacts with B according to the following reaction

$$A + B \rightarrow C \qquad -r_A = 2\times10^{-7}\, c_A \ \ \mathrm{mol/s.m^3}$$

Method of Solution: The physical problem is sketched in Fig. E6.2*a*. The mole balance of A for part (*a*) leads to

$$\frac{\partial c_A}{\partial t} = D_{AB}\frac{\partial^2 c_A}{\partial z^2} \tag{1}$$

The boundary and initial conditions for Eq. (1) are:

$$\text{I.C.} \qquad c_A(z,0) = 0 \qquad \text{for } z > 0 \tag{2}$$

$$\text{B.C. 1} \qquad c_A(0,t) = c_{A_0} \qquad \text{for } t \geq 0 \tag{3}$$

$$\text{B.C. 2} \qquad \frac{\partial c_A}{\partial z}\bigg|_{z=L} = 0 \qquad \text{for } t \geq 0 \tag{4}$$

For part (*b*), moles of A are consumed by the liquid B while diffusing in it. Assuming that the concentration of the product C is negligible, so that the diffusion coefficient remains unchanged, the mole balance of A results in

$$\frac{\partial c_A}{\partial t} = D_{AB}\frac{\partial^2 c_A}{\partial z^2} + kc_A \tag{5}$$

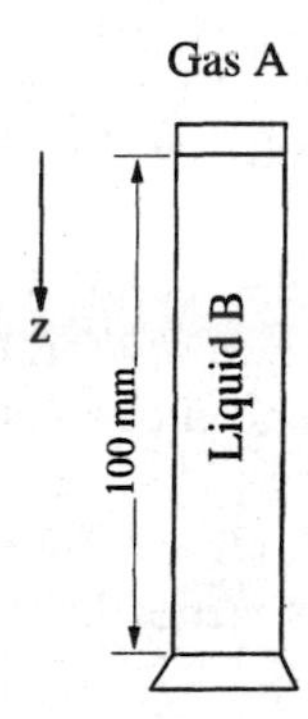

Figure E6.2*a* Diffusion of *A* through *B*.

Initial and boundary conditions remain the same [Eqs. (2)-(4)].

Once the concentration profile of A is known, the molar flux of A entering the liquid B can be calculated from Fick's law for both parts (a) and (b):

$$N_{Az}(t) = -D_{AB}\frac{\partial c_A}{\partial z}\Big|_{z=0} \tag{6}$$

The Crank-Nicolson implicit formula (6.74) is used for the solution of this problem:

$$-\left(\frac{\alpha\Delta t}{\Delta x^2}\right)u_{i-1,n+1} + 2\left(1 + \frac{\alpha\Delta t}{\Delta x^2}\right)u_{i,n+1} - \left(\frac{\alpha\Delta t}{\Delta x^2}\right)u_{i+1,n+1} - (\Delta t)f_{i,n+1}$$

$$= \left(\frac{\alpha\Delta t}{\Delta x^2}\right)u_{i-1,n} + 2\left(1 - \frac{\alpha\Delta t}{\Delta x^2}\right)u_{i,n} + \left(\frac{\alpha\Delta t}{\Delta x^2}\right)u_{i+1,n} + (\Delta t)f_{i,n} \tag{6.79}$$

Because the function f is linear with respect to u, Eq. (6.79) represents a set of linear algebraic equations that may be solved by the matrix inversion method, at each time step.

When a Neumann or Robbins condition is specified, a forward or backward finite difference approximation of the first derivative of order $O(h^2)$ is applied at the start or end point of the x-direction, respectively.

Program Description: The MATLAB function *parabolic1D.m* is developed to solve the parabolic partial differential equation in an unsteady-state one-dimensional problem. The boundary conditions are passed to the function in the same format as that of Example 6.1, with the exception that they are given in only the x-direction. The function also needs the initial condition, u_0, which is a vector containing the values of the dependent variable for all x at time $t = 0$.

The first part of the function is initialization, which checks the inputs and sets the values required in the calculations. The solution of the equation follows next and consists of an outer loop on time interval. At each time interval, the matrix of coefficients and the vector of constants of the set of Eq. (6.79) is formed. The function then solves the set of linear algebraic equations that gives the value of dependent variable in this time interval. This procedure continues until the limit time is reached. If the problem in hand is nonhomogeneous, the name of the MATLAB function containing the function f should be given as the 8th input argument. Because this function is assumed to be linear with respect to u, the set of algebraic equations (6.79) remain linear. The function corrects the matrix of coefficients and the vector of constant in this case accordingly.

The main program *Example6_2.m* asks the user to input all the necessary parameters for solving the problem from the keyboard. It then calls the function *parabolic1D.m* to solve the partial differential equation and finally plots the contour line graph of concentration profiles and the plot of molar flux of A entering the container versus time.

The function *Ex6_2_func.m* contains the rate law equation for the reaction of part (b). It is important to note that the first to third input arguments to this function have to be u, x, and t, respectively, even if one of them is not used in the function f.

Program

Example6_2.m

```
% Example6_2.m
% Solution to Example 6.2. This program calculates and plots
% the concentration profiles of a gas A diffusing in liquid B
% by solving the unsteady-state mole balance equation using
% the function PARABOLIC1D.M.

clear
redo = 1;
clc
disp(' Solution of parabolic partial differential equation.')
while redo
   disp(' ')
   h = input(' Depth of the container (m) = ');
   tmax = input(' Maximum time (s) = ');
   p = input(' Number of divisions in z-direction = ');
   q = input(' Number of divisions in t-direction = ');
   Dab = input(' Diffusion coefficient of A in B = ');
   disp(' ')
   disp(' 1 - No reaction between A and B')
   disp(' 2 - A reacts with B')
   react = input(' Enter your choice : ') - 1;
   if react
      disp(' ')
      k = input(' Rate constant = ');
      f = input(' Name of the file containing the rate law = ');
   end
   disp(' '), disp(' ')
   disp(' Boundary conditions:')
   disp(' ')
   ca0 = input(' Concentration of A at interface (mol/m3) = ');
   bc(1,1) = 1;
   bc(1,2) = ca0;
   disp(' ')
   disp(' Condition at Bottom of the container:')
   disp('    1 - Dirichlet')
   disp('    2 - Neumann')
   disp('    3 - Robbins')
   bc(2,1) = input(' Enter your choice : ');
   if bc(2,1) < 3
      bc(2,2) = input(' Value = ');
   else
      disp(' u'' = (beta) + (gamma)*u')
      bc(2,2) = input(' Constant    (beta)  = ');
      bc(2,3) = input(' Coefficient (gamma) = ');
   end
```

```
   u0 = [ca0; zeros(p,1)];
   % Calculating concentration profile
   if react
      [z,t,ca] = parabolic1D(p,q,h/p,tmax/q,Dab,u0,bc,f,k);
   else
      [z,t,ca] = parabolic1D(p,q,h/p,tmax/q,Dab,u0,bc);
   end
   % Calculating the flux of A
   Naz = -Dab*diff(ca(1:2,:))/diff(z(1:2));
   % Plotting concentration profile
   tt=[]; % Making time matrix from time vector
   for kk = 1 : p+1
      tt = [tt; t];
   end
   zz = [];   % Making height matrix from height vector
   for kk = 1 : q+1
      zz = [zz z'];
   end
   figure(1)
   [a,b]=contour(zz*1000,ca/ca0,tt/3600/24,[0:5:tmax/3600/24]);
   clabel(a,b,[10:10:tmax/3600/24])
   xlabel('z (mm)')
   ylabel('C_A/C_A_0')
   title('t (days)')
   % Plotting the unsteady-state flux
   figure(2)
   loglog(t/3600/24,Naz*3600*24)
   xlabel('t (days)')
   ylabel(' N_{Az} (mol/m^2.day)')
   disp(' ')
   redo = input(' Repeat calculations (0/1) ? ');
   clc
end
```

Ex6_2_func.m

```
function f = Ex6_2_func(ca,x,t,k)
% Function Ex6_2_func.M
% This function introduces the reaction rate equation
% used in Example 6.2.

f = -k*ca;
```

parabolic1D.m

```
function [x,t,u] = parabolic1D(nx,nt,dx,dt,alpha,u0,bc,func,...
      varargin)
%PARABOLIC1D solution of a one-dimensional parabolic partial
%   differential equation
%
```

```
%    [X,T,U]=PARABOLIC1D(NX,NT,DX,DT,ALPHA,U0,BC) solves the
%    homogeneous parabolic equation by Crank-Nicolson implicit
%    formula where
%       X = vector of x values
%       T = vector of T values
%       U = matrix of dependent variable [U(X,T)]
%       NX = number of divisions in x-direction
%       NT = number of divisions in t-direction
%       DX = x-increment
%       DY = t-increment
%       ALPHA = coefficient of equation
%       U0 = vector of U-distribution at T=0
%       BC is a matrix of 2x2 or 2x3 containing the types
%       and values of boundary conditions in x-direction.
%       The order of appearing boundary conditions are lower
%       x and upper x in rows 1 and 2 of the matrix BC,
%       respectively. The first column of BC determines the
%       type of condition:
%          1 for Dirichlet condition, followed by the set
%          value of U in the second column.
%          2 for Neumann condition, followed by the set value
%          of U' in the second column.
%          3 for Robbins condition, followed by the constant
%          and the coefficient of U in the second and third
%          column, respectively.
%
%    [X,T,U]=PARABOLIC1D(NX,NT,DX,DT,ALPHA,U0,BC,F,P1,P2,...) solves
%    the nonhomogeneous parabolic equation where F(U,X,T) is a
%    constant or linear function with respect to U, described in
%    the M-file F.M. The extra parameters P1, P2, ... are passed
%    directly to the function F(U,X,T,P1,P2,...).
%
%    See also PARABOLIC2D, PARABOLIC

% (c) N. Mostoufi & A. Constantinides
% January 1, 1999

% Initialization
if nargin < 7
   error(' Invalid number of inputs.')
end

nx = fix(nx);
x = [0:nx]*dx;
nt = fix(nt);
t = [0:nt]*dt;
r = alpha*dt/dx^2;
```

```
u0 = (u0(:).')'; % Make sure it's a column vector
if length(u0) ~= nx+1
 error(' Length of the vector of initial condition is not correct.')
end

[a,b]=size(bc);
if a ~= 2
   error(' Invalid number of boundary conditions.')
end
if b < 2 | b > 3
   error(' Invalid boundary condition.')
end
if b == 2 & max(bc(:,1)) <= 2
   bc = [bc zeros(2,1)];
end

u(:,1) = u0;
c = zeros(nx+1,1);
% Iteration on t
for n = 2:nt+1

   % Lower x boundary condition
   switch bc(1,1)
   case 1
      A(1,1) = 1;
      c(1) = bc(1,2);
   case {2, 3}
      A(1,1) = -3/(2*dx) - bc(1,3);
      A(1,2) = 2/dx;
      A(1,3) = -1/(2*dx);
      c(1) = bc(1,2);
   end

   % Internal points
   for i = 2:nx
      A(i,i-1) = -r;
      A(i,i) = 2*(1+r);
      A(i,i+1) = -r;
      c(i) = r*u(i-1,n-1) + 2*(1-r)*u(i,n-1) + r*u(i+1,n-1);
      if nargin >= 8   % Nonhomogeneous equation
         intercept = feval(func,0,x(i),t(n),varargin{:});
         slope = feval(func,1,x(i),t(n),varargin{:}) - intercept;
         A(i,i) = A(i,i) - dt*slope;
         c(i) =c(i)+dt*feval(func,u(i,n-1),x(i),t(n-1), ...
            varargin{:}) +dt*intercept;
      end
   end
```

```
   % Upper x boundary condition
   switch bc(2,1)
   case 1
      A(nx+1,nx+1) = 1;
      c(nx+1) = bc(2,2);
   case {2, 3}
      A(nx+1,nx+1) = 3/(2*dx) - bc(2,3);
      A(nx+1,nx) = -2/dx;
      A(nx+1,nx-1) = 1/(2*dx);
      c(nx+1) = bc(2,2);
   end

   u(:,n) = inv(A)*c;   % Solving the set of equations
end
```

Input and Results

```
>>Example6_2

 Solution of parabolic partial differential equation.

 Depth of the container (m) = 0.1
 Maximum time (s) = 70*3600*24
 Number of divisions in z-direction = 10
 Number of divisions in t-direction = 500
 Diffusion coefficient of A in B = 2e-9

 1 - No reaction between A and B
 2 - A reacts with B
 Enter your choice : 1

 Boundary conditions:

 Concentration of A at interface (mol/m3) = 0.01

 Condition at Bottom of the container:
   1 - Dirichlet
   2 - Neumann
   3 - Robbins
 Enter your choice : 2
 Value = 0

 Repeat calculations (0/1) ? 1

 Depth of the container (m) = 0.1
 Maximum time (s) = 70*3600*24
```

```
Number of divisions in z-direction = 10
Number of divisions in t-direction = 500
Diffusion coefficient of A in B = 2e-9

1 - No reaction between A and B
2 - A reacts with B
Enter your choice : 2

Rate constant = 2e-7
Name of the file containing the rate law = 'Ex6_2_func'

Boundary conditions:

Concentration of A at interface (mol/m3) = 0.01

Condition at Bottom of the container:
   1 - Dirichlet
   2 - Neumann
   3 - Robbins
Enter your choice : 2
Value = 0

Repeat calculations (0/1) ? 0
```

Discussion of Results: *Part* (*a*) The unsteady-state concentration profile is plotted in Fig. E6.2*b*. The steady-state concentration profile is $c_A = 0.01$ mol/m^3 at all levels. The unsteady-state mole flux of A entering the container is shown in Fig. E6.2*c*. This flux decreases with time and reaches zero at steady-state.

Part (*b*) The unsteady-state concentration profile is plotted in Fig. E6.2*d*. Like part (*a*), the steady-state concentration profile is

$$\frac{c_A}{c_{A_0}} = \frac{\cosh\beta[1 - (z/L)]}{\cosh\beta}$$

where $\beta = \sqrt{kL^2/D_{AB}}$.

The unsteady-state mole flux of A entering the container is shown in Fig. E6.2*e*. This flux decreases with time at the beginning. However, it reaches the steady-state value of 1.3×10^{-5} mol/m^{3}day. This happens because A is constantly consumed by B in the container. In fact, the steady-state flux at top of the container is equal to the consumption of A in the container by reaction:

$$-D_{AB}\frac{\partial c_A}{\partial z}\bigg|_{z=0} = \int_0^L (-r_A)\,dz = \frac{D_{AB}c_{A_0}}{L}\beta\tanh\beta \qquad t \to \infty$$

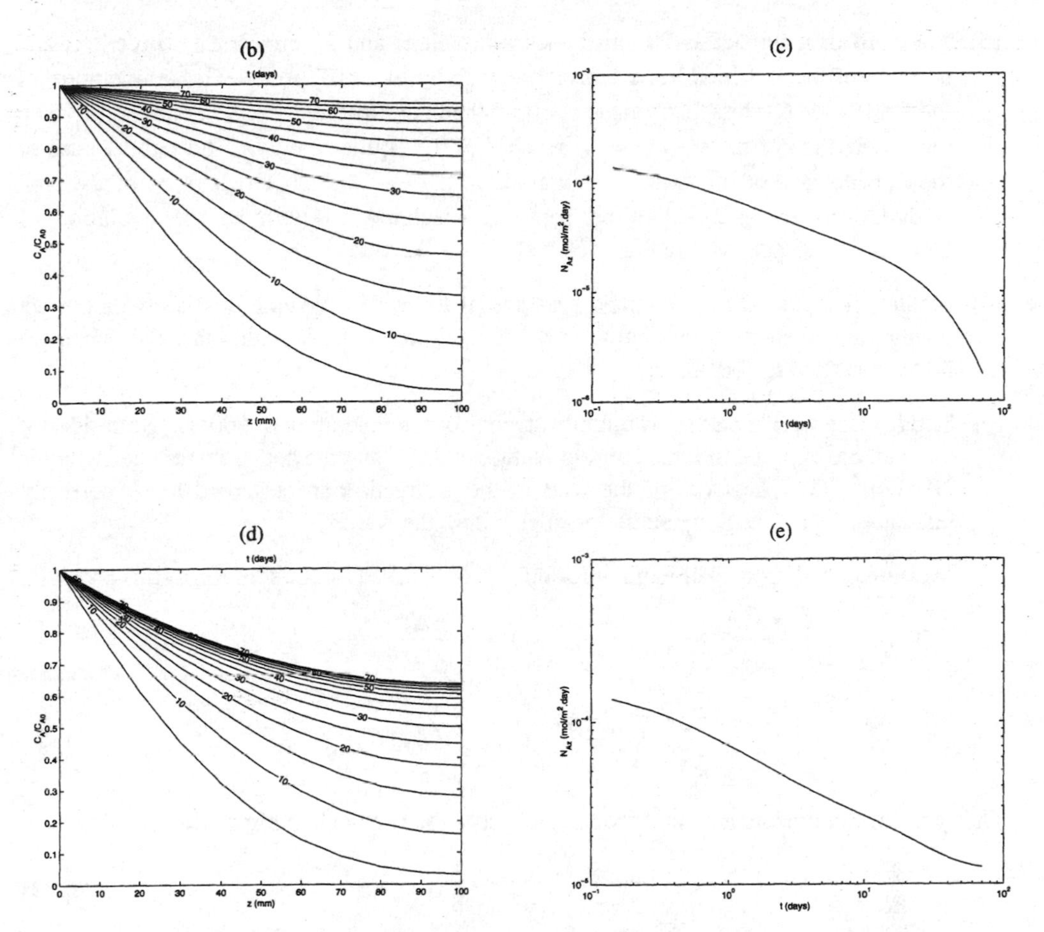

Figure E6.2 Unsteady-state concentration and flux profiles with and without reaction. (*b*) Concentration profile of *A* with no reaction. (*c*) Flux profile of *A* with no reaction. (*d*) Concentration profile of *A* with reaction. (*e*) Flux profile of *A* with reaction.

Example 6.3: Two-Dimensional Parabolic Partial Differential Equation for Heat Transfer. Write a general MATLAB function to determine the numerical solution of the parabolic partial differential equation

$$\frac{\partial u}{\partial t} = \alpha\left(\frac{\partial^2 u}{\partial x^2} + \frac{\partial^2 u}{\partial y^2}\right) + f(x, y, t) \tag{6.65}$$

by explicit method. Apply this function to solve the following problems ($u = T$):

(*a*) The wall of a furnace is 20 cm thick (x-direction) and 50 cm long (y-direction) and is made of brick, which has a thermal diffusivity of 2×10^{-7} m^2/s. The temperature of the wall is 25°C when the furnace is off. When the furnace is fired, the temperature on the inside face of the wall ($x = 0$) reaches 500°C quite rapidly. The temperature of the outside face of the wall is maintained at 25°C. The other two faces of the wall (y-direction) are assumed to be perfectly insulated. Determine the evolution of temperature profiles within the brick wall.

(*b*) Insulation is placed on the outside surface of the wall. Assume this is also a perfect insulation and show the evolution of the temperature profiles within the wall when the furnace is fired to 500°C.

(*c*) The furnace wall of part (*a*) is initially at a uniform temperature of 500°C. Both sides of the wall are exposed to forced air circulation at 25°C and the heat transfer coefficient is 20 W/m^2 °C. The faces of the wall in the y-direction are assumed to be perfectly insulated. Show the temperature profiles within the wall.

Method of Solution: The explicit formula (6.66) is used for the solution of this problem:

$$u_{i,j,n+1} = \left(\frac{\alpha\Delta t}{\Delta x^2}\right)(u_{i+1,j,n} + u_{i-1,j,n}) + \left(\frac{\alpha\Delta t}{\Delta y^2}\right)(u_{i,j+1,n} + u_{i,j-1,n})$$

$$+ \left(1 - 2\frac{\alpha\Delta t}{\Delta x^2} - 2\frac{\alpha\Delta t}{\Delta y^2}\right)u_{i,j,n} + (\Delta t)f_{i,j,n} \tag{6.66}$$

The value of the time increment Δt for a stable solution is calculated from Eq. (6.72):

$$\frac{\alpha\Delta t}{\Delta x^2 + \Delta y^2} \leq \frac{1}{8} \tag{6.72}$$

When Neumann or Robbins conditions are specified, for example, (for Neumann condition $\gamma = 0$):

$$\frac{\partial u}{\partial x}\bigg|_{0,y,t} = \beta + \gamma u(0,y,t)$$

a forward difference approximation of the condition is used at this boundary

$$\frac{1}{2\Delta x}(-u_2 + 4u_1 - 3u_0) = \beta + \gamma u_0$$

from which the dependent variable at the boundary, u_0, can be obtained:

$$u_0 = \frac{-2\beta\Delta x - u_2 + 4u_1}{3 + 2\gamma\Delta x}$$

Similarly, if the Neumann or Robbins condition is at the upper boundary, the dependent variable can be calculated from

$$u_{p+1} = \frac{-2\beta\Delta x + u_{p-1} + 4u_p}{-3 + 2\gamma\Delta x}$$

The same discussion applies for y-direction boundaries.

Program Description: The MATLAB function *parabolic2D.m* is written for solution of the parabolic partial differential equation in an unsteady-state two-dimensional problem. The boundary conditions are passed to the function in the same format as that of Example 6.1. Initial condition, u_0, is a matrix of the values of the dependent variable for all x and y at time $t = 0$. If the problem at hand is nonhomogeneous, the name of the MATLAB function containing the function f should be given as the 10th input argument.

The function starts with the initialization section, which checks the inputs and sets the values required in the calculations. The solution of the equation follows next and consists of an outer loop on time interval. At each time interval, values of the dependent variable for inner grid points are being calculated based on Eq. (6.66), followed by calculation of the grid points on the boundaries according to the formula developed in the previous section. The values of the dependent variable at corner points are assumed to be the average of their adjacent points on the converging boundaries.

In the main program *Example6_3.m*, all the necessary parameters for solving the problem are introduced from the keyboard. The program then asks for initial and boundary conditions, builds the matrix of initial conditions, and calls the function *parabolic2D.m* to solve the partial differential equation. It is possible to repeat the same problem with different initial and boundary conditions.

The last part of the program is visualization of the results. There are two ways to look at the results. One way is dynamic visualization, which is an animation of the temperature profile evolution of the wall. This method may be time consuming because it makes individual frames of the temperature profiles at each time interval and then shows them one after another using *movie* command. Instead, the user may select the other option, which is to see a summary of the results in nine succeeding chronological frames.

Program

Example6_3.m

```
% Example6_3.m
% Solution to Example 6.3. This program calculates and plots
% the temperature profiles of a furnace wall by solving the
% two-dimensional unsteady-state energy balance equation using
% the function PARABOLIC2D.M.

clear
```

```
bcdialog = [' Lower x boundary condition:'
    ' Upper x boundary condition:'
    ' Lower y boundary condition:'
    ' Upper y boundary condition:'];

clc
disp(' Solution of two-dimensional parabolic')
disp(' partial differential equation.')
disp(' ')
width = input(' Width of the plate (x-direction) (m) = ');
length = input(' Length of the plate  (y-direction) (m) = ');
tmax = input(' Maximum time (hr) = ')*3600;
p = input(' Number of divisions in x-direction = ');
q = input(' Number of divisions in y-direction = ');
r = input(' Number of divisions in t-direction = ');
alpha = input(' Thermal diffusivity of the wall = ');

redo = 1;
while redo
   clc
   clf
   T0 = input(' Initial temperature of the wall (deg C) = ');
   u0 = T0*ones(p+1,q+1);  % Matrix of initial condition

   disp(' ')
   disp(' Boundary conditions:')
   for k = 1:4
      disp(' ')
      disp(bcdialog(k,:))
      disp(' 1 - Dirichlet')
      disp(' 2 - Neumann')
      disp(' 3 - Robbins')
      bc(k,1) = input(' Enter your choice : ');
      if bc(k,1) < 3
         bc(k,2) = input(' Value = ');
      end
      switch bc(k,1)
      case 3
         disp(' u'' = (beta) + (gamma)*u')
         bc(k,2) = input(' Constant    (beta)  = ');
         bc(k,3) = input(' Coefficient (gamma) = ');
      case 1
         switch k
         case 1
            u0(1,:) = bc(k,2)*ones(1,q+1);
         case 2
            u0(p+1,:) = bc(k,2)*ones(1,q+1);
         case 3
            u0(:,1) = bc(k,2)*ones(p+1,1);
         case 4
            u0(:,q+1) = bc(k,2)*ones(p+1,1);
```

```
        end
     end
  end

  % Calculating concentration profile
  [x,y,t,T] = parabolic2D(p,q,r,width/p,length/q,tmax/r,...
     alpha,u0,bc);
  r = max(size(t))-1; % Time step may be changed by the solver

disp(' ')
  disp(' Which version of MATLAB are you using?')
  disp(' 0 - The Student Edition')
  disp(' 1 - The Complete Edition')
  ver = input(' Choose either 0 or 1: ');
  maxt = max(max(max(T)));
  mint = min(min(min(T)));
  switch ver
  case 0
     for kr = 1:3
        for kc = 1:3
           m1 = (kr-1)*3+kc;
           m2 = fix(r/8*(m1-1)+1);
           subplot(3,3,m1), surf(y/length,x/width,T(:,:,m2))
           view(135,45)
           axis([0 1 0 1 0 maxt])
           if kr == 2 & kc == 1
              zlabel('Temperature (deg C)')
           end
           if kr == 3 & kc == 2
              xlabel('y/Length')
              ylabel('x/Width')
           end
           ttl = [num2str(t(m2)/3600) 'h'];
           title(ttl)
        end
     end
  case 1
     disp(' ')
     disp(' Are you patient enough to see a movie of temperature')
     mv = input(' profile evolution (0/1)? ');
    if mv
      % Making movie of temperature profile evolution
      M = moviein(r);
      for k = 1:r+1
         surf(y/length,x/width,T(:,:,k))
           axis([0 1 0 1 0 maxt])
           view(135,45)
           shading interp
           ylabel('x/Width')
           xlabel('y/Length')
           zlabel('Temperature (deg C)')
```

```
            M(:,k) = getframe;
         end
         movie(M,5)
      else % Show results in 9 succeeding frames
       for kr = 1:3
          for kc = 1:3
             m1 = (kr-1)*3+kc;
             m2 = fix(r/8*(m1-1)+1);
             subplot(3,3,m1), surf(y/length,x/width,T(:,:,m2))
             view(135,45)
             axis([0 1 0 1 0 maxt])
             if kr == 2 & kc == 1
                zlabel('Temperature (deg C)')
             end
             if kr == 3 & kc == 2
                xlabel('y/Length')
                ylabel('x/Width')
             end
             ttl = [num2str(t(m2)/3600) 'h'];
             title(ttl)
          end
       end
      end
    end
    disp(' ')
   redo = input(' Repeat with different initial and boundary conditions
(0/1)? ');
end
```

parabolic2D.m

```
function [x,y,t,u] = parabolic2D(nx,ny,nt,dx,dy,dt,alpha,...
   u0,bc,func,varargin)
%PARABOLIC2D solution of a two-dimensional parabolic partial
%   differential equation
%
%   [X,Y,T,U]=PARABOLIC2D(NX,NY,NT,DX,DY,DT,ALPHA,U0,BC) solves
%   the homogeneous parabolic equation by Crank-Nicolson implicit
%   formula where
%      X = vector of x values
%      Y = vector of y values
%      T = vector of T values
%      U = 3D array of dependent variable [U(X,Y,T)]
%      NX = number of divisions in x-direction
%      NY = number of divisions in y-direction
%      NT = number of divisions in t-direction
%      DX = x-increment
%      DY = y-increment
%      DT = t-increment (leave empty to use the default value)
%      ALPHA = coefficient of equation
%      U0 = matrix of U-distribution at T=0 [U0(X,Y)]
%      BC is a matrix of 4x2 or 4x3 containing the types and values
```

```
%       of boundary conditions in x- and y-directions. The order of
%       appearing boundary conditions are lower x, upper x, lower y,
%       and upper y in rows 1 to 4 of the matrix BC, respectively.
%       The first column of BC determines the type of condition:
%          1 for Dirichlet condition, followed by the set
%          value of U in the second column.
%          2 for Neumann condition, followed by the set value
%          of U' in the second column.
%          3 for Robbins condition, followed by the constant
%          and the coefficient of U in the second and third
%          column, respectively.
%
%    [X,Y,T,U]=PARABOLIC2D(NX,NY,NT,DX,DY,DT,ALPHA,U0,BC,F,P1,P2,...)
%    solves the nonhomogeneous parabolic equation where F(U,X,Y,T)
%    is a function described in the M-file F.M. The extra
%    parameters P1, P2, ... are passed directly to the function
%    F(U,X,Y,T,P1,P2,...).
%
%    See also PARABOLIC1D, PARABOLIC

% (c) N. Mostoufi & A. Constantinides
% January 1, 1999

% Initialization
if nargin < 9
   error(' Invalid number of inputs.')
end

nx = fix(nx);
x = [0:nx]*dx;
ny = fix(ny);
y = [0:ny]*dy;

% Checking dt for stability(use 1/16 instead of 1/8 of Eq. 6.72)
tmax = dt*nt;
if isempty(dt) | dt > (dx^2+dy^2)/(16*alpha)
   dt = (dx^2+dy^2)/(16*alpha);
   nt = tmax/dt+1;
   fprintf('\n dt is adjusted to %6.2e (nt=%3d)',dt,fix(nt))
end

nt = fix(nt);
t = [0:nt]*dt;
rx = alpha*dt/dx^2;
ry = alpha*dt/dy^2;

[r0,c0] = size(u0);
if r0 ~= nx+1 | c0 ~= ny+1
   error(' Size of the matrix of initial condition is not correct.')
end
```

```
[a,b]=size(bc);
if a ~= 4
   error(' Invalid number of boundary conditions.')
end
if b < 2 | b > 3
   error(' Invalid boundary condition.')
end
if b == 2 & max(bc(:,1)) <= 2
   bc = [bc zeros(4,1)];
end

% Solution of differential equation
u(:,:,1) = u0;
for n = 1:nt              % Iteration on t
   for i = 2:nx           % Iteration on x
      for j = 2:ny        % Iteration on y
         u(i,j,n+1) = rx*(u(i+1,j,n)+u(i-1,j,n))+ ry*(u(i,j+1,n)...
            + u(i,j-1,n)) + (1-2*rx-2*ry)*u(i,j,n);
         if nargin >= 10
            u(i,j,n+1) = u(i,j,n+1) +...
               dt*feval(func,u(i,j,n),x(i),y(j),t(n),varargin{:});
         end
      end
   end

   % Lower x boundary condition
   switch bc(1,1)
   case 1
      u(1,2:ny,n+1) = bc(1,2) * ones(1,ny-1,1);
   case {2, 3}
      u(1,2:ny,n+1) = (-2*bc(1,2)*dx + 4*u(2,2:ny,n+1) ...
         - u(3,2:ny,n+1)) / (2*bc(1,3)*dx + 3);
   end

   % Upper x boundary condition
   switch bc(2,1)
   case 1
      u(nx+1,2:ny,n+1) = bc(2,2) * ones(1,ny-1,1);
   case {2, 3}
      u(nx+1,2:ny,n+1) = (-2*bc(2,2)*dx -4*u(nx,2:ny,n+1)...
         +u(nx-1,2:ny,n+1)) / (2*bc(2,3)*dx - 3);
   end

   % Lower y boundary condition
   switch bc(3,1)
   case 1
      u(2:nx,1,n+1) = bc(3,2) * ones(nx-1,1,1);
   case {2, 3}
      u(2:nx,1,n+1) = (-2*bc(3,2)*dy + 4*u(2:nx,2,n+1)...
         - u(2:nx,3,n+1)) / (2*bc(3,3)*dy + 3);
   end
```

```
    % Upper y boundary condition
    switch bc(4,1)
    case 1
       u(2:nx,ny+1,n+1) = bc(4,2) * ones(nx-1,1,1);
    case {2, 3}
       u(2:nx,ny+1,n+1) = (-2*bc(4,2)*dy -4*u(2:nx,ny,n+1)...
           +u(2:nx,ny-1,n+1)) / (2*bc(4,3)*dy - 3);
    end
end

% Corner nodes
u(1,1,:) = (u(1,2,:) + u(2,1,:)) / 2;
u(nx+1,1,:) = (u(nx+1,2,:) + u(nx,1,:)) / 2;
u(1,ny+1,:) = (u(1,ny,:) + u(2,ny+1,:)) / 2;
u(nx+1,ny+1,:) = (u(nx+1,ny,:) + u(nx,ny+1,:)) / 2;
```

Input and Results

```
>>Example6_3

 Solution of two-dimensional parabolic
 partial differential equation.

 Width of the plate (x-direction) (m) = 0.2
 Length of the plate  (y-direction) (m) = 0.5
 Maximum time (hr) = 12
 Number of divisions in x-direction = 8
 Number of divisions in y-direction = 8
 Number of divisions in t-direction = 30
 Thermal diffusivity of the wall = 2e-7
```

Part (a)

```
 Initial temperature of the wall (deg C) = 25

 Boundary conditions:

 Lower x boundary condition:
 1 - Dirichlet
 2 - Neumann
 3 - Robbins
 Enter your choice : 1
 Value = 500

 Upper x boundary condition:
 1 - Dirichlet
 2 - Neumann
 3 - Robbins
 Enter your choice : 1
 Value = 25
```

```
Lower y boundary condition:
1 - Dirichlet
2 - Neumann
3 - Robbins
Enter your choice : 2
Value = 0

Upper y boundary condition:
1 - Dirichlet
2 - Neumann
3 - Robbins
Enter your choice : 2
Value = 0

Which version of MATLAB are you using?
0 - The Student Edition
1 - The Complete Edition
Choose either 0 or 1: 1

Are you patient enough to see a movie of temperature
profile evolution (0/1)? 0

Repeat with different initial and boundary conditions (0/1)? 1
```

Part (b)

```
Initial temperature of the wall (deg C) = 25

Boundary conditions:

Lower x boundary condition:
1 - Dirichlet
2 - Neumann
3 - Robbins
Enter your choice : 1
Value = 500

Upper x boundary condition:
1 - Dirichlet
2 - Neumann
3 - Robbins
Enter your choice : 2
Value = 0

Lower y boundary condition:
1 - Dirichlet
2 - Neumann
3 - Robbins
Enter your choice : 2
Value = 0
```

```
Upper y boundary condition:
1 - Dirichlet
2 - Neumann
3 - Robbins
Enter your choice : 2
Value = 0

Which version of MATLAB are you using?
0 - The Student Edition
1 - The Complete Edition
Choose either 0 or 1: 1

Are you patient enough to see a movie of temperature
profile evolution (0/1)? 0

Repeat with different initial and boundary conditions (0/1)? 1
```

Part (c)

```
Initial temperature of the wall (deg C) = 500

Boundary conditions:

Lower x boundary condition:
1 - Dirichlet
2 - Neumann
3 - Robbins
Enter your choice : 3
u' = (beta) + (gamma)*u
Constant    (beta)  = -25*20
Coefficient (gamma) = 20

Upper x boundary condition:
1 - Dirichlet
2 - Neumann
3 - Robbins
Enter your choice : 3
u' = (beta) + (gamma)*u
Constant    (beta)  = 25*20
Coefficient (gamma) = -20

Lower y boundary condition:
1 - Dirichlet
2 - Neumann
3 - Robbins
Enter your choice : 2
Value = 0

Upper y boundary condition:
1 - Dirichlet
```

```
2 - Neumann
3 - Robbins
Enter your choice : 2
Value = 0

Which version of MATLAB are you using?
0 - The Student Edition
1 - The Complete Edition
Choose either 0 or 1: 1

Are you patient enough to see a movie of temperature
profile evolution (0/1)? 0

Repeat with different initial and boundary conditions (0/1)? 0
```

Discussion of Results: *Part* (*a*) Heat transfers from the inside of the furnace (left boundary), where the temperature is 500°C, towards the outside (right boundary), where the temperature is maintained at 25°C. Therefore, the temperature profile progresses from the left of the wall toward the right, as shown in Fig. E6.3*a*. If the integration is continued for a sufficiently long time, the profile will reach the steady-state, which for this case is a straight plane connecting the two Dirichlet conditions. This is easily verified from the analytical solution of the steady-state problem:

$$\frac{\partial^2 T}{\partial x^2} + \frac{\partial^2 T}{\partial y^2} = 0$$

Because the two faces of the wall in the *y*-direction are insulated, this becomes essentially a one-dimensional problem that yields the equation of a straight line:

$$T = -2375x + 500$$

calculated using the Dirichlet conditions.

Part (*b*) In this case, the insulation installed on the outside surface of the furnace wall causes the temperature within the wall to continue rising, as shown in Fig. E6.3*b*. The steady-state temperature profile would be T = 500°C throughout the solid wall. This is also verifiable from the analytical solution of the steady-state problem in conjunction with the imposed boundary conditions.

Part (*c*) The cooling of the wall occurs from both sides, and the temperature profile moves, symmetrically, as shown in Fig. E6.3*c*. The final temperature would be 25°C.

The reader is encouraged to repeat this example and choose the movie option to see the temperature profile evolutions dynamically. It should be noted that the rate of evolution of the temperature profile on the screen is not the same as that of the heat transfer process itself.

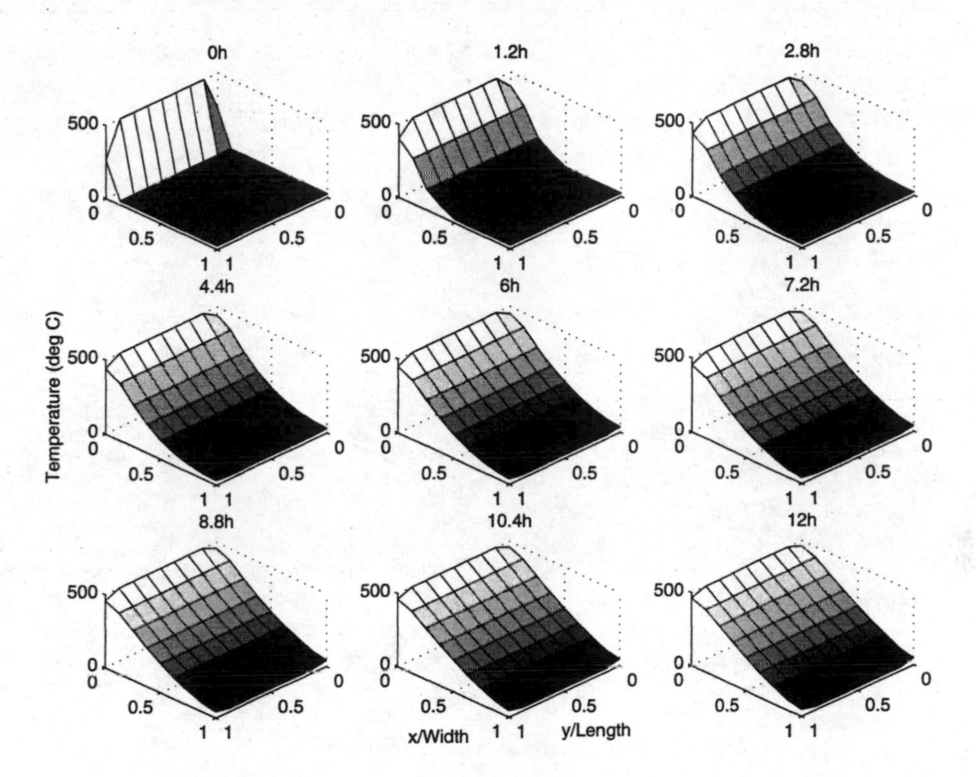

Figure E6.3*a* Evolution of temperature within the wall of the furnace with no insulation. The length and width have been normalized to be in the range of (0, 1).

6.4.3 Hyperbolic Partial Differential Equations

Second-order partial differential equations of the hyperbolic type occur principally in physical problems connected with vibration processes. For example the one-dimensional wave equation

$$\rho \frac{\partial^2 u}{\partial t^2} = T_0 \frac{\partial^2 u}{\partial x^2} + f(x, t) \tag{6.86}$$

describes the transverse motion of a vibrating string that is subjected to tension T_0 and external force $f(x, t)$. In the case of constant density ρ, the equation is written in the form

$$\frac{\partial^2 u}{\partial t^2} = a^2 \frac{\partial^2 u}{\partial x^2} + F(x, t) \tag{6.87}$$

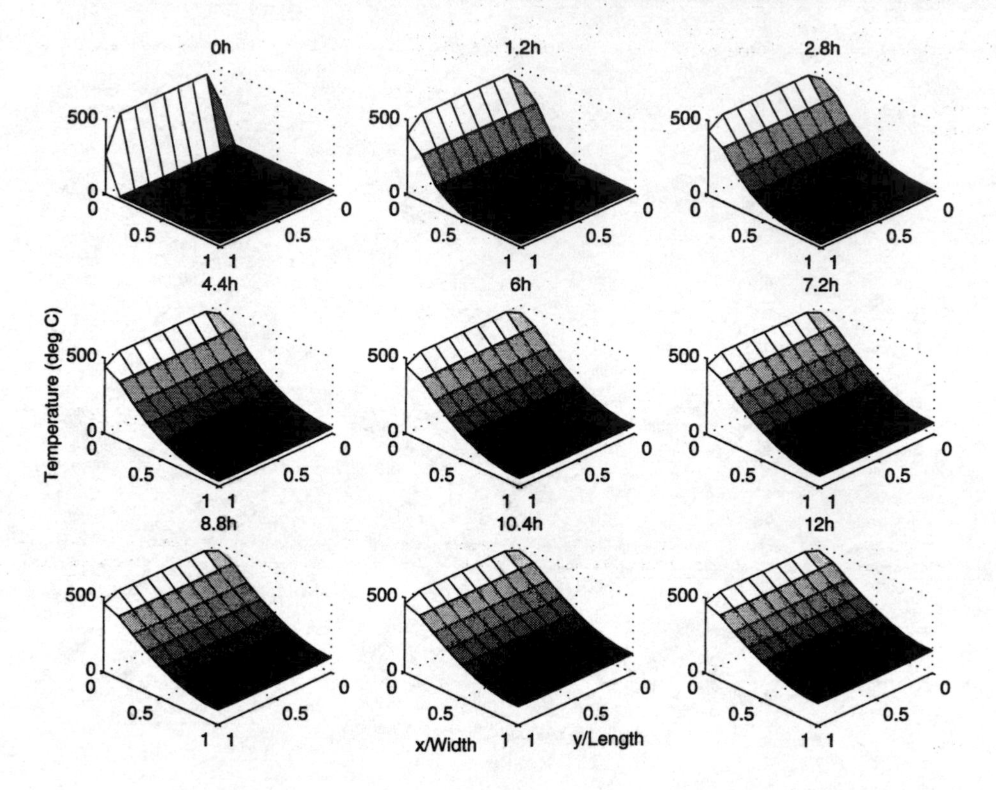

Figure E6.3*b* Evolution of temperature within the wall of the furnace with insulation. The length and width have been normalized to be in the range of (0, 1).

where

$$a^2 = \frac{T_0}{\rho}$$

$$F(x, t) = \frac{1}{\rho} f(x, t)$$

If no external force acts on the string, Eq. (6.87) becomes a homogeneous equation:

$$\frac{\partial^2 u}{\partial t^2} = a^2 \frac{\partial^2 u}{\partial x^2} \tag{6.88}$$

The two-dimensional extension of Eq. (6.87) is

$$\frac{\partial^2 u}{\partial t^2} = a^2 \left(\frac{\partial^2 u}{\partial x^2} + \frac{\partial^2 u}{\partial y^2} \right) + F(x, y, t) \tag{6.89}$$

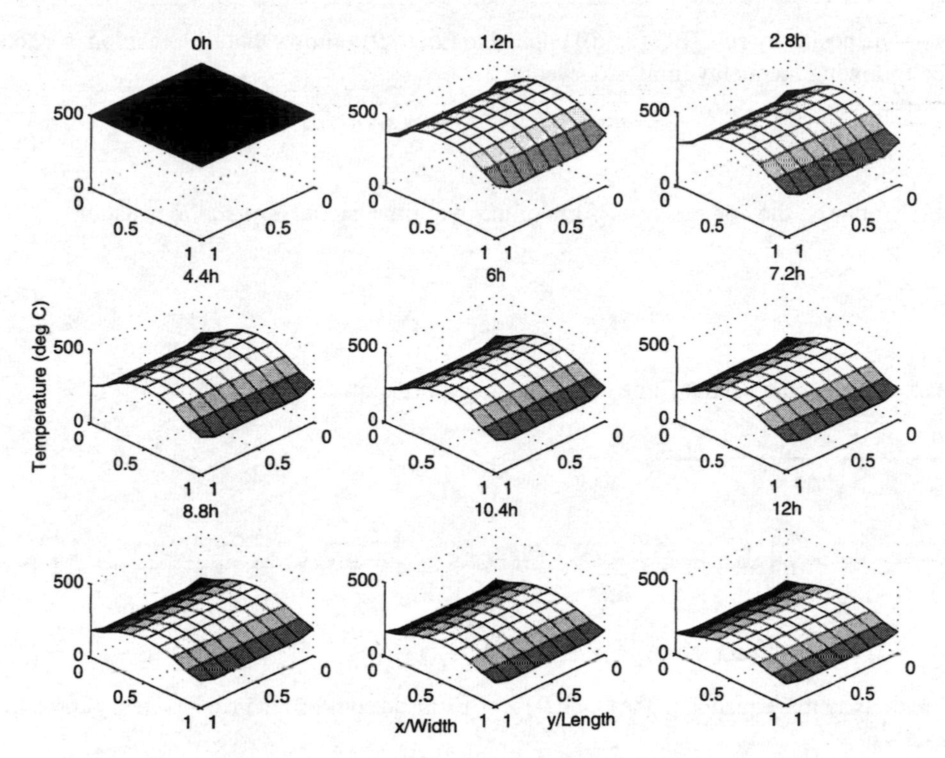

Figure E6.3c Evolution of temperature within the wall of the furnace with cooling from both sides. The length and width have been normalized to be in the range of (0, 1).

which describes the vibration of a membrane subjected to tension T_0 and external force $f(x, y, t)$.

To find the numerical solution of Eq. (6.88) we expand each second-order derivative in terms of central finite differences to obtain

$$\frac{u_{i,n+1} - 2u_{i,n} + u_{i,n-1}}{\Delta t^2} = a^2\left(\frac{u_{i+1,n} - 2u_{i,n} + u_{i-1,n}}{\Delta x^2}\right) + O(\Delta x^2 + \Delta t^2) \qquad (6.90)$$

Rearranging to solve for $u_{i,n+1}$,

$$u_{i,n+1} = 2\left(1 - \frac{a^2\Delta t^2}{\Delta x^2}\right)u_{i,n} + \frac{a^2\Delta t^2}{\Delta x^2}(u_{i+1,n} + u_{i-1,n}) - u_{i,n-1} + O(\Delta x^2 + \Delta t^2) \qquad (6.91)$$

This is an *explicit* numerical solution of the hyperbolic equation (6.88).

The positivity rule [Eq. (6.56)] applied to Eq. (6.91) shows that this solution is stable if the following inequality limit is obeyed:

$$\frac{a^2\Delta t^2}{\Delta x^2} \le 1 \tag{6.92}$$

Similarly, the homogeneous form of the two-dimensional hyperbolic equation:

$$\frac{\partial^2 u}{\partial t^2} = a^2\left(\frac{\partial^2 u}{\partial x^2} + \frac{\partial^2 u}{\partial y^2}\right) \tag{6.93}$$

is expanded using central finite difference approximation to yield

$$\frac{u_{i,j,n+1} - 2u_{i,j,n} + u_{i,j,n-1}}{\Delta t^2} = a^2\left(\frac{u_{i+1,j,n} - 2u_{i,j,n} + u_{i-1,j,n}}{\Delta x^2}\right) + a^2\left(\frac{u_{i,j+1,n} - 2u_{i,j,n} + u_{i,j-1,n}}{\Delta y^2}\right) + O(\Delta x^2 + \Delta y^2 + \Delta t^2) \tag{6.94}$$

Rearranging this equation to the explicit form, using an equidistant grid in x- and y-directions, results in

$$u_{i,j,n+1} = 2\left[1 - 2\left(\frac{a^2\Delta t^2}{\Delta x^2}\right)\right]u_{i,j,n} - u_{i,j,n-1} + \frac{a^2\Delta t^2}{\Delta x^2}(u_{i+1,j,n} + u_{i-1,j,n} + u_{i,j+1,n} + u_{i,j-1,n}) \tag{6.95}$$

This solution is stable when

$$\frac{a^2\Delta t^2}{\Delta x^2} \le \frac{1}{2} \tag{6.96}$$

Implicit methods for solution of hyperbolic partial differential equations can be developed using the *variable-weight* approach, where the space partial derivatives are weighted at $(n+1)$, n, and $(n-1)$. The implicit formulation of Eq. (6.88) is

$$\frac{u_{i,n+1} - 2u_{i,n} + u_{i,n-1}}{\Delta t^2} = \frac{a^2}{\Delta x^2}[\theta(u_{i+1,n+1} - 2u_{i,n+1} + u_{i-1,n+1}) + (1 - 2\theta)(u_{i+1,n} - 2u_{i,n} + u_{i-1,n}) + \theta(u_{i+1,n-1} - 2u_{i,n-1} + u_{i-1,n-1})] \tag{6.97}$$

where $0 \le \theta \le 1$. When $\theta = 0$, Eq. (6.97) reverts back to the explicit method [Eq. (6.91)]. When $\theta = ½$, Eq. (6.97) is a Crank-Nicolson-type approximation. Implicit methods yield tridiagonal sets of linear algebraic equations whose solutions can be obtained using Gauss elimination methods.

6.4.4 Irregular Boundaries and Polar Coordinate Systems

The finite difference approximations of partial differential equations developed in this chapter so far were based on regular Cartesian coordinate systems. Quite often, however, the objects, whose properties are being modeled by the partial differential equations, may have circular, cylindrical, or spherical shapes, or may have altogether irregular boundaries. The finite difference approximations may be modified to handle such cases.

Let us first consider an object which is well described by Cartesian coordinates everywhere except near the boundary, which is of irregular shape, as shown in Fig. 6.9. There are two methods of treating the curved boundary. One simple method is to reshape the boundary to pass through the grid point closest to it. For example, in Fig. 6.9, the point (i, j) can be assumed to be on the boundary instead of the point $(i + 1, j)$ on the original boundary. Although it is a simple method, the approximation of the boundary introduces an error in the calculations, especially at the boundary itself.

A more precise method of expressing the finite difference equation at the irregular boundary is to modify it accordingly. We may use a Taylor series expansion of the dependent value at the point (i, j) in the x-direction to get (see Fig. 6.9)

$$u_{i+1,j} = u_{i,j} + (\alpha\Delta x)\frac{\partial u}{\partial x}\bigg|_{i,j} + \frac{(\alpha^2\Delta x^2)}{2!}\frac{\partial^2 u}{\partial x^2}\bigg|_{i,j} + O(\Delta x^3) \tag{6.98}$$

$$u_{i-1,j} = u_{i,j} - (\Delta x)\frac{\partial u}{\partial x}\bigg|_{i,j} + \frac{(\Delta x^2)}{2!}\frac{\partial^2 u}{\partial x^2}\bigg|_{i,j} + O(\Delta x^3) \tag{6.99}$$

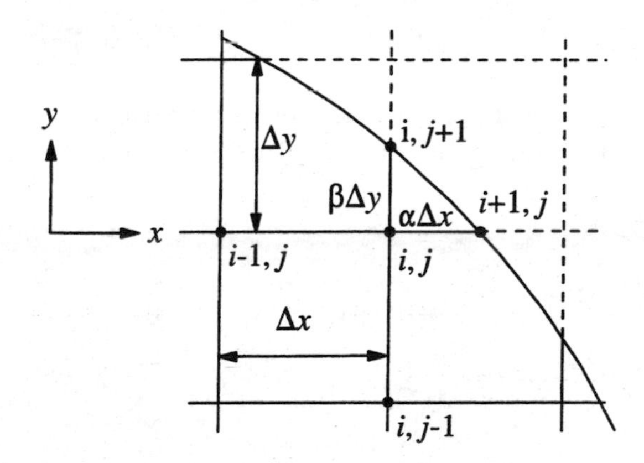

Figure 6.9 Finite difference grid for irregular boundaries.

Eliminating $(\partial^2 u/\partial x^2)$ from Eqs. (6.98) and (6.99) results in

$$\frac{\partial u}{\partial x}\bigg|_{i,j} = \frac{1}{\Delta x}\left[\frac{1}{\alpha(1+\alpha)}\right][u_{i+1,j} - (1 - \alpha^2)u_{i,j} - \alpha^2 u_{i-1,j}] \tag{6.100}$$

and eliminating $(\partial u/\partial x)$ from Eqs. (6.98) and (6.99) gives

$$\frac{\partial^2 u}{\partial x^2}\bigg|_{i,j} = \frac{1}{\Delta x^2}\left[\frac{2}{\alpha(1+\alpha)}\right][u_{i+1,j} - (1 + \alpha)u_{i,j} + \alpha u_{i-1,j}] \tag{6.101}$$

Similarly, in the y-direction:

$$\frac{\partial u}{\partial y}\bigg|_{i,j} = \frac{1}{\Delta y}\left[\frac{1}{\beta(1+\beta)}\right][u_{i,j+1} - (1 - \beta^2)u_{i,j} - \beta^2 u_{i,j-1}] \tag{6.102}$$

$$\frac{\partial^2 u}{\partial y^2}\bigg|_{i,j} = \frac{1}{\Delta y^2}\left[\frac{2}{\beta(1+\beta)}\right][u_{i,j+1} - (1 + \beta)u_{i,j} + \beta u_{i,j-1}] \tag{6.103}$$

When $\alpha = \beta = 1$, Eqs. (6.100)-(6.103) become identical to those developed earlier in this chapter for regular Cartesian coordinate systems. Therefore, for objects with irregular boundaries, the partial differential equations would be converted to algebraic equations using Eqs. (6.100)-(6.103). For points adjacent to the boundary, the parameters α and β would assume values that reflect the irregular shape of the boundary, and for internal points away from the boundary, the value of α and β would be unity.

Eqs. (6.100)-(6.103) can be used at the boundaries with Dirichlet condition where the dependent variable at the boundary is known. Treatment of Neumann and Robbins conditions where the normal derivative at the curved or irregular boundary is specified is more complicated. Considering again Fig. 6.9, the normal derivative of the dependent variable at the boundary can be expressed as

$$\frac{\partial u}{\partial \boldsymbol{n}}\bigg|_{i+1,j} = \frac{\partial u}{\partial x}\bigg|_{i+1,j}\cos\gamma + \frac{\partial u}{\partial y}\bigg|_{i+1,j}\sin\gamma \tag{6.104}$$

where $\boldsymbol{n}$ is the unit vector normal to the boundary and γ is the angle between the vector $\boldsymbol{n}$ and x-axis. The derivatives with respect to x and y in Eq. (6.104) can be approximated by Taylor series expansions

$$\frac{\partial u}{\partial x}\bigg|_{i+1,j} = \frac{\partial u}{\partial x}\bigg|_{i,j} + (\alpha\Delta x)\frac{\partial^2 u}{\partial x^2}\bigg|_{i,j} \tag{6.105}$$

$$\frac{\partial u}{\partial y}\bigg|_{i+1,j} = \frac{\partial u}{\partial y}\bigg|_{i,j} + (\alpha\Delta x)\frac{\partial^2 u}{\partial x \partial y}\bigg|_{i,j} \tag{6.106}$$

The derivatives at the grid point (i, j) should be known in order to calculate the normal derivative at the boundary. For the particular configuration of Fig. 6.9, we may use backward finite differences to evaluate the derivatives in Eqs. (6.105) and (6.106) (see Table 6.3):

$$\frac{\partial u}{\partial x}\bigg|_{i,j} = \frac{1}{\Delta x}(u_{i,j} - u_{i-1,j}) \tag{6.107}$$

$$\frac{\partial^2 u}{\partial x^2}\bigg|_{i,j} = \frac{1}{\Delta x^2}(u_{i,j} - 2u_{i-1,j} + u_{i-2,j}) \tag{6.108}$$

$$\frac{\partial u}{\partial y}\bigg|_{i,j} = \frac{1}{\Delta y}(u_{i,j} - u_{i,j-1}) \tag{6.109}$$

$$\frac{\partial u^2}{\partial x \partial y} = \frac{1}{\Delta x \Delta y}(u_{i,j} - u_{i-1,j} - u_{i,j-1} + u_{i-1,j-1}) \tag{6.110}$$

Combining Eqs. (6.105), (6.107), and (6.108) gives

$$\frac{\partial u}{\partial x}\bigg|_{i+1,j} = \frac{1}{\Delta x}[(1+\alpha)u_{i,j} - (1+2\alpha)u_{i-1,j} + \alpha u_{i-2,j}] \tag{6.111}$$

and combination of Eqs. (6.106), (6.109), and (6.110) results in

$$\frac{\partial u}{\partial y}\bigg|_{i+1,j} = \frac{1}{\Delta y}[(1+\alpha)u_{i,j} - \alpha u_{i-1,j} - (1+\alpha)u_{i,j-1} + \alpha u_{i-1,j-1}] \tag{6.112}$$

Replacing Eqs. (6.111) and (6.112) into Eq. (6.104) provides the normal derivative which can be used when dealing with Neumann or Robbins conditions at an irregular boundary.

Similarly in the y-direction:

$$\frac{\partial u}{\partial \boldsymbol{n}}\bigg|_{i,j+1} = \frac{\partial u}{\partial x}\bigg|_{i,j+1}\cos\gamma + \frac{\partial u}{\partial y}\bigg|_{i,j+1}\sin\gamma \tag{6.113}$$

where

$$\frac{\partial u}{\partial x}\bigg|_{i,j+1} = \frac{1}{\Delta x}[(1+\beta)u_{i,j} - \beta u_{i,j-1} - (1+\beta)u_{i-1,j} + \beta u_{i-1,j-1}] \tag{6.114}$$

$$\frac{\partial u}{\partial y}\bigg|_{i,j+1} = \frac{1}{\Delta y}[(1+\beta)u_{i,j} - (1+2\beta)u_{i,j-1} + \beta u_{i,j-2}] \tag{6.115}$$

It is important to remember that Eqs. (6.111), (6.112), (6.114), and (6.115) are specific to the configuration shown in Fig. 6.9. For other possible configurations, forward differences, or a combination of forward and backward differences (in different directions), may be used to treat the derivative boundary condition.

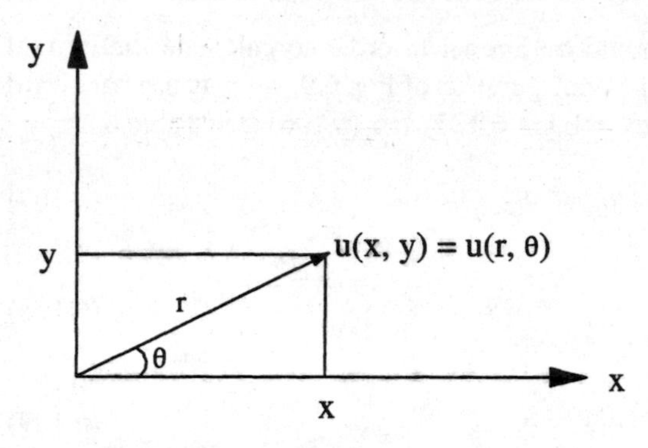

Figure 6.10 Transformation to polar coordinates.

Cylindrical-shaped objects are more conveniently expressed in polar coordinates. The transformation from Cartesian coordinate to polar coordinate systems is performed using the following relationships, which are based on Fig. 6.10:

$$x = r\cos\theta \qquad y = r\sin\theta$$
$$r = \sqrt{x^2 + y^2} \qquad \theta = \tan^{-1}\frac{y}{x} \tag{6.116}$$

The Laplacian operator in polar coordinates becomes

$$\frac{\partial^2 u}{\partial x^2} + \frac{\partial^2 u}{\partial y^2} = \frac{\partial^2 u}{\partial r^2} + \frac{1}{r}\frac{\partial u}{\partial r} + \frac{1}{r^2}\frac{\partial^2 u}{\partial \theta^2} \tag{6.117}$$

Fick's second law of diffusion [Eq. (6.10)] in polar coordinates is

$$\frac{\partial c_A}{\partial t} = D_{AB}\left(\frac{\partial^2 c_A}{\partial r^2} + \frac{1}{r}\frac{\partial c_A}{\partial r} + \frac{1}{r^2}\frac{\partial^2 c_A}{\partial \theta^2} + \frac{\partial^2 c_A}{\partial z^2}\right) \tag{6.118}$$

Using the finite difference grid for polar coordinates shown in Fig. 6.11, the partial derivatives are approximated by

$$\left.\frac{\partial^2 u}{\partial r^2}\right|_{i,j} = \frac{1}{\Delta r^2}(u_{i,j+1} - 2u_{i,j} + u_{i,j-1}) \tag{6.119}$$

$$\left.\frac{\partial^2 u}{\partial \theta^2}\right|_{i,j} = \frac{1}{\Delta \theta^2}(u_{i+1,j} - 2u_{i,j} + u_{i-1,j}) \tag{6-120}$$

$$\left.\frac{\partial u}{\partial r}\right|_{i,j} = \frac{1}{2\Delta r}(u_{i,j+1} - u_{i,j-1}) \tag{6-121}$$

where j and i are counters in r- and θ-directions, respectively. Partial derivatives in z- and t-dimensions (not shown in Fig. 6.11) would be similarly expressed through the use of

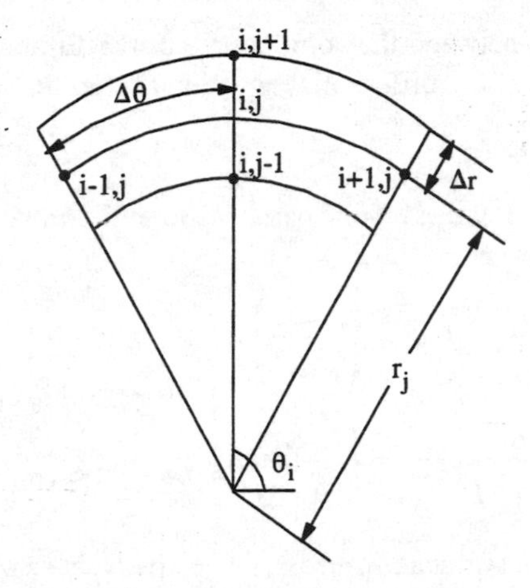

Figure 6.11 Finite difference grid for polar coordinates.

additional subscripts.

6.4.5 Nonlinear Partial Differential Equations

The discussion in this chapter has focused on linear partial differential equations that yield sets of linear algebraic equations when expressed in finite difference approximations. On the other hand, if the partial differential equation is nonlinear, for instance,

$$u\frac{\partial^2 u}{\partial x^2} + u\frac{\partial^2 u}{\partial y^2} = f(u) \tag{6.122}$$

The resulting finite difference discretization would generate sets of nonlinear algebraic equations. The solution of this problem would require the application of Newton's method for simultaneous nonlinear equations (see Chap. 1).

6.5 Stability Analysis

In this section, we discuss the stability of finite difference approximations using the well-known von Neumann procedure. This method introduces an initial error represented by a finite Fourier series and examines how this error propagates during the solution. The von Neumann method applies to initial-value problems; for this reason it is used to analyze the stability of the explicit method for parabolic equations developed in Sec. 6.4.2 and the explicit method for hyperbolic equations developed in Sec. 6.4.3.

Define the error $\epsilon_{m,n}$ as the difference between the solution $u_{m,n}$ of the finite difference approximation and the exact solution $\bar{u}_{m,n}$ of the differential equation at step (m, n):

$$\epsilon_{m,n} \equiv u_{m,n} - \bar{u}_{m,n} \tag{6.123}$$

The explicit finite difference solution (6.58) of the parabolic partial differential equation (6.18) can be written for $u_{m,n+1}$ and $\bar{u}_{m,n+1}$ as follows:

$$u_{m,n+1} = \left(\frac{\alpha\Delta t}{\Delta x^2}\right) u_{m+1,n} + \left(1 - 2\frac{\alpha\Delta t}{\Delta x^2}\right) u_{m,n} + \left(\frac{\alpha\Delta t}{\Delta x^2}\right) u_{m-1,n} + R_{E_{m,n+1}} \tag{6.124}$$

$$\bar{u}_{m,n+1} = \left(\frac{\alpha\Delta t}{\Delta x^2}\right) \bar{u}_{m+1,n} + \left(1 - 2\frac{\alpha\Delta t}{\Delta x^2}\right) \bar{u}_{m,n} + \left(\frac{\alpha\Delta t}{\Delta x^2}\right) \bar{u}_{m-1,n} + T_{E_{m,n+1}} \tag{6.125}$$

where $R_{E_{m,n+1}}$ and $T_{E_{m,n+1}}$ are the roundoff and truncation errors, respectively, at step $(m, n+1)$.

Combining Eqs. (6.123)-(6.125) we obtain

$$\epsilon_{m,n+1} - \left(\frac{\alpha\Delta t}{\Delta x^2}\right) \epsilon_{m+1,n} - \left(1 - 2\frac{\alpha\Delta t}{\Delta x^2}\right) \epsilon_{m,n} - \left(\frac{\alpha\Delta t}{\Delta x^2}\right) \epsilon_{m-1,n} = R_{E_{m,n+1}} - T_{E_{m,n+1}} \tag{6.126}$$

This is a *nonhomogeneous finite difference equation in two dimensions*, representing the propagation of error during the numerical solution of the parabolic partial differential equation (6.18). The solution of this finite difference equation is rather difficult to obtain. For this reason, the von Neumann analysis considers the *homogeneous* part of Eq. (6.126):

$$\epsilon_{m,n+1} - \left(\frac{\alpha\Delta t}{\Delta x^2}\right) \epsilon_{m+1,n} - \left(1 - 2\frac{\alpha\Delta t}{\Delta x^2}\right) \epsilon_{m,n} - \left(\frac{\alpha\Delta t}{\Delta x^2}\right) \epsilon_{m-1,n} = 0 \tag{6.127}$$

which represents the propagation of the error introduced at the initial point ($n = 0$) only and ignores truncation and roundoff errors that enter the solution at $n > 0$.

The solution of the homogeneous finite difference equation may be written in the separable form

$$\epsilon_{m,n} = c\, e^{\gamma n \Delta t} e^{i\beta m \Delta x} \tag{6.128}$$

where $i = \sqrt{-1}$ and c, γ, and β are constants. At $n = 0$:

$$\epsilon_{m,0} = c\, e^{i\beta m \Delta x} \tag{6.129}$$

which is the error at the initial point. Therefore, the term $e^{\gamma\Delta t}$ is the *amplification factor* of the initial error. In order for the original error not to grow as n increases, the amplification factor must satisfy the *von Neumann condition for stability*:

$$|e^{\gamma\Delta t}| \le 1 \tag{6.130}$$

The amplification factor can have complex values. In that case, the modulus of the complex numbers must satisfy the above inequality, that is,

$$|r| \le 1 \tag{6.131}$$

Therefore the stability region in the complex plane is a circle of radius = 1, as shown in Fig. 6.12.

The amplification factor is determined by substituting Eq. (6.128) into Eq. (6.127) and rearranging to obtain

$$e^{\gamma \Delta t} = \left(1 - 2\frac{\alpha \Delta t}{\Delta x^2}\right) + \frac{\alpha \Delta t}{\Delta x^2}\left(e^{i\beta \Delta x} + e^{-i\beta \Delta x}\right) \tag{6.132}$$

Using the trigonometric identities

$$\frac{e^{i\beta \Delta x} + e^{-i\beta \Delta x}}{2} = \cos \beta \Delta x \tag{6.133}$$

and

$$1 - \cos \beta \Delta x = 2\sin^2 \frac{\beta \Delta x}{2} \tag{6.134}$$

Eq. (6.132) becomes

$$e^{\gamma \Delta t} = 1 - \left(4\frac{\alpha \Delta t}{\Delta x^2}\right)\left(\sin^2 \frac{\beta \Delta x}{2}\right) \tag{6.135}$$

Combining this with the von Neumann condition for stability, we obtain the stability bound

$$0 \le \left(\frac{\alpha \Delta t}{\Delta x^2}\right)\left(\sin^2 \frac{\beta \Delta x}{2}\right) \le \frac{1}{2} \tag{6.136}$$

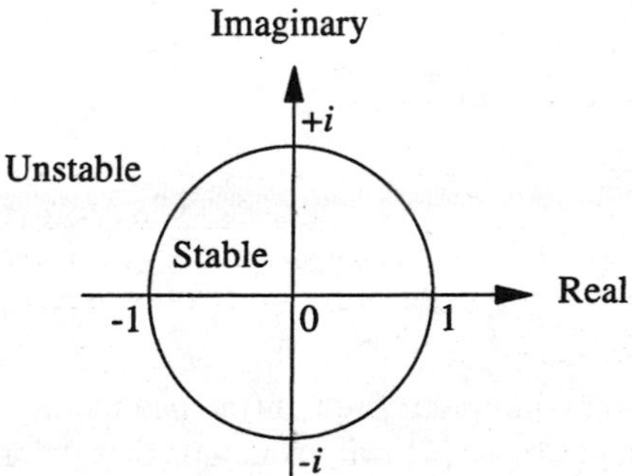

Figure 6.12 Stability region in the complex plane.

The $\sin^2(\beta\Delta x/2)$ term has its highest value equal to unity; therefore:

$$0 \leq \frac{\alpha\Delta t}{\Delta x^2} \leq \frac{1}{2} \tag{6.137}$$

is the limit for conditional stability for this method. It should be noted that this limit is identical to that obtained by using the positivity rule (Sec. 6.4.2).

The stability of the explicit solution (6.91) of the hyperbolic equation (6.88) can be similarly analyzed using the von Neumann method. The homogeneous equation for the error propagation of that solution is

$$\epsilon_{m,n+1} - 2\left(1 - \frac{a^2\Delta t^2}{\Delta x^2}\right)\epsilon_{m,n} - \frac{a^2\Delta t^2}{\Delta x^2}\left(\epsilon_{m+1,n} + \epsilon_{m-1,n}\right) + \epsilon_{m,n-1} = 0 \tag{6.138}$$

Substitution of the solution (6.128) into (6.138) and use of the trigonometric identities (6.133) and (6.134) give the amplification factor as

$$e^{\gamma\Delta t} = \left(1 - 2\frac{a^2\Delta t^2}{\Delta x^2}\sin^2\frac{\beta\Delta x}{2}\right) \pm \sqrt{\left(1 - 2\frac{a^2\Delta t^2}{\Delta x^2}\sin^2\frac{\beta\Delta x}{2}\right)^2 - 1} \tag{6.139}$$

The above amplification factor satisfies inequality (6.131) in the complex plane, that is, when

$$\left(1 - 2\frac{a^2\Delta t^2}{\Delta x^2}\sin^2\frac{\beta\Delta x}{2}\right)^2 - 1 \leq 0 \tag{6.140}$$

which converts to the following inequality:

$$\frac{a^2\Delta t^2}{\Delta x^2} \leq \frac{1}{\sin^2(\beta\Delta x/2)} \tag{6.141}$$

The $\sin^2(\beta\Delta x/2)$ term has its highest value equal to unity; therefore,

$$\frac{a^2\Delta t^2}{\Delta x^2} \leq 1 \tag{6.142}$$

is the conditional stability limit for this method.

In a similar manner the stability of other explicit and implicit finite difference methods may be examined. This has been done by Lapidus and Pinder [4], who conclude that "most explicit finite difference approximations are conditionally stable, whereas most implicit approximations are unconditionally stable."

6.6 Introduction to Finite Element Methods

The finite element methods are powerful techniques for the numerical solution of differential equations. Their theoretical formulation is based on the variational principle. The minimization of the functional of the form

$$J(U) = \int_D \phi\left(U, \frac{\partial U}{\partial x}, \frac{\partial U}{\partial y}, x, y \right) dD \tag{6.143}$$

must satisfy the Euler-Lagrange equation

$$\frac{\partial}{\partial x}\left[\frac{\partial \phi}{\partial(\partial U/\partial x)} \right] + \frac{\partial}{\partial y}\left[\frac{\partial \phi}{\partial(\partial U/\partial y)} \right] - \frac{\partial \phi}{\partial U} = 0 \tag{6.144}$$

which is a partial differential equation with certain natural boundary conditions.

It has been shown that many differential equations that originate from the physical sciences have equivalent variational formulations.[4] This is the basis for the well-known Rayleigh-Ritz procedure which in turn forms the basis for the finite element methods.

An equivalent formulation of finite element methods can be developed using the concept of weighted residuals. In Sec. 5.6.3, we discussed the method of weighted residuals in connection with the solution of the two-point boundary-value problem. In that case we chose the solution of the ordinary differential equation as a polynomial *basis function* and caused the integral of weighted residuals to vanish:

$$\int_{x_0}^{x_f} W_k R_m(x)\, dx = 0 \tag{5.129}$$

We now extend this method to the solution of partial differential equations where the desired solution $u(.)$ is replaced by a piecewise polynomial approximation of the form

$$u(.) = \sum_{j=1}^{N} a_j \phi_j(.) \tag{6.145}$$

[4] For a complete discussion of the variational formulation of the finite element method, see Vichnevetsky [3] and Vemuri and Karplus [6].

The set of functions $\{\phi_j(.) \mid j = 1, 2, \ldots, N\}$ are the basis functions and the $\{a_j \mid j = 1, 2, \ldots, N\}$ are undetermined coefficients. The integral of weighted residuals is made to vanish:

$$\int_t \int_V W_j(.) R(.)\, dV dt = 0 \tag{6.146}$$

The choice of basis functions $\phi_j(.)$ and weighted functions $W_j(.)$ determines the particular finite element method. The *Galerkin* method [13] chooses the basis and weighted functions to be identical to each other. The *orthogonal collocation* method uses the Dirac delta function for weights and orthogonal polynomials for basis functions. The *subdomain* method chooses the weighted function to be unity in the subregion V_i, for which it is defined, and zero elsewhere. A complete discussion of the finite element methods is outside the scope of this book. The interested reader is referred to Lapidus and Pinder [4], Reddy [7], Huebner et al. [8], and Pepper and Heinrich [9] for detailed developments of these methods.

MATLAB has a powerful toolbox for solution of linear and nonlinear partial differential equations which is called *Partial Differential Equation* (or *PDE*) *TOOLBOX*. This toolbox uses the finite element method for solution of partial differential equations in two space dimensions. The basic equation of this toolbox is the equation

$$-\nabla.(c\nabla u) + au = f \tag{6.147}$$

where c, a, and f are complex-valued functions in the solution domain and may also be a function of u. The toolbox can also solve the following equations:

$$d\frac{\partial u}{\partial t} - \nabla.(c\nabla u) + au = f \tag{6.148}$$

and

$$d\frac{\partial^2 u}{\partial t^2} - \nabla.(c\nabla u) + au = f \tag{6.149}$$

where d, c, a, and f are complex-valued functions in the solution domain and also can be functions of time. The symbol ∇ is the vector differential operator (not to be confused with ∇, the backward difference operator).

In the PDE toolbox Eqs. (6.147)-(6.149) are named elliptic, parabolic, and hyperbolic, respectively, regardless of the values of the coefficients and boundary conditions.

In order to solve a partial differential equation using the PDE toolbox, one may simply use the *graphical user interface* by employing the *pdetool* command. In this separate environment, the user is able to define the two-dimensional geometry, introduce the boundary conditions, solve the partial differential equation, and visualize the results. In special cases where the problem is complicated or nonstandard, the user may wish to solve it using command-line functions. Some of these functions (solvers only) are listed in Table 6.4.

Table 6.4 Partial differential equation solvers in MATLAB's PDE TOOLBOX

Solver	Description
adaptmesh	Adaptive mesh generation and solution of elliptic partial differential equation
assempde	Assembles and solves the elliptic partial differential equation
hyperbolic	Solves hyperbolic partial differential equation
parabolic	Solves parabolic partial differential equation
pdenonlin	Solves nonlinear elliptic partial differential equation
poisolv	Solves the Poisson equation on a rectangular grid

PROBLEMS

6.1 Modify *elliptic.m* in Example 6.1 to solve for the three-dimensional problem

$$\frac{\partial^2 u}{\partial x^2} + \frac{\partial^2 u}{\partial y^2} + \frac{\partial^2 u}{\partial z^2} = 0$$

Apply this function to calculate the distribution of the dependent variable within a solid body which is subject to the following boundary conditions:

$$u(0,y,z) = 100 \qquad u(1,y,z) = 100$$
$$u(x,0,z) = 0 \qquad u(x,1,z) = 0$$
$$u(x,y,0) = 50 \qquad u(x,y,1) = 50$$

6.2 Solve Laplace's equation with the following boundary conditions and discuss the results:

$$u(0,y) = 100 \qquad \left.\frac{\partial u}{\partial x}\right|_{10,y} = 10$$
$$\left.\frac{\partial u}{\partial y}\right|_{x,0} = 0 \qquad \left.\frac{\partial u}{\partial y}\right|_{x,1} = 0$$

6.3 The ambient temperature surrounding a house is 50°F. The heat in the house had been turned off; therefore, the internal temperature is also at 50°F at $t = 0$. The heating system is turned on and raises the internal temperature to 70°F at the rate of 4°F/h. The ambient temperature remains at 50°F. The wall of the house is 0.5 ft thick and is made of material that has an average thermal diffusivity $\alpha = 0.01$ ft²/h and a thermal conductivity $k = 0.2$ Btu/(h.ft².°F). The heat transfer coefficient on the inside of the wall is $h_{in} = 1.0$ Btu/(h.ft².°F), and the heat transfer coefficient on the outside is h_{out} = 2.0 Btu/(h.ft^2.°F). Estimate how long it will take to reach a steady-state temperature distribution across the wall.

6.4 Develop the finite difference approximation of Fick's second law of diffusion in polar coordinates. Write a MATLAB program that can be used to solve the following problem [10]:

A wet cylinder of agar gel at 278 K with a uniform concentration of urea of 0.1 kgmol/m^3 has a diameter of 30.48 mm and is 38.1 mm long with flat parallel ends. The diffusivity is 4.72×10^{-10} m^2/s. Calculate the concentration at the midpoint of the cylinder after 100 h for the following cases if the cylinder is suddenly immersed in turbulent pure water:

(a) For radial diffusion only
(b) Diffusion that occurs radially and axially.

6.5 Express the two-dimensional parabolic partial differential equation

$$\frac{\partial u}{\partial t} = \alpha\left(\frac{\partial^2 u}{\partial x^2} + \frac{\partial^2 u}{\partial y^2}\right)$$

in an explicit finite difference formulation. Determine the limits of conditional stability for this method using

(a) The von Neumann stability
(b) The positivity rule.

6.6 Consider a first-order chemical reaction being carried out under isothermal steady-state conditions in a tubular-flow reactor. On the assumptions of laminar flow and negligible axial diffusion, the material balance equation is

$$-v_0\left[1 - \left(\frac{r}{R}\right)^2\right]\frac{\partial c}{\partial z} + D\left(\frac{\partial^2 c}{\partial r^2} + \frac{1}{r}\frac{\partial c}{\partial r}\right) - kc = 0$$

where v_0 = velocity of central stream line
R = tube radius
k = reaction rate constant
c = concentration of reactant
D = radial diffusion constant
z = axial distance along the length of tube
r = radial distance from center of tube.

Upon defining the following dimensionless variables:

$$\lambda = \frac{kz}{v_0} \qquad C = \frac{c}{c_0} \qquad \alpha = \frac{D}{kR^2} \qquad U = \frac{r}{R}$$

the equation becomes

$$(1 - U^2)\frac{\partial C}{\partial \lambda} = \alpha\left(\frac{\partial^2 C}{\partial U^2} + \frac{1}{U}\frac{\partial C}{\partial U}\right) - C$$

where c_0 is the entering concentration of the reactant to the reactor.

(a) Choose a set of appropriate boundary conditions for this problem. Explain your choice.
(b) What class of PDE is the above equation (hyperbolic, parabolic, or elliptic)?
(c) Set up the equation for numerical solution using finite difference approximations.
(d) Does your choice of finite differences result in an explicit or implicit set of equations? Give the details of the procedure for the solution of this set of equations.
(e) Discuss stability considerations with respect to the method you have chosen.

6.7 A 12-in-square membrane (no bending or shear stresses), with a 4-in-square hole in the middle, is fastened at the outside and inside boundaries as shown in Fig. P6.7 [11]. If a highly stretched membrane is subject to a pressure p, the partial differential equation for the deflection w in the z-direction is

$$\frac{\partial^2 w}{\partial x^2} + \frac{\partial^2 w}{\partial y^2} = -\frac{p}{T}$$

where T is the tension (pounds per linear inch). For a highly stretched membrane, the tension T may be assumed constant for small deflections. Utilizing the following values of pressure and tension:

$$p = 5 \text{ psi} \qquad \text{(uniformly distributed)}$$
$$T = 100 \text{ lb/in}$$

(a) Express the differential equation in finite difference form to obtain the deflection w of the membrane.
(b) List all the boundary conditions needed for the numerical solution of the problem. Utilize some or all of these boundary conditions to simplify the finite difference equations of part (a).
(c) Solve the equation numerically.

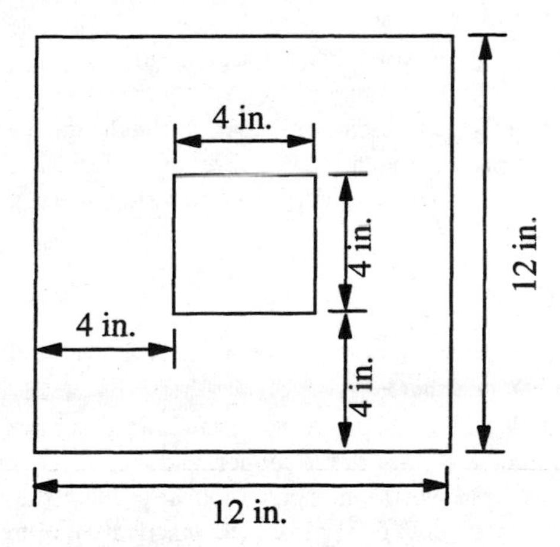

Figure P6.7 Stretched membrane.

6.8 Figure P6.8 shows a cross section of a long cooling fin of width W, thickness t, and thermal conductivity k that is bonded to a hot wall, maintaining its base (at $x = 0$) at a temperature T_w [12]. Heat is conducted steadily through the fin in the plane of Fig. P6.8 so that the fin temperature T obeys Laplace's equation, $\partial^2 T/\partial x^2 + \partial^2 T/\partial y^2 = 0$. (Temperature variations along the length of the fin in the z-direction are ignored.)

Heat is lost from the sides and tip of the fin by convection to the surrounding air (radiation is neglected at sufficiently low temperatures) at a local rate $q = h(T_s - T_a)$ Btu/(h.ft^2). Here, T_s and T_a, in degrees Fahrenheit, are the temperatures at a point on the fin surface and of the air, respectively. If the surface of the fin is vertical, the heat transfer coefficient h obeys the dimensional correlation $h = 0.21(T_s - T_a)^{1/3}$.

(a) Set up the equations for a numerical solution of this problem to determine the temperature at a finite number of points within the fin and at the surface.

(b) Describe in detail the step-by-step procedure for solving the equation of part (a) and evaluating the temperature within the fin and at the surface.

(c) Solve the problem numerically using the following quantities:

$$T_w = 200^\circ\text{F} \qquad T_a = 70^\circ\text{F}$$
$$t = 0.25 \text{ in} \qquad k = 25.9 \text{ Btu}/(\text{h.ft.}^\circ\text{F})$$
$$w = 0.5 \text{ in}$$

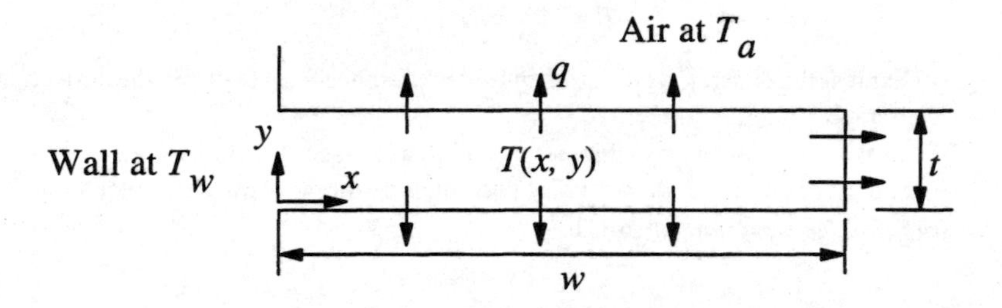

Figure P6.8 Cooling fin.

6.9 Consider a steady-state plug flow reactor of length z through which a substrate is flowing with a constant velocity v with no dispersion effects. The reactor is made up of a series of collagen membranes, each impregnated with two enzymes catalyzing the sequential reaction [14]:

$$A \xrightarrow{\text{Enzyme 1}} B \xrightarrow{\text{Enzyme 2}} C$$

The membranes in the reactor are arranged in parallel, as shown in Fig. P6.9. The nomenclature for this problem is shown in Table P6.9.

For a substrate molecule to encounter the immobilized enzymes, it must diffuse across a *Nernst* diffusion layer on the surface of the support and then some distance into the membrane. The coupled reaction takes place in the membrane and the product, the unreacted intermediate, and substrate diffuse back into the bulk fluid phase. No inactivation of the enzymes occurs, and it is assumed that the enzymes behave independently of each other.

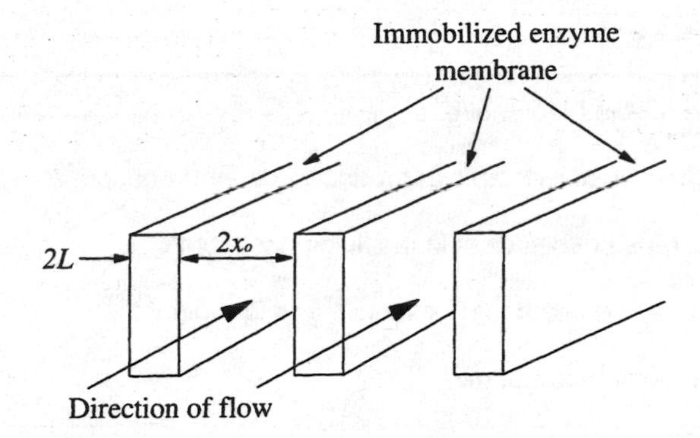

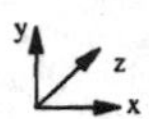

Figure P6.9 Biocatalytic reactor.

Since the membrane can accommodate only a finite number of enzymes molecules per unit weight, it becomes necessary to introduce a control parameter ϵ that measures the ratio of molar concentration of enzyme 1 to molar concentration of Enzymes 1 plus 2. It is implicitly assumed that the binding sites on collagen do not discriminate between the enzymes. Thus, when both enzymes are present, the maximum reaction velocities reduce to ϵV_1 and $(1 - \epsilon)V_2$. The control ϵ is constrained between the bounds of 0 (only Enzyme 2 present) and 1 (only Enzyme 1 present).

The reaction rates for the two sequential reactions are given by the Michaelis-Menten relationship:

$$R_1 = \frac{\epsilon V_1 Y_{Am}}{K_{M1} + C_{Af} Y_{Am}}$$

$$R_2 = \frac{(1 - \epsilon) V_2 Y_{Bm}}{K_{M2} + C_{Af} Y_{Bm}}$$

Material balance for the species A, B, and C in the membrane yield the following differential equations:

$$\frac{D}{L^2} \frac{\partial^2 Y_{Am}}{\partial X^2} - R_1 = 0$$

$$\frac{D}{L^2} \frac{\partial^2 Y_{Bm}}{\partial X^2} + R_2 - R_1 = 0$$

$$\frac{D}{L^2} \frac{\partial^2 Y_{Cm}}{\partial X^2} + R_2 = 0$$

Table P6.9 Nomenclature for problem 6.9

C_{At}	=	concentration of A in feed, mol/L
D	=	molecular diffusivity of reactants or products in membrane, cm^2/s
k_L	=	overall mass transfer coefficient in the fluid phase, cm/s
K_{M1}, K_{M2}	=	Michaelis-Menten constant for Enzymes 1 and 2, mol/L
L	=	half thickness of membrane, mils
v	=	superficial fluid velocity in reactor, cm/s
V_1, V_2	=	maximum reaction velocity for Enzymes 1 and 2, mol/(L.s)
X	=	variable axial distance from center of membrane to surface, cm
x_0	=	half distance between two consecutive membranes, cm
X	=	x/L, dimensionless distance
Y_{Ab}, Y_{Bb}, Y_{Cb}	=	bulk concentration of species A, B, or C divided by the feed concentration of A (C_{Af}), dimensionless
Y_{Am}, Y_{Bm}, Y_{Cm}	=	membrane concentration of species A, B, or C divided by the feed concentration of A (C_{Af}), dimensionless
Y_{As}, Y_{Bs}, Y_{Cs}	=	surface concentration of species A, B, or C divided by the feed concentration of A (C_{Af}), dimensionless
z	=	variable longitudinal distance from entrance of reactor, cm
ϵ	=	control, ratio of molar concentration of enzyme 1 to total molar concentration of enzymes 1 plus 2, dimensionless
θ	=	z/v, space time, s

In the bulk fluid phase, the material balances for species A, B, and C can be defined as follows:

$$\frac{dY_{Ab}}{d\theta} + \frac{k_L}{x_0}(Y_{Ab} - Y_{As}) = 0$$

$$\frac{dY_{Bb}}{d\theta} + \frac{k_L}{x_0}(Y_{Bb} - Y_{Bs}) = 0$$

$$\frac{dY_{Cb}}{d\theta} + \frac{k_L}{x_0}(Y_{Cb} - Y_{Cs}) = 0$$

Since each membrane is symmetric about $X = 0$, the boundary conditions at $X = 0$ and $X = 1$ become

$$\frac{\partial Y_{Am}}{\partial X} = \frac{\partial Y_{Bm}}{\partial X} = \frac{\partial Y_{Cm}}{\partial X} = 0 \quad \text{at } X = 0$$

$$\left.\begin{aligned} Y_{Am} &= Y_{As} \\ Y_{Bm} &= Y_{Bs} \\ Y_{Cm} &= Y_{Cs} \end{aligned}\right\} \quad \text{at } X = 1$$

The surface concentrations are determined by equating the surface flux to the bulk transport flux, that is,

$$\frac{D}{L}\left(\frac{\partial Y_{Am}}{\partial X}\right)_{X=1} = k_L(Y_{Ab} - Y_{As})$$

$$\frac{D}{L}\left(\frac{\partial Y_{Bm}}{\partial X}\right)_{X=1} = k_L(Y_{Bb} - Y_{Bs})$$

$$\frac{D}{L}\left(\frac{\partial Y_{Cm}}{\partial X}\right)_{X=1} = k_L(Y_{Cb} - Y_{Cs})$$

Finally at the entrance of the reactor, that is, at $\theta = 0$:

$$Y_{Ab} = 1 \qquad Y_{Bb} = 0 \qquad Y_C = 0$$

Develop a numerical procedure for solving the above set of equations and write a computer program to calculate the concentration profiles in the membranes and in the bulk fluid for the following set of kinetic and transport parameters:

$$\begin{aligned} V_1 &= 4.4\times10^{-3}\ \text{mol/L.s} & V_2 &= 12.0\times10^{-3}\ \text{mol/L.s} \\ K_{M1} &= 0.022\ \text{mol/L} & K_{M2} &= 0.010\ \text{mol/L} \\ D &= 5.7\times10^{-8}\ \text{cm}^2/\text{s} & C_{Af} &= 1.0\ \text{mol/L} \\ k_L &= 1.2\times10^{-4}\ \text{cm/s} & x_0 &= 23\ \text{mils} \\ \epsilon &- 0.75 & 2L &= 3\ \text{mils} \end{aligned}$$

6.10 Coulet et al. [15] have developed a glucose sensor that has glucose oxidase enzyme immobilized as a surface layer on a highly polymerized collagen membrane. In this system, glucose (analyte) is converted to hydrogen peroxide, which is subsequently detected on the membrane face (that is not exposed to the analyte solution) by an amperometric electrode. The hydrogen peroxide flux is a direct measure of the sensor response [16].

The physical model and coordinates system are shown in Fig. P6.10. The local analyte concentration at the enzyme surface is low so that the reaction kinetics are adequately described by a first-order law. This latter assumption ensures that the electrode response is proportional to the analyte concentration.

The governing dimensionless equation describing analyte transport within the membrane is

$$\frac{\partial C}{\partial \xi} = \frac{\partial^2 C}{\partial \zeta^2}$$

where the dimensionless time ξ and penetration ζ variables are defined as

$$\xi = \frac{Dt}{\delta^2}$$

$$\zeta = \frac{x}{\delta}$$

where δ is the membrane thickness and D the diffusion coefficient. The initial and boundary conditions are

$$C = 0 \qquad \xi = 0 \qquad 0 \le \zeta \le 1$$

$$C = 1 \qquad \xi > 0 \qquad \zeta = 1$$

$$\frac{\partial C}{\partial \zeta} = -\mu C \qquad \xi > 0 \qquad \zeta = 1$$

where μ is the *Damkoehler* number, defined as

$$\mu = \frac{k''\delta}{D}$$

The surface rate constant k'' is related to the surface concentration of the enzyme $[E'']$, the turnover number k_{cat}, and the intrinsic Michaelis-Menten constant K_m by

$$k'' = \frac{k_{cat}[E'']}{K_m}$$

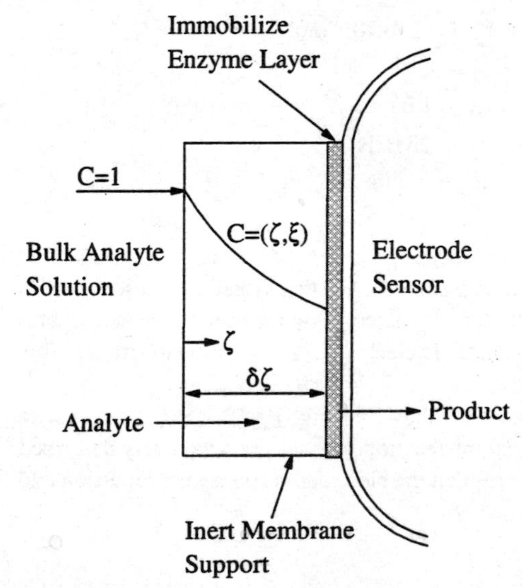

Figure P6.10 Schematic description of an anisotropic enzyme electrode. The membrane (exaggerated) has active enzyme deposited as a surface layer at the electrode sensor interface. The product flux is the result of the reaction involving analyte diffusing through the membrane.

(a) Predict the electrode response as a function of the dimensionless time ξ for a 0.3-mm-thick membrane with the analyte diffusion coefficient $D = 2\times10^{-6}$ cm²/s and immobilized enzyme with the surface rate constant $k'' = 0.24$ cm/h.

(b) Repeat part (a) for the reaction kinetics defined by the Michaelis-Menten law.

6.11 The radial dispersion coefficient of solids in a fluidized bed can be evaluated by the injection of tracer particles at the center of the fluidized bed and monitoring the unsteady-state dispersion of these particles [17]. Assuming instantaneous axial mixing of solids and radial mixing occurring by dispersion, the governing partial differential equation of the model, in cylindrical coordinates, is

$$\frac{\partial C}{\partial t} = D_{sr}\frac{1}{r}\frac{\partial}{\partial r}\left(r\frac{\partial C}{\partial r}\right)$$

where C is the concentration of the tracer, t is time, r is the radial position, and D_{sr} is the radial solid dispersion coefficient. The appropriate initial and boundary conditions are

$$t = 0 \qquad 0 \le r \le a \qquad C = 100\%$$

$$t > 0 \qquad r = 0 \qquad \frac{\partial C}{\partial r} = 0$$

$$t > 0 \qquad r = R \qquad \frac{\partial C}{\partial r} = 0$$

where a is the radius of the tracer injection tube and R is the radius of the column. The analytical solution of the dispersion equation, subject to these conditions, is

$$\frac{C}{C_\infty} = 1 + \frac{2}{a}\sum_{i=0}^{\infty}\frac{J_1(\lambda_i a)J_0(\lambda_i r)}{\lambda_i[J_0(\lambda_i R)]^2}\exp[-D_{sr}\lambda_i^2 t]$$

where C_∞ is the concentration of the tracer at the steady-state condition, J is the Bessel function of the first kind, and λ_i is calculated from

$$J_1(\lambda_i R) = 0$$

(a) Use the analytical solution of the dispersion equation to plot the unsteady-state concentration profiles of the tracer.

(b) Solve the dispersion equation numerically and compare it with the exact solution.

Additional data:

$$2R = 0.27\ \text{m} \qquad 2a = 19\ \text{mm} \qquad D_{sr} = 2\times10^{-4}\ \text{m}^2/\text{s}$$

REFERENCES

1. Bird, R. B., Stewart, W. E., and Lightfoot, E. N., *Transport Phenomena*, Wiley, New York, 1960.

2. Tychonov, A. N., and Samarski, A. A., *Partial Differential Equations of Mathematical Physics*, Holden-Day, San Francisco, 1964.

3. Vichnevetsky, R., *Computer Methods for Partial Differential Equations*, vol. I, Prentice Hall, Englewood Cliffs, NJ, 1981.

4. Lapidus, L., and Pinder, G. F., *Numerical Solution of Partial Differential Equations in Science and Engineering*, Wiley, New York, 1982.

5. Finlayson, B. A., *Nonlinear Analysis in Chemical Engineering*, McGraw-Hill, New York, 1980.

6. Vemuri, V., and Karplus, W. J., *Digital Computer Treatment of Partial Differential Equations*, Prentice Hall, Englewood Cliffs, NJ, 1981.

7. Reddy, J. N., *An Introduction to the Finite Element Method*, 2nd ed., McGraw-Hill, New York, 1993.

8. Huebner, K. H., Thornton, E. A., and Byrom, T. G., *The Finite Element Method for Engineers*, 3rd ed., Wiley, New York, 1995.

9. Pepper, D. W. and Heinrich, J, C., *The Finite Element Method: Basic Concepts and Applications*, Hemisphere, Washington, DC, 1992.

10. Geankoplis, C. J., *Transport Processes and Unit Operations*, 3rd ed., Prentice Hall, Englewood Cliffs, NJ, 1993

11. James, M. L., Smith, G. M., and Wolford J. C., *Applied Numerical Methods for Digital Computation with FORTRAN and CSMP*, 2nd ed., Harper & Row, New York, 1977.

12. Carnahan, B., Luther, H. A., and Wilkes, J. O., *Applied Numerical Methods*, Wiley, New York, 1969.

13. Fairweather, G., *Finite Element Galerkin Methods for Differential Equations*, Marcel Dekker, New York, 1978.

14. Fernandes, P. M., Constantinides, A., Vieth, W. R., and Venkatasubramanian, K., "Enzyme Engineering: Part V. Modeling and Optimizing Multi-Enzyme Reactor Systems," *Chemtech*, July 1975, p. 438.

15. Coulet, P. R., Sternberg, R., and Thevenot, D. R., "Electrochemical Study of Reactions at Interfaces of Glucose Oxidase Collagen Membranes," *Biochim. Biophys. Acta*, vol. 612, 1980, p. 317.

16. Pedersen, H., and Chotani, G. K., "Analysis of a Theoretical Model for Anisotropic Enzyme Membranes: Application to Enzyme Electrodes," *Appl. Biochem. Biotech.*, vol. 6, 1981, p. 309.

17. Berruti, F., Scott, D. S., and Rhodes, E., "Measurement and Modelling Lateral Solid Mixing in a Three-Dimensional Batch Gas-Solid Fluidized Bed Reactor," *Canadian J. Chem. Eng.*, vol. 64, 1986, p. 48.

CHAPTER 7

Linear and Nonlinear Regression Analysis

7.1 Process Analysis, Mathematical Modeling, and Regression Analysis

Engineers and scientists are often required to analyze complex physical or chemical systems and to develop mathematical models which simulate the behavior of such systems. *Process analysis* is a term commonly used by chemical engineers to describe the study of complex chemical, biochemical, or petrochemical processes. More recently coined phrases such as *systems engineering* and *systems analysis* are used by electrical engineers and computer scientists to refer to analysis of electric network and computer systems. No matter what the phraseology is, the principles applied are the same.

According to Himmelblau and Bischoff [1]: "Process analysis is the application of scientific methods to the recognition and definition of problems and the development of procedures for their solution. In more detail, this means (1) mathematical specification of the problem for the given physical solution, (2) detailed analysis to obtain mathematical models, and (3) synthesis and presentation of results to ensure full comprehension."

In the heart of successful process analysis is the step of *mathematical modeling*. The objective of modeling is to construct, from theoretical and empirical knowledge of a process, a mathematical formulation that can be used to predict the behavior of this process. Complete understanding of the mechanism of the chemical, physical, or biological aspects of the process under investigation is not usually possible. However, some information on the mechanism of the system may be available; therefore, a combination of empirical and theoretical methods can be used. According to Box and Hunter [2]: "No model can give a precise description of what happens. A working theoretical model, however, supplies information on the system under study over important ranges of the variables by means of equations which reflect at least the major features of the mechanism."

The engineer in the process industries is usually concerned with the operation of existing plants and the development of new processes. In the first case, the control, improvement, and optimization of the operation are the engineer's main objectives. In order to achieve this, a quantitative representation of the process, a model, is needed that would give the relationship between the various parts of the system. In the design of new processes, the engineer draws information from theory and the literature to construct mathematical models that may be used to simulate the process (see Fig. 7.1). The development of mathematical models often requires the implementation of an experimental program in order to obtain the necessary information for the verification of the models. The experimental program is originally designed based on the theoretical considerations coupled with *a priori* knowledge of the process and is subsequently modified based on the results of regression analysis.

Regression analysis is the application of mathematical and statistical methods for the analysis of the experimental data and the fitting of the mathematical models to these data by the estimation of the unknown parameters of the models. The series of statistical tests, which normally accompany regression analysis, serve in model identification, model verification, and efficient design of the experimental program.

Strictly speaking, a mathematical model of a dynamic system is a set of equations that can be used to calculate how the *state* of the system evolves through time under the action of the *control variables*, given the state of the system at some initial time. The state of the system is described by a set of variables known as *state variables*. The first stage in the development of a mathematical model is to identify the state and control variables.

The control variables are those that can be directly controlled by the experimenter and that influence the way the system changes from its initial state to that of any later time. Examples of control variables in a chemical reaction system may be the temperature, pressure, and/or concentration of some of the components. The state variables are those that describe the state of the system and that are not under direct control. The concentrations of reactants and products are state variables in chemical systems. The distinction between state and control

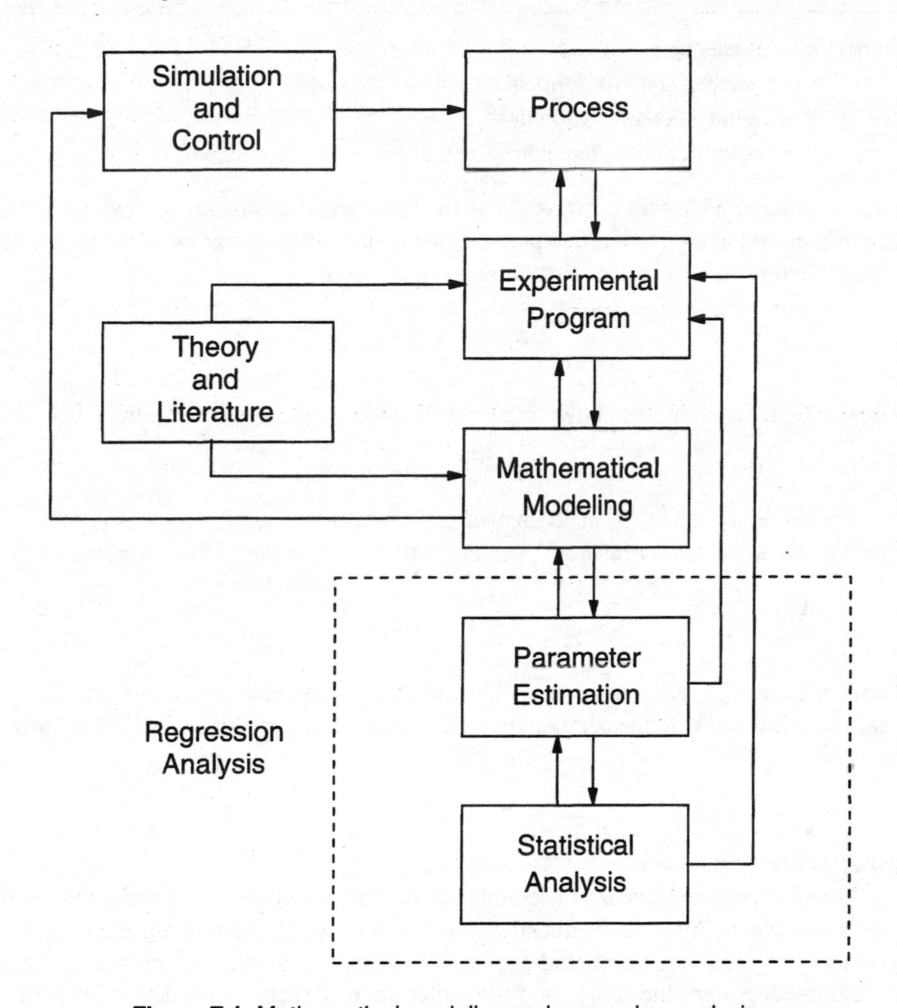

Figure 7.1 Mathematical modeling and regression analysis.

variables is not always fixed but can change when the method of operating the system changes. For example, if temperature is not directly controlled, it becomes a state variable.

The equations comprising the mathematical model of the process are called the *performance equations*. These equations should show the effect of the control variables on the evolution of the state variables. The performance equation may be a set of differential equations and/or a set of algebraic equations. For example, a set of ordinary differential equations describing the dynamics of a process may have the general form:

$$\frac{d\boldsymbol{y}}{dx} = \boldsymbol{g}(x, \boldsymbol{y}, \boldsymbol{\theta}, \boldsymbol{b}) \tag{7.1}$$

where x = independent variable
$\boldsymbol{y}$ = vector of state (dependent) variables
$\boldsymbol{\theta}$ = vector of control variables
$\boldsymbol{b}$ = vector of parameters whose values must be determined.

In this chapter, we concern ourselves with the methods of estimating the parameter vector $\boldsymbol{b}$ using regression analysis. For this purpose, we assume that the vector of control variables $\boldsymbol{\theta}$ is fixed; therefore, the mathematical model simplifies to

$$\frac{d\boldsymbol{y}}{dx} = \boldsymbol{g}(x, \boldsymbol{y}, \boldsymbol{b}) \tag{7.2}$$

In their integrated form, the above set of performance equations convert to

$$\boldsymbol{y} = \boldsymbol{f}(x, \boldsymbol{b}) \tag{7.3}$$

For regression analysis, mathematical models are classified as *linear* or *nonlinear* with respect to the *unknown parameters*. For example, the following differential equation:

$$\frac{dy}{dt} = ky \tag{7.4}$$

which we classified earlier as linear with respect to the dependent variable (see Chap. 5), is *nonlinear* with respect to the parameter k. This is clearly shown by the integrated form of Eq. (7.4):

$$y = y_0 e^{kt} \tag{7.5}$$

where y is highly nonlinear with respect to k.

Most mathematical models encountered in engineering and the sciences are nonlinear in the parameters. Attempts at linearizing these models, by rearranging the equations and regrouping the variables, were common practice in the precomputer era, when graph paper and the straightedge were the tools for fitting models to experimental data. Such primitive techniques have been replaced by the implementation of *linear* and *nonlinear regression* methods on the computer.

The theory of linear regression has been expounded by statisticians and econometricians, and a rigorous statistical analysis of the regression results has been developed. Nonlinear regression is an extension of the linear regression methods used iteratively to arrive at the values of the parameters of the nonlinear models. The statistical analysis of the nonlinear regression results is also an extension of that applied in linear analysis but does not possess the rigorous theoretical basis of the latter.

In this chapter, after giving a brief review of statistical terminology, we develop the basic algorithm of linear regression and then show how this is extended to nonlinear regression. We develop the methods in matrix notation so that the algorithms are equally applicable to fitting single or multiple variables and to using single or multiple sets of experimental data.

7.2 Review of Statistical Terminology Used in Regression Analysis

It is assumed that the reader has a rudimentary knowledge of statistics. This section serves as a review of the statistical definitions and terminology needed for understanding the application of linear and nonlinear regression analysis and the statistical treatment of the results of this analysis. For a more complete discussion of statistics, the reader should consult a standard text on statistics, such as Bethea [3] and Ostle et al. [4].

7.2.1 Population and Sample Statistics

A *population* is defined as a group of similar items, or events, from which a sample is drawn for test purposes; the population is usually assumed to be very large, sometimes infinite. A *sample* is a random selection of items from a population, usually made for evaluating a variable of that population. The *variable* under investigation is a characteristic property of the population.

A *random variable* is defined as a variable that can assume any value from a set of possible values. A *statistic* or *statistical parameter* is any quantity computed from a sample; it is characteristic of the sample, and it is used to estimate the characteristics of the population variable.

Degrees of freedom can be defined as the number of observations made in excess of the minimum theoretically necessary to estimate a statistical parameter or any unknown quantity.

Let us use the notation N to designate the total number of items in the population under study, where $0 \le N \le \infty$, and n to specify the number of items contained in the sample taken from that population, where $0 \le n \le N$. The variable being investigated will be designated as X; it may have discrete values, or it may be a continuous function, in the range $-\infty < x < \infty$. For specific populations, these ranges may be more limited, as will be mentioned below.

For the sake of example and in order to clarify these terms, let us consider for study the entire human population and examine the age of this population. The value of N, in this case, would be approximately 6 billion. The age variable would range from 0 to possibly 150 years. Age can be considered either as a continuous variable, because all ages are possible, or more commonly as a discrete variable, because ages are usually grouped by year. In the continuous case, the age variable takes an infinite number of values in the range $0 < x \le 150$, and in the discrete case it takes a finite number of values x_j, where $j = 1, 2, 3, \ldots, M$ and $M \le 150$. Assume that a random sample of n persons is chosen from the total population (say $n = 1$ million) and the age of each person in the sample is recorded.

The frequency at which each value of the variable (age, in the above example) may occur in the population is not the same; some values (ages) will occur more frequently than others. Designating m_j as the number of times the value of x_j occurs, we can define the concept of *probability of occurrence* as

$$Pr\{X = x_j\} = \begin{Bmatrix} \text{Probability of} \\ \text{occurrence of} \\ x_j \end{Bmatrix} = \frac{\text{number of occurrences of } x_j}{\text{total number of observations}}$$

$$= \lim_{n \to N} \frac{m_j}{n} = p(x_j) \tag{7.6}$$

For a discrete random variable, $p(x_j)$ is called the *probability function,* and it has the following properties:

$$0 \le p(x_j) \le 1$$
$$\sum_{j=1}^{M} p(x_j) = 1 \tag{7.7}$$

The shape of a typical probability function is shown in Fig. 7.2*a*.

For a continuous random variable, the probability of occurrence is measured by the continuous function $p(x)$, which is called the *probability density function*, so that

$$Pr\{x < X \le x + dx\} = p(x)\,dx \tag{7.8}$$

The probability density function has the following properties:

$$0 \le p(x) \le 1$$
$$\int_{-\infty}^{\infty} p(x)\,dx = 1 \tag{7.9}$$

The smooth curve obtained from plotting $p(x)$ versus x (Fig. 7.3*a*) is called the *continuous probability density distribution.*

The *cumulative distribution function* is defined as the probability that a random variable X will not exceed a given value x, that is:

$$Pr\{X \le x\} = P(x) = \int_{-\infty}^{x} p(x)\,dx \tag{7.10}$$

The equivalent of Eq. (7.10) for a discrete random variable is

$$Pr\{X \le x_i\} = P(x_i) = \sum_{j=1}^{i} p(x_j) \tag{7.11}$$

The cumulative distribution functions for discrete and continuous random variables are illustrated in Figs. 7.2*b* and 7.3*b*, respectively.

It is obvious from the integral of Eq. (7.10) that the cumulative distribution function is obtained from calculating the area under the density distribution function. The three area segments shown in Fig. 7.3*a* correspond to the following three probabilities:

$$Pr\{X \le x_a\} = \int_{-\infty}^{x_a} p(x)\,dx \tag{7.12}$$

$$Pr\{x_a < X \le x_b\} = \int_{x_a}^{x_b} p(x)\,dx \tag{7.13}$$

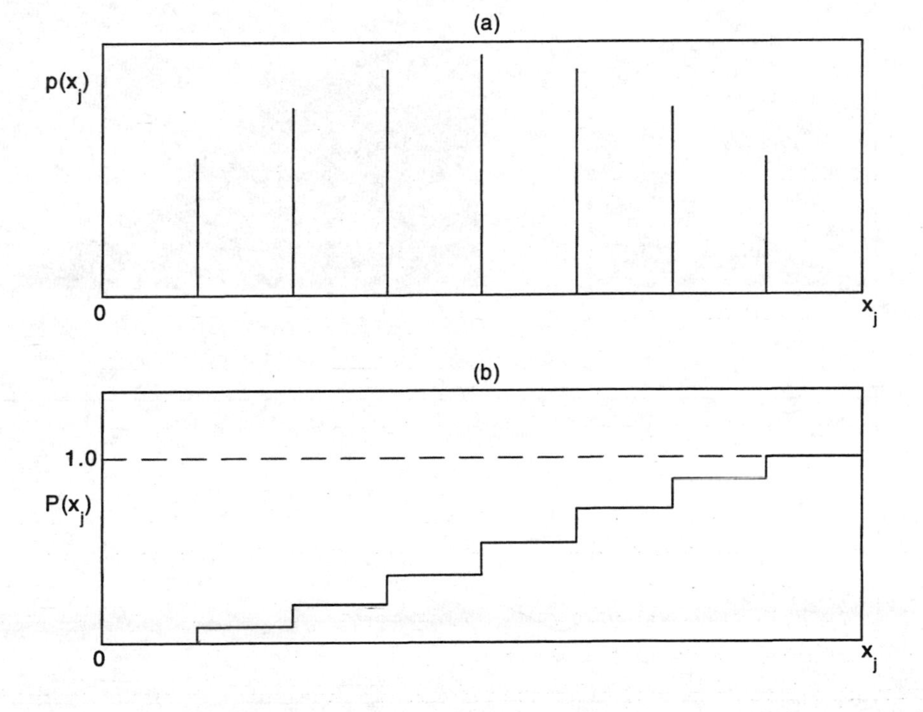

Figure 7.2 (*a*) Probability function and (*b*) cumulative distribution function for discrete random variable.

$$Pr\{X > x_a\} = \int_{x_a}^{\infty} p(x)\,dx \tag{7.14}$$

The *population mean*, or *expected value*, of a discrete random variable is defined as

$$\mu = E[X] = \sum_{j=1}^{M} x_j p(x_j) \tag{7.15}$$

and that of a continuous random variable as

$$\mu = E[X] = \int_{-\infty}^{\infty} x\,p(x)\,dx \tag{7.16}$$

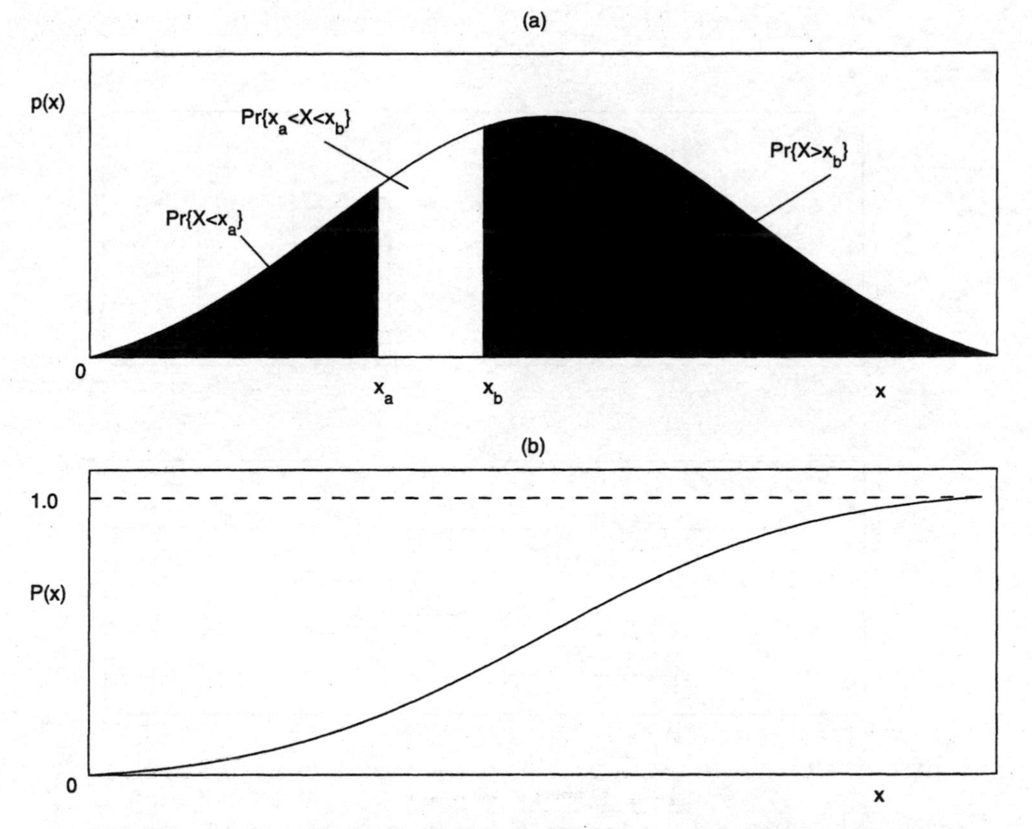

Figure 7.3 (*a*) Probability density function and (*b*) cumulative distribution function for a continuous random variable.

The usefulness of the concept of *expectation*, as defined above, is that it corresponds to our intuitive idea of *average*, or equivalently to the center of gravity of the probability density distribution along the x-axis. It is easy to show that combining Eqs. (7.15) and (7.6) yields the arithmetic average of the random variable for the entire population:

$$\mu = E[X] = \frac{\sum_{i=1}^{N} x_i}{N} \tag{7.17}$$

In addition, the integral of Eq. (7.16) can be recognized from the field of mechanics as the *first noncentral moment of X*.

The *sample mean*, or *arithmetic average*, of a sample of observations is the value obtained by dividing the sum of observations by their total number:

$$\bar{x} = \frac{\sum_{i=1}^{n} x_i}{n} \tag{7.18}$$

The expected value of the sample mean is given by

$$E[\bar{x}] = E\left[\frac{\sum_{i=1}^{n} x_i}{n}\right] = \frac{1}{n}\sum_{i=1}^{n} E[x_i] = \frac{1}{n}\sum_{i=1}^{n} \mu = \mu \tag{7.19}$$

that is, the sample mean is an *unbiased estimate* of the population mean.

In MATLAB the built-in function *mean*($\boldsymbol{x}$) calculates the mean value of the vector $\boldsymbol{x}$ [Eq. (7.18)]. If $\boldsymbol{x}$ is a matrix, *mean*($\boldsymbol{x}$) returns a vector of mean values of each column.

The *population variance* is defined as the expected value of the square of the deviation of the random variable X from its expectation:

$$\begin{aligned} \sigma^2 &= V[X] \\ &= E[(X - E[X])^2] \\ &= E[(X - \mu)^2] \end{aligned} \tag{7.20}$$

For a discrete random variable, Eq. (7.20) is equivalent to

$$\sigma^2 = \sum_{j=1}^{M} (x_j - \mu)^2 p(x_j) \tag{7.21}$$

When combined with Eq. (7.6), Eq. (7.21) becomes

$$\sigma^2 = \frac{\sum_{i=1}^{N} (x_i - \mu)^2}{N} \tag{7.22}$$

which is the arithmetic average of the square of the deviations of the random variable from its mean. For a continuous random variable, Eq. (7.20) is equivalent to

$$\sigma^2 = \int_{-\infty}^{\infty} (x - \mu)^2 p(x)\,dx \tag{7.23}$$

which is the *second central moment of X* about the mean.

It is interesting and useful to note that Eq. (7.20) expands as follows:[1]

$$\begin{aligned} V[X] &= E[(X - E[X])^2] = E[X^2 + (E[X])^2 - 2XE[X]] \\ &= E[X^2] + E[(E[X])^2] - 2E[XE[X]] \\ &= E[X^2] + (E[X])^2 - 2(E[X])^2 \\ &= E[X^2] - (E[X])^2 \\ &= E[X^2] - \mu^2 \end{aligned} \tag{7.24}$$

The positive square root of the population variance is called the *population standard deviation*:

$$\sigma = +\sqrt{\sigma^2} \tag{7.25}$$

The *sample variance* is defined as the arithmetic average of the square of the deviations of x_i from the population mean μ:

$$s^2 = \frac{\sum_{i=1}^{n} (x_i - \mu)^2}{n} \tag{7.26}$$

However, since μ is not usually known, $\bar{x}$ is used as an estimate of μ, and the sample variance is calculated from

$$s^2 = \frac{\sum_{i=1}^{n} (x_i - \bar{x})^2}{n - 1} \tag{7.27}$$

[1] The expected value of a constant is that constant. The expected value of X is a constant; therefore, $E[E[X]] = E[X]$.

where the degrees of freedom have been reduced to $(n - 1)$, because the calculation of the sample mean consumes one degree of freedom. The sample variance obtained from Eq. (7.27) is an unbiased estimate of population variance, that is,

$$E[s^2] = \sigma^2 \tag{7.28}$$

The positive square root of the sample variance is called the sample *standard deviation*:

$$s = +\sqrt{s^2} \tag{7.29}$$

In MATLAB, the built-in function *std*(**x**) calculates the standard deviation of the vector **x** [Eq. (7.29)]. If **x** is a matrix, *std*(**x**) returns a vector of standard deviations of each column.

The covariance of two random variables X and Y is defined as the expected value of the product of the deviations of X and Y from their expected values:

$$Cov[X, Y] = E[(X - E[X])(Y - E[Y])] \tag{7.30}$$

Eq. (7.30) expands to

$$\begin{aligned} Cov[X, Y] &= E[XY - YE[X] - XE[Y] + E[X]E[Y]] \\ &= E[XY] - E[X]E[Y] \end{aligned} \tag{7.31}$$

The covariance is a measurement of the association between the two variables. If large positive deviations of X are associated with large positive deviations of Y, and likewise large negative deviations of the two variables occur together, then the covariance will be positive. Furthermore, if positive deviations of X are associated with negative deviations of Y, and vice versa, then the covariance will be negative. On the other hand, if positive and negative deviations of X occur equally frequently with positive and negative deviations of Y, then the covariance will tend to zero.

In MATLAB, the built-in function *cov*(**x**, **y**) calculates the covariance of the vectors of the same length **x** and **y** [Eq. (7.30)]. If **x** is a matrix where each row is an observation and each column a variable, *cov*(**x**) returns the covariance matrix.

The variance of X, defined earlier in Eq. (7.20), is a special case of the covariance of the random variable with itself:

$$\begin{aligned} Cov[X, X] &= E[(X - E[X])(X - E[X])] \\ &= E[(X - E[X])^2] = V[X] \end{aligned} \tag{7.32}$$

The magnitude of the covariance depends on the magnitude and units of X and Y and could conceivably range from $-\infty$ to ∞. To make the measurement of covariance more manageable, the two dimensionless standardized variables are formed:

$$\frac{X - E[X]}{\sqrt{V[X]}} \quad \text{and} \quad \frac{Y - E[Y]}{\sqrt{V[Y]}}$$

The covariance of the standardized variables is known as the *correlation coefficient*:

$$\rho_{XY} = Cov\left[\frac{X - E[X]}{\sqrt{V[X]}}, \frac{Y - E[Y]}{\sqrt{V[Y]}}\right] \tag{7.33}$$

Using the definition of covariance reduces the correlation coefficient to

$$\rho_{XY} = \frac{Cov[X, Y]}{\sqrt{V[X]V[Y]}} \tag{7.34}$$

If $\rho_{XY} = 0$, we say that X and Y are *uncorrelated*, and this implies that

$$Cov[X, Y] = 0 \tag{7.35}$$

We know from probability theory that if X and Y are *independent variables*, then

$$p\{x, y\} = p_x(x)p_y(y) \tag{7.36}$$

from which it follows that

$$E[XY] = E[X]E[Y] \tag{7.37}$$

Combining Eqs. (7.37) and (7.31) shows that

$$Cov[X, Y] = 0 \tag{7.38}$$

and from Eq. (7.34)

$$\rho_{XY} = 0 \tag{7.39}$$

Thus independent variables are uncorrelated.

In MATLAB, the built-in function $corrcoef(\boldsymbol{x}, \boldsymbol{y})$ calculates the matrix of the correlation coefficients of the vectors of the same length $\boldsymbol{x}$ and $\boldsymbol{y}$ [Eq. (7.34)]. If $\boldsymbol{x}$ is a matrix where each row is an observation and each column a variable, $corrcoef(\boldsymbol{x})$ also returns the correlation coefficients matrix.

The population and sample statistics discussed above are summarized in Table 7.1.

Table 7.1 Summary of population and sample statistics

Statistics	Population: Continuous variable	Population: Discrete variable	Sample
Mean	$\mu = E[X] = \sum_{j=1}^{M} x_j p(x_j)$	$\mu = E[X] = \int_{-\infty}^{\infty} x\,p(x)\,dx$	$\bar{x} = \frac{1}{n}\sum_{i=1}^{n} x_i$
Variance	$\sigma^2 = V[X] = E[(X - E[X])^2]$	$\sigma^2 = V[X] = E[(X - E[X])^2]$	$s^2 = \frac{1}{n}\sum_{i=1}^{n}(x_i - \mu)^2$
			or
	$= \int_{-\infty}^{\infty}(x - \mu)^2 p(x)\,dx$	$= \sum_{j=1}^{M}(x_j - \mu)^2 p(x_j)$	$s^2 = \frac{1}{n-1}\sum_{i=1}^{n}(x_i - \bar{x})^2$
Standard deviation	$\sigma = +\sqrt{\sigma^2}$	$\sigma = +\sqrt{\sigma^2}$	$s = +\sqrt{s^2}$
Covariance	$Cov[X,Y] = E[(X - E[X])(Y - E[Y])]$		
Correlation coefficient	$\rho_{XY} = \frac{Cov[X,Y]}{\sqrt{V[X]V[Y]}}$		

7.2.2 Probability Density Functions and Probability Distributions

There are many different probability density functions encountered in statistical analysis. Of particular interest to regression analysis are the *normal*, χ^2, *t*, and *F distributions*, which will be discussed in this section. The *normal* or *Gaussian* density function has the form:

$$p(x) = \frac{1}{\sqrt{2\pi}\sigma} \exp\left[-\frac{1}{2}\left(\frac{x-\mu}{\sigma}\right)^2\right] \tag{7.40}$$

where $-\infty < x < \infty$. The cumulative distribution function of the normal density function is

$$P(x) = \frac{1}{\sqrt{2\pi}\sigma} \int_{-\infty}^{x} \exp\left[-\frac{1}{2}\left(\frac{x-\mu}{\sigma}\right)^2\right] dx \tag{7.41}$$

which involves an integral that does not have an explicit form and must be integrated numerically (see Chap. 4). The normal probability distributions are illustrated in Fig. 7.4.

The expected value of a variable that has a normal distribution is

$$E[X] = \mu \tag{7.42}$$

and the variance is

$$V[X] = \sigma^2 \tag{7.43}$$

For this reason, the normal density function is usually abbreviated as $N(\mu, \sigma^2)$, and the notation

$$X \sim N(\mu, \sigma^2) \tag{7.44}$$

means that the variable X has normal distribution with expected value μ and variance σ^2.

A normal density function can be transformed to the *standard normal density function* by the substitution

$$u = \frac{x-\mu}{\sigma} \tag{7.45}$$

which transforms Eq. (7.40) to

$$\phi(u) = \frac{1}{\sqrt{2\pi}} \exp\left[-\frac{u^2}{2}\right] \tag{7.46}$$

The expected value of the standardized variable u is

$$E[u] = 0 \tag{7.47}$$

and the variance is

$$V[u] = 1 \tag{7.48}$$

Therefore

$$u \sim N(0,1) \tag{7.49}$$

The standard normal density function $\phi(u)$ and its cumulative distribution function

$$\Phi(u) = \frac{1}{\sqrt{2\pi}} \int_{-\infty}^{u} \exp\left[-\frac{u^2}{2}\right] du \tag{7.50}$$

are shown in Fig. 7.5.

The function $\phi(u)$ is symmetrical about zero; therefore, the area in the left tail, below $-u$, is equal to the area in the right tail, above $+u$ (shaded area in Fig. 7.5). The unshaded area between -1.960 and 1.960 is equivalent to 95 percent of the total area under the density function. This area is designated as $(1 - \alpha)$ and the area under each tail as $\alpha/2$. Application of Eqs. (7.12)-(7.14) shows that

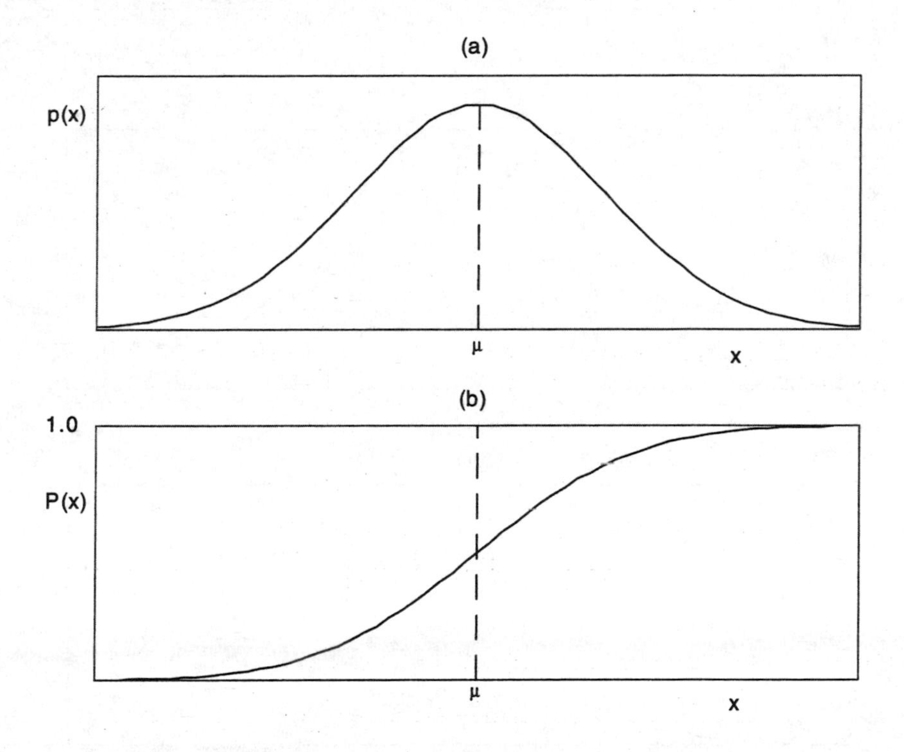

Figure 7.4 (*a*) Normal probability density function and (*b*) normal cumulative distribution function for a continuous random variable.

$$Pr\{u \le -1.960\} = \frac{\alpha}{2} = 0.025$$

$$Pr\{-1.960 < u \le 1.960\} = 1 - \alpha = 0.95 \tag{7.51}$$

$$Pr\{u > 1.960\} = \frac{\alpha}{2} = 0.025$$

If a set of normally distributed variables X_j, where

$$X_k \sim N(\mu_k, \sigma_k^2) \tag{7.52}$$

is linearly combined to form another variable Y, where

$$Y = \sum_k a_k X_k \tag{7.53}$$

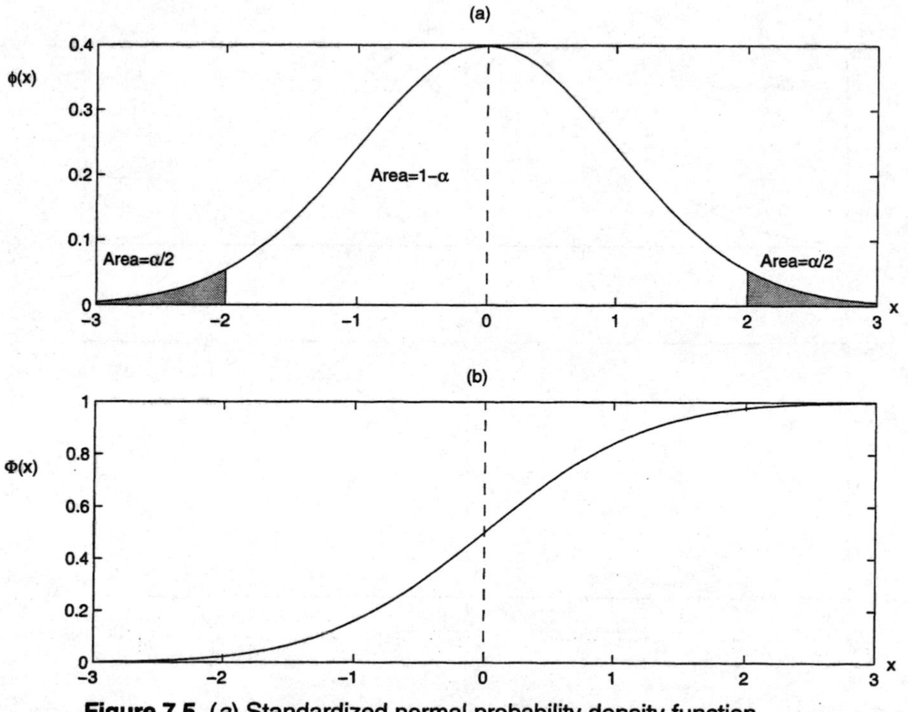

Figure 7.5 (*a*) Standardized normal probability density function.
(*b*) Standardized normal cumulative distribution function.

then Y is also normally distributed, that is,

$$Y \sim N\left(\sum_k a_k \mu_k , \sum_k a_k^2 \sigma_k^2 \right) \tag{7.54}$$

The sample mean [Eq. (7.18)] of a normally distributed population is a linear combination of normally distributed variables; therefore, the sample mean itself is normally distributed:

$$\bar{x} \sim N\left(\mu , \frac{\sigma^2}{n} \right) \tag{7.55}$$

It follows then, from Eqs. (7.45) and (7.49), that

$$\frac{\bar{x} - \mu}{\sqrt{\sigma^2 / n}} \sim N(0, 1) \tag{7.56}$$

If we wish to test the hypothesis that a sample, whose mean is $\bar{x}$, could come from a normal distribution of mean μ and known variance σ^2, the procedure is easy, because the variable $(\bar{x} - \mu)/\sqrt{\sigma^2/n}$ is normally distributed as $N(0, 1)$ and can readily be compared with tabulated values. However, if σ^2 is unknown and must be estimated from the sample variance s^2, then *Student's t distribution*, which is described later in this section, is needed.

Now consider a sequence X_j of identically distributed, independent random variables (not necessarily normally distributed) whose second-order moment exists. Let

$$E[X_k] = \mu \tag{7.57}$$

and

$$E[(X_k - \mu)^2] = V[X_k] = \sigma^2 \tag{7.58}$$

for every k. Consider the random variable Z_n defined by

$$Z_n = X_1 + X_2 + \dots + X_n \tag{7.59}$$

where

$$E[Z_n] = n\mu \tag{7.60}$$

and, by independence of X_j:

$$E[(Z_n - n\mu)^2] = n\sigma^2 \tag{7.61}$$

Let

$$\hat{Z}_n = \frac{Z_n - n\mu}{\sigma\sqrt{n}} \tag{7.62}$$

then the distribution of $\hat{Z}_n$ approaches the standard normal distribution, that is,

$$\lim_{n \to \infty} P_n(z) = \frac{1}{\sqrt{2\pi}} \int_{-\infty}^{z} \exp\left[-\frac{z^2}{2}\right] dz \tag{7.63}$$

This is the *central limit theorem*, a proof of which can be found in Sienfeld and Lapidus [5]. This is a very important theorem of statistics, particularly in regression analysis where experimental data are being analyzed. The experimental error is a composite of many separate errors whose probability distributions are not necessarily normal distributions. However, as the number of components contributing to the error increases, the central limit theorem justifies the assumption of normality of the error.

Suppose we have a set of ν independent observations, $x_1, \ldots, x_j$ from a normal distribution $N(\mu, \sigma^2)$. The standardized variables

$$u_i = \frac{x_i - \mu}{\sigma} \tag{7.64}$$

will also be independent and have distribution $N(0, 1)$. The variable $\chi^2(\nu)$ is defined as the sum of the squares of u_i:

$$\chi^2(\nu) = \sum_{i=1}^{\nu} u_i^2 = \sum_{i=1}^{\nu} \frac{(x_i - \mu)^2}{\sigma^2} \tag{7.65}$$

The $\chi^2(\nu)$ variable has the so-called χ^2 *(chi-square) distribution function,* which is given by

$$p(\chi^2) = \frac{1}{2^{\nu/2}\Gamma(\nu/2)} e^{-\chi^2/2}(\chi^2)^{(\nu/2)-1} \tag{7.66}$$

where $\chi^2 \geq 0$ and

$$\Gamma(\nu/2) = \int_0^{\infty} e^{-x} x^{(\nu/2)-1} dx \tag{7.67}$$

The χ^2 distribution is a function of the degrees of freedom ν, as shown in Fig. 7.6. The distribution is confined to the positive half of the χ^2-axis, as the u_i^2 quantities are always positive.

The expected value of χ^2 variable is

$$\mu = E[\chi^2] = \int_0^{\infty} \chi^2 p(\chi^2)\, d\chi^2 = \nu \tag{7.68}$$

and its variance is

$$\sigma^2 = V[\chi^2] = \int_0^{\infty} (\chi^2)^2 p(\chi^2)\, d\chi^2 = 2\nu \tag{7.69}$$

The χ^2 distribution tends toward the normal distribution $N(\nu, 2\nu)$ as ν becomes large. The χ^2 distribution is widely used in statistical analysis for testing independence of variables and the fit of probability distributions to experimental data.

We saw earlier that the sample variance was obtained from Eq. (7.27):

$$s^2 = \frac{\sum_{i=1}^{n} (x_i - \bar{x})^2}{n - 1} \tag{7.27}$$

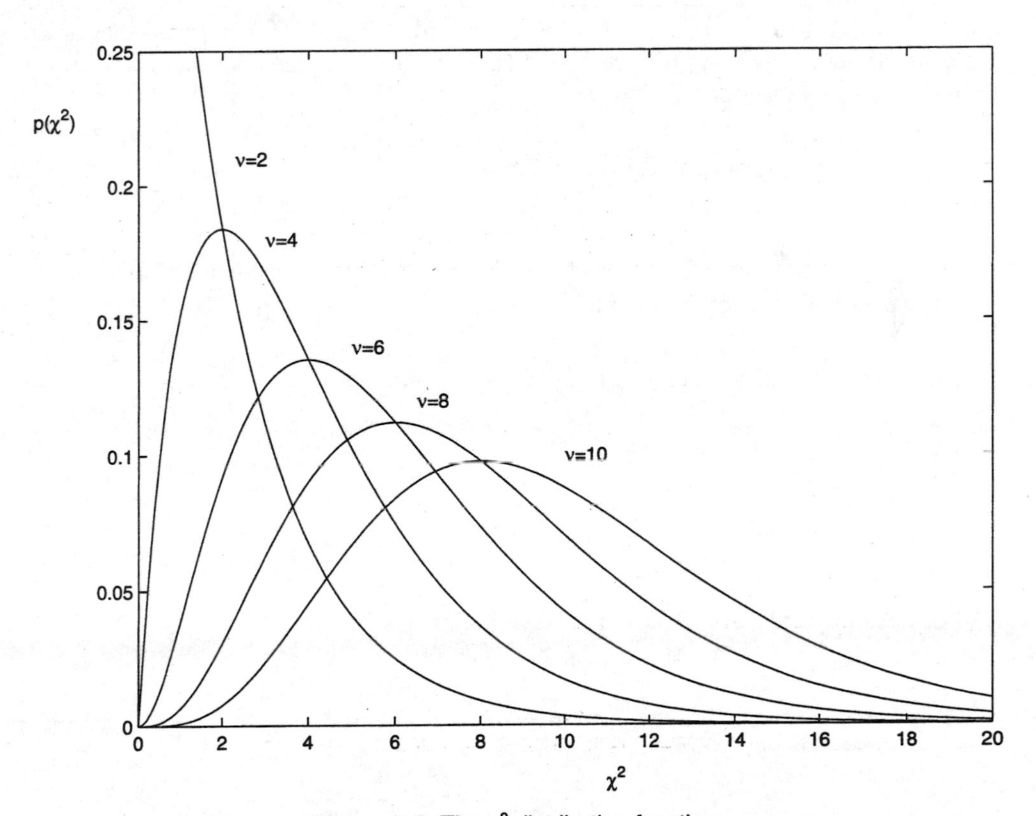

Figure 7.6 The χ^2 distribution function.

with $(n - 1)$ degrees of freedom. When $\bar{x}$ is assumed to be equal to μ then

$$s^2 = \frac{\sum_{i=1}^{n} (x_i - \mu)^2}{n - 1} \tag{7.70}$$

Combining Eqs. (7.65) and (7.70) shows that

$$\chi^2 = (n - 1)\frac{s^2}{\sigma^2} \tag{7.71}$$

with $\nu = (n - 1)$ degrees of freedom. This equation will be very useful in Sec. 7.2.3 in constructing confidence intervals for the population variance.

Let us define a new random variable t, so that

$$t = \frac{u}{\sqrt{\chi^2/\nu}} \tag{7.72}$$

where $u \sim N(0, 1)$ and χ^2 is distributed as chi-square with ν degrees of freedom. It is assumed that u and χ^2 are independent of each other. The variable t is called *Student's t* and has the probability density function

$$p(t) = \frac{1}{\sqrt{\nu\pi}} \frac{\Gamma[(\nu + 1)/2]}{\Gamma(\nu/2)} \left(1 + \frac{t^2}{\nu}\right)^{-(\nu + 1)/2} \tag{7.73}$$

with ν degrees of freedom. The shape of the t density function is shown in Fig. 7.7.

The expected value of the t variable is

$$\mu_t = E[t] = \int_{-\infty}^{\infty} t\,p(t)\,dt = 0 \qquad \text{for } \nu > 1 \tag{7.74}$$

and the variance is

$$\sigma_t^2 = V[t] = \int_{-\infty}^{\infty} t^2 p(t)\,dt = \frac{\nu}{\nu - 2} \qquad \text{for } \nu > 2 \tag{7.75}$$

The t distribution tends toward the normal distribution as ν becomes large.

Combining Eq. (7.72) with (7.56) and (7.71) gives

$$t = \frac{(\bar{x} - \mu)/\sqrt{\sigma^2/n}}{\sqrt{s^2/\sigma^2}} = \frac{\bar{x} - \mu}{\sqrt{s^2/n}} \tag{7.76}$$

The quantity on the right-hand side of Eq. (7.76) in independent of σ and has a t distribution. Therefore, the t distribution provides a test of significance for the deviation of a sample mean from its expected value when the population variance is unknown and must be estimated from the sample variance.

Finally, we define the ratio

$$F(\nu_1, \nu_2) = \frac{\chi_1^2/\nu_1}{\chi_2^2/\nu_2} \tag{7.77}$$

where χ_1^2 and χ_2^2 are two independent random variables of chi-square distribution with ν_1 and ν_2 degrees of freedom, respectively. The variable $F(\nu_1, \nu_2)$ has the F distribution density function with ν_1 and ν_2 degrees of freedom:

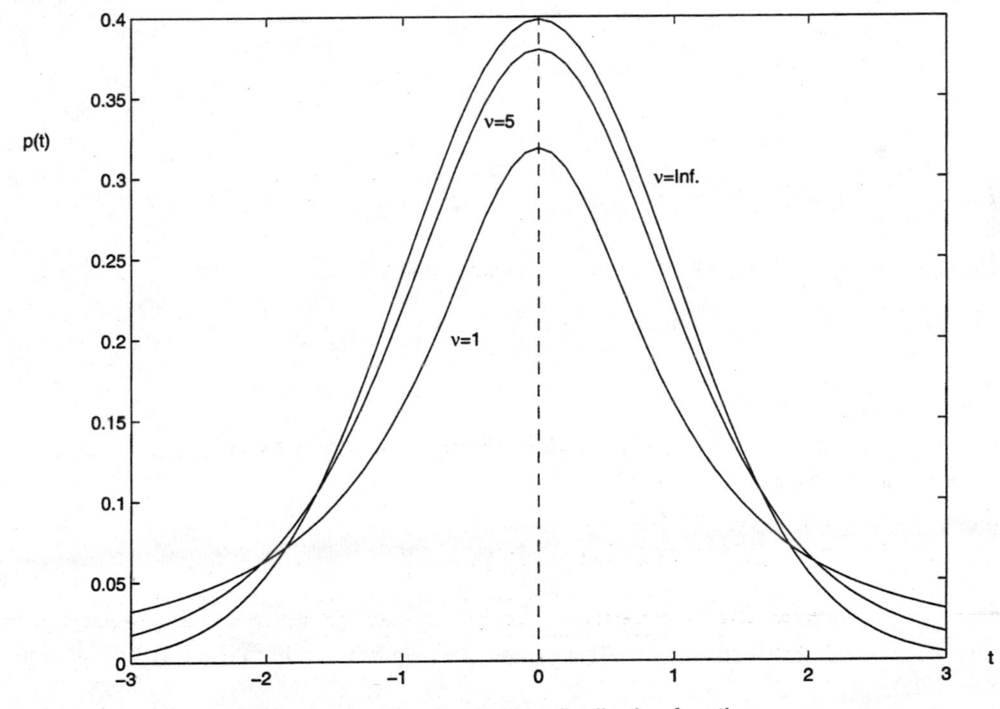

Figure 7.7 The Student's *t* distribution function.

$$p(F) = \frac{\left(\frac{\nu_1}{\nu_2}\right)^{\nu_1/2} F^{(\nu_1/2)-1}\left(1 + \frac{\nu_1}{\nu_2}F\right)^{-(\nu_1+\nu_2)/2}}{\int_0^1 x^{(\nu_1/2)-1}(1 - x)^{(\nu_2/2)-1}\,dx} \tag{7.78}$$

The F distribution is very useful in the analysis of variance of populations. Consider two normally distributed independent random samples:

$$x_{1,1}\,,\, x_{1,2}\,,\, \dots\,,\, x_{1,n_1}$$

and

$$x_{2,1}\,,\, x_{2,2}\,,\, \dots\,,\, x_{2,n_1}$$

The first sample which has a sample variance s_1^2, is from a population with mean μ_1 and variance σ_1^2. The second sample which has a sample variance s_2^2 is from a population with mean μ_2 and variance σ_2^2. Using Eq. (7.71), we see that

$$\chi_1^2 = (n_1 - 1)\frac{s_1^2}{\sigma_1^2} \tag{7.79}$$

and

$$\chi_2^2 = (n_2 - 1)\frac{s_2^2}{\sigma_2^2} \tag{7.80}$$

Combining Eq. (7.77) with (7.79) and (7.80) shows that

$$F(n_1 - 1, n_2 - 1) = \frac{\chi_1^2/(n_1 - 1)}{\chi_2^2/(n_2 - 1)} = \frac{s_1^2/\sigma_1^2}{s_2^2/\sigma_2^2} \tag{7.81}$$

with $(n_1 - 1)$ and $(n_2 - 1)$ degrees of freedom. Furthermore, if the two populations have the same variance, that is, if $\sigma_1^2 = \sigma_2^2$, then

$$F(n_1 - 1, n_2 - 1) = \frac{s_1^2}{s_2^2} \tag{7.82}$$

Therefore, the F distribution provides a means of comparing variances, as will be seen in Sec. 7.2.3.

7.2.3 Confidence Intervals and Hypothesis Testing

The concept of *confidence interval* is of considerable importance in regression analysis. A confidence interval is a range of values defined by an upper and a lower limit, the *confidence limits*. This range is constructed in such a way that we can say with certain confidence that the true value of the statistic being examined lies within this range. The level of confidence is chosen at $100(1 - \alpha)$ percent, where α is usually small, say, 0.05 or 0.01. For example, when $\alpha = 0.05$, the confidence level is 95 percent. We demonstrate the concept of confidence interval by first constructing such an interval for the standard normal distribution, extending the concept to other distributions, and then calculating specific confidence intervals for the mean and variance.

We saw earlier that the standard normal variable u has a density function $\phi(u)$ [Eq. (7.46)] and a cumulative distribution function $\Phi(u)$ [Eq. (7.50)] and is distributed with $N(0, 1)$. Applying Eqs. (7.12) and (7.13) to standard normal distribution:

$$Pr\{u \le u_{\alpha/2}\} = \int_{-\infty}^{u_{\alpha/2}} \phi(u)\,du = \Phi(u_{\alpha/2}) = \frac{\alpha}{2} \tag{7.83}$$

$$Pr\{u \le u_{1-\alpha/2}\} = \int_{-\infty}^{u_{1-\alpha/2}} \phi(u)\,du = \Phi(u_{1-\alpha/2}) = 1 - \frac{\alpha}{2} \tag{7.84}$$

and

$$\begin{aligned} Pr\{u_{\alpha/2} < u \le u_{1-\alpha/2}\} &= \int_{u_{\alpha/2}}^{u_{1-\alpha/2}} \phi(u)\,du \\ &= \Phi(u_{1-\alpha/2}) - \Phi(u_{\alpha/2}) \\ &= 1 - \alpha \end{aligned} \tag{7.85}$$

The inequality

$$u_{\alpha/2} < u \le u_{1-\alpha/2} \tag{7.86}$$

defines the $100(1 - \alpha)$ precent interval for the variable u. If $\alpha = 0.05$, then the 95 percent confidence interval for the standard normal variable is

$$-1.96 < u \le 1.96 \tag{7.87}$$

Let us now determine a confidence interval for the mean of a normally distributed population. We saw earlier that the sample mean $\bar{x}$ of a normally distributed population is also normally distributed:

$$\bar{x} \sim N\left(\mu, \frac{\sigma^2}{n}\right) \tag{7.55}$$

and that this can be converted to the standard normal distribution so that

$$\frac{\bar{x} - \mu}{\sqrt{\sigma^2/n}} \sim N(0, 1) \tag{7.56}$$

Since the quantity $\left[(\bar{x} - \mu)/\sqrt{\sigma^2/n}\right]$ is equivalent to u, Eq. (7.85) can be written as

$$Pr\left\{u_{\alpha/2} < \frac{\bar{x} - \mu}{\sqrt{\sigma^2/n}} \leq u_{(1-\alpha/2)}\right\} = 1 - \alpha \tag{7.88}$$

or rearranged to

$$Pr\left\{\bar{x} - u_{(1-\alpha/2)}\sqrt{\frac{\sigma^2}{n}} \leq \mu < \bar{x} - u_{\alpha/2}\sqrt{\frac{\sigma^2}{n}}\right\} = 1 - \alpha \tag{7.89}$$

The inequality[2]

$$\bar{x} - u_{(1-\alpha/2)}\sqrt{\frac{\sigma^2}{n}} \leq \mu < \bar{x} + u_{(1-\alpha/2)}\sqrt{\frac{\sigma^2}{n}} \tag{7.90}$$

is the $100(1 - \alpha)$ percent confidence interval for the population mean. For $\alpha = 0.05$, the 95 percent confidence interval of the mean of a normally distributed population is

$$\bar{x} - 1.96\sqrt{\frac{\sigma^2}{n}} \leq \mu < \bar{x} + 1.96\sqrt{\frac{\sigma^2}{n}} \tag{7.91}$$

where $\bar{x}$ is the sample mean and σ^2 is the population variance. This simply says that we can state with 95 percent confidence that the true value of the population mean is in the range defined by the inequality (7.91).

If the population variance σ^2 is not known, it will be estimated from the sample variance s^2. Replacing σ^2 with s^2 in the quantity $[(\bar{x} - \mu)/\sqrt{\sigma^2/n}]$, we obtain the variable

$$t = \frac{\bar{x} - \mu}{\sqrt{s^2/n}} \tag{7.76}$$

---

[2] Note that the density distribution of u is symmetrical around $u = 0$, so that $u_{\alpha/2} = -u_{1-\alpha/2}$. This substitution has been made in obtaining (7.90).

which has a Student's t distribution with $\nu = (n - 1)$ degrees of freedom, as shown in Sec. 7.2.2. The confidence interval in this case is obtained from

$$Pr\left\{t_{\alpha/2} < \frac{\bar{x} - \mu}{\sqrt{s^2/n}} \le t_{1-\alpha/2}\right\} = 1 - \alpha \tag{7.92}$$

which rearranges to

$$Pr\left\{\bar{x} - t_{1-\alpha/2}\sqrt{\frac{s^2}{n}} \le \mu < \bar{x} - t_{\alpha/2}\sqrt{\frac{s^2}{n}}\right\} = 1 - \alpha \tag{7.93}$$

to yield the $100(1 - \alpha)$ percent confidence interval[3]

$$\bar{x} - t_{1-\alpha/2}\sqrt{\frac{s^2}{n}} \le \mu < \bar{x} + t_{1-\alpha/2}\sqrt{\frac{s^2}{n}} \tag{7.94}$$

In Sec. 7.2.2 we showed that the sample variance s^2 and the population variance σ^2 were related through the χ^2 distribution:

$$\chi^2 = (n - 1)\frac{s^2}{\sigma^2} \tag{7.71}$$

with $\nu = (n - 1)$ degrees of freedom. This relation can now be used to construct the confidence interval for the variance from

$$Pr\left\{\chi^2_{\alpha/2} < (n - 1)\frac{s^2}{\sigma^2} \le \chi^2_{1-\alpha/2}\right\} = 1 - \alpha \tag{7.95}$$

which gives the $100(1 - \alpha)$ percent confidence interval for σ^2 as

$$\frac{(n - 1)s^2}{\chi^2_{1-\alpha/2}} \le \sigma^2 < \frac{(n - 1)s^2}{\chi^2_{\alpha/2}} \tag{7.96}$$

This discussion leads us to the concept of *hypothesis testing*. This consists of making an assumption about the distribution function of a random variable, very often about the numerical values of the statistical parameters of the distribution function (mean and variance), and deciding whether those values of the parameters are consistent with our sample of observations on that random variable.

For example, suppose that a sample of $n_1 = 10$ observations has a sample mean, $\bar{x}_2 = 2.0$, and a sample variance, $s_1^2 = 4$. Let us make the assumption that this sample came from a population that has a normal distribution with $\mu = 0$ and σ_1^2 unknown, that is, $X \sim N(0, \sigma_1^2)$.

---

[3] The density distribution of the t variable is symmetrical around $t = 0$, so that $t_{\alpha/2} = -t_{(1-\alpha/2)}$. This substitution has been made in obtaining (7.94).

In order to test this assumption, we formalize it by stating the *null hypothesis*:

$$H_0 : \mu = \mu_0 = 0 \tag{7.97}$$

and the *alternative hypothesis*:

$$H_A : \mu = \mu_A \neq 0 \tag{7.98}$$

We recall from Eq. (7.76) that the quantity $\left[(\bar{x}_1 - \mu)/\sqrt{s_1^2/n_1}\right]$ has a t distribution, and from Eq. (7.92) that

$$Pr\{t_{\alpha/2} < t \le t_{1-\alpha/2}\} = 1 - \alpha \tag{7.99}$$

For 95 percent probability and $\nu = (n_1 - 1) = 9$ degrees of freedom, the above equation is

$$Pr\{-2.045 < t \le 2.045\} = 0.95 \tag{7.100}$$

The region defined by Eq. (7.100) is shown in Fig. 7.8 as the *region of acceptance*, wheras the regions outside this range are labeled *regions of rejection*. Based on the assumption that the null hypothesis is true, if the statistic calculated from the experimental sample falls outside the region of acceptance, the null hypothesis is rejected and H_A is accepted. Otherwise, H_0 is accepted and H_A is rejected. In this example, we calculate t_0:

$$t_0 = \frac{\bar{x}_1 - \mu_0}{\sqrt{s_1^2/n_1}} = \frac{2.0 - 0}{\sqrt{\frac{4}{10}}} = 3.16 \tag{7.101}$$

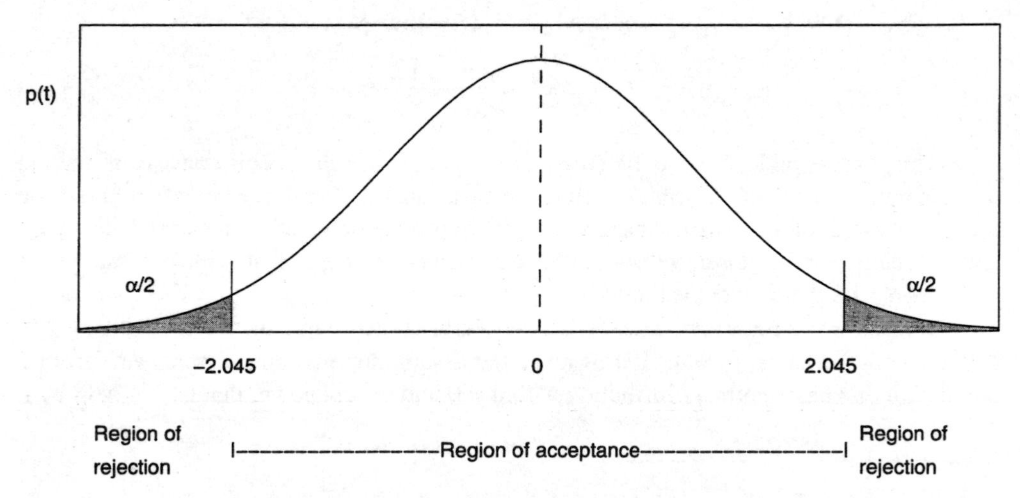

Figure 7.8 Hypothesis test for the mean.

We see that t_0 is outside the region of acceptance defined by Eq. (7.100); therefore, we reject the null hypothesis.

We can generalize this test by saying that if

$$\left| \frac{\bar{x} - \mu_0}{\sqrt{s^2/n}} \right| > t_{(1-\alpha/2)} \tag{7.102}$$

then the null hypothesis that $\mu = \mu_0$ must be rejected. This is the well-known *two-side t test,* which is used extensively in regression analysis to test the values of regression parameters.

Let us now examine the variance of the sample. We draw a second sample of $n_2 = 21$ observations and find that $\bar{x}_2 = 2.0$ and $s_2^2 = 3$. We ask the question: "Is the second sample taken from the same population as the first sample, or from one that has a different variance than the first?" We state that null hypothesis:

$$H_0 : \frac{\sigma_1^2}{\sigma_2^2} = 1 \tag{7.103}$$

and the alternative hypothesis:

$$H_A : \frac{\sigma_1^2}{\sigma_2^2} \neq 1 \tag{7.104}$$

We recall from Eq. (7.81) that the ratio $(s_1^2/\sigma_1^2)/(s_2^2/\sigma_2^2)$ has an F distribution with (ν_1, ν_2) degrees of freedom. From the probability distribution function

$$Pr\{F_{\alpha/2}(\nu_1, \nu_2) < F \leq F_{(1-\alpha/2)}(\nu_1, \nu_2)\} = 1 - \alpha \tag{7.105}$$

To test the null hypothesis at the 95 percent confidence level, obtain the values of $F_{0.025}(9, 20)$ and $F_{0.975}(9, 20)$ for this example from the F distribution tables.[4] Therefore, the interval of acceptance is given by

$$Pr\left\{0.272 < \frac{s_1^2/\sigma_1^2}{s_2^2/\sigma_2^2} \leq 2.84\right\} = 0.95 \tag{7.106}$$

The null hypothesis assumes that $\sigma_1^2 = \sigma_2^2$; therefore, the inequality becomes

$$0.272 < \frac{s_1^2}{s_2^2} \leq 2.84 \tag{7.107}$$

[4] The value of $F_{\alpha/2}(\nu_1, \nu_2)$ is obtained from the relationship: $F_{\alpha/2}(\nu_1, \nu_2) = 1/F_{(1-\alpha/2)}(\nu_2, \nu_1)$.

For this example:

$$\frac{s_1^2}{s_2^2} = \frac{4}{3} = 1.33 \tag{7.108}$$

therefore, the null hypothesis can be accepted.

This is the *two-side F test* used in the analysis of variance of regression results to test the adequacy of a model in fitting the experimental data (see Sec. 7.5).

Hypothesis testing is an involved procedure, which we have briefly introduced here. It is outside the scope of this chapter to discuss hypothesis testing in more depth. The interested reader is referred to Bethea [3] and Ostle et al. [4] for further discussion.

7.3 Linear Regression Analysis

Most mathematical models in engineering and science are nonlinear in the parameters. However, for a complete understanding of nonlinear regression methods, it is necessary to develop first the linear regression case and show how this extends to nonlinear models.

The exact representation of a linear relationship may be shown as

$$y = \alpha + \beta x \tag{7.109}$$

where y represents the true value of the dependent variable, x is the true value of the independent variable, β is the slope of the line, and α is the y-intercept of the line. This deterministic relationship is not useful in this form because it requires knowledge of the true values of y and x. Instead, the linear model is rewritten in terms of the observations of the values of the variables

$$\boldsymbol{Y}^* = \alpha + \beta \boldsymbol{X} + \boldsymbol{u} \tag{7.110}$$

where $\boldsymbol{Y}^*$ is the vector of observations of the dependent variable, $\boldsymbol{X}$ is the vector of observations of the independent variable, and $\boldsymbol{u}$ is the vector of *disturbance terms*. The purpose of the $\boldsymbol{u}$ term is to characterize the discrepancies that emerge between the true values and the observed values of the variables. These discrepancies can be attributed mainly to experimental error. Later in this section, $\boldsymbol{u}$ will be assumed to be a stochastic variable with some specified probability distribution.

Eq. (7.110) can be extended to include more than one independent variable:

$$\boldsymbol{Y}^* = \beta_1 \boldsymbol{X}_1 + \beta_2 \boldsymbol{X}_2 + \ldots + \beta_k \boldsymbol{X}_k + \boldsymbol{u} \tag{7.111}$$

where $\boldsymbol{X}_1, \boldsymbol{X}_2, \ldots, \boldsymbol{X}_k$ are the vectors of observations of k independent variables. To allow for a y-intercept, the vector $\boldsymbol{X}_1$ can be taken as a vector whose components are all unity; thus, β_1 becomes the parameter specifying the value of the y-intercept.

Eq. (7.111) can be condensed to matrix form

$$Y^* = X\beta + u \tag{7.112}$$

where Y^* = $(n \times 1)$ vector of observations of the dependent variable
X = $(n \times k)$ matrix of observations of the independent variables
β = $(k \times 1)$ vector of parameters
u = $(n \times 1)$ vector of disturbance terms
n = number of observations.

Given a set of n observations in the Y variable and in each of the k independent variables, the problem now is to obtain an estimate of the β vector.

The basic assumptions made in the derivation of the method for estimating the parameters are the following:

1. The disturbance terms, represented by the vector u, are random variables with zero expectation, that is,

$$E[u] = \bar{u} = 0 \tag{7.113}$$

Because the variable u is the sum of errors from several sources, the central limit theorem implies that the distribution of u tends toward the normal distribution as the number of factors contributing to u increases.

$$\begin{aligned} V[u_1] &= E[(u_1 - \bar{u}_1)^2] = \sigma^2 \\ V[u_2] &= E[(u_2 - \bar{u}_2)^2] = \sigma^2 \\ &\vdots \\ V[u_n] &= E[(u_n - \bar{u}_n)^2] = \sigma^2 \end{aligned} \tag{7.114}$$

2. The variance of the distribution of u is constant and independent of X, that is, In addition, the values of u for each set of observations are independent of one another, that is,

$$E[u_i u_j] = E[u_i]E[u_j] \qquad i \neq j \tag{7.115}$$

From assumption 1 and from Eqs. (7.31) and (7.37), we also conclude that the covariance of u is zero:

$$Cov[u_i, u_j] = 0 \qquad i \neq j \tag{7.116}$$

The variance-covariance matrix is defined as

$$Var\text{-}Cov[\boldsymbol{u}] = \begin{bmatrix} V[u_1] & Cov[u_1, u_2] & \dots & Cov[u_1, u_n] \\ Cov[u_2, u_1] & V[u_2] & \dots & Cov[u_2, u_n] \\ \dots & \dots & \dots & \dots \\ Cov[u_n, u_1] & Cov[u_n, u_2] & \dots & V[u_n] \end{bmatrix}$$

$$= E[(\boldsymbol{u} - E[\boldsymbol{u}])(\boldsymbol{u} - E[\boldsymbol{u}])'] \tag{7.117}$$

Combining Eqs. (7.113), (7.114), (7.116), and (7.117), we obtain

$$Var\text{-}Cov[\boldsymbol{u}] = E[\boldsymbol{u}\boldsymbol{u}']$$

$$= \begin{bmatrix} \sigma^2 & 0 & \dots & 0 \\ 0 & \sigma^2 & \dots & 0 \\ \dots & \dots & \dots & \dots \\ 0 & 0 & \dots & \sigma^2 \end{bmatrix} = \sigma^2 \boldsymbol{I} \tag{7.118}$$

In summary, Eq. (7.118) says that each $\boldsymbol{u}$ distribution has the same variance, and that all distributions are pairwise uncorrelated.

3. The matrix $\boldsymbol{X}$ is a set of fixed numbers, that is, the values of $\boldsymbol{X}$ do not contain error.
4. The rank of the matrix $\boldsymbol{X}$ is equal to k, and $k < n$. The first part of this assumption ensures that k variables are linearly independent. The second part requires that the number of observations exceeds the number of parameters to be estimated. This is essential in order to have the necessary degrees of freedom for parameter estimation.
5. The vector $\boldsymbol{u}$ has a multivariate normal distribution

$$\boldsymbol{u} = N(0, \sigma^2 \boldsymbol{I}) \tag{7.119}$$

These assumptions are not overly restrictive. Since the value of $\boldsymbol{u}$ is due to many factors acting in opposite directions, it should be expected that small values of $\boldsymbol{u}$ occur more frequently than large values, and that $\boldsymbol{u}$ is a variable with a probability distribution centered at zero and having a finite variance σ^2. This is true when the form of Eq. (7.112) is close to the correct relationship. Because of the many factors involved, the central limit theorem would further suggest that $\boldsymbol{u}$ has a normal distribution, which gives the parameter estimates the desirable property of being maximum-likelihood estimates. Later on in the discussion, it will be shown that the regression method can handle cases where σ^2 is not constant, and where $\boldsymbol{u}$ is not independent of $\boldsymbol{X}$.

7.3.1 The Least Squares Method

Let us consider the hypothesized linear model

$$Y^* = X\beta + u \tag{7.112}$$

Let b denote a k-element vector that is an estimate of the parameter vector β. We use this estimate to define a vector of residuals:

$$\epsilon = Y^* - Xb = Y^* - Y \tag{7.120}$$

These residuals are the differences between the experimental observations Y^* and the calculated values of Y using the estimated vector b. A common way for evaluation of the unknown vector b is the *least squares method,* which minimizes the sum of the squared residuals Φ:

$$\Phi = \epsilon'\epsilon = (Y^* - Xb)'(Y^* - Xb) \tag{7.121}$$

In order to calculate the vector b, which minimizes Φ, we take the partial derivative of Φ with respect to b and set it equal to zero:

$$\frac{\partial \Phi}{\partial b} = (-X)'(Y^* - Xb) + (Y^* - Xb)'(-X) = 0 \tag{7.122}$$

We simplify this utilizing the matrix-vector identity $\mathbf{A}'y = y'\mathbf{A}$:

$$-2X'(Y^* - Xb) = 0 \tag{7.123}$$

Eq. (7.123) can be further rearranged to yield

$$(X'X)b = X'Y^* \tag{7.124}$$

The above constitute a set of simultaneous linear algebraic equations, called the *normal equations.* The matrix $(X'X)$ is a $(k \times k)$ symmetric matrix. Assumption 4 made earlier guarantees that $(X'X)$ is nonsingular; therefore, its inverse exists. Thus, the normal equations can be solved for the vector b:

$$b = (X'X)^{-1}X'Y^* \tag{7.125}$$

The values of the elements of vector b can be obtained readily from Eq. (7.125), because the right-hand side of this equation contains the matrix of observations of the independent variables X and the vector of observations of the dependent variable Y^*, all of which are known.

Polynomial regression may be considered as a special case of linear regression. In such a case the relationship between independent and dependent variables is expressed by the following $(k - 1)$ degree polynomial:

$$y = b_1 + b_2 x + b_3 x^2 + \dots + b_k x^{k-1} \tag{7.126}$$

We may consider $x^0, x^1, x^2, \dots, x^{k-1}$ as independent variables X_1 to X_k and construct the matrix $\boldsymbol{X}$ for the polynomial regression as

$$\boldsymbol{X} = \begin{bmatrix} 1 & x_1 & x_1^2 & \dots & x_1^{k-1} \\ 1 & x_2 & x_2^2 & \dots & x_2^{k-1} \\ & & \dots\dots & & \\ 1 & x_n & x_n^2 & \dots & x_n^{k-1} \end{bmatrix} \tag{7.127}$$

The vector of coefficients of the polynomial (7.126) is then calculated from Eq. (7.125).

In MATLAB, the function *polyfit* does the polynomial regression. The statement *polyfit(X, Y, N)* evaluates the coefficients of the *N*th order polynomial fitted to the data points given in the vectors *X* (independent variable) and *Y* (dependent variable). Note that *polyfit* returns the coefficients in the descending order, which is the opposite of what is shown in Eq. (7.126).

7.3.2 Properties of the Estimated Vector of Parameters

The vector $\boldsymbol{b}$ is an estimate of $\boldsymbol{\beta}$, which minimizes the sum of the squared residuals, irrespective of any distribution properties of the residuals. In addition, $\boldsymbol{b}$ is an unbiased estimate of $\boldsymbol{\beta}$. To show this, we combine Eqs. (7.125) and (7.112):

$$\begin{aligned} \boldsymbol{b} &= (\boldsymbol{X}'\boldsymbol{X})^{-1}\boldsymbol{X}'(\boldsymbol{X}\boldsymbol{\beta} + \boldsymbol{u}) \\ &= (\boldsymbol{X}'\boldsymbol{X})^{-1}(\boldsymbol{X}'\boldsymbol{X})\boldsymbol{\beta} + (\boldsymbol{X}'\boldsymbol{X})^{-1}\boldsymbol{X}'\boldsymbol{u} \\ &= \boldsymbol{\beta} + (\boldsymbol{X}'\boldsymbol{X})^{-1}\boldsymbol{X}'\boldsymbol{u} \end{aligned} \tag{7.128}$$

and take the expected value of $\boldsymbol{b}$:

$$E[\boldsymbol{b}] = E[\boldsymbol{\beta}] + (\boldsymbol{X}'\boldsymbol{X})^{-1}\boldsymbol{X}'E[\boldsymbol{u}] \tag{7.129}$$

but because $E[\boldsymbol{u}] = 0$ (assumption 1) and $\boldsymbol{\beta}$ is constant, then

$$E[\boldsymbol{b}] = \boldsymbol{\beta} \tag{7.130}$$

that is, the expected value of $\boldsymbol{b}$ is $\boldsymbol{\beta}$.

Furthermore, the variance of $\boldsymbol{b}$ can be obtained as follows. Rearranging Eq. (7.128):

$$\boldsymbol{b} - \boldsymbol{\beta} = (\boldsymbol{X}'\boldsymbol{X})^{-1}\boldsymbol{X}'\boldsymbol{u} \tag{7.131}$$

and utilizing Eq. (7.130), we obtain

$$\boldsymbol{b} - E[\boldsymbol{b}] = (\boldsymbol{X}'\boldsymbol{X})^{-1}\boldsymbol{X}'\boldsymbol{u} \tag{7.132}$$

From the definition of the variance-covariance matrix [Eq. (7.117)]:

$$Var\text{-}Cov[\boldsymbol{b}] = E[(\boldsymbol{b} - E[\boldsymbol{b}])(\boldsymbol{b} - E[\boldsymbol{b}])'] \tag{7.133}$$

Using Eq. (7.132) in Eq. (7.133),[5]

$$\begin{aligned} Var\text{-}Cov[\boldsymbol{b}] &= E[(\boldsymbol{X}'\boldsymbol{X})^{-1}\boldsymbol{X}'\boldsymbol{u}\boldsymbol{u}'\boldsymbol{X}(\boldsymbol{X}'\boldsymbol{X})^{-1}] \\ &= (\boldsymbol{X}'\boldsymbol{X})^{-1}\boldsymbol{X}'E[\boldsymbol{u}\boldsymbol{u}']\boldsymbol{X}(\boldsymbol{X}'\boldsymbol{X})^{-1} \end{aligned} \tag{7.134}$$

but from Eq. (7.118) $E[\boldsymbol{uu}'] = \sigma^2\boldsymbol{I}$; therefore the variance-covariance of $\boldsymbol{b}$ simplifies to

$$Var\text{-}Cov[\boldsymbol{b}] = \sigma^2(\boldsymbol{X}'\boldsymbol{X})^{-1} \tag{7.135}$$

where σ^2 is the variance of $\boldsymbol{u}$, as defined by Eq. (7.114).

The elements of the matrix $(\boldsymbol{X}'\boldsymbol{X})^{-1}$ are designated as a_{ij}. Therefore, the variance of b_i is given by

$$V[b_i] = \sigma^2 a_{ii} \tag{7.136}$$

and the covariance of b_i with b_j by

$$Cov[b_i, b_j] = \sigma^2 a_{ij} \tag{7.137}$$

Therefore, if the variance of $\boldsymbol{u}$ is known, or can be estimated, then the variance-covariance of the estimated parameter vector $\boldsymbol{b}$ can be calculated.

It can be seen from Eq. (7.134) that the variance-covariance matrix of $\boldsymbol{b}$ can still be calculated even if assumption 2 is not made. In that case, the matrix $E[\boldsymbol{uu}']$ would not be a diagonal matrix.

We can now draw an important conclusion regarding the distribution of $\boldsymbol{b}$. Eq. (7.128) shows that $\boldsymbol{b}$ is a linear combination of $\boldsymbol{u}$. If $\boldsymbol{u}$ is a multivariate normal distribution (assumption 5, Sec. 7.3), then $\boldsymbol{b}$ is also a multivariate normal distribution, that is,

$$\boldsymbol{b} \sim N(\boldsymbol{\beta}, \sigma^2(\boldsymbol{X}'\boldsymbol{X})^{-1}) \tag{7.138}$$

[5] Note that $(\boldsymbol{X}'\boldsymbol{X})$ is a symmetric matrix; therefore, its inverse $(\boldsymbol{X}'\boldsymbol{X})^{-1}$ is also symmetric. The transpose of a symmetric matrix is the same as the original matrix.

For each individual parameter:

$$b_i \sim N(\beta_i, \sigma^2 a_{ii}) \tag{7.139}$$

where a_{ii} is the ith element on the principal diagonal of $(\boldsymbol{X}'\boldsymbol{X})^{-1}$. The normal distribution can be converted to the standard normal distribution

$$\frac{b_i - \beta_i}{\sigma \sqrt{a_{ii}}} \sim N(0, 1) \tag{7.140}$$

The variance σ^2 of the distribution term is not usually known unless a large number of repetitive experiments have been performed. The value of σ^2 can be estimated from

$$s^2 = \frac{\epsilon' \epsilon}{n - k} \tag{7.141}$$

where $\epsilon' \epsilon$ is the sum of squared residuals [see Eq. (7.121)], and $(n - k)$ is the number of degree of freedom. If there is no lack of fit of the model to the data (see analysis of variances, Sec. 7.5), then s^2 is an unbiased estimate of σ^2, that is,

$$E[s^2] = \sigma^2 \tag{7.142}$$

If lack of fit cannot be tested, using s^2 as an estimate of σ^2 implies an assumption that the model is correct.

We saw earlier that the ratio of s^2/σ^2 has a chi-square distribution:

$$\chi^2 = \nu \frac{s^2}{\sigma^2} \tag{7.71}$$

and that the t variable is given by

$$t = \frac{u}{\sqrt{\chi^2/\nu}} = \frac{N(0,1)}{\sqrt{\chi^2/\nu}} \tag{7.72a}$$

We can, therefore, combine Eqs. (7.140), (7.71), and (7.72*a*) to form the t variable

$$t = \frac{(b_i - \beta_i)/\sigma\sqrt{a_{ii}}}{\sqrt{s^2/\sigma^2}} = \frac{b_i - \beta_i}{s\sqrt{a_{ii}}} \sim t(n-k) \tag{7.143}$$

Eq. (7.143) shows that the quantity $[(b_i - \beta_i)/s\sqrt{a_{ii}}\,]$ has a t distribution with $(n - k)$ degrees of freedom. This is a very important equation, because it enables us to construct confidence intervals of the parameters from quantities that can be calculated from the regression analysis. For example, the 100(1 − α) percent confidence interval for parameter β_i can be obtained from

$$Pr\left\{t_{\alpha/2} < \frac{b_i - \beta_i}{s\sqrt{a_{ii}}} \le t_{1-\alpha/2}\right\} = 1 - \alpha \tag{7.144}$$

which yields the interval

$$b_i - t_{1-\alpha/2}s\sqrt{a_{ii}} \le \beta_i < b_i + t_{(1-\alpha/2)}s\sqrt{a_{ii}} \tag{7.145}$$

The above are *individual parameter confidence intervals*. Fig. 7.9*a* demonstrates these intervals for β_1 and β_2 in a two-parameter model.

Furthermore, Eq. (7.143) enables us to perform the t test for hypothetical values of β (see Sec. 7.2.3). For example, if it is suspected that the value of β_i is not significantly different than zero, then null hypothesis can be stated as

$$H_0 : \beta_i = 0 \tag{7.146}$$

and the alternative hypothesis as

$$H_A : \beta_i \ne 0 \tag{7.147}$$

When Eq. (7.146) is substituted in Eq. (7.143), the resulting expression

$$t = \frac{b_i}{s\sqrt{a_{ii}}} \tag{7.148}$$

is calculated. If this value of t lies within the region of acceptance given by the t distribution for a two-sided test at the required confidence level, then the null hypothesis that $\beta_i = 0$ is accepted. This is a very useful test in deciding the significance of a parameter in a model and in helping the experimenter discriminate between competing models.

In most mathematical models, the covariance between parameters, as measured by Eq. (7.137), is nonzero, that is, the parameters are correlated with each other. Careful experimental design may reduce, but never completely eliminate, this correlation. The individual confidence intervals calculated by Eq. (7.145) do not reflect the covariance. To do so, it is necessary to construct the *joint confidence region* of parameters. Using the multivariate normal distribution of $\boldsymbol{b}$ [Eq. (7.138)], we form the standardized normal variable:

$$\{\sigma^2(\boldsymbol{X}'\boldsymbol{X})^{-1}\}^{-1/2}(\boldsymbol{b} - \boldsymbol{\beta}) \sim N(0, \boldsymbol{I}) \tag{7.149}$$

We recall that the chi-square variable is the sum of the squares of standardized normal variables [Eq. (7.65)]; therefore we can form the χ_1^2 variable from (7.149):

$$\chi_1^2 = \frac{(\boldsymbol{b} - \boldsymbol{\beta})'(\boldsymbol{X}'\boldsymbol{X})(\boldsymbol{b} - \boldsymbol{\beta})}{\sigma^2} \tag{7.150}$$

with k degrees of freedom. We can also form another chi-square variable from the ratio of s^2 and σ^2 [see Eq. (7.71)]:

$$\chi_2^2 = (n - k)\frac{s^2}{\sigma^2} \tag{7.151}$$

From Eq. (7.77) we recall that

$$F(\nu_1, \nu_2) = \frac{\chi_1^2/\nu_1}{\chi_2^2/\nu_2} \tag{7.77}$$

Therefore, combining Eqs. (7.150), (7.151), and (7.77) we obtain

$$F(k, n - k) = \frac{[(\boldsymbol{b} - \boldsymbol{\beta})'(\boldsymbol{X}'\boldsymbol{X})(\boldsymbol{b} - \boldsymbol{\beta})]/k\sigma^2}{[(n - k)s^2]/[(n - k)\sigma^2]} = \frac{(\boldsymbol{b} - \boldsymbol{\beta})'(\boldsymbol{X}'\boldsymbol{X})(\boldsymbol{b} - \boldsymbol{\beta})}{ks^2} \tag{7.152}$$

Finally, the joint $100(1 - \alpha)$ percent confidence region for all the parameters can be obtained from

$$Pr\left\{\frac{(\boldsymbol{b} - \boldsymbol{\beta})'(\boldsymbol{X}'\boldsymbol{X})(\boldsymbol{b} - \boldsymbol{\beta})}{ks^2} \le F_{1-\alpha}(k, n - k)\right\} = 1 - \alpha \tag{7.153}$$

where $F_{1-\alpha}(k, n - k)$ is the $(1 - \alpha)$ point of the F distribution with k and $(n - k)$ degrees of freedom. The inequality in (7.153) defines a hyperellipsoidal region in the k-dimensional parameter space. For a two-parameter model, this joint confidence region is shown in Fig. 7.9*b* as an elongated tilted ellipse.

In the rare case where the parameters are uncorrelated, the matrix $(\boldsymbol{X}'\boldsymbol{X})^{-1}$ is diagonal, the axes of the confidence ellipsoid would be parallel to the coordinates of the parameter space, and the individual parameter confidence intervals would hold for each parameter independently. However, since the parameters are usually correlated, the extent of the correlation can be measured from the *correlation coefficient matrix*, $\boldsymbol{R}$. This is obtained by applying Eq. (7.34) to the variance-covariance matrix (7.135):

$$\rho_{b_i b_j} = \frac{Cov[b_i, b_j]}{\sqrt{V[b_i]V[b_j]}}$$

$$= \frac{a_{ij}}{\sqrt{a_{ii}a_{jj}}} = r_{ij} \tag{7.154}$$

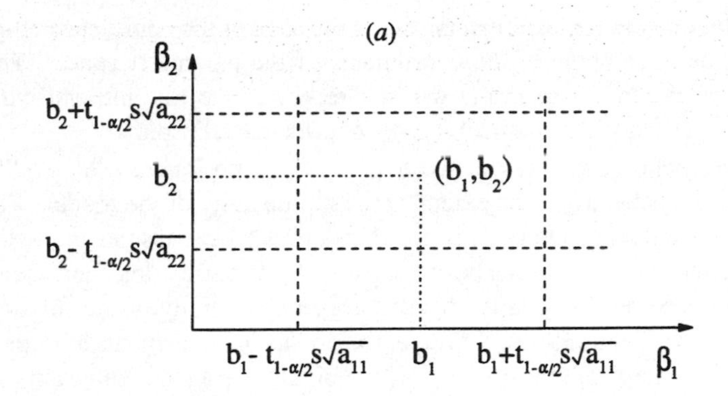

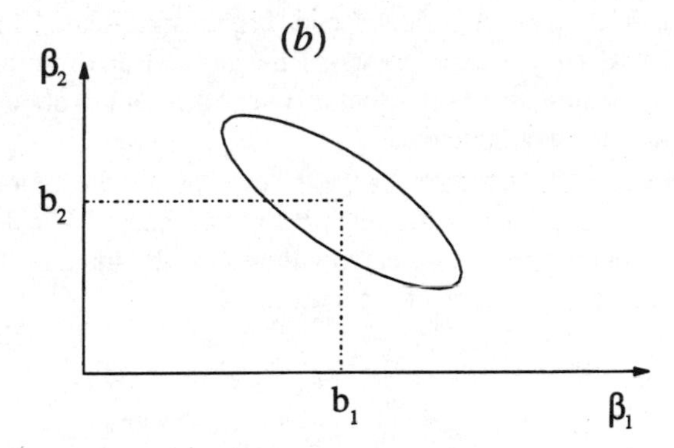

Figure 7.9 Confidence intervals for parameters.
(*a*) Individual confidence intervals.
(*b*) Joint confidence region.

Matrix $\boldsymbol{R}$ is a $(k \times k)$ symmetric matrix, and its elements r_{ij} have all their values in the range $-1.0 \le r_{ij} \le 1.0$.

A typical correlation coefficient matrix may look like this:

$$\boldsymbol{R} = \begin{bmatrix} 1.0 & 0.98 & -0.56 & 0.92 \\ 0.98 & 1.0 & 0.85 & -0.97 \\ -0.56 & 0.85 & 1.0 & 0.35 \\ 0.92 & -0.97 & 0.35 & 1.0 \end{bmatrix} \tag{7.155}$$

A negative correlation between two parameters implies that the errors that cause the estimate of one to be high also cause the estimate of the other to be low. The higher the correlation between two parameters, the closer the value of $|r_{ij}|$ is to 1.0. Consequently, the diagonal elements r_{ii}, which measure the correlation of each parameter with itself, are equal to +1.0.

The correlation between parameters causes the axes of the confidence ellipsoids of the linear model to be at an angle to the coordinates of the parameter space. Therefore, the individual parameter confidence limits will not represent the true interval within which a parameter b_i may lie and still remain within the confidence ellipsoid.

In nonlinear models, the confidence hyperspace is no longer a hyperellipsoid. The amount of distortion depends on the extent of the nonlinearity of the model. Therefore, the calculation of the confidence intervals is not as rigorous an exercise as in the linear model. Still, a lot of valuable information can be extracted from the correlation coefficient matrix that approximates the maximum-likelihood hyperspace in the vicinity of the solution where the model is nearly linear. If the absolute values of the off-diagonal elements of $\boldsymbol{R}$ are close to 1.0, the parameters associated with those elements are highly correlated with each other. Davies [6] tests the values of r_{ij} against a normal distribution with zero mean, that is, no correlation. He classifies the correlation as "significant" and "highly significant" if the value of r_{ij} is higher than the 0.05 and 0.01 significance points of the normal distribution, respectively. High correlation between parameters implies that it is very difficult to obtain separate estimates of these parameters with the available data.

The eigenvectors w of the matrix $\boldsymbol{R}$ give the direction of the major and minor axes of the hyperellipsoidal confidence region of the parameter space. The length of the axes are proportional to the square root of the eigenvalues λ of the matrix. Box [7] calculated the values of the parameters at the ends of the axes by

$$\bar{b}_i = b_i \pm w_{ri}\{\lambda_r(s^2 a_{ii})kF_{(1-\alpha)}(k, n - k)\}^{1/2} \tag{7.156}$$

where $r = 1, 2, \ldots, k$

k = number of parameters

n = number of points used in estimating b_i

$F_{1-\alpha}(k, n - k)$ = value of the F distribution with k and $(n - k)$ degrees of freedom.

Subsequently, he uses these parameter values to calculate the sum of squares at each end of the axes and to compare them with the sum of squares at the center of the hyperellipsoid. This sum-of-squares search, which is based on a linear model, may give vital information for nonlinear models as well. In the case where the solution has only converged on a *local* minimum sum of squares, it is very likely that the search in the direction of one of the axes will produce a lower sum of squares. In such a case, the regression must be repeated, starting from a different initial position, so that the local minimum may be bypassed.

7.4 Nonlinear Regression Analysis

We have stated this earlier in the chapter, and we state it again: The mathematical models encountered in engineering and science are often nonlinear in their parameters. Consider, for example, the analysis of a chemical reaction such as

$$A \xrightarrow{k_1} B \xrightarrow{k_2} C + D$$

$$C + A \xrightarrow{k_3} E + F$$

where the rate of formation of each component may be written as

$$\frac{dC_A}{dt} = -k_1 C_A - k_3 C_A^n C_C^m$$

$$\frac{dC_B}{dt} = k_1 C_A - k_2 C_B$$

$$\frac{dC_C}{dt} = k_2 C_B - k_3 C_A^n C_C^m \tag{7.157}$$

$$\frac{dC_E}{dt} = k_3 C_A^n C_C^m$$

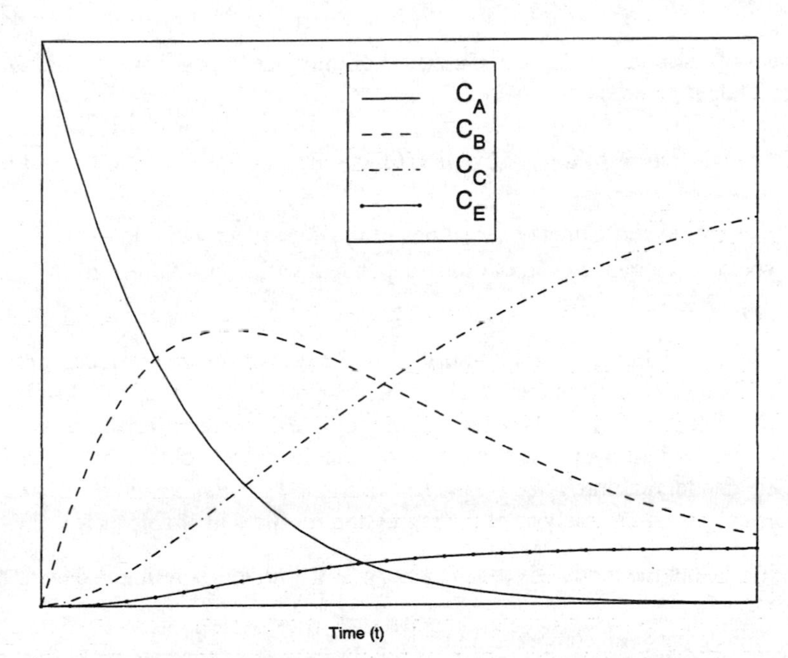

Figure 7.10 Simulated data for batch reactor experiment.

This is only one possible formulation of the reaction mechanism. It contains five unknown parameters, k_1, k_2, k_3, n, and m, which must be calculated by fitting the model to experimental data. Suppose that experiments for this chemical system are carried out in a batch reactor and data of the form shown in Fig. 7.10 are collected. Because experimental data are available for all four dependent variables, C_A, C_B, C_C, and C_E, multiple nonlinear regression can be performed by simultaneously fitting all four equations of (7.157) to the data.

A model consisting of differential equations, such as Eq. (7.157), may be shown in the form:

$$\frac{d\boldsymbol{Y}}{dx} = \boldsymbol{g}(x, \boldsymbol{Y}, \boldsymbol{b}) \tag{7.158}$$

where $d\boldsymbol{Y}/dx$ = vector of derivatives of $\boldsymbol{Y}$
$\boldsymbol{g}$ = vector of functions
x = independent variable
$\boldsymbol{Y}$ = vector of dependent variables
$\boldsymbol{b}$ = vector of parameters.

We assume that if the boundary conditions are given and if the vector $\boldsymbol{b}$ can be estimated, then the differential equations (7.158) can be integrated numerically or analytically to give the integrated results, which are

$$\boldsymbol{Y} = f(x, \boldsymbol{b}) \tag{7.159}$$

For the simple case where the model consists of only one dependent variable, the sum of squared residuals is given by

$$\Phi = \epsilon'\epsilon = (\boldsymbol{Y}^* - \boldsymbol{Y})'(\boldsymbol{Y}^* - \boldsymbol{Y}) \tag{7.160}$$

where $\boldsymbol{Y}^*$ = vector of experimental observations of the dependent variable
$\boldsymbol{Y}$ = vector of calculated values of the dependent variables obtained from Eq. (7.159).

There are several techniques for minimization of the sum of squared residuals described by Eq. (7.160). We review some of these methods in this section. The methods developed in this section will enable us to fit models consisting of multiple dependent variables, such as the one described earlier, to multiresponse experimental data, in order to obtain the values of the parameters of the model that minimize the *overall (weighted) sum of squared residuals.* In addition, a thorough statistical analysis of the regression results will enable us to

1. Decide whether the model gives satisfactory fit within the experimental error of the data.
2. Discriminate between competing models.

3. Measure the accuracy of the estimation of the parameters by constructing the confidence region in the parameter space.
4. Measure the correlation between parameters by examining the correlation coefficient matrix.
5. Perform tests to verify that repeated experimental data come from the same population of experiments.
6. Perform tests to verify whether the residuals between the data and the model are randomly distributed.

MATLAB does the single nonlinear regression calculation by applying the function *curvefit,* which comes with the *Optimization TOOLBOX* of MATLAB. The statement $\boldsymbol{b}$ = *curvefit('file_name'*, $\boldsymbol{b}_0$, $\boldsymbol{x}$, $\boldsymbol{y}$) starts the regression calculations at the vector of initial guesses of the parameters $\boldsymbol{b}_0$ and uses the least squares technique to find the vector of parameters $\boldsymbol{b}$ that best fit the nonlinear expression, introduced in the MATLAB function *file_name.m,* to the data $\boldsymbol{y}$. Inputs to the function *file_name* should be the vector of parameters $\boldsymbol{b}$ and the vector of independent variable $\boldsymbol{x}$. The function *file_name* should return the vector of dependent variable $\boldsymbol{y}$. The default algorithm is Marquardt (see Sec. 7.4.4). A Gauss-Newton method (see Sec. 7.4.2) may be selected via the *options* input to the function.

7.4.1 The Method of Steepest Descent

A simple method, which has been used to arrive at the minimum sum of squares of a nonlinear model, is that of *steepest descent.* We know that the gradient of a scalar function is a vector that gives the direction of the greatest increase of the function at any point. In the steepest descent method, we take advantage of this property by moving in the opposite direction to reach a lower function value. Therefore, in this method, the initial vector of parameter estimates is corrected in the direction of the negative gradient of Φ:

$$\Delta \boldsymbol{b} = -K\left(\frac{\partial \Phi}{\partial \boldsymbol{b}}\right) \tag{7.161}$$

Where K is a suitable constant factor and $\Delta \boldsymbol{b}$ is the correction vector to be applied to the estimated value of $\boldsymbol{b}$ to obtain a new estimate of the parameter vector:

$$\boldsymbol{b}^{(m+1)} = \boldsymbol{b}^{(m)} + \Delta \boldsymbol{b} \tag{7.162}$$

where m is the iteration counter. Combining Eqs. (7.160) and (7.161) results in

$$\Delta \boldsymbol{b} = 2K\boldsymbol{J}'(\boldsymbol{Y}^* - \boldsymbol{Y}) \tag{7.163}$$

where $\boldsymbol{J}$ is the Jacobian matrix of partial derivatives of $\boldsymbol{Y}$ with respect to $\boldsymbol{b}$ evaluated at all n points where experimental observations are available:

$$J = \begin{bmatrix} \frac{\partial Y_1}{\partial b_1} & \cdots & \frac{\partial Y_1}{\partial b_k} \\ & \cdots\cdots & \\ \frac{\partial Y_n}{\partial b_1} & \cdots & \frac{\partial Y_n}{\partial b_k} \end{bmatrix} \tag{7.164}$$

The steepest descent method has the advantage that guarantees moving toward the minimum sum of squares without diverging, provided that the value of K, which determines the step size, is small enough. The value of K may be a constant throughout the calculations, changed arbitrarily at each calculation step, or obtained from optimization of the step size [8]. However, the rate of convergence to the minimum decreases as the search approaches this minimum, and the method loses its attractiveness because of this shortcoming.

7.4.2 The Gauss-Newton Method

Once again, we restate that in the least squares method, our objective is to find the vector of parameters $\boldsymbol{b}$ such that it minimizes the sum of squared residuals Φ. Thus, the vector $\boldsymbol{b}$ may be found by taking the partial derivative of Φ with respect to $\boldsymbol{b}$ and setting it to zero:

$$\frac{\partial \Phi}{\partial \boldsymbol{b}} = 0 \tag{7.165}$$

Because $\boldsymbol{Y}$ is nonlinear with respect to the parameters, Eq. (7.165) will yield a nonlinear equation that would be difficult to solve for $\boldsymbol{b}$. This problem was alleviated by Gauss, who determined that fitting nonlinear functions by least squares can be achieved by an iterative method involving a series of linear approximations. At each stage of the iteration, linear squares theory can be used to obtain the next approximation.

This method, known as the *Gauss-Newton method*, converts the nonlinear problem into a linear one by approximating the function $\boldsymbol{Y}$ by a Taylor series expansion around an estimated value of the parameter vector $\boldsymbol{b}$:

$$\boldsymbol{Y}(x, \boldsymbol{b}) = \boldsymbol{Y}(x, \boldsymbol{b}^{(m)} + \Delta \boldsymbol{b}) = \boldsymbol{Y}(x, \boldsymbol{b}^{(m)}) + \frac{\partial \boldsymbol{Y}}{\partial \boldsymbol{b}}\Big|_{\boldsymbol{b}^{(m)}} \Delta \boldsymbol{b} = \boldsymbol{Y} + \boldsymbol{J} \Delta \boldsymbol{b} \tag{7.166}$$

where the Taylor series has been truncated after the second term. Eq. (7.166) is linear in $\Delta \boldsymbol{b}$. Therefore, the problem has been transformed from finding $\boldsymbol{b}$ to that of finding the correction to $\boldsymbol{b}$, that is, $\Delta \boldsymbol{b}$, which must be added to an estimate of $\boldsymbol{b}$ to minimize the sum of squared residuals. To do this we replace $\boldsymbol{Y}$ in Eq. (7.160) with the right-hand side of Eq. (7.166) to get

$$\Phi = (\boldsymbol{Y}^* - \boldsymbol{Y} - \boldsymbol{J} \Delta \boldsymbol{b})'(\boldsymbol{Y}^* - \boldsymbol{Y} - \boldsymbol{J} \Delta \boldsymbol{b}) \tag{7.167}$$

Taking the partial derivative of Φ with respect to $\Delta \boldsymbol{b}$, setting it equal to zero, and solving for $\Delta \boldsymbol{b}$, we obtain

$$\Delta \boldsymbol{b} = (\boldsymbol{J}'\boldsymbol{J})^{-1}\boldsymbol{J}'(\boldsymbol{Y}^* - \boldsymbol{Y}) \tag{7.168}$$

The Gauss-Newton method applies to both the one-variable model and the multiple regression case (see Sec. 7.4.5). The algorithm of the Gauss-Newton method involves the following steps:

2. Assume initial guesses for the parameter vector $\boldsymbol{b}$.
3. If the model is in the form of differential equation(s), then use the vector $\boldsymbol{b}$ and the boundary condition(s) to integrate the equation(s) to obtain the profile(s) of $\boldsymbol{Y}$. If the model is in the form of algebraic equation(s), then simply use the vector $\boldsymbol{b}$ to evaluate $\boldsymbol{Y}$ from the equation(s).
4. Evaluate the Jacobian matrix $\boldsymbol{J}$ from the equation(s) of the model.
5. Use Eq. (7.168) to obtain the correction vector $\Delta \boldsymbol{b}$.
6. Evaluate the new estimate of the parameter vector from Eq. (7.162):

$$\boldsymbol{b}^{(m+1)} = \boldsymbol{b}^{(m)} + \Delta \boldsymbol{b} \tag{7.162}$$

7. It is also possible to apply the relaxation factor in order to prevent the calculation from diverging (see Sec. 1.8).
8. Repeat steps 2-5 until either (or both) of the following conditions are satisfied:
   a. Φ does not change appreciably.
   b. $\Delta \boldsymbol{b}$ becomes very small.

The Gauss-Newton method is based on the linearization of a nonlinear model; therefore, this method is expected to work well if the model is not highly nonlinear, or if the initial estimate of the parameter vector is near the minimum sum squares. The contours of constant Φ in the parameter space of a linear model are ellipsoids (Fig. 7.11*a*). For a nonlinear model, these contours are distorted (Fig. 7.11*b*), but in the vicinity of the minimum Φ, the contours are very nearly elliptical. Therefore, the Gauss-Newton method is quite effective if the initial starting point for the search is in the nearly elliptical region. On the other hand, this method may diverge if the starting point is in the highly distorted region of the parameter hyperspace.

7.4.3 Newton's Method

Eq. (7.165) represents a set of nonlinear equations; therefore, Newton's method may be applied to solve this set of nonlinear equations. First, let us expand Φ by Taylor series up to the third term:

$$\Phi(x,b) = \Phi(x,b^{(m)}) + \left(\frac{\partial\Phi}{\partial b}\right)^{(m)} \Delta b + \frac{1}{2}\Delta b' \left(\frac{\partial^2\Phi}{\partial b^2}\right)^{(m)} \Delta b \tag{7.169}$$

Taking the partial derivative of both sides of Eq. (7.169) with respect to b gives

$$\frac{\partial\Phi}{\partial b} = \left(\frac{\partial\Phi}{\partial b}\right)^{(m)} + \left(\frac{\partial^2\Phi}{\partial b^2}\right)^{(m)} \Delta b \tag{7.170}$$

The first derivative of Φ with respect to b can be calculated by differentiating Eq. (7.160):

$$\left(\frac{\partial\Phi}{\partial b}\right)^{(m)} = -2J'(Y^* - Y) \tag{7.171}$$

and the second derivative of Φ with respect to b is called the *Hessian matrix* of the second-order partial derivatives of Φ with respect to b evaluated at all n points where experimental observations are available:

$$H = \begin{bmatrix} \dfrac{\partial^2\Phi}{\partial b_1^2} & \dfrac{\partial^2\Phi}{\partial b_1 \partial b_2} & \cdots & \dfrac{\partial^2\Phi}{\partial b_1 \partial b_k} \\ \dfrac{\partial^2\Phi}{\partial b_2 \partial b_1} & \dfrac{\partial^2\Phi}{\partial b_2^2} & \cdots & \dfrac{\partial^2\Phi}{\partial b_2 \partial b_k} \\ \cdots & \cdots & \cdots & \\ \dfrac{\partial^2\Phi}{\partial b_k \partial b_1} & \dfrac{\partial^2\Phi}{\partial b_k \partial b_2} & \cdots & \dfrac{\partial^2\Phi}{\partial b_k^2} \end{bmatrix} \tag{7.172}$$

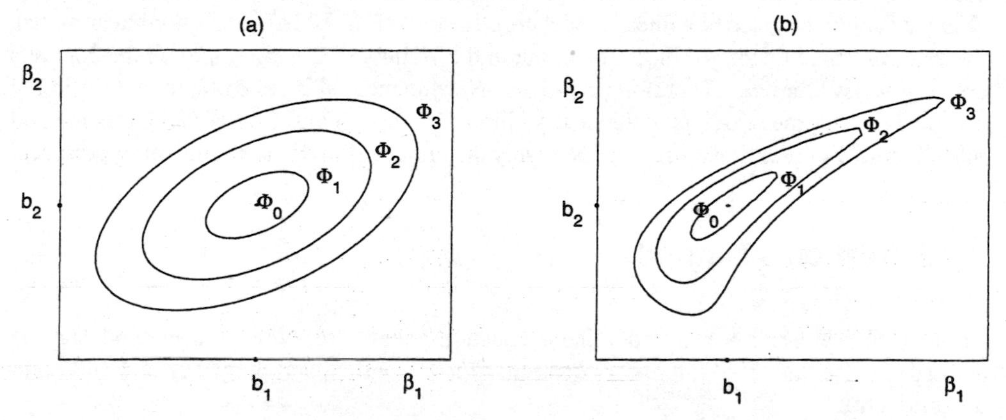

Figure 7.11 Contours for constant sum of squares in parameter space. (*a*) Linear model. (*b*) Nonlinear model.

By applying the necessary condition of having a local minimum of Φ, Eq. (7.165), into Eq. (7.170) and combining with Eqs. (7.171) and (7.172), we can evaluate the correction vector $\Delta \boldsymbol{b}$:

$$\Delta \boldsymbol{b} = 2\boldsymbol{H}^{-1}\boldsymbol{J}'(\boldsymbol{Y}^* - \boldsymbol{Y}) \tag{7.173}$$

It is interesting to note that in the case of single parameter regression, Eq. (7.165) becomes

$$\frac{d\Phi}{db} = \Phi' = 0$$

and Eq. (7.173) simplifies to

$$\Delta b = -\frac{\Phi'^{(m)}}{\Phi''^{(m)}}$$

which is the Newton-Raphson solution of the nonlinear equation $\Phi' = 0$.

The calculation procedure for Newton's method is almost the same as that of Gauss-Newton method with the exception that the vector of corrections to the parameters is calculated from Eq. (7.173). If Φ is quadratic with respect to $\boldsymbol{b}$ (that is, linear regression), Newton's method converges in only one step. Like all other methods applying Newton's technique for the solution of the set of nonlinear equations, a relaxation factor may be used along with Eq. (7.173) when correcting the parameters.

7.4.4 The Marquardt Method[6]

Marquardt [9] has developed an interpolation technique between the Gauss-Newton and the steepest descent methods. This interpolation is achieved by adding the diagonal matrix $(\lambda \boldsymbol{I})$ to the matrix $(\boldsymbol{J}'\boldsymbol{J})$ in Eq. (7.168):

$$\Delta \boldsymbol{b} = (\boldsymbol{J}'\boldsymbol{J} + \lambda \boldsymbol{I})^{-1}\boldsymbol{J}'(\boldsymbol{Y}^* - \boldsymbol{Y}) \tag{7.174}$$

The value of λ is chosen, at each iteration, so that the corrected parameter vector will result in a lower sum of squares in the following iteration. It can be easily seen that when the value of λ is small in comparison with the elements of matrix $(\boldsymbol{J}'\boldsymbol{J})$, the Marquardt method approaches the Gauss-Newton method; when λ is very large, this method is identical to steepest descent, with the exception of a scale factor that does not affect the direction of the parameter correction vector but that gives a small step size.

According to Marquardt, it is desired to minimize Φ in the maximum neighborhood over which the linearized function will give adequate representation of the nonlinear function.

---

[6] Also known as the Levenberg-Marquardt method.

Therefore, the method for choosing λ must give small values of λ when the Gauss-Newton method would converge efficiently and large values of λ when the steepest descent method is necessary.

The Marquardt method may likewise be applied to Newton's method. In this case, the diagonal matrix $\lambda \boldsymbol{I}$ is added to the Hessian matrix in Eq. (7.173):

$$\Delta \boldsymbol{b} = 2(\boldsymbol{H} + \lambda \boldsymbol{I})^{-1} \boldsymbol{J}'(\boldsymbol{Y}^* - \boldsymbol{Y}) \quad (7.175)$$

The Marquardt method consists of the following steps:

1. Assume initial guesses for the parameter vector $\boldsymbol{b}$.
2. Assign a large value, say 1000, to λ. This means that in the first iteration the steepest descent method is predominant and would assure that the method is moving toward the lower sum of squared residuals.
3. Evaluate the Jacobian matrix $\boldsymbol{J}$ from the equation(s) of the model. Also evaluate the Hessian matrix $\boldsymbol{H}$ if using Newton's method.
4. Use either Eq. (7.174) or Eq. (7.175) to obtain the correction vector $\Delta \boldsymbol{b}$.
5. Evaluate the new estimate of the parameter vector from Eq. (7.162).

$$\boldsymbol{b}^{(m+1)} = \boldsymbol{b}^{(m)} + \Delta \boldsymbol{b} \quad (7.162)$$

6. Calculate the new value of Φ. If $\Phi^{(m+1)} < \Phi^{(m)}$, reduce the value of λ, by a factor of 4, for example. If $\Phi^{(m+1)} > \Phi^{(m)}$, keep the old parameters [$\boldsymbol{b}^{(m+1)} = \boldsymbol{b}^{(m)}$] and increase the value of λ, by a factor of 2, for example.
7. Repeat steps 3-6 until either (or both) of the following conditions are satisfied:
   a. Φ does not change appreciably.
   b. $\Delta \boldsymbol{b}$ becomes very small.

7.4.5 Multiple Nonlinear Regression

In the previous four sections, the sum of squared residuals that was minimized was that given by Eq. (7.160). This was the sum of squared residuals determined from fitting one equation to measurements of one variable. However, most mathematical models may involve simultaneous equations in multiple dependent variables. For such a case, when more than one equation is fitted to multiresponse data, where there are v dependent variables in the model, the *weighted sum of squared residuals* is given by

$$\Phi = \sum_{j=1}^{v} w_j \epsilon_j' \epsilon_j = \sum_{j=1}^{v} w_j \phi_j$$

$$= \sum_{j=1}^{v} w_j (\boldsymbol{Y}_j^* - \boldsymbol{Y}_j)'(\boldsymbol{Y}_j^* - \boldsymbol{Y}_j) \quad (7.176)$$

where w_j = weighting factor corresponding to the jth dependent variable
ϕ_j = sum of squared residuals corresponding to the jth dependent variable.

To minimize Φ by the Gauss-Newton method, we first linearize the models using Eq. (7.166) and combine with Eq. (7.176) to obtain

$$\Phi = \sum_{j=1}^{v} w_j (\boldsymbol{Y}_j^* - \boldsymbol{Y}_j - \boldsymbol{J}_j \Delta \boldsymbol{b})'(\boldsymbol{Y}_j^* - \boldsymbol{Y}_j - \boldsymbol{J}_j \Delta \boldsymbol{b}) \tag{7.177}$$

Taking the partial derivative of Φ with respect to $\Delta \boldsymbol{b}$, setting it equal to zero, and solving for $\Delta \boldsymbol{b}$ we obtain

$$\Delta \boldsymbol{b} = \left[\sum_{j=1}^{v} w_j (\boldsymbol{J}_j{}' \boldsymbol{J}_j)\right]^{-1} \left[\sum_{j=1}^{v} w_j \boldsymbol{J}_j{}' (\boldsymbol{Y}_j^* - \boldsymbol{Y}_j)\right] \tag{7.178}$$

Eq. (7.178) gives the correction of the parameter vector when fitting multiple dependent variables simultaneously. Eq. (7.178) becomes identical to Eq. (7.168) when $v = 1$, that is, when only one dependent variable is fitted. When using the Marquardt method, the correction of the parameter vector is calculated from

$$\Delta \boldsymbol{b} = \left[\lambda \boldsymbol{I} + \sum_{j=1}^{v} w_j (\boldsymbol{J}_j{}' \boldsymbol{J}_j)\right]^{-1} \left[\sum_{j=1}^{v} w_j \boldsymbol{J}_j{}' (\boldsymbol{Y}_j^* - \boldsymbol{Y}_j)\right] \tag{7.179}$$

The weighting factors w_j are determined as follows: The basic assumption in the derivation of the regression algorithm was that the variance σ^2 of the distribution of the error in the measurements was constant throughout the profile of a single dependent variable. However, in the case of multiple regression, it is very unlikely that the variance σ_j^2 of all the curves will be the same. Therefore, in order to form an unbiased weighted sum of squared residuals, the individual sum of squares must be multiplied by a weighting factor that is proportional to $1/\sigma_j^2$. The equation for evaluating the weighting factors is given by

$$w_j = \frac{1/\sigma_j^2}{\dfrac{1}{\sum_{i=1}^{v} n_i}\left[\sum_{i=1}^{v} \sum_{l=1}^{n_i} \dfrac{1}{\sigma_i^2}\right]} \tag{7.180}$$

where σ_j^2 or σ_i^2 = variance for each curve
n_i = number of experimental points available for each curve
v = number of variables being fitted.

The denominator of Eq. (7.180) accounts for the possibility that each curve may have a different number of experimental points n_i and weighs that accordingly. If the assumption that σ_j^2 is constant within one curve does not hold, then Eq. (7.180) can be extended so that weighting factor can be calculated at each point with the appropriate value of σ_j^2.

In most cases, the values of σ_j^2 would not be known; however, the estimates of these variances s_j^2 can be obtained from repeated experiments, and the values of s_j^2 are then used in Eq. (7.180) to calculate the weighting factors. In the worst case, where no repeated experiments are made and no *a priori* knowledge of σ_j^2 is available, then the values of w_j must be guessed. Otherwise, the nonlinear regression algorithm would introduce a bias toward fitting more satisfactorily the curve with the highest ϕ_j and partially ignoring the curves with low ϕ_j.

The nonlinear regression can also be extended to fit multiple experimental values of the dependent variable at each value of the independent variable. This can be done by changing Eq. (7.176) so that the squared residuals are also summed up within each group of points. Finally, if the value of the variance of the error is proportional to the value of the dependent variable, the residual in the sum-of-squares calculation must be divided by the theoretical (calculated) value of the dependent variable at each point in the calculation.

7.5 Analysis of Variance and Other Statistical Tests of the Regression Results

The t test on parameters, described in Sec. 7.3.2, is useful in establishing whether a model contains an insignificant parameter. This information can be used to make small adjustments to models and thus discriminate between models that vary from each other by one or two parameters. This test, however, does not give a criterion for testing the adequacy of this model. The residual sum of squares, calculated by Eq. (7.160), contains two components. One is due to the scatter in the experimental data and the other is due to the lack of fit of the model. In order to test the adequacy of the fit of a model, the sum of squares must be partitioned into its components. This procedure is called analysis of variance, which is summarized in Table 7.2. To maintain generality, we examine a set of nonlinear data and assume the availability of multiple values of the dependent variable y_{ij} at each point of the independent variable x_i (see Fig. 7.12).

In Table 7.2, p is the number of points of the independent variable at which there are experimental (observed) values of the dependent variable, n_i are the numbers of repeated experiments available at each point of the independent variable, $\bar{y}$ is the mean value of each group of repeated experiments, y_i are the calculated values of the dependent variable, y^*_{ij} are the experimental values of the dependent variable, and k is the number of parameters being estimated. It should be realized that the total sum of squares shown in Table 7.2:

$$Total\ SS = \sum_{i}^{p} \sum_{j}^{n_i} (y_{ij}^* - y_i)^2 \tag{7.181}$$

is merely a generalization of Eq. (7.160) to apply to both linear and nonlinear models and an extension of that relationship to account for the presence of repeated experimental data.

The ratio of the variances s_1^2/s_2^2 has an F distribution with ν_1 and ν_2 degrees of freedom. This ratio must be tested against the F statistic in order to test the hypothesis that the experimental points are adequately represented by the predicted line. For a good fit, this ratio should be small, that is, to accept the hypothesis the following must be true:

$$\frac{s_1^2}{s_2^2} < F_{1-\alpha}(\nu_1, \nu_2) \tag{7.182}$$

This would mean that the component of the variance due to the lack of fit is small when compared with the variance of the of the experimental error. In that case, the model adequately represents the data. It is obvious that if the experimental data have a large scatter, then s_2^2 is high, and the requirements on the model are less stringent. Stating this more simply, almost anything can be fitted through very noisy or sloppy data, but the value of such a model would be marginal.

If more than one model is found to satisfy the above test, the choice of the best one can be facilitated by performing an F test between the values s_1^2 of pairs of models. If the fit of any one of these models is significantly better than that of the others, it will be discovered by this test.

Table 7.2 Analysis of variance

Source of variance	Sum of squares	Degrees of freedom	Variance
Lack of fit	$\sum_{i=1}^{p} n_i(\bar{y}_i - y_i)^2$	$\nu_1 = p - k$	s_1^2
Experimental error	$\sum_{i=1}^{p}\sum_{j=1}^{n_i} (y_{ij}^* - \bar{y}_i)^2$	$\nu_2 = \left(\sum_{i=1}^{p} n_i\right) - p$	s_2^2
Total	$\sum_{i=1}^{p}\sum_{j=1}^{n_i} (y_{ij}^* - y_i)^2$	$\nu = \left(\sum_{i=1}^{p} n_i\right) - k$	s^2

Furthermore, the F test may be used to determine if an experiment whose results deviate from those of other experiments performed under identical conditions should be grouped together with the other ones. To do this, the model is first fitted to each experiment separately. The individual sum of squares from each regression are pooled together as follows:

$$s_{pooled}^2 = \frac{\sum (\text{individual sum of squares})}{\sum (\text{degrees of freedom})} \tag{7.183}$$

Then the model is fitted to the grouped set of experiments to find the variance $s^2_{grouped}$. Finally, an F test is performed between the pooled and grouped variances. If the inequality

$$\frac{s_{grouped}^2}{s_{pooled}^2} < F_{1-\alpha} \tag{7.184}$$

is not satisfied, it means that the model fits the experiments better individually than when grouped together.

A final test can be performed to investigate the lack of fit of the model. In the least squares regression, the assumption is made that the model being fitted is the correct one, and that the observations deviate from the model in a random fashion. The residuals between the

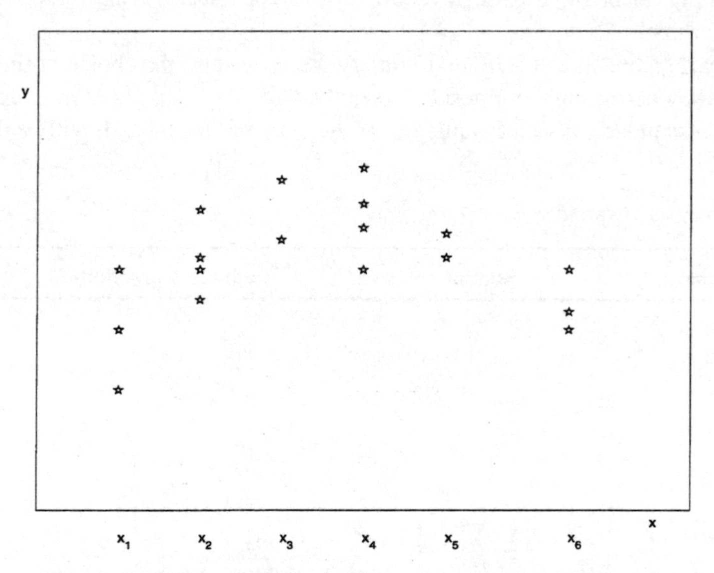

Figure 7.12 Set of nonlinear data where multiple values of the dependent variable are available from repeated experiments ($p = 6$, $n_1 = 3$, $n_2 = 4$, $n_3 = 2$, $n_4 = 4$, $n_5 = 2$, $n_6 = 3$).

observations and the model can be either positive or negative, but if these are truly random, the sign of the residuals should change in a random fashion. The randomness (or lack of fit) can be detected visually by plotting the residuals ($y^* - y$) versus the independent variable x and also versus the dependent variable y. Figs. 7.13-7.16 show several different cases of distribution of residuals. Fig. 7.13 demonstrates the case where the values of the residuals are randomly distributed around zero. This seems to be a satisfactory fit that would probably pass the randomness test (*runs test*) described later in this section.

On the other hand, Fig. 7.14 shows a definite trend in the value of the residuals from positive to negative. The model gives a low prediction of y for low values of x and a high prediction of y for high values of x. A correction to the model to remedy this trend seems to be warranted. At first sight, Fig. 7.15 may seem to give a case of a well-fitting model, but careful examination of the residuals shows that there is an oscillation pattern in the distribution of these residuals around zero. The addition of a term that introduces oscillatory behavior in the model may considerably improve the fit of the model to the data.

Another case is demonstrated in Fig. 7.16, which shows that the value of the residuals grow proportionately to the value of y. In such a case, it would be more appropriate to normalize the residuals by dividing them by the appropriate value of y, that is:

$$\bar{\epsilon} = \frac{\epsilon}{y} \tag{7.185}$$

and then minimize the sum of normalized squared residuals:

$$\bar{\Phi} = \bar{\epsilon}'\bar{\epsilon} \tag{7.186}$$

The randomness of the distribution of the residuals can be quantified, measured, and tested by the so-called *runs test.* In this test, the total number of positive residuals is represented by n_1 and that of negative residuals by n_2. The number of times the sequence of residuals changes sign is r, which is called the number of runs. The distribution of r is approximated by the normal distribution. Brownlee [10] finds the mean and standard deviation of this variable to be

$$\bar{r} = \frac{2n_1 n_2}{n_1 + n_2} + 1 \tag{7.187}$$

$$\sigma = \sqrt{\frac{2n_1 n_2(2n_1 n_2 - n_1 - n_2)}{(n_1 + n_2)^2(n_1 + n_2 - 1)}} \tag{7.188}$$

The standardized form of the variable is

$$Z = \frac{r - \bar{r}}{\sigma} \tag{7.189}$$

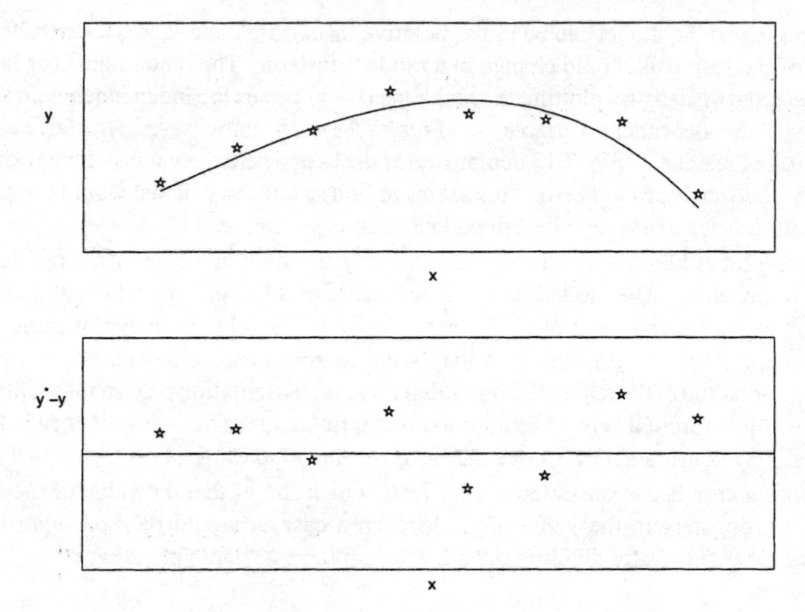

Figure 7.13 Analysis of residuals showing a random trend.

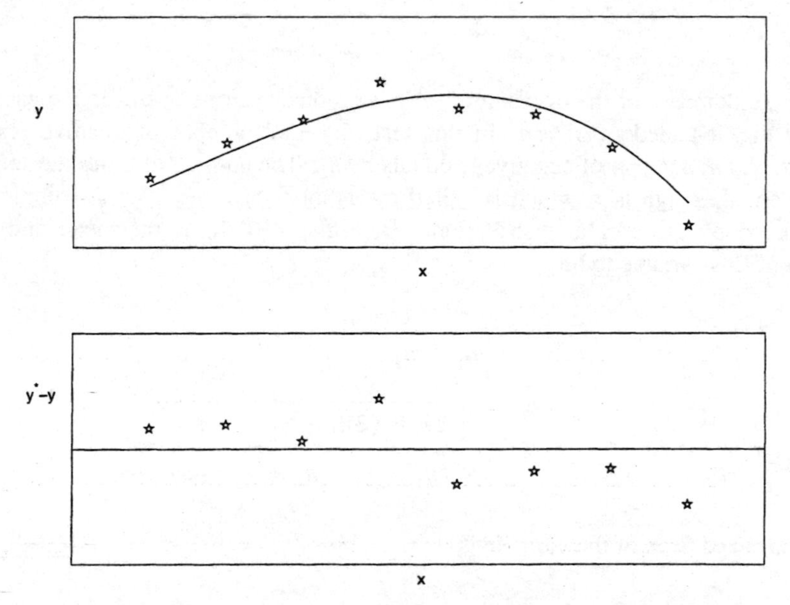

Figure 7.14 Analysis of residuals showing trend from positive to negative.

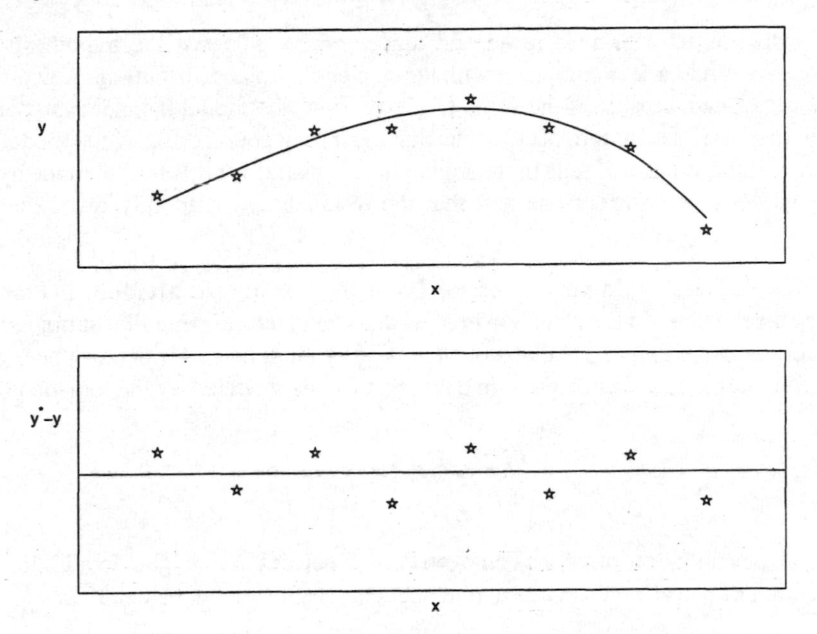

Figure 7.15 Analysis of residuals showing oscillatory trend.

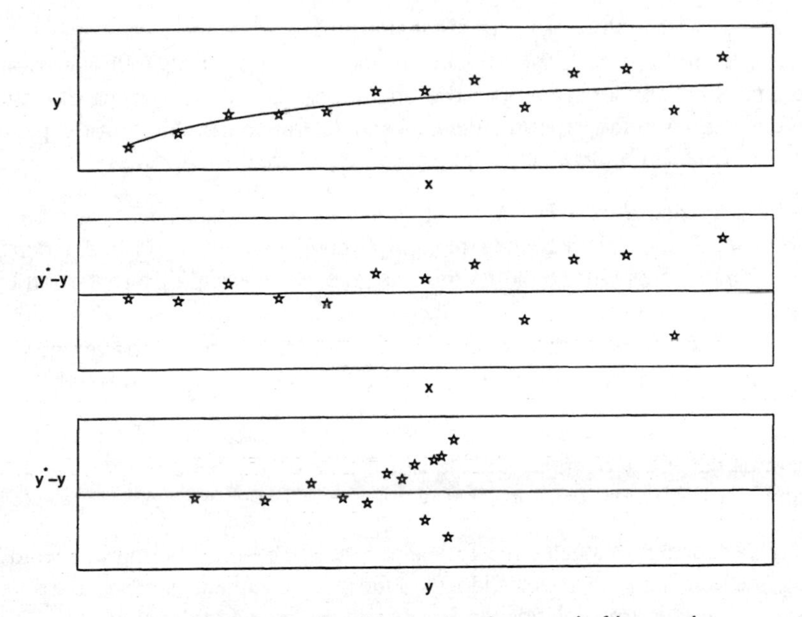

Figure 7.16 Analysis of residuals showing trend of increasing residuals in proportion to *y*.

which is distributed with zero mean and unit variance. To test the hypothesis that the deviations are random, Z is compared with the standard normal distribution. A two-sided test must be performed, because if the value of Z is too low, the model is inadequate; and if Z is too high, then the data contain oscillations that must be accounted for by the model. On the other hand, if the value of Z falls in the region of acceptance for this test, then the hypothesis that the model is the correct one and that the residuals are randomly distributed can be accepted.

Example 7.1: Nonlinear Regression Using the Marquardt Method. In Prob. 5.5, we described the kinetics of a fermentation process that manufactures penicillin antibiotics. When the microorganism *Penicillium chrysogenum* is grown in a batch fermentor under carefully controlled conditions, the cells grow in a rate that can be modeled by the logistic law

$$\frac{dy_1}{dt} = b_1 y_1 \left(1 - \frac{y_1}{b_2}\right)$$

where y_1 is the concentration of the cell expressed as percent dry weight. In addition, the rate of production of penicillin has been mathematically quantified by the equation

$$\frac{dy_2}{dt} = b_3 y_1 - b_4 y_2$$

where y_2 is the concentration of penicillin in units/mL.

The experimental data in Table E7.1 were obtained from two penicillin fermentation runs conducted at essentially identical operating conditions. Using the Marquardt method, fit the above two equations to the experimental data and determine the values of the parameters b_1, b_2, b_3, and b_4, which minimize the weighted sum of squared residuals.

Method of Solution: The Marquardt method using the Gauss-Newton technique, described in Sec. 7.4.4, and the concept of multiple nonlinear regression, covered in Sec. 7.4.5, have been combined together to solve this example. Numerical differentiation by forward finite differences is used to evaluate the Jacobian matrix defined by Eq. (7.164).

The initial conditions of the model equations were chosen to be the average values of the corresponding experimental data at $t = 0$; that is, $y_1(0) = 0.29$ and $y_2\,(0) = 0.0$.

Program Description: Two separate MATLAB functions are written for evaluating the fitting parameters and performing the statistical tests on the parameters. These functions, *NLR.m* and *statistics.m*, are described below:

NLR.m: This function evaluates the fitting parameters by the Marquardt method. At the beginning, the function examines the length of the input arguments and sets the default value, if necessary. The experimental independent and dependent variables should be introduced to the function by matrices of the same size (column vectors in the case of single independent

Table E7.1 Experimental data for penicillin fermentation

	Run No. 1		Run No. 2	
Time (hours)	Cell concentration (percent dry weight)	Penicillin concentration (units/mL)	Cell concentration (percent dry weight)	Penicillin concentration (units/mL)
0	0.40	0	0.18	0
10		0	0.12	0
22	0.99	0.0089	0.48	0.0089
34		0.0732	1.46	0.0062
46	1.95	0.1446	1.56	0.2266
58		0.523	1.73	0.4373
70	2.52	0.6854	1.99	0.6943
82		1.2566	2.62	1.2459
94	3.09	1.6118	2.88	1.4315
106		1.8243	3.43	2.0402
118	4.06	2.217	3.37	1.9278
130		2.2758	3.92	2.1848
142	4.48	2.8096	3.96	2.4204
154		2.6846	3.58	2.4615
166	4.25	2.8738	3.58	2.283
178		2.8345	3.34	2.7078
190	4.36	2.8828	3.47	2.6542

variable), each column of which corresponding to a dependent variable. For example, the first column of x and y matrices contain the data points of the first variable, the second column those of the second variable, and so on. It is not important to give the independent value in order (ascending or descending), but obviously it is important to give matrices of independent and dependent variables such that each element of the latter corresponds to the same element of the former. Because there may be different number of experimental data points for each independent variable, these numbers should be given to the function as a vector. The structure of matrices $\boldsymbol{x}$ and $\boldsymbol{y}$ and relations between these matrices are illustrated in Fig. E7.1*a*.

The function *NLR.m* can handle both algebraic and ordinary differential equations as the model equations. If the model consists of only one type of equation, an empty matrix should be passed to the function in the place of the file name for the other one. It is important to note that both functions evaluating algebraic and differential equations return the function or derivative values as a column vector. The functions should also perform the calculations on an element-by-element manner; that is, using dotted operators (".*", "./", and ".^") where necessary. The boundary conditions passed to the function have to be the initial conditions of the dependent variables.

For multiple regression, the variances are used in Eq. (7.180) to determine the unbiased weighting factors, w_j, which are in turn used in Eq. (7.176) to determine the unbiased weighted sum of squared residuals. In the case where repeated experimental data are available, the program searches for repeated experimental points and evaluates the variance of each dependent variable by dividing the sum of squared differences by the number of degrees of freedom of each dependent variable (see Table 7.2). The degrees of freedom of each variable

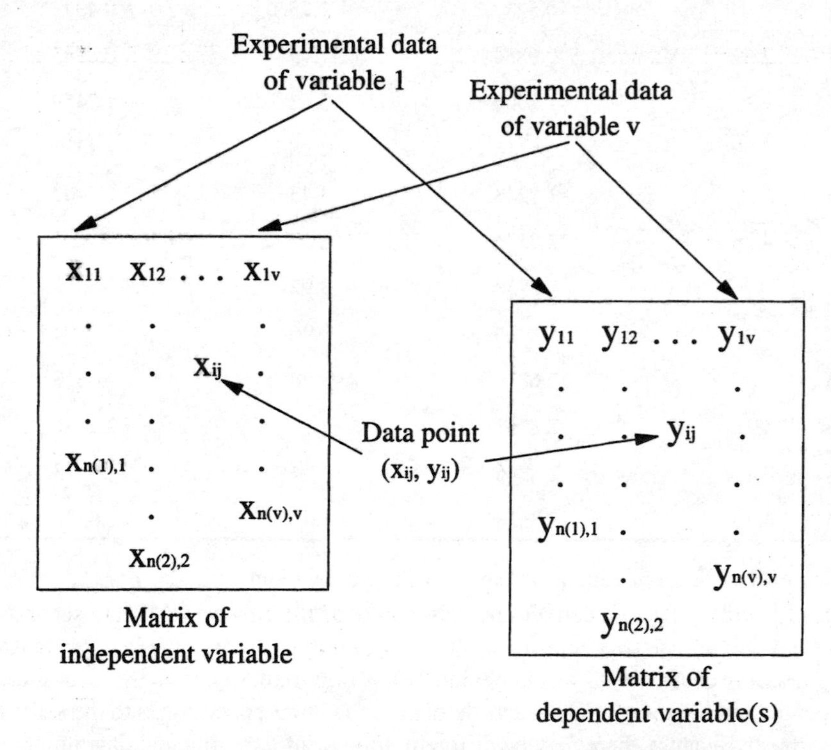

Figure E7.1*a* Structure of the matrices *x* (independent variable) and *y* (dependent variables) and their relation. Each column of the matrix corresponds to a dependent variable. The number of dependent variables is shown by *v*, and the number of experimental data points for each dependent variable is given in the vector *n*.

is calculated by the total number of experimental points available for the variable minus the number of means of groups of repeated data that have been calculated:

$$\textit{degrees of freedom} = \left(\sum_{i=1}^{p} n_i \right) - p$$

In the statistical analysis of the lack of fit, the degrees of freedom is $(p - k)$, where k is the number of parameters associated with that variable. However, in the case where several dependent variables are fitted, and the same parameter appears in the model equation of more than one variable, it is not so easy to decide exactly how many parameters correspond to each variable. For this reason, the *NLR.m* computer program apportions the parameters equally to each variable (that is, k/v degrees of freedom are deducted from each curve).

The main iterative procedure consists of two nested loops. The inner loop evaluates the parameters, whereas weighting factors, evaluated in the outer loop, are assumed constant. Starting with the initial guess of parameters, the function calculates the sum of squared residuals from Eq. (7.176). If the model contains ordinary differential equations, the function first solves the equations by the MATLAB function *ode23* and then evaluates by interpolation the calculated dependent variables corresponding to the vector of the independent variable. We use this procedure because the input may contain repeated experimental data, but *ode23* can handle only ascending or descending vector of independent variables. The Jacobian matrix is then determined numerically, followed by evaluating the vector of corrections to the parameters from Eq. (7.179) for the Marquardt method or its equivalent for the Gauss-Newton method. If applicable, the value of λ is adjusted according to the new sum of squared residuals, and the iterative procedure continues until

$$\left| \frac{\Delta b_i}{b_i} \right| < \epsilon_b \qquad \text{if} \;\; b_i \neq 0$$

or

$$|\Delta b_i| < \epsilon_b \qquad \text{if} \;\; b_i = 0$$

or

$$|\Delta \Phi_i| < \epsilon_\Phi$$

When the inner iteration loop is complete, the function reevaluates the weighting factors from Eq. (7.141) and repeats the above procedure until the increment of all the weighting factors satisfy the convergence condition

$$|\Delta w_j| < \epsilon_w$$

By default, $\epsilon_b = \epsilon_w = 0.001$ and $\epsilon_\Phi = 1\times10^{-6}$, and their values may be changed through the 10th input argument (tol) to the function.

statistics.m: This function performs statistical analysis of the data and the regression results calculated in *NLR.m*. First, the data are analyzed, and for each set of data the following statistical information is calculated: total number of points, degrees of freedom, sum of squares, variance, and standard deviation. Next, statistical analysis of the regression results is performed. The standard deviations of the parameters are calculated from Eq. (7.136), where the value of σ^2 is approximated from Eq. (7.141), and a_{ii} are the diagonal elements of the matrix

$$A = \left(\sum_{j=1}^{v} w_j J_j{}' J_j \right)^{-1}$$

In order to calculate the 100(1 - α) percent confidence intervals of the parameters, first the value of $t_{1-\alpha/2}$ is calculated by the Newton-Raphson method, using the following iteration formula:

$$t_{(1-\alpha/2)}^{(m+1)} = t_{(1-\alpha/2)}^{(m)} + \frac{\int_{t_{(1-\alpha/2)}^{(m)}}^{\infty} p(x, n-k)\, dx - \alpha/2}{p(t_{(1-\alpha/2)}^{(m)}, n-k)}$$

where $p(t,\nu)$ is the Student's t density function defined in Eq. (7.73). The confidence limits of the parameters then can be evaluated by Eq. (7.145). The Student's t density function is given in the function *stud.m*.

A significance test (t test) is performed, as described in Sec. 7.2.3 [(Eq. (7.102)], on the parameters to test the null hypothesis that any one of the parameters might be qual to zero. The 95% confidence intervals of each measured variable are calculated. The variance-covariance matrix and the matrix of correlation coefficients of the parameters are calculated according to Eqs. (7.135) and (7.154), respectively. The analysis of variance of the regression results is performed as shown in Table 7.2. Finally, the randomness tests are applied to the residuals to test for the randomness of the distribution of these residuals.

The results of the statistical analysis are stored in an output file so that they may be viewed and edited by the user.

Example7_1.m: The main program is written to get all the required inputs from the keyboard, perform regression analysis, and display the results both numerically and graphically. It is written in such a general form that it can be used in any other regression calculation as well as the problem in this example.

Ex7_1_func.m: The model equations of this problem are given in this MATLAB function. Note that this function returns values of the derivatives as a column vector.

Program

Example7_1.m

```
% Example7_1.m
% Solution to Example 7.1. This program gets the data
% from the keyboard, and/or from a data file, performs the
% nonlinear regression (using NLR.M) and the statistical
% analysis of the results (using STATISTICS.M). The program
% gives the results of the calculations numerically and
% graphically.

clear
clc
close all
disp(' Nonlinear Regression Analysis'), disp(' ')
fout = input(' Name of output file for storing results = ');
if exist(fout)
   error(' Output file already exists. Use another name.')
end
% Input experimental data
disp(' ')
disp(' Experimental data input:')
disp('    1 - Enter data from keyboard (point-by-point)')
disp('    2 - Enter data from keyboard (in vector form)')
disp('    3 - Read data from data file (prepared earlier)')
datain = input(' Enter your choice : ');
disp(' ')
switch datain
case 1
   fdata = input(' Name of file for storing the data = ');
   if exist ([fdata,'.mat'])
      error(' Data file already exists. Use another name.')
   end
   v = input(' Number of dependent variables = ');
   for m = 1:v
      countpoints=0;
      fprintf('\n Variable %2d\n',m)
      datasets=input(' How many data sets for this variable? = ');
      for nset = 1:datasets
      fprintf('\n    Data set %2d\n',nset)
      npoints = input('     How many points in this set? = ');
      disp(' ')
         for np = 1:npoints
            countpoints=countpoints+1;
            fprintf('     Point %2d\n',np)
            xm = input('     Enter independent value = ');
            ym = input('     Enter dependent value   = ');
            x(countpoints,m)=xm;
            y(countpoints,m)=ym;
         end
```

```
      end
      n(m) = countpoints;
   end
   eval(['save ',fdata ' n x y'])
case 2
   fdata = input(' Name of file for storing the data = ');
   if exist ([fdata,'.mat'])
      error(' Data file already exists. Use another name.')
   end
   v = input(' Number of independent variables = ');
   for m = 1:v
      fprintf('\n Variable %2d\n',m)
      xm = input(' Vector of independent variable = ');
      x(1:length(xm),m) = (xm(:).')';
      ym = input(' Vector of dependent variable = ');
      y(1:length(ym),m) = (ym(:).')';
      n(m) = length(ym);
   end
   eval(['save ',fdata ' n x y'])
case 3
   fname = input(' Name of file containing the data = ');
   eval(['load ', fname])
   [r,v] = size(y);
end

disp(' ')
namex = input(' Name of independent variable = ');
for m = 1:v
   nam = input([' Name of dependent variable ', int2str(m), ' = ']);
   namey(m,1:length(nam))= nam;
end

% Input type of equation(s)
disp(' ')
disp(' Type of equation(s)')
disp('   1 - Algebraic equation')
disp('   2 - Ordinary differential equation')
disp('   3 - Both 1 and 2')
eqin = input(' Enter your choice : ');
disp(' ')
switch eqin
case 1        % Algebraic equation
  fnctn=input(' Name of M-file containing algebraic equation(s)=');
  ODEfile = []; x0 = []; y0 = [];
case 2        % ODE
  ODEfile = input(' Name of M-file containing differential
equation(s) = ');
  disp(' ')
  x0=input(' Value of independent variable at boundary condition=
');
```

```
  y0=input(' Value(s) of dependent variable(s) at boundary condition
= ');
  fnctn = [];
otherwise    % Algebraic and ODE
  fnctn=input(' Name of M-file containing algebraic equation(s)= ');
  ODEfile=input(' Name of M-file containing differential equation(s)
= ');
  disp(' ')
  x0=input(' Value of independent variable at boundary condition=
');
  y0=input(' Value(s) of dependent variable(s) at boundary condition
= ');
end
% Input method of solution
disp(' ')
disp(' Method of solution')
disp('   1 - Marquardt')
disp('   2 - Gauss-Newton')
method = input(' Enter your choice : ');

disp(' ')
b0 = input(' Vector of initial guess of fitting parameters = ');
disp(' ')
trace = input(' Show results of each iteration (0/1) ? ');
if trace == 0
   disp(' ')
   disp(' Please wait for final results')
end

% Regression
[b, yc, w, JTJ] = NLR(b0, n, x, y, fnctn, ODEfile, x0, y0, ...
   method, [], trace);
if trace == 1
   disp(' ')
   disp(' Please wait for final results')
end
% Statistical properties
[sd, cl] = statistics(b, n, x, y, yc, w, JTJ, 95, fout);

% Displaying final results
disp(' ')
disp(' ***************************************************************')
disp('                        Final Results')
disp(' ***************************************************************')
disp(' ')
disp('  No.   Parameter    Standard      95% Confidence interval ')
disp('                     deviation     for the parameters      ')
disp('                                   lower value  upper value')
for m = 1:length(b)
   fprintf('  %2d    %10.4e  %10.4e   %10.4e  %10.4e\n',...
      m,b(m),sd(m),cl(m,:))
```

```
end
disp(' ')
disp(' ***********************************************************')

% Plotting the results
for m = 1:v
   figure(m)
   [xx,loc] = sort(x(1:n(m),m));
   plot(x(1:n(m),m),y(1:n(m),m),'o',xx,yc(loc,m))
   xlabel(namex)
   ylabel(namey(m,:))
end
```

NLR.m

NOTE: The program *NLR.m* is not listed here because of its length. The user is encouraged to examine the program using the MATLAB Editor.

statistics.m

NOTE: The program *statistics.m* is not listed here because of its length. The user is encouraged to examine the program using the MATLAB Editor.

stud.m

```
function p = stud(t, nu)
% Student t distribution
p = 1/sqrt(nu*pi) * gamma((nu+1)/2)/gamma(nu/2) * ...
   (1 + t.^2/nu).^(-(nu + 1)/2);
```

Ex7_1_func.m

```
function dy = Ex7_1_func(x,y,flag,b)
% Function Ex7_1_func.M
% Model equations for Example 7.1.
dy = [b(1)*y(1)*(1-y(1)/b(2));
      b(3)*y(1)-b(4)*y(2)];
```

Input and Results

```
>>Example7_1

Nonlinear Regression Analysis

 Name of output file for storing the results = 'Ex7_1_results'

 Experimental data input:
   1 - Enter data from keyboard (point-by-point)
   2 - Enter data from keyboard (in vector form)
   3 - Read data from data file (prepared earlier)
```

```
Enter your choice : 1

Name of data file for storing the data = 'Ex7_1_data'
Number of dependent variables = 2

Variable  1
How many data sets for this variable? = 2

  Data set  1
  How many points in this set? = 9

  Point  1
  Enter independent value = 0
  Enter dependent value   = 0.40
  Point  2
  Enter independent value = 22
  Enter dependent value   = 0.99
  Point  3
  Enter independent value = 46
  Enter dependent value   = 1.95
  Point  4
  Enter independent value = 70
  Enter dependent value   = 2.52
  Point  5
  Enter independent value = 94
  Enter dependent value   = 3.09
  Point  6
  Enter independent value = 118
  Enter dependent value   = 4.06
  Point  7
  Enter independent value = 142
  Enter dependent value   = 4.48
  Point  8
  Enter independent value = 166
  Enter dependent value   = 4.25
  Point  9
  Enter independent value = 190
  Enter dependent value   = 4.36

  Data set  2
  How many points in this set? = 17

  Point  1
  Enter independent value = 0
  Enter dependent value   = 0.18
  Point  2
  Enter independent value = 10
```

```
Enter dependent value   = 0.12
Point  3
Enter independent value = 22
Enter dependent value   = 0.48
Point  4
Enter independent value = 34
Enter dependent value   = 1.46
Point  5
Enter independent value = 46
Enter dependent value   = 1.56
Point  6
Enter independent value = 58
Enter dependent value   = 1.73
Point  7
Enter independent value = 70
Enter dependent value   = 1.99
Point  8
Enter independent value = 82
Enter dependent value   = 2.62
Point  9
Enter independent value = 94
Enter dependent value   = 2.88
Point 10
Enter independent value = 106
Enter dependent value   = 3.43
Point 11
Enter independent value = 118
Enter dependent value   = 3.37
Point 12
Enter independent value = 130
Enter dependent value   = 3.92
Point 13
Enter independent value = 142
Enter dependent value   = 3.96
Point 14
Enter independent value = 154
Enter dependent value   = 3.58
Point 15
Enter independent value = 166
Enter dependent value   = 3.58
Point 16
Enter independent value = 178
Enter dependent value   = 3.34
Point 17
Enter independent value = 190
Enter dependent value   = 3.47
```

```
Variable  2
How many data sets for this variable? = 2

  Data set  1
  How many points in this set? = 17

  Point  1
  Enter independent value = 0
  Enter dependent value   = 0
  Point  2
  Enter independent value = 10
  Enter dependent value   = 0
  Point  3
  Enter independent value = 22
  Enter dependent value   = 0.0089
  Point  4
  Enter independent value = 34
  Enter dependent value   = 0.0732
  Point  5
  Enter independent value = 46
  Enter dependent value   = 0.1446
  Point  6
  Enter independent value = 58
  Enter dependent value   = 0.5230
  Point  7
  Enter independent value = 70
  Enter dependent value   = 0.6854
  Point  8
  Enter independent value = 82
  Enter dependent value   = 1.2566
  Point  9
  Enter independent value = 94
  Enter dependent value   = 1.6118
  Point 10
  Enter independent value = 106
  Enter dependent value   = 1.8243
  Point 11
  Enter independent value = 118
  Enter dependent value   = 2.2170
  Point 12
  Enter independent value = 130
  Enter dependent value   = 2.2758
  Point 13
  Enter independent value = 142
  Enter dependent value   = 2.8096
  Point 14
  Enter independent value = 154
```

```
Enter dependent value   = 2.6846
Point 15
Enter independent value = 166
Enter dependent value   = 2.8738
Point 16
Enter independent value = 178
Enter dependent value   = 2.8345
Point 17
Enter independent value = 190
Enter dependent value   = 2.8828

Data set  2
How many points in this set? = 17

Point  1
Enter independent value = 0
Enter dependent value   = 0
Point  2
Enter independent value = 10
Enter dependent value   = 0
Point  3
Enter independent value = 22
Enter dependent value   = 0.0089
Point  4
Enter independent value = 34
Enter dependent value   = 0.0642
Point  5
Enter independent value = 46
Enter dependent value   = 0.2266
Point  6
Enter independent value = 58
Enter dependent value   = 0.4373
Point  7
Enter independent value = 70
Enter dependent value   = 0.6943
Point  8
Enter independent value = 82
Enter dependent value   = 1.2459
Point  9
Enter independent value = 94
Enter dependent value   = 1.4315
Point 10
Enter independent value = 106
Enter dependent value   = 2.0402
Point 11
Enter independent value = 118
Enter dependent value   = 1.9278
```

```
  Point 12
  Enter independent value = 130
  Enter dependent value   = 2.1848
  Point 13
  Enter independent value = 142
  Enter dependent value   = 2.4204
  Point 14
  Enter independent value = 154
  Enter dependent value   = 2.4615
  Point 15
  Enter independent value = 166
  Enter dependent value   = 2.2830
  Point 16
  Enter independent value = 178
  Enter dependent value   = 2.7078
  Point 17
  Enter independent value = 190
  Enter dependent value   = 2.6542

Name of independent variable = 'Time, hours'
Name of dependent variable 1 = 'Cell concentration, % dry weight'
Name of dependent variable 2 = 'Penicillin concentration, units/mL'

Type of equation(s)
  1 - Algebraic equation
  2 - Ordinary differential equation
  3 - Both 1 and 2
Enter your choice : 2

Name of M-file containing differential equation(s) = 'Ex7_1_func'

Value of independent variable at boundary condition = 0
Value(s) of dependent variable(s) at boundary condition = [0.29 0]

Method of solution
  1 - Marquardt
  2 - Gauss-Newton
Enter your choice : 1

Vector of initial guess of fitting parameters = [0.1,4,0.02,0.02]

Show results of each iteration (0/1) ? 1


Iteration on weights =    1
Variable    Weight
    1       2.4241e-001
    2       1.5793e+000
```

```
Starting values
Parameter    Value
     1       1.0000e-001
     2       4.0000e+000
     3       2.0000e-002
     4       2.0000e-002
Sum of squares = 6.9923e+001
Lambda = 1.0000e+003

Iteration on parameters =    1
Parameter    Value
     1       6.8498e-002
     2       3.9970e+000
     3       7.3256e-003
     4       1.1040e-002
Sum of squares = 1.5563e+001
Lambda = 2.5000e+002

Iteration on parameters =    2
Parameter    Value
     1       3.9016e-002
     2       3.9926e+000
     3       1.2042e-002
     4       1.5427e-002
Sum of squares = 8.3942e+000
Lambda = 6.2500e+001

Iteration on parameters =    3
Parameter    Value
     1       4.2387e-002
     2       3.9898e+000
     3       1.8434e-002
     4       2.5628e-002
Sum of squares = 2.7515e+000
Lambda = 1.5625e+001

Iteration on parameters =    4
Parameter    Value
     1       4.2834e-002
     2       3.9817e+000
     3       1.7434e-002
     4       2.1875e-002
Sum of squares = 2.2976e+000
Lambda = 3.9063e+000

Iteration on parameters =    5
```

```
Parameter   Value
     1      4.2922e-002
     2      3.9623e+000
     3      1.7713e-002
     4      2.2475e-002
Sum of squares = 2.2858e+000
Lambda = 9.7656e-001

Iteration on parameters =   6
Parameter   Value
     1      4.3208e-002
     2      3.9368e+000
     3      1.7425e-002
     4      2.1858e-002
Sum of squares = 2.2832e+000
Lambda = 2.4414e-001

Iteration on parameters =   7
Parameter   Value
     1      4.3257e-002
     2      3.9278e+000
     3      1.7413e-002
     4      2.1794e-002
Sum of squares = 2.2830e+000
Lambda = 6.1035e-002

Iteration on parameters =   8
Parameter   Value
     1      4.3260e-002
     2      3.9270e+000
     3      1.7409e-002
     4      2.1783e-002
Sum of squares = 2.2830e+000
Lambda = 1.5259e-002


Iteration on weights =   2
Variable   Weight
    1      3.7508e-001
    2      1.4779e+000


Starting values
Parameter   Value
     1      4.3260e-002
     2      3.9270e+000
     3      1.7409e-002
```

```
     4       2.1783e-002
Sum of squares = 2.5720e+000
Lambda = 1.0000e+003

Iteration on parameters =    1
Parameter    Value
     1       4.2944e-002
     2       3.9270e+000
     3       1.7618e-002
     4       2.2063e-002
Sum of squares = 2.5706e+000
Lambda = 2.5000e+002

Iteration on parameters =    2
Parameter    Value
     1       4.2913e-002
     2       3.9269e+000
     3       1.7686e-002
     4       2.2174e-002
Sum of squares = 2.5705e+000
Lambda = 6.2500e+001

Iteration on parameters =    3
Parameter    Value
     1       4.2916e-002
     2       3.9267e+000
     3       1.7682e-002
     4       2.2165e-002
Sum of squares = 2.5705e+000
Lambda = 1.5625e+001


Iteration on weights =    3
Variable    Weight
    1       3.7808e-001
    2       1.4756e+000


Starting values
Parameter    Value
     1       4.2916e-002
     2       3.9267e+000
     3       1.7682e-002
     4       2.2165e-002
Sum of squares = 2.5770e+000
Lambda = 1.0000e+003
```

```
Iteration on parameters =   1
Parameter   Value
    1       4.2916e-002
    2       3.9267e+000
    3       1.7682e-002
    4       2.2165e-002
Sum of squares = 2.5770e+000
Lambda = 2.0000e+003

Please wait for final results

**********************************************************
      Statistical analysis of the experimental data
**********************************************************

Unweighted statistics

Variable No.                      1
Total points                     26
Degrees of freedom                9
Sum of squares                1.387
Variance                     0.1541
Standard deviation           0.3925

Variable No.                      2
Total points                     34
Degrees of freedom               17
Sum of squares                0.402
Variance                    0.02365
Standard deviation           0.1538

Weighted statistics

Total points                     60
Total degrees of freedom         26
Weighted sum of squares       1.117
Weighted variance           0.04298
Weighted stand. dev.         0.2073

**********************************************************
         Statistical analysis of the regression
**********************************************************

 No.   Parameter    Standard       95% Confidence interval
                    deviation      for the parameters
                                   lower value  upper value

  1    4.2916e-002  2.3579e-003    3.8192e-002  4.7639e-002
```

2	3.9267e+000	1.2089e-001	3.6845e+000	4.1689e+000
3	1.7682e-002	2.6767e-003	1.2320e-002	2.3044e-002
4	2.2165e-002	4.7113e-003	1.2727e-002	3.1603e-002

Degrees of freedom = 56

Total (weighted) sum of squared residuals = 2.577

Combined (weighted) residual variance (s^2) = 0.04602

Significance tests

No.	Parameter	t-calculated	Is parameter significantly different from zero?
1	4.2916e-002	1.8200e+001	Yes
2	3.9267e+000	3.2482e+001	Yes
3	1.7682e-002	6.6061e+000	Yes
4	2.2165e-002	4.7047e+000	Yes

Confidence limits of regressed variables

Measured variable	Degrees of freedom	Residual variance	95% Confidence limit for each measured variable
1	24	1.2171e-001	7.2005e-001
2	32	3.1186e-002	3.5971e-001

Covariance analysis

Variance-covariance matrix: s^2*Inverse(J transpose J)

5.5599e-006	-1.5722e-004	-4.6564e-006	-7.7490e-006
-1.5722e-004	1.4614e-002	1.2934e-004	2.8603e-004
-4.6564e-006	1.2934e-004	7.1645e-006	1.2369e-005
-7.7490e-006	2.8603e-004	1.2369e-005	2.2196e-005

Matrix of correlation coefficients

1	-0.5515	-0.7378	-0.6975
-0.5515	1	0.3997	0.5022
-0.7378	0.3997	1	0.9809
-0.6975	0.5022	0.9809	1

Analysis of variance

Source of variance	Sum of squares	Degrees of freedom	Variance

```
 Lack of fit           1.4596e+000          30          4.8654e-002

 Experimental error    1.1174e+000          26          4.2979e-002

 Total                 2.5771e+000          56          4.6019e-002

Randomness test

 Variable                                1
 Number of positive residuals            9
 Number of negative residuals            8
 Number of runs (changes of sign)        8
 Z =                                    -0.7395

 Random at 95% level of confidence

 Variable                                2
 Number of positive residuals            8
 Number of negative residuals            9
 Number of runs (changes of sign)        3
 Z =                                    -3.254

 Not random at 95% level of confidence

**********************************************************
              End of statistical analysis
**********************************************************

**********************************************************
The results of the statistical analysis have been stored
in the output file you specified.  This file is located
in the default directory you have operated from.  You may
open and view this file using any editor.
**********************************************************

**********************************************************
                  Final Results
**********************************************************

 No.   Parameter    Standard      95% Confidence interval
                    deviation     for the parameters
                                  lower value  upper value
  1    4.2916e-002  2.3579e-003   3.8192e-002  4.7639e-002
  2    3.9267e+000  1.2089e-001   3.6845e+000  4.1689e+000
  3    1.7682e-002  2.6767e-003   1.2320e-002  2.3044e-002
  4    2.2165e-002  4.7113e-003   1.2727e-002  3.1603e-002
**********************************************************
```

Discussion of Results: The experimental and calculated values of y_1 and y_2 are shown in Figs. E7.1*b*. The final values of the parameters are calculated as

$$b_1 = 0.0429 \qquad b_2 = 3.9244 \qquad b_3 = 0.0177 \qquad b_4 = 0.0221$$

The program starts out with the weighting factors calculated from the variances of the experimental data and the Marquardt method converges after eight iterations on the parameters. The weighting factors are then adjusted and the function repeats the iteration on the parameters and converges in three iterations. Finally, the weighing factors are changed for the third time and the method converges in one iteration. The reader is encouraged to repeat the calculations with different methods and different starting guesses of parameters. A bad guess may cause the Gauss-Newton method to diverge. If the Marquardt method is chosen, the method does not diverge. However, a bad guess may require a large number of iterations to converge, and may possibly exceed the maximum limit of iterations defined in the function.

The program gives a complete statistical analysis of the experimental data and the regression results. For the experimental data, the program calculates the following statistics for each dependent variable being fitted: total points, degrees of freedom, sum of squares, variance, and standard deviation. These statistics are shown for both unweighted and weighted data. For the regression results, the program evaluates the standard deviation and 95% confidence intervals of the parameters, it performs a significance test (t test) on each parameter to determine whether the value of that parameter is different than zero, it calculates the 95% confidence limit for each fitted variable, it lists the variance-covariance matrix and the matrix of correlation coefficients of the parameters, performs a complete analysis of variance, and does a randomness test on the distribution of the residuals.

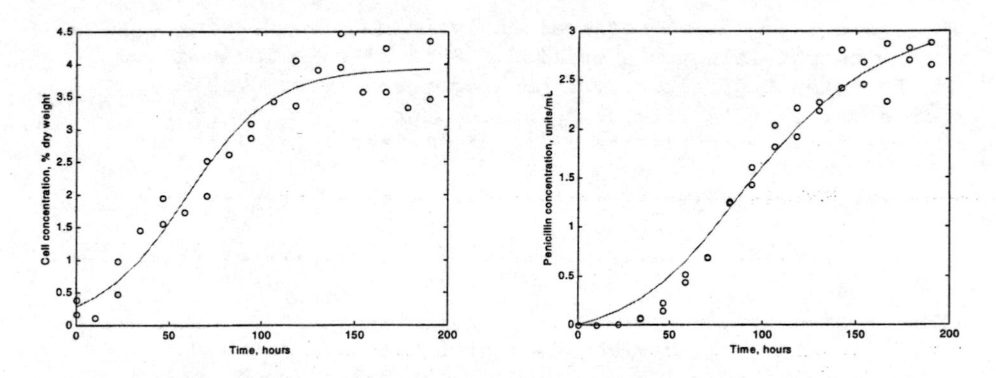

Figure E7.1*b* Results of the regression of cell and penicillin experimental data.

PROBLEMS

7.1 The heat capacity of gases and liquids is a function of temperature. Various forms of polynomial equations have been used to represent this functionality. Two such equations are shown below:

$$c_p = a + bT + cT^2 + dT^3$$

$$c_p = a + bT + cT^{-2}$$

Using the heat capacity data of Table P7.1, determine the coefficients of these equations. Discuss your results and recommend which equation gives the best representation of the data.

Table P7.1 Simulated heat capacity data

Temperature (°C)	Heat capacity (J/gmol °C) Set no. 1	Heat capacity (J/gmol °C) Set no. 2	Heat capacity (J/gmol °C) Set no. 3
100	29.38	30.04	28.52
200	29.88	29.08	29.79
300	30.42	30.18	31.41
400	30.98	30.14	31.18
500	31.57	32.27	31.16
600	32.15	31.79	32.81
700	32.73	32.97	32.38
800	33.29	32.56	34.26
900	33.82	34.24	34.72
1000	34.31	35.27	33.69

7.2 A mathematical model of the fermentation of the bacterium *Pseudomonas ovalis*, which produces gluconic acid, was given in Prob. 5.6. This model, which describes the dynamics of the logarithmic growth phase, can be summarized as follows:

Rate of cell growth:

$$\frac{dy_1}{dt} = b_1 y_1 \left(1.0 - \frac{y_1}{b_2} \right)$$

Rate of gluconolactone formation:

$$\frac{dy_2}{dt} = \frac{b_3 y_1 y_4}{b_4 + y_4} - 0.9082 b_5 y_2$$

Rate of gluconic acid formation:

$$\frac{dy_3}{dt} = b_5 y_2$$

Rate of glucose consumption:

$$\frac{dy_4}{dt} = -1.011\left(\frac{b_3 y_1 y_4}{b_4 + y_4}\right)$$

where y_1 = concentration of cell
y_2 = concentration of gluconolactone
y_3 = concentration of gluconic acid
y_4 = concentration of glucose
b_1-b_5 = parameters of the system that are functions of temperature and pH.

Using the batch fermentation data given in Table P7.2, determine the values of the parameters, b_1-b_5, at the three different temperatures of 25°, 28°, and 30°C.

7.3 Accurate vapor-liquid equilibrium measurements can be used to compute liquid-phase activity coefficients and excess Gibbs free energies. Consider the data in Table P7.3 for benzene-2,2,4-trimethylpentane (B-TMP) mixtures at constant temperature of 55°C.

(a) Assume that the gas phase is ideal and neglect any fugacity and Poynting corrections for the liquid phase. Calculate the activity coefficients for *B* and TMP and the molar excess Gibbs free energy at each temperature point. The vapor pressure of pure *B* at 55°C is 327.05 mmHg and of pure TMP at 55°C is 178.08 mmHg.

(b) If a *three-constant* Redlich-Kister expansion for the excess molar Gibbs free energy is assumed [see Denbigh, Ref. 11, p. 286], evaluate the constants that appear using the data of part (a); that is, find "best fits" for A_0, A_1, and A_2 (Denbigh notation).

(c) Calculate the activity coefficients from your expressions in part (b) for *B* and TMP. (*Hint*: First derive an expression for the activity coefficients assuming a three-constant Redlich-Kister expansion.)

(d) Plot the theoretical excess molar Gibbs free energy and the theoretical activity coefficients with the experimental data as well.

7.4 Use the data of Prob. 5.7 to fit the Lotka-Volterra predator-prey equations (shown below) in order to obtain accurate estimates of the parameters of the model. Modify the Lotka-Volterra equations as recommended in Prob. 5.8, and determine the parameters of your new models. Compare the results of the statistical analysis for each model, and choose the set of equations that gives the best representation of the data.

The Lotka-Volterra equations:

$$\frac{dN_1}{dt} = \alpha N_1 - \beta N_1 N_2$$

$$\frac{dN_2}{dt} = -\gamma N_2 + \beta N_1 N_2$$

Table P7.2 Data obtained by varying the temperature at pH 7.0

Time (hour)	Cell concentration (UOD/ml)	Gluconolactone concentration (mg/ml)	Gluconic acid concentration (mg/ml)	Glucose concentration (mg/ml)
Experiment No. 34, batch fermentation data at 25.0°C				
0.0	0.56	1.28	0.16	45.00
1.0	0.86	2.20	1.56	43.00
2.0	1.60	3.50	5.00	38.00
3.0	2.60	5.60	9.50	33.00
4.0	3.20	7.00	16.00	25.00
5.0	3.30	7.80	24.50	16.50
6.0	3.50	7.20	32.00	9.00
7.0	3.40	6.30	45.50	4.00
8.0	3.40	3.20	45.80	2.00
Experiment no. 33, batch fermentation data at 28.0°C				
0.0	0.66	-	-	48.00
1.0	1.00	1.96	0.15	45.00
2.0	1.60	6.67	7.00	37.50
3.0	2.60	10.50	15.00	28.00
4.0	3.20	10.50	25.00	18.00
5.0	3.30	7.58	35.00	8.00
6.0	3.30	2.05	42.50	3.00
7.0	3.30	1.90	45.50	-
Experiment no. 32, batch fermentation data at 30.0°C				
0.0	0.80	1.34	0.95	44.50
1.0	1.50	4.00	4.96	37.50
2.0	2.60	7.50	16.10	25.00
3.0	3.50	8.00	32.10	9.00
4.0	3.50	5.00	43.70	3.00
5.0	3.50	2.42	44.50	2.00

Source: V. R. Rai, "Mathematical Modeling and Optimization of the Gluconic Acid Fermentation," Ph.D. dissertation, Rutgers - The State University of New Jersey, Piscataway, NJ., 1973.

Table P7.3 Vapor-liquid equilibrium data

Liquid phase mole fraction x_B	Vapor phase mole fraction y_B	Equilibrium total pressure P (mmHg)
0.0819	0.1869	202.74
0.2192	0.4065	236.86
0.3584	0.5509	266.04
0.3831	0.5748	270.73
0.5256	0.6786	293.36
0.8478	0.8741	324.66
0.9872	0.9863	327.39

Source: S. Weissman and S. E. Wood, "Vapor-Liquid Equilibrium of Benzene-2,2,4-trimethylpentane Mixtures," *J. Chem. Phys.*, vol. 32, 1960, p.1153.

7.5 Svirbely and Blaner [12] modeled a set of chemical reactions represented by

$$A + B \xrightarrow{k_1} C + F$$

$$A + C \xrightarrow{k_2} D + F$$

$$A + D \xrightarrow{k_3} E + F$$

using the following differential equations

$$\frac{dA}{dt} = -k_1AB - k_2AC - k_3AD$$

$$\frac{dB}{dt} = -k_1AB$$

$$\frac{dC}{dt} = k_1AB - k_2AC$$

$$\frac{dD}{dt} = k_2AC - k_3AD$$

$$\frac{dE}{dt} = k_3AD$$

Estimate the coefficients k_1, k_2, and k_3 (all positive) from the data of Table P7.5, and the following initial conditions:

$$C(0) = D(0) = 0.0$$

$$A(0) = 0.02090 \text{ mol/L}$$
$$B(0) = (1/3)A(0)$$

The estimates reported in the article were:

$$k_1 = 14.7$$
$$k_2 = 1.53$$
$$k_3 = 0.294$$

Could estimates be obtained if the initial conditions were not known?

Table P7.5 Experimental reaction data

Time (min)	Ax10³ (mol/L)	Time (min)	Ax10³ (mol/L)
4.50	15.40	76.75	8.395
8.67	14.22	90.00	7.891
12.67	13.35	102.00	7.510
17.75	12.32	108.00	7.370
22.67	11.81	147.92	6.646
27.08	11.39	198.00	5.883
32.00	10.92	241.75	5.322
36.00	10.54	270.25	4.960
46.33	9.780	326.25	4.518
57.00	9.157	418.00	4.075
69.00	8.594	501.00	3.715

7.6 Choose one of the equations given in Prob. 7.1 and perform the following:
(a) Fit the equation to the specific heat data given in Prob. 7.1.
(b) Add random error (r.e.) to the data, where this error is in the range $-1.0 \le \text{r.e.} \le 1.0$, and fit the equation to the noisy data.
(c) Repeat part (b) with $-5.0 \le \text{r.e.} \le 5.0$.
(d) Repeat part (b) with $-10.0 \le \text{r.e.} \le 10.0$.
Compare the results of the statistical analysis in parts (a) to (d). What conclusions do you draw?

7.7 Solids mixing in fluidized beds is assumed to be described by the dispersion model which is a diffusion-type model. Based on this model, axial dispersion of solids is represented by the differentia equation

$$\frac{\partial C}{\partial t} = D\frac{\partial^2 C}{\partial z^2}$$

where C is the concentration of tagged particles at axial position z and time t, and D is the axial dispersion coefficient of the solids. In an experiment in a gas-solid fluidized bed, a certain amount of tagged particles (1000 units) is injected at the top of the bed ($z = 350$ mm) at the beginning of the experiment. The tagged particles are not added or taken out during the experiment. Therefore, the diffusion equation in this case should be solved subject to the following initial and boundary conditions:

$$\text{at } t = 0, \quad \begin{cases} C = C_0 = 1000 & \text{for } z = L = 350\,mm \\ C = 0 & \text{for } z < L \end{cases}$$

$$\text{at } t \geq 0, \quad \frac{\partial C}{\partial z}\Big|_{z=L} = 0$$

$$\text{at } t \geq 0, \quad \frac{\partial C}{\partial z}\Big|_{z=0} = 0$$

Concentration of tagged particles at different heights and times are measured during the experiment and are given in Table P7.7.

Table P7.7 Experimental concentration of tagged particles as a function of time and height

t (s)	z = 300 mm	z = 250 mm	z = 200 mm
0.0	0	0	0
0.1	20	5	0
0.2	63	22	6
0.3	72	40	24
0.4	57	53	36
0.5	30	56	38
0.6	29	32	48
0.8	26	25	40
1.0	25	23	25
1.5	17	21	16
2.0	29	20	23

Develop a MATLAB function to calculate the dispersion coefficient in the above partial differential equation from $C(t, z)$ data by using a least squares technique. Apply this function to the data of Table P7.7 to evaluate the dispersion coefficient at the conditions of this experiment.

REFERENCES

1. Himmelblau, D. M., and Bischoff, K. B., *Process Analysis and Simulation: Deterministic Systems*, Wley, New York, 1968.

2. Box, G. E. P. and Hunter, W. G., "A Useful Method for Model-Building," *Technometrics*, vol. 4, no. 3, 1962, p. 301.

3. Bethea, R. M., *Statistical Methods for Engineers and Scientists*, 3rd ed., Marcel Dekker, New York, 1995.

4. Ostle, B., Turner, K. V., Hicks, C. R., and McEleath, G. W., *Engineering Statistics: The Industrial Experience*, Durburg Press, Belmont, 1996.

5. Seinfeld, J. H., and Lapidus, L., *Mathematical Methods in Chemical Engineering*, vol. 3, *Process Modeling, Estimation, and Identification*, Prentice Hall, Englewood Cliffs, NJ, 1974.

6. Davies, O. L., *Statistical Methods in Research and Production*, Hafner, New York, 1957.

7. Box, G. E. P., "Fitting Empirical Data," *Ann. N.Y. Acad. Sci.*, vol. 86, 1960, p. 792.

8. Edgar, T. F., and Himmelblau, D. M., *Optimization of Chemical Processes*, McGraw-Hill, New York, 1988.

9. Marquardt, D. W., "An Algorithm for Least Squares Estimation of Nonlinear Parameters," *J. Soc. Ind. Appl. Math.*, vol. 11, 1963, p. 431.

10. Brownlee, K. A., *Statistical Theory and Methodology in Science and Engineering*, 2nd ed., Wiley, New York, 1965.

11. Denbigh, K., *The Principles of Chemical Equilibrium*, 3rd ed., Cambridge University Press, Cambridge, U.K., 1971.

12. Svirbely, W. J. and Blaner, J. A., *J. Amer. Chem. Soc.*, vol. 83, 1961, p. 4118. (Problem reproduced from D. M. Himmelblau, *Process Analysis by Statistical Methods*, Wiley, New York, 1970).

APPENDIX **A**

Introduction to MATLAB

This appendix is intended to help the reader get started with MATLAB and should not be considered as a complete reference. We have introduced in this section the MATLAB features that are essential to understanding the software developed throughout the text. A good detailed MATLAB tutorial is given by Hanselman and Littlefield [1, 2]. It is assumed that the reader reads this appendix and practices each command while sitting at a computer running MATLAB. For this reason, the outputs to each command are not printed here. Many commands may fail to convey the intended lesson if the user is not practicing them on the computer. In the following, "»" is MATLAB's prompt. You do not need to type it.

A.1 BASIC OPERATIONS AND COMMANDS

The four elementary arithmetic operations in MATLAB are done by the operators +, -, *, and /, and ^ stands for power operator:

```
»2+3*4^(1-1/5)
```

The operator \ is for left division. For example, try

```
»2/4
»2\4
```

MATLAB easily handles complex and infinite numbers:

```
»sqrt(-1), 1/0
```

Both *i* and *j* stand for the complex number $\sqrt{-1}$ unless another value is assigned to them. Also the variable *pi* represents the ratio of the circumference of a circle to its diameter (i.e., 3.141592653...). If an expression cannot be evaluated, MATLAB returns *NaN*, which stands for Not-a-Number:

```
»0*log(0)
```

The equality sign is used to assign values to variables:

```
»a = 2
»b = 3*a
```

If no name is introduced, result of the expression is saved in a variable named *ans*:

```
»a+b
```

If you do not want to see the result of the command, put a semicolon at the end of it:

```
»a+b;
```

You can see the value of the variable by simply typing it:

```
»a
```

MATLAB is case sensitive. This means MATLAB distinguishes between upper and lower case variables:

```
»A
```

In MATLAB all computations are done in double precision. However, the result of calculation is normally shown with only 5 digits. The *format* command may be used to switch between different output display formats:

»c = *exp*(*pi*)
»*format long*, c
»*format short e*, c
»*format long e*, c
»*format short*, c

Use the command *who* to see names of the variables, currently available in the workspace:

»*who*

and to see a list of variables together with information about their size, density, etc., use the command *whos*:

»*whos*

In order to delete a variable from the memory use the *clear* command:

»*clear* a, *who*

Using *clear* alone deletes all the variables from the workspace:

»*clear*, *who*

The *clc* command clears the command window and homes the cursor:

»*clc*

Remember that by using the up arrow key you can see the commands you have entered so far in each session. If you need to call a certain command that has been used already, just type its first letter (or first letters) and then use the up arrow key to call that command.

Several navigational commands from DOS and UNIX may be executed from the MATLAB Command Window, such as *cd*, *dir*, *mkdir*, *pwd*, *ls*. For example:

»*cd* d:\matlab\toolbox
»*cd* 'c:\Program Files\Numerical Methods\Chapter1'

The single quotation mark (') is needed in the last command because of the presence of blank spaces in the name of the directory.

A.2 VECTORS, MATRICES, AND MULTIDIMENSIONAL ARRAYS

MATLAB is designed to make operations on matrices as easy as possible. Most of the variables in MATLAB are considered as matrices. A scalar number is a 1×1 matrix and a vector is a 1×*n* (or *n*×1) matrix. Introducing a matrix is also done by an equality sign:

```
»m = [1  2  3; 4, 5, 6]
```

Note that elements of a row may be separated either by a space or a comma, and the rows may be separated by a semicolon or carriage return (i.e., a new line). Elements of a matrix can be called or replaced individually:

```
»m(1,3)
»m(2,1) = 7
```

Matrices may combine together to form new matrices:

```
»n = [m; m]
»o = [n, n]
```

The transpose of a matrix results from interchanging its rows and columns. This can be done by putting a single quote after a matrix:

```
»m = [m; 7, 8, 9]'
```

A very useful syntax in MATLAB is the colon operator that produces a row vector:

```
»v = -1:4
```

The default increment is 1, but the user can change it if required:

```
»w = [-1:0.5:4; 8:-1:-2; 1:11]
```

A very common use of the colon notation is to refer to rows, columns, or a part of the matrix:

```
»w(:,5)
»w(1,:)
»w(2:3,4:7)
»w(2,8:end)
```

Multidimentional arrays (i.e., arrays with more than two dimensions) is a new feature in MATLAB 5. Let us add the third dimension to the matrix w:

```
»w(:,:,2) = ones(3,11)
```

and the fourth dimension

»w(1,1,1,2) = 5; *whos*

The third dimension is called page and no generic name is given to the higher dimensions.

There are many built-in array construction functions. For example:

»*ones*(2)	generates a 2×2 matrix of ones
»*ones*(2,3)	generates a 2×3 matrix of ones
»*zeros*(2,3,2)	generates a 2×3×2 array of zeros
»*eye*(3)	generates a 3×3 identity matrix
»*rand*(4,2)	generates a 4×2 matrix of random entries
»*linspace*(-1,5,7)	generates a 7-element row vector of equally spaced numbers between -1 and 5
»*logspace*(-1,2,8)	generates an 8-element row vector of logarithmically equally spaced points between 10^{-1} and 10^2

Two useful array functions are *size,* which gives the size of the array, and *length,* which gives the maximum length of the array:

»*size*(w)
»*length*(w)

A.2.1 Array Arithmetic

Multiplying a scalar to an array, multiplies all the elements by the scalar:

»a = [1, 2, 4; 2:4; 4:0.5:5], 2*a

Only two arrays of the same size may be added or subtracted:

»b = *ones*(3); a-b
»c = *ones*(3,1); a+c

Adding a scalar to an array results in adding the scalar to all the elements of the array:

»a+2

Vector and matrix multiplication requires that the sizes match:

»a*b
»a*c
»c*a

»c'*a

To perform an operation on an array element-by-element, use a "." before the operator:

»a.*b
»a^2
»a.^2
»1./a

Some useful matrix functions are:

»*det*(a)	determinant of a square matrix
»*inv*(a)	inverse of matrix
»*rank*(a)	rank of matrix (see Chap. 2)
»*eig*(a)	eigenvalues and eigenvectors of square matrix
»*poly*(a)	characteristic polynomial of matrix (see Chap. 2)
»*svd*(a)	singular value decomposition

A.3 GRAPHICS

A.3.1 2-D Graphs

Functions with one independent variable can easily be visualized in MATLAB:

»x = *linspace*(0,2); y = x.**exp*(-x);

»*plot*(y)	plots *y* versus their index
»*plot*(x,y)	plots *y* against *x*
»*grid*	adds grid lines to the current axes
»*grid*	removes the grid lines
»*xlabel*('x')	adds text below the x-axis
»*ylabel*('y')	adds text besides the y-axis
»*title*('y = x*exp(-x)')	adds text at the top of the graph
»*gtext*('anywhere')	places text with mouse
»*text*(1,0.2,'(1,0.2)')	places text at the specific point

You can use symbols instead of lines. You can also plot more than one function in a graph:

»*plot*(x,y,'.',x,x.**sin*(x))

Also, more than one graph can be shown in different frames:

»*subplot*(2,1,1), *plot*(x,x.***cos*(x))
»*subplot*(2,1,2), *plot*(x,x.***sin*(x))

Axis limits can be seen and modified using the *axis* command:

»*axis*
»*axis*([0, 1.5, 0, 1.5])

Before continuing, clear the graphic window:

»*clf*

Now let us see the comet-like trajectory of the function:

»*shg*, *comet*(x,y)

The *shg* command brings up the current graphic window. It is possible to use more than one graphic window by using *figure*(n) command, where n is a positive integer.

Another easy way to plot a function is:

»*fplot*('x***exp*(-x)',[0, 2])

The function to be plotted may also be a user-defined function (see Sec. A.4).

Other useful two-dimensional plotting facilities are:

»*semilogx*(x,y)	semilogarithmic plot, logarithmic x-axis
»*semilogy*(x,y)	semilogarithmic plot, logarithmic y-axis
»*loglog*(x,y)	full logarithmic plot, both x- and y-axis logarithmic
»*area*(x,y)	filled area plot
»*polar*(x,y)	polar coordinate plot
»*bar*([2:5])	bar graph

A.3.2 3-D Graphs

There are several commands for visualizing three-dimensional (3-D) functions. A 3-D curve can be shown by the *plot3* command:

»t = 0:0.01:3***pi*;
»*plot3*(t,*sin*(t),*cos*(t))
»*xlabel*('t'), *ylabel*('sin t'), *zlabel*('cos t')

Surfaces can be shown in many ways. Here are two of interest:

```
»[x, y] = meshgrid(-pi:pi/10:pi,-pi:pi/10:pi);
»z = cos(x).*cos(y);
»mesh(x,y,z)
»surf(x,y,z)
»view(30,60)
```

You may make your graph look better, when required, by shading:

```
»shading interp
```

To see the color scale, use *colorbar*

```
»colorbar
```

A.3.3 2½-D Graphs

The so-called 2½-D graph is used for visualizing a 3-D graph on a 2-D system of coordinates. Let us use x, y, and z variables from the previous section. We can show different z-levels on a x-y system of coordinates by its contour lines:

```
»contour(x,y,z)
```

It is possible to show only the levels required:

```
»contour(x,y,z,[-0.9:0.3:0.9])
```

and show the level values on the contour lines:

```
»[c,h] = contour(x,y,z,[-0.9:0.2:0.9]);
»clabel(c,h)
```

Another method is to watch the graph from top z-axis and assign different colors to different z-values:

```
»pcolor(x,y,z)
»colorbar
»shading interp
```

The *quiver* command can also be useful in visualizing vector fields such as velocity profiles.

A.4 SCRIPTS AND FUNCTIONS

The programs written in the language of MATLAB should be saved with the extension of "*m*", from where the name of *m-file* comes. If you use the editor of MATLAB, it automatically saves your files with the "*m*" extension. Otherwise, be sure to save them with the "*m*" extension. M-files can be in the form of *scripts* and *functions* and could be executed in the MATLAB workspace.

A script is simply a series of MATLAB commands that could have been entered in the workspace. When typing the name of the script, the commands will be executed in their sequential order as if they were individually typed in the workspace. For example, let us calculate the volume of an ideal gas as a function of pressure and temperature. Type the following commands in the editor and save it as "*myscript.m*":

```
% A sample script file
disp(' Calculating the volume of an ideal gas.')
R = 8314;                          % Gas constant
t = input(' Vector of temperatures (K) = ');
p = input(' Pressure (bar) = ')*1e5;
v = R*t/p;                         % Ideal gas law
% Plotting the results
plot(t,v)
xlabel('T (K)')
ylabel('V (m^3/kmol)')
title('Ideal gas volume vs temperature')
```

The symbol % indicates that this line contains comments. The % sign, and what comes after it in that line, will be ignored at the time of execution. Return to the MATLAB command window and type:

»*myscript*

Input the required data and see the results.

A practical method for beginners to create a script is using *diary*. You can start creating a diary by:

»*diary* mydiary

Then you start typing your statements in the workspace one by one. For example, you can type the statements of the script developed above, see the results at each step, and make corrections if necessary. When you got your desired results, close the diary:

»*diary off*

Now you can develop a script by editing the file "mydiary" (no extension is added by MATLAB), deleting the unnecessary lines, and saving it as a m-file.

You can develop your own function and execute it just like other built-in functions in MATLAB. A function takes some data as input, performs required calculations, and returns the results of calculations back to you. As an example, let us write a function to do the ideal gas volume calculations that we have already done in a script. We make this function more general so that it would be able to calculate the volume at multiple pressures and multiple temperatures:

```
function v = myfunction(t,p)
% Function "myfunction.m"
% This function calculates the specific volume of an ideal gas

R = 8314;                    % Gas constant
for k = 1:length(p)
  v(k,:) = R*t/p(k);         % Ideal gas law
end
```

This function must be saved as "*myfunction.m*". You can now use this function in the workspace, in a script, or in another function. For example:

```
»p = 1:10; t = 300:10:400;
»vol = myfunction(t,p);
»surf(t,p,vol)
»view(135,45), colorbar
```

The first line of a function is called function declaration line and should start with the word function followed by the output argument(s), equality sign, name of the function, and input argument(s), as illustrated in the example. The first set of continuous comment lines immediately after the function declaration line is the help for the function and can be reviewed separately:

```
»help myfunction
```

A.4.1 Flow Control

MATLAB has several flow control structures that allow the program to make decisions or control its execution sequence. These structures are *for*, *if*, *while*, and *switch* which we describe briefly below:

if . . . (else . . .) end – The *if* command enables the program to make decision about what commands to execute:

```
x = input(' x = ');
if x >= 0
  y = x^2
end
```

You can also define an *else* clause, which is executed if the condition in the *if* statement is not true:

```
x = input(' x = ');
if x >= 0
  y = x^2
else
  y = -x^2
end
```

for ... end – The *for* command allows the script to cause a command, or a series of commands, to be executed several times:

```
k = 0;
for x = 0:0.2:1
  k = k+1
  y(k) = exp(-x)
end
```

while ... end – The while statement causes the program to execute a group of commands until some condition is no longer true:

```
x = 0;
while x<1
  y = sin(x)
  x = x+0.1;
end
```

switch . . . case . . . end – When a variable may have several values and the program has to execute different commands based on different values of the variable, a *switch-case* structure is easier to use than a nested *if* structure:

```
a = input('a = ');
switch a
case 1
  disp('One')
case 2
  disp('Two')
```

```
case 3
  disp('Three')
end
```

Two useful commands in programming are *break* and *pause*. You can use the *break* command to jump out of a loop before it is completed. The *pause* command will cause the program to wait for a key to be pressed before continuing:

```
k = 0;
for x = 0:0.2:1
  if k>3
    break
  end
  k = k+1
  y(k) = exp(-x)
  pause
end
```

A.5 Data Export and Import

There are different ways you can save your data in MATLAB. Let us first generate some data:

```
»a = magic(3); b = magic(4);
```

The following command saves all the variables in the MATLAB workspace in the file "*f1.mat*":

```
»save f1
```

If you need to save just some of the variables, list their names after the file name. The following saves only a in the file "*f2.mat*":

```
»save f2 a
```

The files generated above have the extension "*.mat*" and could be retrieved only by MATLAB. To use your data elsewhere you may want to save your data as text:

```
»save f3 b -ascii
```

Here, the file "*f3*" is a text file with no extension. You can also use *fprintf* command to save your data into a file using a desired format.

You can load your data into the MATLAB workspace using the *load* command. If the file to be loaded is generated by MATLAB (carrying ".*mat*" extension), the variables will appear in the workspace with their name at the time they were saved:

```
»clear
»load f1
»whos
```

However, if the file is a text file, the variables will appear in the workspace under the name of the file:

```
»load f3, whos
```

A.6 WHERE TO FIND HELP

As a beginner, you may want to see a tutorial about MATLAB. This is possible by typing *demo* at the command line to see the available demonstrations. In the MATLAB demo window you may choose the subject you are interested in and then follow the lessons.

If you know the name of the function you want help on you can use the *help* command:

```
»help sign
```

Typing *help* alone lists the names of all directories in the MATLAB search path. Also, if you type a directory name in the place of the file name, MATLAB lists contents of the directory (if the directory is a directory in the MATLAB path search and contains a *contents.m* file):

```
»help
»help matlab\general
```

If you are not sure of the function name, you can try to find the name using the *lookfor* command:

```
»lookfor absolute
```

Extensive MATLAB help and manuals may be found on the following websites:

http://www.mathworks.com
http://www.owlnet.rice.edu/~ceng303
http://www.indiana.edu/~statmath/smdoc/matlab.html

REFERENCES

1. Hanselman, D., and Littlefield, B., *The Student Edition of MATLAB 5, Version 5 User's Guide*, Prentice Hall, Upper Saddle River, NJ, 1997.

2. Hanselman, D., and Littlefield, B., *Mastering MATLAB 5: A Comprehensive Tutorial and Reference*, Prentice Hall, Upper Saddle River, NJ, 1998.

Index

A

B

N

O

P

Q

R

S

T

U

V

W

X

Z

THE AUTHORS

We sincerely hope that you have enjoyed reading this book and using the software that accompanies it. For updates of the software and other news about the book please visit our website: http://sol.rutgers.edu/~constant. If you have any questions or comments, you will be able to e-mail us via the website.

Alkis Constantinides is Professor and Chairman of the Department of Chemical and Biochemical Engineering at Rutgers, The State University of New Jersey. He was born in Cyprus, where he lived until he graduated from high school. In 1959 he came to the United States to attend Ohio State University, Columbus, and received the B.S. and M.S. degrees in chemical engineering in 1964. For the next two years he worked at Exxon Research and Engineering Company in Florham Park, NJ. In 1969, he received the Ph.D. degree in chemical engineering from Columbia University, New York, NY. He then joined the Department of Chemical Engineering at Rutgers University where he helped establish the biochemical engineering curriculum of the department.

Professor Constantinides has 30 years experience teaching graduate and undergraduate courses in chemical and biochemical engineering. His research interests are in the fields of computer applications in chemical and biochemical engineering, process modeling and optimization, artificial intelligence, biotechnology, fermentations, and enzyme engineering. Professor Constantinides has industrial experience in process development and design of large petrochemical plants and in pilot plant research. He has served as consultant to industry in the areas of fermentation processes, enzyme engineering, application of artificial intelligence in chemical process planning, design and economics of chemical processes, technology assessment, modeling, and optimization. He is the author of the textbook *Applied Numerical Methods with Personal Computers*, published by McGraw-Hill in 1987. He is the editor and co-editor of three volumes of *Biochemical Engineering,* published by the NY Academy of Sciences, and the author of more than 50 papers in professional journals. He served as the Director of the Graduate Program in Chemical and Biochemical Engineering from 1976 to 1985. In addition to being the Chairman, he is also the Director of the Microcomputer Laboratory of the department.

Professor Constantinides is the recipient of Rutgers University's prestigious Warren I. Susman Award for Excellence in Teaching (1991) and the 1998 Teaching Excellence Award given by the Graduating Senior Class of the Chemical and Biochemical Engineering Department.

Alkis Constantinides is a member of the American Institute of Chemical Engineers and the American Chemical Society.

Navid Mostoufi is Assistant Professor of Chemical Engineering at the University of Tehran, Iran. He was born in Abadan, Iran. He received the B.S. and M.S. degrees in chemical engineering from the University of Tehran. From 1989 to 1994 he worked as process engineer with Chagalesh Consulting Engineers and Farazavaresh Consulting Engineers, Tehran. In 1999 he received the Ph.D. degree in chemical engineering from École Polytechnique de Montréal and then joined the Department of Chemical Engineering in the Faculty of Engineering, University of Tehran. His areas of active investigation are multiphase reactors and numerical methods. Professor Mostoufi has five publications in *Chemical Engineering Science* and other major journals. He is a member of the Iranian Society for Chemical Engineering and the Iranian Petroleum Institute.

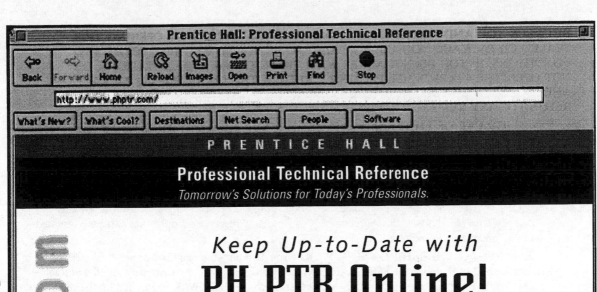

LICENSE AGREEMENT AND LIMITED WARRANTY

READ THE FOLLOWING TERMS AND CONDITIONS CAREFULLY BEFORE OPENING THIS CD PACKAGE. THIS LEGAL DOCUMENT IS AN AGREEMENT BETWEEN YOU AND PRENTICE-HALL, INC. (THE "COMPANY"). BY OPENING THIS SEALED CD PACKAGE, YOU ARE AGREEING TO BE BOUND BY THESE TERMS AND CONDITIONS. IF YOU DO NOT AGREE WITH THESE TERMS AND CONDITIONS, DO NOT OPEN THE CD PACKAGE. PROMPTLY RETURN THE UNOPENED CD PACKAGE AND ALL ACCOMPANYING ITEMS TO THE PLACE YOU OBTAINED THEM FOR A FULL REFUND OF ANY SUMS YOU HAVE PAID.

1. **GRANT OF LICENSE:** In consideration of your purchase of this book, and your agreement to abide by the terms and conditions of this Agreement, the Company grants to you a nonexclusive right to use and display the copy of the enclosed software program (hereinafter the "SOFTWARE") on a single computer (i.e., with a single CPU) at a single location so long as you comply with the terms of this Agreement. The Company reserves all rights not expressly granted to you under this Agreement.

2. **OWNERSHIP OF SOFTWARE:** You own only the magnetic or physical media (the enclosed CD) on which the SOFTWARE is recorded or fixed, but the Company and the software developers retain all the rights, title, and ownership to the SOFTWARE recorded on the original CD copy(ies) and all subsequent copies of the SOFTWARE, regardless of the form or media on which the original or other copies may exist. This license is not a sale of the original SOFTWARE or any copy to you.

3. **COPY RESTRICTIONS:** This SOFTWARE and the accompanying printed materials and user manual (the "Documentation") are the subject of copyright. The individual programs on the CD are copyrighted by the authors of each program. Some of the programs on the CD include separate licensing agreements. If you intend to use one of these programs, you must read and follow its accompanying license agreement. If you intend to use the trial version of Internet Chameleon, you must read and agree to the terms of the notice regarding fees on the back cover of this book. You may not copy the Documentation or the SOFTWARE, except that you may make a single copy of the SOFTWARE for backup or archival purposes only. You may be held legally responsible for any copying or copyright infringement which is caused or encouraged by your failure to abide by the terms of this restriction.

4. **USE RESTRICTIONS:** You may not network the SOFTWARE or otherwise use it on more than one computer or computer terminal at the same time. You may physically transfer the SOFTWARE from one computer to another provided that the SOFTWARE is used on only one computer at a time. You may not distribute copies of the SOFTWARE or Documentation to others. You may not reverse engineer, disassemble, decompile, modify, adapt, translate, or create derivative works based on the SOFTWARE or the Documentation without the prior written consent of the Company.

5. **TRANSFER RESTRICTIONS:** The enclosed SOFTWARE is licensed only to you and may not be transferred to any one else without the prior written consent of the Company. Any unauthorized transfer of the SOFTWARE shall result in the immediate termination of this Agreement.

6. **TERMINATION:** This license is effective until terminated. This license will terminate automatically without notice from the Company and become null and void if you fail to comply with any provisions or limitations of this license. Upon termination, you shall destroy the Documentation and all copies of the SOFTWARE. All provisions of this Agreement as to warranties, limitation of liability, remedies or damages, and our ownership rights shall survive termination.

7. **MISCELLANEOUS:** This Agreement shall be construed in accordance with the laws of the United States of America and the State of New York and shall benefit the Company, its affiliates, and assignees.

8. **LIMITED WARRANTY AND DISCLAIMER OF WARRANTY:** The Company warrants that the SOFTWARE, when properly used in accordance with the Documentation, will operate in substantial conformity with the description of the SOFTWARE set forth in the Documentation. The Company does not warrant that the SOFTWARE will meet your requirements or that the operation of the SOFTWARE will be uninterrupted or error-free. The Company warrants that the media on which the SOFTWARE is delivered shall be free from defects in materials and workmanship under normal use for a period of thirty (30) days from the date of your purchase. Your only remedy and the Company's only obligation under these limited warranties is, at the Company's option, return of the warranted item for a refund of any amounts paid by you or replacement of the item. Any replacement of SOFTWARE or media under the warranties shall not extend the original warranty period. The limited warranty set forth above shall not apply to any SOFTWARE which the Company determines in good faith has been subject to misuse, neglect, improper installation, repair, alteration, or damage by you. EXCEPT FOR THE EXPRESSED WARRANTIES SET FORTH ABOVE, THE COMPANY DISCLAIMS ALL WARRANTIES, EXPRESS OR IMPLIED, INCLUDING WITHOUT LIMITATION, THE IMPLIED WARRANTIES OF MERCHANTABILITY AND FITNESS FOR A PARTICULAR PURPOSE. EXCEPT FOR THE EXPRESS WARRANTY SET FORTH ABOVE, THE COMPANY DOES NOT WARRANT, GUARANTEE, OR MAKE ANY REPRESENTATION REGARDING THE USE OR THE RESULTS OF THE USE OF THE SOFTWARE IN TERMS OF ITS CORRECTNESS, ACCURACY, RELIABILITY, CURRENTNESS, OR OTHERWISE.

IN NO EVENT, SHALL THE COMPANY OR ITS EMPLOYEES, AGENTS, SUPPLIERS, OR CONTRACTORS BE LIABLE FOR ANY INCIDENTAL, INDIRECT, SPECIAL, OR CONSEQUENTIAL DAMAGES ARISING OUT OF OR IN CONNECTION WITH THE LICENSE GRANTED UNDER THIS AGREEMENT, OR FOR LOSS OF USE, LOSS OF DATA, LOSS OF INCOME OR PROFIT, OR OTHER LOSSES, SUSTAINED AS A RESULT OF INJURY TO ANY PERSON, OR LOSS OF OR DAMAGE TO PROPERTY, OR CLAIMS OF THIRD PARTIES, EVEN IF THE COMPANY OR AN AUTHORIZED REPRESENTATIVE OF THE COMPANY HAS BEEN ADVISED OF THE POSSIBILITY OF SUCH DAMAGES. IN NO EVENT SHALL LIABILITY OF THE COMPANY FOR DAMAGES WITH RESPECT TO THE SOFTWARE EXCEED THE AMOUNTS ACTUALLY PAID BY YOU, IF ANY, FOR THE SOFTWARE.

SOME JURISDICTIONS DO NOT ALLOW THE LIMITATION OF IMPLIED WARRANTIES OR LIABILITY FOR INCIDENTAL, INDIRECT, SPECIAL, OR CONSEQUENTIAL DAMAGES, SO THE ABOVE LIMITATIONS MAY NOT ALWAYS APPLY. THE WARRANTIES IN THIS AGREEMENT GIVE YOU SPECIFIC LEGAL RIGHTS AND YOU MAY ALSO HAVE OTHER RIGHTS WHICH VARY IN ACCORDANCE WITH LOCAL LAW.

ACKNOWLEDGMENT

YOU ACKNOWLEDGE THAT YOU HAVE READ THIS AGREEMENT, UNDERSTAND IT, AND AGREE TO BE BOUND BY ITS TERMS AND CONDITIONS. YOU ALSO AGREE THAT THIS AGREEMENT IS THE COMPLETE AND EXCLUSIVE STATEMENT OF THE AGREEMENT BETWEEN YOU AND THE COMPANY AND SUPERSEDES ALL PROPOSALS OR PRIOR AGREEMENTS, ORAL, OR WRITTEN, AND ANY OTHER COMMUNICATIONS BETWEEN YOU AND THE COMPANY OR ANY REPRESENTATIVE OF THE COMPANY RELATING TO THE SUBJECT MATTER OF THIS AGREEMENT.

Should you have any questions concerning this Agreement or if you wish to contact the Company for any reason, please contact in writing at the address below.

Robin Short
Prentice Hall PTR
One Lake Street
Upper Saddle River, New Jersey 07458

About the CD-ROM

This CD-ROM was prepared using the ISO 9660 format, therefore it can be used by all three computing platforms: WINDOWS, Macintosh, and UNIX.

Hardware requirements

PC computer, Macintosh computer, or UNIX workstation with a CD-ROM drive.

Software requirements

For PC computers: WINDOWS (3.1, 95, 98, or NT).

For Macintosh computers: System 7 with File Exchange activated, or System 8 with PC Exchange activated in order to be able to read the CD-ROM. If you have not activated File Exchange, please do so via the Control Panel. In addition, Macintosh computers need *zipit* or *StuffIt Expander* to uncompress the files.

For UNIX workstations: Any UNIX operating system.

MATLAB: The programs contained on this CD-ROM have been written in the MATLAB 5.0 language and will execute in the MATLAB command environment in all three operating systems (WINDOWS, Macintosh, and UNIX). Version 5.0 or higher, including the Student Edition, is sufficient to run the programs.

NOTE for users of MATLAB 5.2: The original MATLAB Version 5.2 had a "bug" which causes some programs to function incorrectly. A patch which corrects this problem is available on the website of Math Works, Inc.:

http://www.mathworks.com

If you have Version 5.2, you are strongly encouraged to download and install this patch, if you have not done so already.

Installation procedures

For complete installation instructions see the section entitled "Programs on the CD-ROM" (page xv) or the README file on the CD-ROM.

Updates

Updates of the software will be posted on our website at:

http://sol.rutgers.edu/~constant

Disclaimer

Prentice Hall does not offer technical support for this software. However, if there is a problem with the media, you may obtain a replacement copy by e-mailing us with your problem at:

disk_exchange@prenhall.com